Quantum Bio-Informatics III

From Quantum Information to Bio-Informatics

QP–PQ: Quantum Probability and White Noise Analysis*

QP–PQ: Quantum Probability and White Noise Analysis

Vol. 26: Quantum Bio-Informatics III
From Quantum Information to Bio-Informatics
eds. L. Accardi, W. Freudenberg and M. Ohya

Vol. 25: Quantum Probability and Infinite Dimensional Analysis
Proceedings of the 29th Conference
eds. H. Ouerdiane and A. Barhoumi

Vol. 24: Quantum Bio-Informatics II
From Quantum Information to Bio-informatics
eds. L. Accardi, W. Freudenberg and M. Ohya

Vol. 23: Quantum Probability and Related Topics
eds. J. C. García, R. Quezada and S. B. Sontz

Vol. 22: Infinite Dimensional Stochastic Analysis
eds. A. N. Sengupta and P. Sundar

Vol. 21: Quantum Bio-Informatics
From Quantum Information to Bio-Informatics
eds. L. Accardi, W. Freudenberg and M. Ohya

Vol. 20: Quantum Probability and Infinite Dimensional Analysis
eds. L. Accardi, W. Freudenberg and M. Schürmann

Vol. 19: Quantum Information and Computing
eds. L. Accardi, M. Ohya and N. Watanabe

Vol. 18: Quantum Probability and Infinite-Dimensional Analysis
From Foundations to Applications
eds. M. Schürmann and U. Franz

Vol. 17: Fundamental Aspects of Quantum Physics
eds. L. Accardi and S. Tasaki

Vol. 16: Non-Commutativity, Infinite-Dimensionality, and Probability at the Crossroads
eds. N. Obata, T. Matsui and A. Hora

Vol. 15: Quantum Probability and Infinite-Dimensional Analysis
ed. W. Freudenberg

Vol. 14: Quantum Interacting Particle Systems
eds. L. Accardi and F. Fagnola

Vol. 13: Foundations of Probability and Physics
ed. A. Khrennikov

*For the complete list of the published titles in this series, please visit:
www.worldscibooks.com/series/qqpwna_series.shtml

QP–PQ
Quantum Probability and White Noise Analysis
Volume XXVI

Quantum Bio-Informatics III

From Quantum Information to Bio-Informatics

Tokyo University of Science, Japan 11 – 14 March 2009

Editors

Luigi Accardi
Università di Roma "Tor Vergata", Italy

Wolfgang Freudenberg
Brandenburgische Technische Universität Cottbus, Germany

Masanori Ohya
Tokyo University of Science, Japan

World Scientific

NEW JERSEY • LONDON • SINGAPORE • BEIJING • SHANGHAI • HONG KONG • TAIPEI • CHENNAI

Published by

World Scientific Publishing Co. Pte. Ltd.

5 Toh Tuck Link, Singapore 596224

USA office: 27 Warren Street, Suite 401-402, Hackensack, NJ 07601

UK office: 57 Shelton Street, Covent Garden, London WC2H 9HE

Library of Congress Cataloging-in-Publication Data

Quantum bio-informatics III : from quantum information to bio-informatics, Tokyo University of Science, Japan, 11–14 March 2009 / edited by Luigi Accardi, Wolfgang Freudenberg & Masanori Ohya.

p. cm. -- (QP-PQ, quantum probability and white noise analysis ; v. 26)

Includes bibliographical references.

ISBN-13: 978-981-4304-05-4 (hardcover : alk. paper)

ISBN-10: 981-4304-05-0 (hardcover : alk. paper)

1. Bioinformatics--Congresses. 2. Quantum theory--Congresses. I. Accardi, L. (Luigi), 1947– II. Freudenberg, Wolfgang, 1950– III. Ohya, Masanori, 1947– IV. Title: Quantum bio-informatics 3.

QH324.2.Q363 2010

572.80285--dc22

2009049238

British Library Cataloguing-in-Publication Data

A catalogue record for this book is available from the British Library.

Printed by FuIsland Offset Printing (S) Pte Ltd, Singapore

PREFACE

This volume is based on the third international conference of quantum bio-informatics held at the QBI Center of Tokyo University of Sciences.

The purpose of the conference is towards new stage making interdisciplinary bridges in mathematics, physics, information and life sciences, in particular, research for new paradigm for information science and life science on the basis of quantum theory. For this purpose we firstly return to the starting point of bio-informatics and quantum information, fields that are growing rapidly at present, and to seriously attempt mutual interaction between the two, with a view to enumerating and solving the many fundamental problems they entail. Those three years, we have established various relations between quantum information and life science. The papers submitted in this volume are all referred, whose contents are related to one of the following subjects:

(1) Mathematical foundation of quantum information
(2) Quantum probability
(3) Quantum entanglement
(4) Quantum entropy and lifting
(5) Quantum dynamics and white noise analysis
(6) Brain model
(7) Classical and Quantum computing
(8) Experimental studies of quantum computer
(9) Superconductivity and its uses
(10) Bio-Informatics
(11) Genome analysis

Luigi Accardi
Wolfgang Freudenberg
Masanori Ohya

CONTENTS

Quantum Bio-Informatics III

From Quantum Information to Bio-Informatics

Quantum Bio-Informatics III
eds. L. Accardi, W. Freudenberg & M. Ohya

CLASSICAL AND QUANTUM CONDITIONING: MATHEMATICAL AND INFORMATION THEORETICAL ASPECTS

LUIGI ACCARDI

Centro Vito Volterra, Università degli Studi di Roma "Tor Vergata", Italy
E-mail: accardi@volterra.mat.uniroma2.it

The different notions of stochastic independences, introduced in quantum probability open new fascinating possibilities to deepen our intuition on what a *composite system*.
In the present note we propose a general mathematical definition of *composite system* which emphasizes the fact that the naive idea, that a physical system is *composed* of a multiplicity of sub-systems, can be substantiated by a multiplicity of inequivalent mathematical models.
This wealth of possibilities can considerably enrich the present approach to the theory of open systems, with potential implications for the theory of measurement and the theory of complex systems, such as biological or economical ones.
The standard approach to composite system strongly privileges the tensor product construction and the corresponding notion of stochastic independence. But there are a multiplicity of other possibilities whose mathematical and physical investigation is only at the beginning. In particular, to any notion of statistical independence it is canonically associated a corresponding notion of entanglement.

Contents

1. Conditioning: qualitative aspects

Conditioning is one of the crucial notions of probability theory.
It is traditionally considered a quantitative measure of statistical dependence, i.e. of how far are two or more random variables from being statis-

tically independent.
But we know from quantum probability that there are uncountably many notions of statistical independence: for each of them we should expect a different form of conditioning.
Even the extreme case of independence is far from trivial in fact, if we accept the traditional physical interpretation of *classical or boson statistical independence* as *absence of interactions* (think of the EPR type experiments) which will be the physical interpretation of the other notions of statistical independence?
Even within a single fixed probabilistic framework (e.g. usual quantum probability on $\mathcal{B}(\mathcal{H})$) with the usual notion of tensor independence, many inequivalent notions of conditioning are mathematically possible.
It is natural to conjecture that these notions correspond to different physical situations, i.e. that one can characterize the physical situations to which each of these notions (hence the corresponding mathematical formulae) can be applied.
Unfortunately at the moment this is hardly ever the case.
In almost 40 years of activity quantum probability has brought several contributions to the conditioning problem, both from the mathematical and the conceptual point of view.
However what we understood only scratches the surfaces of this deep problem at the crossroad between probability, physics and information theory.
From a model independent point of view, conditioning is the mathematical answer to a deep information theoretical problem:
if my information changes, how will my statistical predictions change?
Thus: conditioning (and its dual aspect of *lifting* see [1]) is equivalent to *information dynamics*, a notion introduced by M. Ohya [9].
Since in physics changes of information occur by effect of measurements, it is not surprising that measurement theory, and more generally the theory of composite systems, is the physical counterpart to the conditioning problem.
More specifically, in measurement theory this situation is the folowing.
Given two systems S and M, how will my statistical predictions on S change, if I acquire some information on M (typically by a measurement)?
An extreme case is given by information theoretical independence:
whichever information I acquire on M my statistical predictions on S do not change.
Information theoretical independence is related to statistical independence, but the two notions should be distinguished.

For example in classical probability statistical independence is a symmetric notion:
S is statistically independent from $M \Leftrightarrow M$ is statistically independent from S
However it is not natural to postulate this property for information theoretical independence. For example if I receive the information that Warren Buffet is buying stocks of General Electric, this might influence my propensity to buy them, however it is quite it is quite unlikely that Buffet might be influenced by my single decision to buy stocks of General Electric.
Thus B can be information independent from A without A being from B.
In physics statistical independence is usually identified with absence of interaction, reflecting the naive idea that interacting systems cannot be independent. This identification helps our intuition to realize that the dichotomy dependence–independence denotes, just like in the case of order and chaos, the two extremes of a continuous hierarchy of possibilities.
Quantum probability provides a mathematical framework for this more subtle scenario: the various notions of quantum independences are an indication of a hierarchy of hidden dependences or, in physical terminology, of hidden interactions.
Here the term *quantum* is essential because the freedom for classical notions of independence is very narrow.

2. Mathematical models of a statistical theory: single system

2.1. *Statistical description of a physical system: algebraic probability spaces*

A physical system is characterized by a family of observables related to it and experimentally measurable. It is mathematically convenient and physically not restrictive to model the possible results of measurements on these observables as a subset of a $*$–algebra $\mathcal{A}$.
The statistical predictions of any theory are then condensed into probability distributions on the values of these observables and, in view of Gleason's theorem, it is not too restrictive to postulate that these values determine a unique state φ on the $*$–algebra $\mathcal{A}$. Thus the natural mathematical model for the statistical description of a physical system leads us to one of the basic notion of quantum probability, that of **algebraic probability space**.
Definition An algebraic probability space is a pair:

$$\{\mathcal{A}, \varphi\}$$

If $\mathcal{A}$ is commutative, then one speaks of classical probability space; if it is non commutative, of a quantum probability space.
In the following, when the commutativity of $\mathcal{A}$ is not specified, it means that the notions and results introduced are valid both for classical and quantum systems.

2.2. *Dynamical systems*

Algebraic probability spaces account for the statistical description of a physical system. The corresponding mathematical tools for their time evolutions are the different notions of dynamical systems. Here we describe two of them.

Definition A Hilbert space dynamical system is defined by a triple

$$\{\mathcal{H}, U_t, \Phi\}$$

where
- $\mathcal{H}$ is a Hilbert space
- U_t is a 1–parameter unitary group (U_t is unitary and $U_{t+s} = U_t U_s$, $U_0 = 1$)
- Φ is a unit vector in $\mathcal{H}$

The system is called stationary if, in addition, for all $t \in \mathbb{R}$

$$U_t \Phi = \Phi$$

An algebraic dynamical system is defined by a triple

$$\{\mathcal{A}, u_t, \varphi\}$$

where
- $\mathcal{A}$
- u_t is a 1–parameter group of automorphisms of $\mathcal{A}$ ($u_{t+s} = u_t u_s$, $u_0 = id_{\mathcal{A}}$)
- φ is a state on $\mathcal{A}$

The system is called stationary if, in addition for all $t \in \mathbb{R}$

$$\varphi \circ u_t = \varphi$$

To every stationary algebraic dynamical system one can canonically associate the stationary Hilbert space dynamical system $\{\mathcal{H}, U_t, \Phi\}$ where $\{\mathcal{H}, \pi, \Phi\}$ is the GNS representation of the pair $\{\mathcal{A}, \varphi\}$, so that

$$\varphi = \langle \Phi, \pi(\,\cdot\,)\Phi \rangle$$

and the unitary U_t is defined by linear continuous extension of the map

$$\pi(a)\Phi \mapsto \pi(u_t(a))\Phi =: U_t \pi(a)\Phi \qquad ; \qquad \mathcal{A}$$

Thus by construction:

$$U_t\Phi = \Phi$$
$$u_t = U_t\pi(\,\cdot\,)U_t^*$$

and

$$U_t\mathcal{A}U_t^* = u_t(\mathcal{A}) \subseteq \mathcal{A}$$

The converse is also true, i.e.: to every stationary Hilbert space dynamical system $\{\mathcal{H}, U_t, \Phi\}$ one can canonically associate the stationary algebraic dynamical system $\{\mathcal{A}, u_t, \varphi\}$ characterized by:

$$\mathcal{A} = \mathcal{B}(\mathcal{H}) \ ; \ \varphi = \langle \Phi, (\,\cdot\,)\Phi \rangle \ ; \ u_t = U_t(\,\cdot\,)U_t^*$$

By Stone's theorem

$$U_t = e^{itH}$$

so that the assignment of U_t is equivalent to the assignment of its generator H, called the Hamiltonian of the system.

3. Two system conditioning: composite systems

The intuitive idea of *interaction* between two systems is that it combines two originally independent dynamical systems creating a third one, which is intuitively thought to be their *composite system*.
Measurement theory is a particular case of the theory of composite systems. During the interaction (act of measurement), system S and apparatus M merge into a composite system (S, M) and the main problem of measurement theory is: *which informations on the system S can be deduced from informations acquired on the system M?*
Recalling the qualitative discussion at the beginning of the present paper, we can say that the main problem of measurement theory is equivalent to the conditioning problem from the composite system (S, M) to its *sub–system M*.
Also open systems are included in composite systems and, as for measurements, there in no clear cut distinction between these two notions. Intuitively one speaks of open system when the role of the two interacting systems is asymmetric and typically one is much larger, or faster, ..., than the other one (e.g. and atom and a field, a system and its environment, system and measurement apparatus, ...). In the following we will appeal to this intuition and frequently we will use terms such as *reservoir, measurement apparatus, environment,* ... to suggest that we are referring to

the *large system.* However from the mathematical point of view we will not introduce specific distinctions between the two types of systems.
The idea of composite system should be made more precise by specifying what do we mean by asserting that *two dynamical systems are combined so to create a third one.*
To this goal one should begin to list some natural physical, i.e. model independent, requirements such as the following ones.
Requirements on the observables (kinematical):
1) one should be able to define a notion of *sub–system* in such a way that S and M can be identified to sub–systems of (S, M) denoted respectively $j_S(\mathcal{A}_S)$ and $j_M(\mathcal{A}_M)$
2) the observable quantities of the the composite system (S, M) should be entirely constructable in terms of observables of S and M (minimality)
Requirements on the state (statistical):
3) We assume that, at time 0 (the time in which the interaction is switched on), the two systems are independent in the (model independent) sense that:
the knowledge of the individual statistics of each system uniquely determines the statistics of any observable of the composite system.
Requirements on the interaction (dynamical):
4) Under assumption (3) above one would like to have a dynamical law which, given the state at time 0, uniquely determines the state at any time t.
The second step should be to give a precise mathematical formulation of these intuitive requirements, i.e. to construct mathematical models for these physical, model independent, requirements.
To this goal recall that we have seen that algebraic probability spaces account for the statistical description of a physical system and dynamical systems for their time evolutions. This means that:
- S is described by an algebraic probability space $\{\mathcal{A}_S, \varphi_S\}$
- M is described by an algebraic probability space $\{\mathcal{A}_M, \varphi_M\}$
Therefore the attempt to make precise the idea of composite system, leads to the following problem:
which algebraic probability space $\{\mathcal{A}, \varphi\}$ shall describe the composite system (S, M)?
The situation described above is the one most commonly considered in physics.
Below we describe two other situations which are met in concrete applications and which, from a mathematical point of view can be reduced to the

previous one even if the physical situations they describe are quite different. The composite system $(S,\ M)$ undergoes an evolution

$$u_t : \mathcal{A} \to \mathcal{A}$$

which is a one-parameter group of $*-$automorphisms of $\mathcal{A}$.
This means that the state φ of $(S,\ M)$ evolves according to the law

$$\varphi_t := \varphi \circ u_t \tag{1}$$

Concerning the initial state, one has to distinguish three possibilities, for each of which there exist physical examples:

i) only the state φ_S of S is known (standard measurement)
ii) only the state φ_M of M is known (open systems, stochastic limit)
iii) both states φ_S and φ_M are known (composite system)

In any of these cases one wants to determine a state φ, of the composite system $(S,\ M)$, with the following properties:

$$\varphi \circ j_S = \varphi_S \quad ; \quad \text{in case } (i) \tag{2}$$

$$\varphi \circ j_M = \varphi_M \quad ; \quad \text{in case } (ii) \tag{3}$$

and in case (iii) both (2) and (3) are required. In this case, following a terminology introduced by M. Ohya (see [9] p. 402), the state φ is called a *compound state* of the states (φ_S, φ_M) through the embeddings j_s, j_M. We will see that the various notions of independence correspond to rules to create a compound state from two given states.

4. Mathematical formulation of the theory of composite systems

Definition A composite algebraic probability space of the two algebraic probability spaces

$$\{\mathcal{A}_S, \varphi_S\}\ (system) \quad ; \quad \{\mathcal{A}_R, \varphi_R\}\ (reservoir)$$

is defined by:
(i) a third algebraic probability space

$$\{\mathcal{A}_{SR}, \varphi_{SR}\}$$

(ii) two isometric embeddings

$$j_S : \mathcal{A}_S \to \mathcal{A} \quad ; \quad j_M : \mathcal{A}_M \to \mathcal{A} \tag{4}$$

such that the algebras $j_S(\mathcal{A}_S)$ and $j_M(\mathcal{A}_M)$ generate $\mathcal{A}$:

$$j_S(\mathcal{A}_S) \vee j_M(\mathcal{A}_M) = \mathcal{A}$$

(iii) the state φ_{SR} is a compound state of φ_S, φ_M in the sense of Ohya through the isometric embeddings j_S, j_M, i.e.

$$\varphi_{SR} \circ j_S = \varphi_S \qquad ; \qquad \varphi_{SR} \circ j_M = \varphi_M$$

(iv) the state φ_{SR} is a generalized product state of φ_S, φ_M in the sense defined below.

Any rule to construct the association $(\varphi_S, \varphi_M) \mapsto \varphi$ will be called **an hidden interaction.**

Below we will give examples to show how to translate these physical requirements into precise mathematical rules.

The interaction is best specified in terms of Hilbert space dynamical systems.

Definition A composite system of the two Hilbert space dynamical systems

$$\{\mathcal{H}_S, H_S, \Phi_S\}\ (system) \qquad ; \qquad \{\mathcal{H}_R, H_R, \Phi_R\}\ (reservoir)$$

is defined by:

(i) a third dynamical system

$$\{\mathcal{H}_{SR}, H_{SR}, \Phi_{SR}\}$$

(ii) isometric embeddings

$$J_S : \mathcal{H}_S \to \mathcal{H}_{SR} \qquad ; \qquad J_R : \mathcal{H}_R \to \mathcal{H}_{SR}$$

such that

$$j_S(\mathcal{A}_S) := J_S \mathcal{A}_S J_S^* \subseteq \mathcal{A}_{SR} \qquad ; \qquad j_R(\mathcal{A}_R) := J_R \mathcal{A}_R J_R^* \subseteq \mathcal{A}_{SR}$$

(iii) Three operators $j_S(H_S)$, $j_R(H_R)$, H_I such that

$$H_{SR} = j_S(H_S) + j_R(H_R) + H_I$$

Notice that, while the relation between the Hamiltonian of the composite system and those of its constituents is very simple, no such simple relation exists for the corresponding dynamics.

5. The standard (Boson) approach to the theory of composite systems

The standard approach to the theory of composite systems is based on the following postulates.

Postulate (I) No kinematical constraints at time zero.

This is a postulate on the structure of the observables of the composite system:

$$\mathcal{A} = \mathcal{A}_S \otimes \mathcal{A}_M \tag{5}$$

Postulate (II) This postulate defines a very special notion of sub–system (tensor sub–system) by specifying the embeddings j_S, j_M (tensor and identity preserving embeddings)

$$j_S(a_S) = a_s \otimes 1_M \quad ; \quad a_S \in \mathcal{A}_S \tag{6}$$

$$j_M(a_M) = 1_S \otimes a_M \quad ; \quad a_M \in \mathcal{A}_{\mathcal{M}} \tag{7}$$

Postulate (III) This postulate specifies in what sense there are no statistical constraints at time zero by choosing a special notion of generalized product state (tensor product state)

$$\varphi = \varphi_S \otimes \varphi_M \tag{8}$$

In the remaining of this Section we shall fix the choices (5) and (6). Under these assumptions the state φ_S of the system S evolves according to the *reduced evolution*:

$$\varphi_{S,t}(a_S) := \varphi_S(E_M \circ u_t \circ j_S(a_S)) \quad ; \quad a_S \in \mathcal{A}_S \tag{9}$$

where $E_M : \mathcal{A} = \mathcal{A}_S \otimes \mathcal{A}_M \to \mathcal{A}_S$ is the linear map characterized by

$$E_M(a_S \otimes a_M) = a_S \varphi_M(a_M) \quad ; \quad a_S \in \mathcal{A}_S \;\; ; \; a_M \in \mathcal{A}_M \tag{10}$$

The one–parameter family of linear maps $P^t : \mathcal{A}_S \to \mathcal{A}_S$, defined by

$$P^t = E_M \circ u_t \circ j_S \;;\; t \geq 0 \tag{11}$$

is called the **reduced evolution** of the system S. Each P^t is a **completely positive** identity preserving map, in particular

$$P^t(1) = 1 \;;\; \forall t \geq 0 \tag{12}$$

The family (P^t) is usually not an evolution in the strict sense, i.e. it does not satisfy the condition

$$P^s P^t = P^{s+t} \tag{13}$$

However, in some limiting and physically meaningful situations, the reduced evolution of the system S becomes indeed an (usually irreversible) evolution (see [2]). In these cases it is called a *Markovian semigroup* and the differential form of the equation (13) is called a *Markovian master equation.*
Notice that all the choices in the above (von Neumann's) construction are non canonical from a mathematical point of view:
- why tensor product of algebras?
- why tensor product of states?
- even accepting tensor product of states and of algebras,
why the embeddings should have the special form

$$j_S(a_S) = a_S \otimes 1_M \ ; \ j_M(a_M) = 1_S \otimes a_M \quad ; \quad a_S \in \mathcal{A}_S \ ; \ a_M \in \mathcal{A}_M$$

Also from a physical point of view these choices are non canonical because they are equivalent to postulate that, at time $t = 0$, there is no kinematical constraint among the observables of the two subsystems: it is clear that such a postulate in many concrete situations is not physically justified.

6. Notions of independence and of composite systems

The tensor construction of a composite system can be equivalently rephrased as follows:
Let S, M, $\mathcal{A}_S$, $\mathcal{A}_M$ be as before. Then the embeddings

$$j_S : \mathcal{A}_S \to \mathcal{A} \quad ; \quad j_M : \mathcal{A}_M \to \mathcal{A}$$

and the state φ on $\mathcal{A}$ are such that the sub–algebras $j_S(\mathcal{A}_S)$ and $j_M(\mathcal{A}_M)$ of $\mathcal{A}$ are φ–Boson–independent with marginals φ_S and φ_M.
This formulation emphasizes the fact that von Neumann's notion of composite system should be more precisely called *Boson (or tensor) composite system.*
Therefore a possible extension of von Neumann construction of composite systems can be obtained replacing Boson–independence by a more general notion of independence.
Definition Given an algebraic probability space $\{\mathcal{A}, \varphi\}$, two sub–algebras $\mathcal{A}_S^0, \mathcal{A}_R^0 \subseteq \mathcal{A}$ are called φ–independent if they generate $\mathcal{A}$,

$$\mathcal{A}_S^0 \vee \mathcal{A}_R^0 = \mathcal{A}$$

and the corresponding restrictions of the state φ

$$\varphi_S := \varphi|_{\mathcal{A}_S^0} \ ; \ \varphi_R := \varphi|_{\mathcal{A}_R^0}$$

uniquely determine the state φ itself in the sense that, for each $n \in \mathbb{N}$ and for any choice of $b_{1,S}, b_{2,S}, \ldots, b_{n,S} \in \mathcal{A}^0_S$ and of $b_{1,R}, b_{2,R}, \ldots, b_{n,R} \in \mathcal{A}^0_R$, the expectation value

$$\varphi(b_{1,S} \cdot b_{1,R} \cdot b_{2,S} \cdot b_{2,R} \ldots b_{n,S} \cdot b_{n,R})$$

is expressed as a function of expectation values of the form

$$\varphi_S(b_{j_1,S}, \ldots, b_{j_\alpha,S}) \quad ; \quad \varphi_R(b_{h_1,R}, \ldots, b_{h_\beta,S})$$

where

$$\alpha, \beta \in \{1, \ldots, n\} \quad ; \quad j_1 < \cdots < j_\alpha \quad ; \quad h_1 < \cdots < h_\beta$$

In this case we say that φ is a **generalized product** of the states φ_S and φ_R.

Any set of rules $\mathcal{I}$ to construct explicitly φ as a function of the states φ_S and φ_R is called a notion of **stochastic independence**.

Remark

Since the moste general element of the algebra $\mathcal{A}^0_S \vee \mathcal{A}^0_R = \mathcal{A}$ is a linear combination of products of the form

$$b_{1,S} \cdot b_{1,R} \cdot b_{2,S} \cdot b_{2,R} \ldots b_{n,S} \cdot b_{n,R}$$

the above definition is a way to make precise the intuitive idea that the statistics of the composite system is uniquely determined by the statistics of the individual component systems. By induction the above definition can be extended to an arbitrary number of component systems.

Remark

To emphasize the connection with (generalized) measurements, we will consider embeddings

$$j_S : \mathcal{A}_S \to \mathcal{A} \quad ; \quad j_M : \mathcal{A}_M \to \mathcal{A}$$

such that

$$j_S(\mathcal{A}_S) := \mathcal{A}^0_S \quad ; \quad j_R(\mathcal{A}_R) := \mathcal{A}^0_R$$

Once given a notion of *generalized product states* , one can define the notion of *generalized entanglement* as usual:

Definition Given two sub–algebras

$$\mathcal{A}^0_S \ , \ \mathcal{A}^0_R \subseteq \mathcal{A}$$

a state ψ on $\mathcal{A}$ is called entangled more precisely $(\mathcal{A}^0_S, \mathcal{A}^0_R)$–entangled if if it does not belong to the closed convex hull of generalized product states. Let us emphasize that, even if products of non commuting operators are in general non positive, the joint probabilities can be positive.

7. Alternatives to the standard description of composite systems

Let S, M, $\mathcal{A}_S$, $\mathcal{A}_M$ be as before and let $\mathcal{I}$ be a notion of statistical independence.
If the embeddings

$$j_S : \mathcal{A}_S \to \mathcal{A} \qquad ; \qquad j_M : \mathcal{A}_M \to \mathcal{A}$$

and the state φ on $\mathcal{A}$ are such that the sub–algebras $j_S(\mathcal{A}_S)$ and $j_M(\mathcal{A}_M)$ of $\mathcal{A}$ are φ–$\mathcal{I}$–independent with marginals φ_S and φ_M then we call these systems $\mathcal{I}$–*composite systems.* To construct concrete examples of this general statement, we have to clarify the connections between stochastic independences and embeddings.

7.1. Φ–*embeddings*

It is known that, given a tensor product of two Hilbert spaces $\mathcal{H}_S \otimes \mathcal{H}_R$ there does not exist a canonical way to embed any of the factors $\mathcal{H}_S, \mathcal{H}_R$ into their product $\mathcal{H}_S \otimes \mathcal{H}_R$.
The simplest embeddings are the isometric ones. Our examples will be built on a very special class of these examples, described in the following Lemma.

Lemma.
For any choice of $\Phi \in \mathcal{H}_R$ with $\|\Phi\| = 1$, the embedding

$$J_S : \xi \in \mathcal{H}_S \to \xi \otimes \Phi \in \mathcal{H}_S \otimes \mathcal{H}_R \tag{14}$$

is isometric and its adjoint is

$$J_S^*(\eta \otimes \psi) = \langle \Phi, \psi \rangle \eta \ ; \quad \eta \in \mathcal{H}_S, \psi \in \mathcal{H}_R \tag{15}$$

so that

$$j_S(a_S) := J_S a_S J_S^* = a_s \otimes P_\Phi \qquad ; \ a_S \in \mathcal{B}(\mathcal{H}_S) \tag{16}$$

where

$$P_\Phi = |\Phi\rangle\langle\Phi|$$

Proof. A simple calculation
Definition.
Embeddings

$$j_S :\in \mathcal{B}(\mathcal{H}_S) \to \mathcal{B}(\mathcal{H}_S) \otimes \mathcal{B}(\mathcal{H}_M)$$

of the form (16) above are called Φ–embeddings.

Remark

Using this one finds

$$\langle J_S a_S J_S^*(\eta \otimes \psi), \eta \otimes \psi\rangle = \langle a_S \eta, \eta\rangle |\langle \Phi, \psi\rangle|^2$$

The following two sections show how different notions of embeddings lead to different notions of independence, hence different notions of *composite systems.*

7.2. *Boolean composite systems*

Let us choose, as in von Neumann case,

$$\mathcal{A} := \mathcal{B}(\mathcal{H}_S) \otimes \mathcal{B}(\mathcal{H}_R)$$
$$\varphi(x) := \varphi_S \otimes \varphi_R(x) := \langle \Phi_S \otimes \Phi_R, x(\Phi_S \otimes \Phi_R)\rangle \quad ; \quad x \in \mathcal{A}$$

where Φ_S and Φ_R are unit vectors in $\mathcal{H}_S$ and $\mathcal{H}_R$ respectively, but choose the embeddings j_S, j_R to be of the form

$$j_S(b_S) := b_S \otimes P_{\Phi_R} \quad ; \quad \Phi_R\text{–embedding} \quad ; \quad b_S \in \mathcal{B}(\mathcal{H}_S)$$
$$j_R(b_R) := P_{\Phi_S} \otimes b_R \quad ; \quad \Phi_S\text{–embedding} \quad ; \quad b_R \in \mathcal{B}(\mathcal{H}_R)$$

Having fixed $n \in \mathbb{N}$ and $b_{1,S}, b_{2,S}, \ldots, b_{n,S} \in \mathcal{A}_S$ $b_{1,R}, b_{2,R}, \ldots, b_{n,R} \in \mathcal{A}_R$ we want to calculate the mixed moments (correlators)

$$\varphi(\bar{b}_{1,S} \cdot \bar{b}_{1,R} \cdot \bar{b}_{2,S} \cdot \bar{b}_{2,R} \ldots \bar{b}_{n,S} \cdot \bar{b}_{n,R})$$

To this goal recall that

$$\bar{b}_{j,S} := j_S(b_{j,S}) = b_{j,S} \otimes P_{\Phi_R} \in \mathcal{A}_S^0 \quad ; \quad b_{j,S} \in \mathcal{A}_S$$
$$\bar{b}_{j,R} := j_S(b_{j,R}) = P_{\Phi_S} \otimes b_{j,R} \in \mathcal{A}_R^0 \quad ; \quad b_{j,R} \in \mathcal{A}_R$$

Therefore one has

$$\varphi(\bar{b}_{1,S} \cdot \bar{b}_{1,R} \cdot \bar{b}_{2,S} \cdot \bar{b}_{2,R} \ldots \bar{b}_{n,S} \cdot \bar{b}_{n,R}) =$$

$$= \varphi([b_{1,S} \otimes P_{\Phi_R}] \cdot [P_{\Phi_S} \otimes b_{1,R}] \cdot [b_{2,S} \otimes P_{\Phi_R}] \cdot [P_{\Phi_S} \otimes b_{2,R}] \ldots [b_{n,S} \otimes P_{\Phi_R}] \cdot [P_{\Phi_S} \otimes b_{n,R})]) =$$

$$= \varphi([b_{1,S} P_{\Phi_S} b_{2,S} P_{\Phi_S} \ldots b_{n,S} P_{\Phi_S}] \otimes [P_{\Phi_R} b_{1,R} P_{\Phi_R} b_{2,R} P_{\Phi_R} \ldots b_{n,R}])$$

Now recall that

$$P_{\Phi_S} b_{2,S} P_{\Phi_S} = \langle \Phi_S, b_{2,S} \Phi_S\rangle P_{\Phi_S} = \varphi_S(b_{2,S}) P_{\Phi_S}$$
$$P_{\Phi_R} b_{1,R} P_{\Phi_R} = \langle \Phi_R, b_{1,R} \Phi_R\rangle P_{\Phi_R} = \varphi_R(b_{1,R}) P_{\Phi_R}$$

Therefore

$$\varphi([b_{1,S}P_{\Phi_S}b_{2,S}P_{\Phi_S}\dots b_{n,S}P_{\Phi_S}]\otimes[P_{\Phi_R}b_{1,R}P_{\Phi_R}b_{2,R}P_{\Phi_R}\dots b_{n,R}]) =$$

$$= \varphi_S(b_{2,S})\varphi_R(b_{1,R})\cdot\varphi([b_{1,S}P_{\Phi_S}b_{3,S}P_{\Phi_S}\dots b_{n,S}P_{\Phi_S}]\otimes[P_{\Phi_R}b_{2,R}P_{\Phi_R}\dots b_{n,R}]) =$$

$$= \dots = \varphi_S(b_{2,S})\varphi_S(b_{3,S})\dots\varphi_R(b_{1,R})\varphi_R(b_{2,R})\cdots\varphi([b_{1,S}P_{\Phi_S}]\otimes[P_{\Phi_R}b_{n,R}])$$

Now recall that, for $x \in \mathcal{A}$,

$$\varphi(x) := \langle\Phi_S\otimes\Phi_R, x(\Phi_S\otimes\Phi_R)\rangle$$

Therefore

$$\begin{aligned}\varphi([b_{1,S}P_{\Phi_S}]\otimes[P_{\Phi_R}b_{n,R}]) &= \langle\Phi_S\otimes\Phi_R, [b_{1,S}P_{\Phi_S}]\otimes[P_{\Phi_R}b_{n,R}](\Phi_S\otimes\Phi_R)\rangle\\ &= \langle\Phi_S b_{1,S}\Phi_S\rangle\langle\Phi_R b_{n,R}\Phi_R\rangle\end{aligned}$$

In conclusion

$$\varphi(\bar{b}_{1,S}\cdot\bar{b}_{1,R}\cdot\bar{b}_{2,S}\cdot\bar{b}_{2,R}\dots\bar{b}_{n,S}\cdot\bar{b}_{n,R}) = \prod_{j=1}^{n}\varphi_S(b_{j,S})\prod_{j=1}^{n}\varphi_R(b_{j,R})$$

Comparing with the boson (or classical) case:

$$\varphi(\bar{b}_{1,S}\cdot\bar{b}_{1,R}\cdot\bar{b}_{2,S}\cdot\bar{b}_{2,R}\dots\bar{b}_{n,S}\cdot\bar{b}_{n,R}) = \varphi(b_{1,S}\cdot b_{2,S}\dots b_{n,S})\varphi(b_{1,R}\cdot\bar{b}_{2,R}\dots b_{n,R})$$

Corresponding to the choices:

$$\bar{b}_{j,S} := j_S(b_{j,S}) = b_{j,S}\otimes 1_R\ ;\ b_{j,S}\in\mathcal{A}_S\ ;\ \bar{b}_{j,R} := j_S(b_{j,R}) = 1_S\otimes b_{j,R}\ ;\ b_{j,R}\in\mathcal{A}_R$$

7.3. *Monotone composite systems*

Choose as in von Neumann case

$$\begin{aligned}\mathcal{A} :&= \mathcal{B}(\mathcal{H}_S)\otimes\mathcal{B}(\mathcal{H}_R)\\ \varphi(x) :&= \varphi_S\otimes\varphi_R(x) := \langle\Phi_S\otimes\Phi_R, x(\Phi_S\otimes\Phi_R)\rangle \qquad ; \quad x\in\mathcal{A}\end{aligned}$$

where Φ_S and Φ_R are unit vectors in $\mathcal{H}_S$ and $\mathcal{H}_R$ respectively. Then choose the embeddings j_S, j_R to be of the form

$$\begin{aligned} j_S(b_S) :&= b_S\otimes P_{\Phi_R} \quad ; \quad \Phi_R\text{–embedding} \quad ; \quad b_S\in\mathcal{B}(\mathcal{H}_S)\\ j_R(b_R) :&= 1_S\otimes b_R \quad ; \quad \text{tensor embedding} \quad ; \quad b_R\in\mathcal{B}(\mathcal{H}_R)\end{aligned}$$

i.e. Notice that $j_S(b_S)$ does not commute with $j_R(1_R)$, in fact:

$$j_S(b_S) = b_S\otimes 1_R \quad ; \quad j_R(b_R) = P_{\Phi_S}\otimes b_R \quad ; \quad j_R(1_R) = P_{\Phi_S}\otimes 1_R$$

Having fixed $n \in \mathbb{N}$ and $b_{1,S}, b_{2,S}, \dots, b_{n,S} \in \mathcal{A}_S$, $b_{1,R}, b_{2,R}, \dots, b_{n,R} \in \mathcal{A}_R$ and with the notations

$$\bar{b}_{j,S} := j_S(b_{j,S}) = b_{j,S} \otimes P_{\Phi_R} \in \mathcal{A}_S^0 \quad , \quad b_{j,S} \in \mathcal{A}_S$$
$$\bar{b}_{j,R} := j_S(b_{j,R}) = 1_S \otimes b_{j,R} \in \mathcal{A}_R^0 \quad , \quad b_{j,R} \in \mathcal{A}_R$$

we want to calculate the mixed moments (correlators)

$$\varphi(\bar{b}_{1,S} \cdot \bar{b}_{1,R} \cdot \bar{b}_{2,S} \cdot \bar{b}_{2,R} \dots \bar{b}_{n,S} \cdot \bar{b}_{n,R}) =$$

$$= \varphi([b_{1,S} \otimes P_{\Phi_R}] \cdot [1_S \otimes b_{1,R}] \cdot [b_{2,S} \otimes P_{\Phi_R}] \cdot [1_S \otimes b_{2,R}] \dots [b_{n,S} \otimes P_{\Phi_R}] \cdot [1_S \otimes b_{n,R})]) =$$

$$= \varphi([b_{1,S} b_{2,S} \dots b_{n,S}] \otimes [P_{\Phi_R} b_{1,R} P_{\Phi_R} b_{2,R} P_{\Phi_R} \dots b_{n,R}])$$

Now recall that

$$P_{\Phi_R} b_{1,R} P_{\Phi_R} = \langle \Phi_R, b_{1,R} \Phi_R \rangle P_{\Phi_R} = \varphi_R(b_{1,R}) P_{\Phi_R}$$

Therefore

$$\varphi([b_{1,S} b_{2,S} \dots b_{n,S}] \otimes [P_{\Phi_R} b_{1,R} P_{\Phi_R} b_{2,R} P_{\Phi_R} \dots P_{\Phi_R} b_{n,R}]) =$$

$$= \varphi_R(b_{1,R}) \cdot \varphi([b_{1,S} b_{2,S} \dots b_{n,S}] \otimes [P_{\Phi_R} b_{2,R} P_{\Phi_R} \dots P_{\Phi_R} b_{n,R}]) =$$

$$= \dots = \varphi_R(b_{1,R}) \varphi_R(b_{2,R}) \dots \varphi_R(b_{n-1,R}) \cdot \varphi([b_{1,S} b_{2,S} \dots b_{n,S}] \otimes [P_{\Phi_R} b_{n,R}])$$

Now recall that, for $x \in \mathcal{A}$

$$\varphi(x) := \langle \Phi_S \otimes \Phi_R, x(\Phi_S \otimes \Phi_R) \rangle$$

Therefore

$$\varphi([b_{1,S} b_{2,S} \dots b_{n,S}] \otimes [P_{\Phi_R} b_{n,R}]) = \langle \Phi_S \otimes \Phi_R, [b_{1,S} b_{2,S} \dots b_{n,S}] \otimes [P_{\Phi_R} b_{n,R}](\Phi_S \otimes \Phi_R) \rangle =$$

$$= \langle \Phi_S, b_{1,S} b_{2,S} \dots b_{n,S} \Phi_S \rangle \langle \Phi_R b_{n,R} \Phi_R \rangle$$

In conclusion

$$\varphi(\bar{b}_{1,S} \cdot \bar{b}_{1,R} \cdot \bar{b}_{2,S} \cdot \bar{b}_{2,R} \dots \bar{b}_{n,S} \cdot \bar{b}_{n,R}) = \varphi_S(b_{1,S} b_{2,S} \dots b_{n,S}) \prod_{j=1}^{n} \varphi_R(b_{j,R})$$

Comparing with the boson (or tensor) case:

$$\varphi(\bar{b}_{1,S} \cdot \bar{b}_{1,R} \cdot \bar{b}_{2,S} \cdot \bar{b}_{2,R} \dots \bar{b}_{n,S} \cdot \bar{b}_{n,R}) = \varphi(b_{1,S} \cdot b_{2,S} \dots b_{n,S}) \varphi(b_{1,R} \cdot b_{2,R} \dots b_{n,R})$$

corresponding to the choices:

$\bar{b}_{j,S} := j_S(b_{j,S}) = b_{j,S} \otimes 1_R$; $b_{j,S} \in \mathcal{A}_S$; $\bar{b}_{j,R} := j_S(b_{j,R}) = 1_S \otimes b_{j,R}$; $b_{j,R} \in \mathcal{A}_R$

one can verify the asymmetry of the notion of monotone independence.

8. Conclusions

We have only considered two examples, but as stated at the beginning of this note, in quantum probability there are uncountably many notions of independence and each of them may turn out to be appropriate to describe different classes of complex physical or biological systems.
An axiomatic characterization of 5 of the most studied notions of quantum probabilistic independence can be found in the paper by Ben Ghorbal and Schürmann [5], which completes a line of research initiated by Schürmann and pursued by several authors. An unification of these notions in terms of generalized embeddings in tensor product spaces is due to Lenczewski [8]. The wide variety of possibilities for such notions is emphasized in the paper [4] and, from a more constructive point of view, in [3]. The equivalence between the constructive and the convolution point of view is proved in [7]. The conditioning problem in the general context of von Neumann algebras is discussed in ([6]).

References

1. Accardi L., Ohya M.: Compound channels, transition expectations and liftings, Applied Mathematics Optimization 39 (1999) 33–59 Volterra preprint N. 75 (1991)
2. Accardi L., Lu Y.G., Volovich I.: Quantum Theory and its Stochastic Limit, Springer Verlag (2002)
3. Accardi L., Crismale V., Lu Y.G.: Constructive universal central limit theorems based on interacting Fock spaces, IDA–QP (Infinite Dimensional Analysis, Quantum Probability and Related Topics) 8 (4) (2005) 631-650
4. Accardi L., Bozejko M.: Interacting Fock spaces and Gaussianization of probability measures, IDA–QP (Infinite Dimensional Analysis, Quantum Probability and Related Topics) 1 (4) (1998) 663–670
5. Ben Ghorbal, Anis; Schürmann, Michael: Non-commutative notions of stochastic independence Mathematical Proceedings of the Cambridge Philosophical Society, 133 (3) (2008) 531-561
6. Stefano Cavallaro, Carlo Cecchini: A unified approach to generalized conditional expectations, operator valued weights and Radon-Nikodym derivatives on von Neumann algebras IDA–QP (Infinite Dimensional Analysis, Quantum Probability and Related Topics) 5 (3) (2002) 409–419
7. Anna Dorota Krystek, Lukasz Jan Wojakowski: Convolution and central limit theorem arising from addition of field operators in one mode type Interacting Fock Spaces, IDA–QP (Infinite Dimensional Analysis, Quantum Probability and Related Topics) 8 (4) (2005) 651–657
8. Romuald Lenczewski: Unification of independence in quantum probability IDA–QP (Infinite Dimensional Analysis, Quantum Probability and Related Topics) 1 (3) (1998) 383
9. Ohya M.: Selected papers, World Scientific

Quantum Bio-Informatics III
eds. L. Accardi, W. Freudenberg & M. Ohya

DYNAMICS AND POTENTIALS

HUZIHIRO ARAKI

Research Institute for Mathematical Sciences,
Kyoto University, Sakyoku, Kyoto 606-8502, Japan,
E-mail: araki@kurims.kyoto-u.ac.jp

(Based on a joint work with Hajime Moriya [[2]])

1. Introduction

In algebraic approach, a quantum system is described by a C*−algebra $\mathcal{A}$ and its dynamics is described by a one-parameter group α_t, $t \in \mathbb{R}$, of $*$−automorphisms of $\mathcal{A}$. For a lattice system such as spins and Fermions on a discrete lattice $\mathbb{L}$ (e.g., $\mathbb{Z}^d$), a widely used method of constructing dynamics $(\mathcal{A}, \alpha_t)$ is as follows.

The interaction energy among physical objects (such as spins and Fermions) in a finite subset I of the lattice $\mathbb{L}$ is denoted $P(I)$ and called the potential. (We denote $I \subset\subset \mathbb{L}$ to indicate that I is a finite subset of $\mathbb{L}$.) According to quantum theory, the time derivative

$$d_\alpha A \equiv \left.\frac{d}{dt}\alpha_t(A)\right|_{t=0} \tag{1}$$

of a physical quantity $A \in \mathcal{A}$ is formally $i[H, A]$ with H representing the total interaction energy (called Hamiltonian). As $\{P(I); \quad I \subset\subset \mathbb{L}\}$ together is the total interaction energy, the time derivative (mathematically a $*$−derivation) would be

$$d_\alpha A = i \sum_{I \subset\subset \mathbb{L}} [P(I), A] \tag{2}$$

The dynamics α_t is then defined by exponentiating d_α. Under some technical conditions (chosen for convenience of the proof) on $\{P(I)\}$, this procedure

$$\{P(I)\} \to d_\alpha \to \alpha_t$$

is proved to make sense for each case.

In the proof of the convergence of the right hand side of (2), a use of a local structure of $\mathcal{A}$ is crucial. For each $I \subset\subset \mathbb{L}$, $\mathcal{A}$ has a subalgebra denoted by $\mathcal{A}(I)$ corresponding to physical quantities in the set I. Furthermore, there is an automorphism Θ satisfying $\Theta^2 = id$ and each $\mathcal{A}(I)$ is $\Theta-$invariant as a set. $\mathcal{A}$ and $\mathcal{A}(I)$ split into a sum of $\Theta-$even and $\Theta-$odd parts:

$$\mathcal{A}_{\pm} = \{A \in \mathcal{A}; \quad \Theta(A) = \pm A\}, \quad \mathcal{A}(I)_{\pm} = \mathcal{A}(I) \cap \mathcal{A}_{\pm}. \tag{3}$$

In fact, for any $A \in \mathcal{A}$,

$$A_{\pm} = \frac{1}{2}(A \pm \Theta(A)) \in \mathcal{A}_{\pm}, \quad A_{+} + A_{-} = A, \tag{4}$$

so that

$$\mathcal{A} = \mathcal{A}_{+} + \mathcal{A}_{-}, \quad \mathcal{A}(I) = \mathcal{A}(I)_{+} + \mathcal{A}(I)_{-}. \tag{5}$$

(This automorphism counts Fermion even-oddness of $A \in \mathcal{A}$.) To make our presentation simpler, we assume that α_t commutes with Θ. (See [[4]] for non-even α_t.) Then $P(I) \in \mathcal{A}(I)_{+}$ and $P(I)$ commutes with $A \in \mathcal{A}(J)$ if $I \cap J = \varnothing$ due to the graded commutation relations. Therefore, for any $A \in \mathcal{A}(J)$, the term with I satisfying $I \cap J = \varnothing$ vanishes in (2) and this property contributes to the convergence of (2).

2. Main results

(1) Existence of Potentials

Let

$$\mathcal{A}_0 = \bigcup_{I \subset\subset \mathbb{L}} \mathcal{A}(I). \tag{6}$$

It is a dense $*-$subalgebra of $\mathcal{A}$. In conventional approach, the $*-$derivation d_α is defined by (2) for all $A \in \mathcal{A}_0$ under some technical condition on potentials $\{P(I); \quad I \subset\subset \mathbb{L}\}$; namely d_α is a $*-$derivation on $\mathcal{A}_0$. By our assumption on α_t, d_α commutes with Θ.

The first of our main results is the following:

Theorem 2.1. *For any $*-$derivation d_α with domain $\mathcal{A}_0$, which commutes with Θ, there exists a system of potentials $P(I) \in \mathcal{A}(I)_{+}$, $I \subset\subset \mathbb{L}$ which gives d_α by the right hand side of (2).*

Furthermore, they satisfy the following convergence property, which guarantees the convergence of (2).

Convergence property: For any $I \subset\subset \mathbb{L}$,

$$S(J) \equiv \sum_K \{P(K) \mid \quad K \cap I \neq \varnothing, \quad K \subset J\} \in \mathcal{A}_+ \tag{7}$$

is a Cauchy net as $J \nearrow \mathbb{L}$, J being a finite subset.

This means that for any $*$−derivation defined on $\mathcal{A}_0$ and commuting with Θ, one can find potentials $\{P(I); \quad I \subset\subset \mathbb{L}\}$ which gives that $*$−derivation via (2). This is a vast improvement over the past situation. It guarantees the existence of potentials for any Θ−even derivation which has $\mathcal{A}_0$ as its definition domain.

For a later use, we define

$$H(I) = \lim_{J \nearrow \mathbb{L}} S(J) \in \mathcal{A}_+. \tag{8}$$

Then $d_\alpha A = i[H(I), A]$ for $A \in \mathcal{A}(I)$.

(2) Unique Choice of Potentials

(2-a) Equivalent Potentials

Two sets of potentials $\{P_1(I); \quad I \subset\subset \mathbb{L}\}$ and $\{P_2(I); \quad I \subset\subset \mathbb{L}\}$ are called equivalent if the $*$−derivations given by (2) coincide. A simple example is given by moving the potential $P_1(I)$ for a finite subset to a part of the potential for subset J, of which I is a proper subset. More explicitly, fix $I_1 \subset\subset \mathbb{L}$ and $I_2 \subset\subset \mathbb{L}$ such that $I_2 \supset I_1$ and $I_2 \neq I_1$. Then construct P_2 out of P_1 by

$$P_2(I) = P_1(I) \quad \text{for } I \neq I_1, I_2,$$
$$P_2(I_1) = 0, \quad P_2(I_2) = P_1(I_2) + P_1(I_1).$$

Then $P_2 \neq P_1$ and they produce the same $*$−derivation by (2).

In order to avoid such situation, one can demand that any of $P(I)$ does not contain a part which can be viewed as a part of $P(J)$ for any $J \subset I, \quad J \neq I$. This can be mathematically formulated by a use of a conditional expectation, which we now introduce.

(2-b) Conditional Expectations

A state φ of $\mathcal{A}$ is called a product state if

$$\varphi(AB) = \varphi(A)\varphi(B)$$

whenever $A \in \mathcal{A}(I)$ and $B \in \mathcal{A}(J)$ for disjoint finite subsets I and J of $\mathbb{L}$. Such a product state exists. For example, a tracial state of $\mathcal{A}$ is an example (a tracial state τ is characterized by $\tau(AB) = \tau(BA)$ for any A

and $B \in \mathcal{A}$) and the vacuum state for Fermions is another example. Both of them are even state in the sense that $\varphi(\Theta(A)) = \varphi(A)$ for all $A \in \mathcal{A}$.

Theorem 2.2. *Let φ be an even product state. Then there exists a map E_I^φ from any $A \in \mathcal{A}$ to $E_I^\varphi(A) \in \mathcal{A}(I)$ for each subset I of $\mathbb{L}$ with the following properties.*

(1) *E_I^φ is linear, positive and unital.*
(2) *For any $A \in \mathcal{A}$ and $B \in \mathcal{A}(I)$,*

$$E_I^\varphi(AB) = E_I^\varphi(A)B, \quad E_I^\varphi(BA) = BE_I^\varphi(A).$$

(3) *E_I^φ is a projection of norm 1 onto $\mathcal{A}(I)$.*
(4) *For any $A \in \mathcal{A}$,*

$$E_I^\varphi(\Theta(A)) = \Theta(E_I^\varphi(A)).$$

Such E_I^φ is uniquely determined by the property $E_I^\varphi(A) \in \mathcal{A}(I)$ and

$$\varphi(B_1 A B_2) = \varphi(B_1 E_I^\varphi(A) B_2)$$

to be satisfied for all B_1 and B_2 in $\mathcal{A}(I)$.

The conditional expectation E_I^φ satisfies

$$E_I^\varphi E_J^\varphi = E_{I\cap J}^\varphi, \quad E_I^\varphi E_{I^c}^\varphi = E_{I^c}^\varphi E_I^\varphi = \varphi,$$
$$\lim_\nu E_{K_\nu}^\varphi(A) = E_K^\varphi(A) \quad \text{if the net } K_\nu \text{ tends to } K,$$
$$\lim_{K\nearrow\mathbb{L}} E_K^\varphi(A) = A.$$

(2-c) <u>Standard Potentials</u>

Definition 2.1. An $\mathcal{A}$–valued function $P(I)$ of finite subsets I of $\mathbb{L}$ is called a standard potential w.r.t. the conditional expectations $\{E_I^\varphi;\ I \subset\subset \mathbb{L}\}$ if it satisfies the following:

(0) $P(I) = 0$ for $I = \varnothing$.
(1) $P(I) = P(I)^* \in \mathcal{A}(I)$.
(2) $E_J^\varphi(P(I)) = 0$ if $J \subset I$ and $J \neq I$.
(3) (Convergence condition) For any fixed $I \subset\subset \mathbb{L}$,

$$\sum_k \{P(K);\quad K \cap I \neq \varnothing,\quad K \subset J\}$$

is a Cauchy net indexed by $J \subset\subset \mathbb{L}$ as $J \nearrow \mathbb{L}$.

The properties (0) and (1) are conventional properties of a potential. The property (3) is mentioned already in Theorem 2.1. The property (2) is a new requirement mentioned verbally at the end of (2-a).

As a consequence of (1) and (2), any standard potential satisfies

$$(2)' \qquad E_J^{\varphi}(P(I)) = \begin{cases} 0 & \text{if } J \not\supset I, \\ P(I) & \text{if } J \supset I. \end{cases}$$

Due to this property, E_J^{φ} can be used to select those terms with $J \supset I$ out of any (finite or infinite) linear combination of $P(I)$.

(2-d) Main Results

Theorem 2.3. *(Existence and uniqueness) For any $*-$derivation d with domain $\mathcal{A}_0$, which commutes with Θ, there exists a system of standard potentials $P(I)$, $I \subset\subset \mathbb{L}$ (standardness referring to any fixed product state φ) which gives d by (2) and such a system $\{P(I)\}$ is unique (for any chosen product state φ).*

If the reference product state φ is changed, one obtains a different system of potentials, which are equivalent.

3. Applications

(3-a) A Banach space of potentials

A cubic lattice is an additive group. Each $i \in \mathbb{L}$ induces an automorphism τ_i of $\mathcal{A}$ satisfying

$$\tau_i \mathcal{A}(I) = \mathcal{A}(I + i).$$

(It represents the translation of positions (in $\mathbb{L}$) of spins and Fermions by i.) We will now consider the dynamics α_t which is translation invariant, namely $\tau_i \alpha_t = \alpha_t \tau_i$. Then we will also use a product state φ which is translation invariant ($\varphi \circ \tau_i = \varphi$), which implies $\tau_i E_I^{\varphi} = E_{I+i}^{\varphi} \tau_i$ and $\tau_i(P(I)) = P(I + i)$ (translation covariance of the potential).

The set of all families $\{P(I);\ \ I \subset\subset \mathbb{L}\}$ of translation covariant standard potentials can be made a normed space by defining

$$(c_1 P_1 + c_2 P_2)(I) = c_1 P_1(I) + c_2 P_2(I),$$
$$\|P\| = \|H_p(\{i\})\|,$$

where $H_p(\{i\})$ is defined by (8) and is independent of a point $\{i\}$ of $\mathbb{L}$ due to the translation covariance of H_p.

Theorem 3.1. *The normed space of translation covariant standard potentials defined above is a separable Banach space.*

This can be used in the proof of equivalence of different criteria for equilibrium state [[2]].

(3-b) <u>*Energy Estimates*</u>

The standardness properties have never been noticed before. However it is sometimes a useful technical tool. The following energy estimates can be proved in a simple and elementary manner by using the property $(2)'$ in (2-c) without any other properties of the potential [[5], Theorem 8.1].

Theorem 3.2. *For any translation invariant standard potential P, the total energies in a finite subset I, defined by*

$$U_p(I) = \sum_{K \subset I} P(K) \qquad \textit{(energy as closed system),}$$

$$H_p(I) = \lim_{J \nearrow \mathbb{L}} \sum \{P(K);\ K \cap I = \varnothing,\ K \subset J\} \quad \textit{(energy as an open system),}$$

have a bound proportional to the size $|I|$ (the number of lattice points in I):

$$\|U_p(I)\| \leq \|H_p(I)\| \leq |I| \, \|P\| \, .$$

References

[1]. Araki, H., Moriya, H.: Local thermodynamical stability of Fermion lattice systems. *Lett. Math. Phys.* 62, 33–45 (2002)

[2]. Araki, H., Moriya, H.: Equilibrium statistical mechanics of Fermion lattice systems. *Rev. Math. Phys.* 15, 93–198 (2003)

[3]. Araki, H.: Conditional expectations relative to a product state and the corresponding standard potentials. *Commun. Math. Phys.* 246, 113–132 (2004)

[4]. Araki, H.: Standard potentials for non-even dynamics. *Commun. Math. Phys.* 262, 161–176 (2006)

[5]. Araki, H.: Dynamics and potentials. *Lett. Math. Phys.* 82, 297-306 (2007)

[6]. Araki, H.: Super-derivations and associated standard super-potentials. *Commun. Math. Phys.* 276, 261–285 (2007)

[7]. Bratelli, O., Robinson, D.W.: Operator Algebras and Quantum Statistical Mechanics. 2. Equilibrium states. Models in Quantum Statistical Mechanics. 2nd ed. *Texts and Monographs in Physics.* Springer, Berlin (1997)

Quantum Bio-Informatics III
eds. L. Accardi, W. Freudenberg & M. Ohya

KOSSAKOWSKI-OHYA TELEPORTATION SCHEME AND ITS APPLICATIONS

M. ASANO, M. OHYA AND Y. TANAKA

Department of Information Sciences,
Faculty of Science and Technology,
Tokyo University of Science, Noda City, Chiba 278, Japan

In many models of perfect quantum teleportation, maximal entangled states are postulated between Alice and Bob. Recently, Kossakowski and Ohya proposed a new scheme of perfect teleportation where a non-maximal entangled state is used [3]. In this paper, we introduce a concrete teleportation model of K-O scheme, called a m-level teleportation. A non-maximal entangled state in this model is realizable in optical techniques. Moreover, we propose another teleportation model designed by using a mathematical formalism of K-O theory. In the second model, a new operation is added to a conventional teleportation process, which is called a filtering and implemented before Alice's measurement. We show effects of the filtering for success probabilities of teleportation.

1. Introduction

Quantum teleportation[1,2] is an important concept in quantum information theory. In order to transmit an information of target qubit from Alice to Bob, an entangled pair of qubits is shared between them. Firstly, Alice implements a quantum Bell measurement on her two qubits, a target qubit and one of the shared qubits. Measured states are described in the forms of Bell states, $\frac{1}{\sqrt{2}}(|00\rangle \pm |11\rangle)$ and $\frac{1}{\sqrt{2}}(|01\rangle \pm |10\rangle)$. Bob's mission after Alice's measurement is to recover the target bit from his shared qubit, and this operation is written as a unitary transformation to a state of the qubit. Conventionally, Bob can complete his mission if the shared entangled state is maximal one as $\frac{1}{\sqrt{2}}(|00\rangle + |11\rangle)$. Maximal entangled states are ideal resources for perfect quantum teleportations, but it is difficult to make a maximal entangled state experimentally. We have to design more practical teleportations that become successful with high probabilities, even if non-maximal entangled states are used.

For this problem, a new scheme proposed by Kossakowski and Ohya is useful [3]. In their mathematical formalism, a process of teleportation is represented as a map from a state of a target qubit to a state of a qubit Bob obtains after Alice's measurement. This map is called a teleportation map, when a teleportation map is linear, its teleportation is perfect. As explained briefly in Sec. 1, Kossakowski and Ohya constructed an ideal teleportation map whose linearity is preserved regardless an intensity of entangled state. In this paper, as applications of K-O formalism, two teleportation models are introduced. The Sec. 2 describes an m-level teleportation model [5] that realizes the K-O teleportation map. This model transmits a quantum state called an m-level state and uses a non-maximal entangled state on n^2-dimensional Hilbert space. If the non-maximal entangled state has a specific property, its teleportation becomes perfect. In Sec. 3, we design another teleportation model with using K-O formalism. This model adds a new operation to a conventional teleportation process. The operation, "filtering" we say here, is implemented to a target qubit before Alice's Bell measurement, and a form of filtering is decided depending on a intensity of entangled pair prepared between Alice and Bob. We show the filtering has potential to improve success probabilities of teleportation.

2. Kossakowski-Ohya Scheme

In this section, we briefly review a representation of teleportation process and an ideal teleportation map, proposed by Kossakowski and Ohya.

In K-O representation, an input state is written as an n-dimensional state vector $\rho = |\phi\rangle\langle\phi|$ in Hilbert space $\mathcal{H}_1 = \mathbb{C}^n$, and a vector of entangled state on $\mathcal{H}_2 \otimes \mathcal{H}_3 = \mathbb{C}^n \otimes \mathbb{C}^n$is written in the form;

$$|\Psi\rangle = \sum_{i=1}^{n} |i\rangle_2 \otimes f\, |i\rangle_3 \,, \tag{1}$$

where $\{|i\rangle_{2,3}\}$ are orthogonal basis in $\mathcal{H}_{2,3}$, and the operator $f \in \mathcal{B}(\mathcal{H}_3)$ acts on basis $|i\rangle_3$. ($rank(f) = n$ is assumed.) The reduced density operator $tr_2(|\Psi\rangle\langle\Psi|)$ is equivalent to the operator ff^*. The eigenvalues of ff^*, $\{\eta_i\}$ are closely related to intensity of entanglement. If $\sqrt{n}f$ is unitary, all η_i are equal to $\frac{1}{n}$, and $|\Psi\rangle$ defined with such the f implies a vector of maximum entangled state.

States Alice measures on a target qubit and one of the entangled pair are described as

$$|\Phi_\alpha\rangle = \sum_{i=1}^{n} g_\alpha^* \,|i\rangle_1 \otimes |i\rangle_2 \;\; (\alpha = 1, ..n^2). \tag{2}$$

The operators $\{g^*_\alpha\} \in \mathcal{B}(\mathcal{H}_1)$ satisfy $tr(g_\alpha g^*_\beta) = \delta_{\alpha,\beta}$ from the orthogonal condition $\langle \Phi_\alpha | \Phi_\beta \rangle = \delta_{\alpha,\beta}$. Before a measurement, a compound system in $\mathcal{H}_1 \otimes \mathcal{H}_2 \otimes \mathcal{H}_3$ is written by

$$|\phi\rangle \otimes |\Psi\rangle = \sum_{\alpha=1}^{n^2} |\Phi_\alpha\rangle \otimes f g_\alpha |\phi\rangle . \tag{3}$$

Conventionally, a quantum measurement is represented by a projection $P_\alpha = |\Phi_\alpha\rangle \langle \Phi_\alpha|$. When Alice measures $|\Phi_\alpha\rangle$ in $\mathcal{H}_1 \otimes \mathcal{H}_2$, Bob obtains a qubit with

$$\tilde{\rho}_\alpha = \frac{T_\alpha(\rho)}{\mathrm{tr}(T_\alpha(\rho))}. \tag{4}$$

The map T_α is a teleportation map defined by

$$\begin{aligned} T_\alpha(\rho) &\equiv tr_{\mathcal{H}_T \otimes \mathcal{H}_A}(P_\alpha \otimes I)\rho \otimes \sigma (P_\alpha \otimes I) \\ &= f g_\alpha \rho g^*_\alpha f^*, \end{aligned} \tag{5}$$

where $\rho = |\phi\rangle \langle \phi|$ and $\sigma = |\Psi\rangle \langle \Psi|$. Note that this teleportation map T_α is not trace preserving. So, a channel $\Lambda^*_\alpha(\rho) = \tilde{\rho}_\alpha$ is nonlinear in general.

Here let us consider a normalization of teleportation map T_α. In order to do this, a dual map $\tilde{T}_\alpha : \mathcal{B}(\mathcal{H}_3) \mapsto \mathcal{B}(\mathcal{H}_1)$ (i.e. $trAT_\alpha(\rho) = tr\tilde{T}_\alpha(A)\rho$, $A \in \mathcal{B}(\mathcal{H}_1)$) is intorduced. Under the condition of $rank(f) = rank(g) = n$, $\tilde{T}_\alpha$ is normalized as

$$\kappa_\alpha^{-\frac{1}{2}} \tilde{T}_\alpha \kappa_\alpha^{-\frac{1}{2}} \equiv \tilde{\Gamma}_\alpha, \tag{6}$$

where $\kappa_\alpha \equiv g^*_\alpha f^* f g_\alpha$. The dual map Γ_α of $\tilde{\Gamma}_\alpha$ is trace preserving. A mapped state is given by

$$\Gamma_\alpha(\rho) \equiv f g_\alpha \kappa_\alpha^{-\frac{1}{2}} \rho \kappa_\alpha^{-\frac{1}{2}} g^*_\alpha f^*.$$

Importantly, $f g_\alpha \kappa_\alpha^{-\frac{1}{2}}$ is unitary even if $\sqrt{n} f$ is not unitary. There always exists a unitary key $W_\alpha \equiv (f g_\alpha \kappa_\alpha^{-\frac{1}{2}})^*$ recovering ρ from $\Gamma_\alpha(\rho)$.

3. *m*-level Teleportation

In an m-level teleportation model [5], we assume the operators $f \in \mathcal{B}(\mathbb{C}^n)$ and $\{g_\alpha\} \in \mathcal{B}(\mathbb{C}^n)$ such that the operator $\kappa_\alpha = g^*_\alpha f^* f g_\alpha$ has a set of eigenstates $\{|e_{\alpha i}\rangle\}$ and a set of eigenvalues $\{k^\alpha_i\}$ that are invariant for any $\alpha(= 1 \cdots n^2)$; one can define $\{|e_{\alpha i}\rangle\} \equiv \{|e_i\rangle\}$ and $\{k^\alpha_i\} \equiv \{\omega_i \in \mathbb{R}\}$. We consider the following state vector as an input;

$$|\phi(m)\rangle = \sum_{i=1}^{m} c_i |f_i\rangle \in \mathcal{H}_1 = \mathbb{C}^n, \tag{7}$$

where $m \leq n$, $\{|f_j\rangle \,|j = 1 \cdots m\} \subseteq \{|e_i\rangle\}$ and the coefficients $c_{i=1,..,m}$ are not zero. The state $\rho(m) \equiv |\phi(m)\rangle \langle\phi(m)|$ is called an m-level state. In the paper of [5], we showed that a perfect teleportation of m-level state can be successful with using a certain non-maximal entangled state, and then, the K-O map Γ_α is realized. For example, in the case of $m = 2$ and $n = 2^s (s \in \mathbb{N})$, one needs a non-entangled state satisfying the condition; for the eigenvalues $\{\omega_i \in \mathbb{R}\}$,

$$\sum_{i=1}^{2^s-1} \omega_i = \sum_{j=1}^{2^s} \omega_j = \frac{1}{2}. \tag{8}$$

We call such the entangled state, a two-level entangled state.

It is expected that in experiments, one can make a two-level entangled state easier than a maximal entangled state. For example, a beam-splitter splits a beam of photon number states $|N\rangle$ into two beams which is described as an entangled state of

$$\begin{aligned} |\Psi\rangle &= \sum_{n=0}^{N} d_n \,|n\rangle \otimes |N-n\rangle \\ d_n &= \mathrm{e}^{\mathrm{i}\phi n}\sqrt{\frac{N!}{n!(N-n)!}}(\sin\frac{\beta}{2})^n(\cos\frac{\beta}{2})^{N-n}, \end{aligned} \tag{9}$$

where N is the number of photon, β is transmission rate and ϕ is the phase difference between the reflected beam and the transmitted beam [6,7]. The operator f in the description of $|\Psi\rangle = \sum_i |i\rangle \otimes f\,|i\rangle$ is given by $f = \sum_n^N d_n \,|N-n\rangle \langle N|$. Thus, eigenvalues of f^*f, namely $\{\omega_i\}$ are equal to $\{|d_i|^2\}$. Figure. 1 shows values of $S_{even} = \sum_{n=even} |d_i|^2$ and $S_{odd} = \sum_{n=odd} |d_i|^2$ for the transmission β at $N = 4$. We can see $S_{even} = S_{odd}$ at certain βs.

4. Quantum Teleportation with Filtering

In this section, we propose another model of teleportation where a new operation called filtering is added to a conventional teleportation process. This model is defined in two-dimensional Hilbert spaces; $\mathcal{H}_{i=1,2,3} = \mathbb{C}^2$.

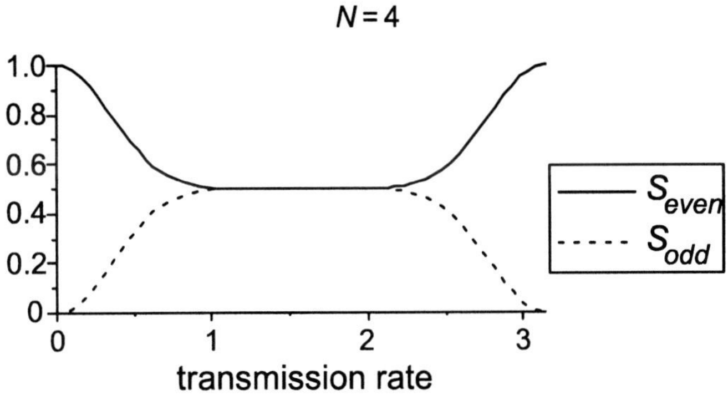

Figure 1. Values of $S_{even} = \sum_{n=even} |d_i|^2$ and $S_{odd} = \sum_{n=odd} |d_i|^2$ for β.

4.1. *Filtering*

An operation of filtering is implemented to a target qubit of ρ before Alice's measurement, and it is described as a map $F : \mathcal{S}(\mathcal{H}_1) \mapsto \mathcal{S}(\mathcal{H}_1)$ as

$$F(\rho) = \frac{\tilde{f}\rho\tilde{f}^*}{\mathrm{tr}(\tilde{\kappa}\rho)}. \tag{10}$$

Here, $\tilde{\kappa} = \tilde{f}^*\tilde{f}$, and the operator $\tilde{f}$ has the form of

$$\tilde{f} = \begin{pmatrix} \omega_2 & 0 \\ 0 & \omega_1 \end{pmatrix}, \tag{11}$$

for the operator

$$f = \begin{pmatrix} \omega_1 & 0 \\ 0 & \omega_2 \end{pmatrix}. \tag{12}$$

(About a realization of filtering F, we explain in Subsection. 4.4.) Alice implements a Bell measurement for the state $F(\rho)$ instead of the original ρ, and after measurement, Bob obtains a qubit with the state of

$$\Lambda_\alpha^*(F(\rho)) = \frac{f g_\alpha \tilde{f}\rho\tilde{f}^* g_\alpha^* f^*}{\mathrm{tr}(\tilde{f}\kappa_\alpha\tilde{f}^*\rho)}, \tag{13}$$

where $\{g_\alpha\}$ are given by

$$g_{\alpha=1} = \frac{1}{\sqrt{2}}\begin{pmatrix} 1 & 0 \\ 0 & 1 \end{pmatrix}, g_{\alpha=2} = \frac{1}{\sqrt{2}}\begin{pmatrix} 1 & 0 \\ 0 & -1 \end{pmatrix},$$

$$g_{\alpha=3} = \frac{1}{\sqrt{2}}\begin{pmatrix} 0 & 1 \\ 1 & 0 \end{pmatrix}, g_{\alpha=3} = \frac{1}{\sqrt{2}}\begin{pmatrix} 0 & 1 \\ -1 & 0 \end{pmatrix}.$$

One can easily check that unitary keys for $\alpha = 3, 4$ do not exist unless $|\omega_1| = |\omega_2| = \frac{1}{\sqrt{2}}$. In contrast, if α is 1 or 2, $\Lambda_\alpha^*(F(\rho))$ is rewritten as

$$\Lambda_\alpha^*(F(\rho)) = g_\alpha \rho g_\alpha, \tag{14}$$

and then, unitary keys for recovering the original ρ from $T_\alpha(F(\rho))$ are given by

$$W_{\alpha=1} = \frac{1}{\sqrt{2}} \begin{pmatrix} 1 & 0 \\ 0 & 1 \end{pmatrix}, W_{\alpha=2} = \frac{1}{\sqrt{2}} \begin{pmatrix} 1 & 0 \\ 0 & -1 \end{pmatrix}. \tag{15}$$

4.2. *Success Probability of Teleportation*

In this model, a teleportation is successful when Alice measures the state of $|\Phi_{\alpha=1,2}\rangle$ on $\mathcal{H}_1 \otimes \mathcal{H}_2$, and its success probability P is given by

$$P = \sum_{\alpha=1,2} \frac{\mathrm{tr}(\tilde{f}\kappa_\alpha \tilde{f}^* \rho)}{\mathrm{tr}(\tilde{\kappa}\rho)}. \tag{16}$$

With introducing components of a target $\rho = |\phi\rangle\langle\phi|$, p and q (i.e. $|\phi\rangle = p|0\rangle + q|1\rangle$), we rewrite P as

$$\begin{aligned} P &= \frac{|\omega_1|^2|\omega_2|^2}{|\omega_2|^2|p|^2 + |\omega_1|^2|q|^2} \\ &= \frac{x(1-x)}{t(1-2x)+x}. \end{aligned} \tag{17}$$

Here, $t = |p|^2$ and $x = |\omega_2|^2$. If $x = \frac{1}{2}$, an intensity of an entangled state σ is maximized, and then P takes invariant $\frac{1}{2}$ for any values of t. If $x < \frac{1}{2}$ which implies σ is non-maximal entangled state, the success probability P takes various values depending on t. Figure. 1 shows values of P for range of $0 \le t \le 0.5$ at $x = 0.25$. It should be noted that the range of $0 \le t \le 0.25$, P has above $\frac{1}{2}$.

In general, $\frac{1}{2} \le P \le 1 - x$ in the range of $0 \le t \le x$.

The above property of P implies that our model makes teleportation successful with high probability of $P \ge \frac{1}{2}$, when target ρ is set in the range of $0 \le t \le x$. Let us put $|p|$ into $|\sin\xi|$ and $|\omega_2|$ into $|\sin\theta|$. Then the range $0 \le t \le x$ is represented as

$$-|\theta| \le \xi \le |\theta|\ , \ \pi - |\theta| \le \xi \le \pi + |\theta|. \tag{18}$$

(See Fig. 2.) Here, note $0 \le |\theta| \le \pi/4$ if $x \le \frac{1}{2}$.

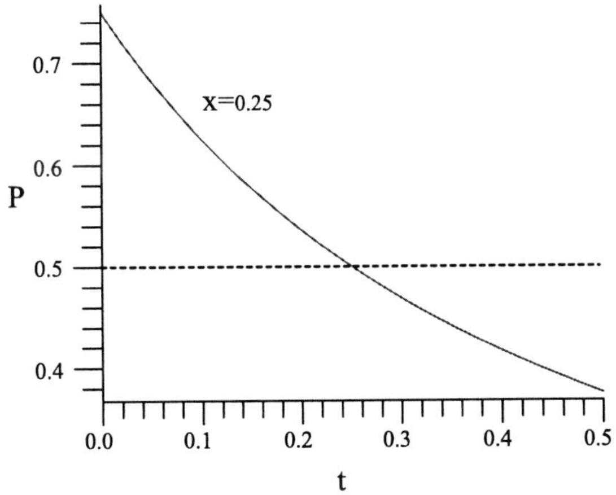

Figure 2. Success probability P at $x = 0.25$

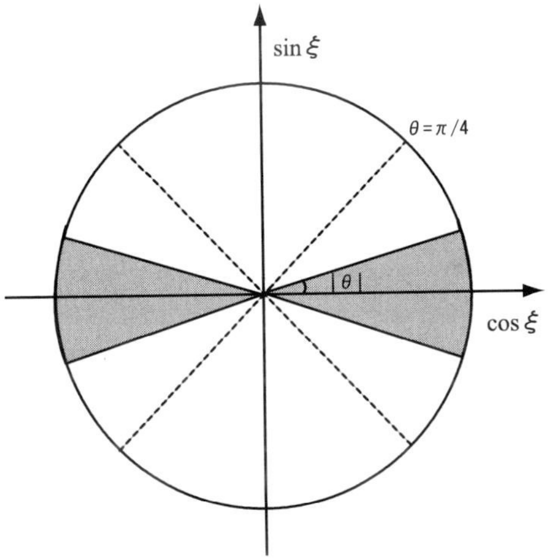

Figure 3. Range of angle ξ for $P \geq \frac{1}{2}$

4.3. *Tuning of Target Qubit*

A target state ρ with success probability $P \geq \frac{1}{2}$ is restricted in a range as seen in Fig. 2. The range width $2|\theta|$ is decided depending on the intensity of entanglement of the state σ. A most simple method to provide high success probability for any state of target is to modify an angle ξ of an original ρ

into $\xi - \zeta$ by a rotation

$$R(\zeta) = \begin{pmatrix} \cos\zeta & -\sin\zeta \\ \sin\zeta & \cos\zeta \end{pmatrix}. \tag{19}$$

An appropriate angle ζ satisfies

$$-|\theta| \leq \xi - \zeta \leq |\theta| \ , \ \pi - |\theta| \leq \xi - \zeta \leq \pi + |\theta|. \tag{20}$$

We call such the operation "tuning". Here, note that for the implementation of tuning, the following two assumptions are needed.
Assumption1: Alice knows the angle θ.
Assumption2: Alice knows that the angle ξ exists in a range of $\mu \leq \xi \leq \nu$, $\pi + \mu \leq \xi \leq \pi + \nu$ with extent of $|\mu - \nu| \leq 2|\theta|$.
Also, Alice has to send the value of tuning angle ζ to Bob for recovering ρ from $T_\alpha(F(\tilde{\rho}))$. ($\tilde{\rho}$ is the state of target after tuning.)

We should remark about tolerance for Eve in our model. Even if Eve obtains the information of ζ, the information of ξ is unknown for Eve in the extent of range $2|\theta|$. In our model, degree of the tolerance is related with the intensity of entanglement of σ quantitatively.

4.4. *Realization of Filtering*

In this sub section, we discuss realization of filtering

$$F(\rho) = \frac{\tilde{f}\rho\tilde{f}^*}{\mathrm{tr}(\tilde{\kappa}\rho)}. \tag{21}$$

With using the components of state vector of target $|\phi\rangle = p\,|0\rangle + q\,|1\rangle$ and the two elements of f, we can rewrite $F(\rho)$ as

$$F(\rho) = \frac{p\omega_2\,|0\rangle + q\omega_1\,|1\rangle}{|p\omega_2|^2 + |q\omega_1|^2}. \tag{22}$$

First, an ancila qubit of

$$|f\rangle = \omega_1\,|0\rangle + \omega_2\,|1\rangle \tag{23}$$

is prepared on $\mathcal{H}_1' = \mathbb{C}^2$. Next, on the compound system of $|\phi\rangle \otimes |f\rangle \in \mathcal{H} \otimes \mathcal{H}_1'$, controlled-Not [4] operation is acted;

$$\begin{aligned} U_{C-N}\,|\phi\rangle \otimes |f\rangle = p\,|0\rangle \otimes (\omega_1\,|0\rangle + \omega_2\,|1\rangle) \\ + q\,|1\rangle \otimes (\omega_1\,|1\rangle + \omega_2\,|0\rangle). \end{aligned} \tag{24}$$

Lastly, Alice measures $|0\rangle\langle 0|$ or $|1\rangle\langle 1|$ on $\mathcal{H}_1'$. If she obtains $|1\rangle\langle 1|$, the state of $F(\rho)$ is provided on $\mathcal{H}_1$. When $|0\rangle\langle 0|$ is obtained, the filtering is failed. The generation of the state $F(\rho)$ by this method is probabilistic. However, the filtering $F(\rho)$ is implemented locally at Alice's site, and Alice can try to construct it repeatedly not breaking resources of entangled states shared between her and Bob.

5. Conclusion

The two different teleportation models were introduced in this paper. Both models are designed with using K-O formalism describing a general teleportation process. The first one, m-level teleportation model realizes the K-O map and can make a perfect teleportation of m-level state with a non-maximal entangled state. On the other hand, the second model with a new operation called filtering does not realize the ideal map, however, improves success probabilities of teleportations. It will be a further problem to find a more effective filtering. Also, it is interesting that in this model, intensity of entanglement is related with tolerance for Eve, quantitatively. Such the relation can not be discussed in conventional perfect teleportation models.

References

1. C. H. Bennett, G. Brassard, C. Crepeau, R. Jozsa, A. Peres and W. K. Wootters, Teleporting an unknown quantum state via dual classical and EPR channels, Phys. Rev. Lett. **70**, 1895 (1993).
2. M. Nielsen and I. Chuang, *Quantum Computation and Quantum Information* (Cambridge University Press, Cambridge, UK, 2000).
3. A. kossakowski, M. Ohya, New scheme of quantum teleportation, Infinite Dimensional Analysis, Quantum Probability and Related Topics Vol. **10**, No. 3 411 (2007).
4. A. Barenco, C. H. Bennett, R. Cleve, D. P. DiVincenzo, N. Margolus, P. Shor, T. Sleator, J. A. Smolin, and H. Weinfurter, Phys. Rev. A **52**, 3457 (1995).
5. M. Asano,M.Ohya and Y. Tanaka, Complete M-level Quantum Teleportation Based on Kossakowski-Ohya Scheme, Quantum Probability and White Noise Analysis Vol. **24** (2009) 19-29.
6. M. S. Kim,W. Son,V. Buzek, and P. L. Kinght, Entanglement by a beam splitter: Nonclassicality as a prerequisite for entanglement. Phys. Rev.A**65**, 032323 (2002).
7. P.T.Cochrane, G.J.Milburn,and W.J.Munro, Teleportation using coupled oscillator states, Phys. Rev. A**62**, 062307 (2000)

Quantum Bio-Informatics III
eds. L. Accardi, W. Freudenberg & M. Ohya

UTILITY AND VALUE OF INFORMATION IN COGNITIVE SCIENCE, BIOLOGY AND QUANTUM THEORY

ROMAN V. BELAVKIN

School of Engineering and Information Sciences
Middlesex University, London NW4 4BT, UK
E-mail: R.Belavkin@mdx.ac.uk

A generalisation of the concepts utility and information value is given for the non-commutative case. In particular, states and utility operators are considered in dual linear spaces equipped with pre-orders, generated by a wedge of utility operators. It is shown that solutions to the information value problem give rise to an isotone Galois connection between the pre-ordered spaces. A particular form of this connection depends on the choice of a functional representing information resource. The properties of information resource are discussed from the point of information value theory, and an example is presented that generalises several known forms of classical and quantum information. Potential areas of application of information value in cognitive science, biology and quantum theory are discussed.

Keywords: Information value; Utility; Quantum information.

1. Introduction

The concepts of probability and expectation first appeared in correspondence between Blaise Pascal and Pierre Fermat, where they discussed how to predict the outcomes of games of chance. It was soon realised how useful these ideas were for predictions and optimisation of decisions under uncertainty, even though axiomatisation of the classical probability and expected utility did not happen until the first half of the 20th century in the works of Kolmogorov [1] and von Neumann and Morgenstern [2] respectively. The influence of these theories has been profound not only in physics, cybernetics and economics, but also in cognitive science and biology.

In cognitive science, computational models are used to study cognition and behaviour of humans and some animals. Bayesian inference and expected utility theories are successfully used in cognitive architectures [3] in their action-selection or decision-making algorithms. In biology and evolutionary theory, game-theoretic concepts, such as the expected pay-off, and

Bayesian analysis have been used to study evolutionary stable strategies [4,5].

Despite these advances, however, there are many open problems both in cognitive science and computational biology. In particular, there are many examples supported by data showing that behaviour of humans and animals deviate significantly from the 'rational' predictions of the expected utility theory. Some of these deviations are related to the so-called paradoxes of decision-making, of which there are plenty [6,7] (e.g. St. Petersburg, Allais, Ellsberg paradoxes to name a few). Other questions are related to optimisation in learning and adaptive systems (e.g. the exploration-exploitation dilemma in machine learning). In evolutionary theory, the relation between genotype and phenotype is not well understood (e.g. the relation between point-mutation and Darwinian fitness of an organism).

The concepts of entropy and information have been applied recently in cognitive science [8] and biology [9] to model learning and adaptation in natural systems. In particular, information value, which amalgamates the concepts of expected utility and information, has been applied in optimisation of learning systems [10,11]. The classical information value theory has been developed by Stratonovich [12], and it is related to the earlier works on information transmission [13,14]. Although the quantum probability and quantum information theories have advanced significantly in the recent years, a quantum generalisation of information value is less understood. Hence, the main motivation of this work is to develop a unified approach to information value both in the classical and non-classical (quantum) setting.

We begin by giving a modern overview of the classical expected utility theory, and then relating it to the theory of dual pre-ordered linear spaces in a general setting that will allow for a natural generalisation of the classical concepts to the non-commutative case. Then we shall introduce the information value as an abstract convex optimisation problem. This will enable us to derive some general properties of its solutions without specifying the form of quantum information. We then show how these properties give rise to a more specific and familiar representation of information. We conclude by a discussion of potential applications of information value and its relation to information geometry.

2. Utility Functions and Operators

A *choice set* is a pair $(\Omega, \lesssim)$, where Ω is a set of arbitrary nature (i.e. a 'master set' of states or the set of outcomes), and $\lesssim \subseteq \Omega \times \Omega$ is a *preference* relation [7] — a complete and transitive binary relation (i.e. $\lesssim$ is a total pre-order on Ω). A subset $\sim \subseteq \lesssim$ of symmetric pairs is the equivalence relation, while the set $< \subseteq \lesssim$ of antisymmetric pairs is a partial order. The quotient set $\Omega/\sim$ (the set of equivalence classes $[\omega] := \{\omega' \in \Omega : \omega' \sim \omega\}$) is a chain (a totally ordered set). The optimal choice is any element $\bar{\omega} \in \Omega$ in the top class $\top \in \Omega/\sim$ (if it exists).

In practical applications, a preference relation is represented by a *utility* function $u : \Omega \to \mathbb{R}$ such that:

$$\omega_1 \lesssim \omega_2 \qquad \Longleftrightarrow \qquad u(\omega_1) \leq u(\omega_2)$$

A utility function admits factorisation $u = x \circ \pi$, where $\pi : \omega \mapsto [\omega] \in \Omega/\sim$ is the canonical projection, and $x : [\omega] \mapsto x([\omega]) \in \mathbb{R}$ is an order embedding of $(\Omega/\sim, <)$ into $(\mathbb{R}, \leq)$. This embedding (and hence a utility representation) is clearly possible for any $(\Omega, \lesssim)$ with countable quotient set. Although there exists examples of preference relations with no utility representation (e.g. an uncountable lexicographic preference relation), we consider such examples pathological for practical applications, where equivalence classes have to be enumerated (e.g. on a digital computer).

The utility defines metric $\rho(\omega_1, \omega_2) = |u(\omega_1) - u(\omega_2)|$ on $\Omega/\sim$, which makes it a separable, complete metric space. The topology induced by the metric on $\Omega/\sim$ coincides with the order topology or the interval topology with closed sets $[\omega_1, \omega_2] := \{\omega \in \Omega : \omega_1 \lesssim \omega \lesssim \omega_2\}$. For convenience of notation, we shall assume further $\Omega = \Omega/\sim$ and consider only the order embedding part x of the utility functions. Thus, we shall assume further that Ω is a separable, complete, metrisable space. Probability measures defined on Borel σ-algebra $\mathfrak{R}(\Omega) \subseteq 2^\Omega$ of such spaces are Radon measures.

Recall that complex Radon measures are in one-to-one correspondence with linear functionals $y(f) := \langle f, y \rangle$ on space $\mathcal{C}_c(\Omega)$ of continuous functions $f : \Omega \to \mathbb{C}$ with compact support [15]. The correspondence is given by the formula:

$$\langle f, y \rangle = \int_\Omega f(\omega)\, dy(\omega) \tag{1}$$

Note that $\mathcal{C}_c(\Omega)$ is an involutive, commutative, associative, linear algebra, the completion of which is the pre-dual C^*-algebra L_* of the Banach space $L(\Omega)$ of all complex finite Radon measures (complex charges).

At this point we can make the following generalisation. Let L_* be a complex Banach space that is a complex linear C^*-algebra with antilinear and isometric involution $x \mapsto x^\dagger$, $x^{\dagger\dagger} = x$, $\|x^\dagger\| = \|x\|$, and complete with respect to a norm having the property $\|x^\dagger x\| = \|x^\dagger\|\|x\|$. Quantum phenomena can be represented by not necessarily commutative elements $x \in L_*$, such as compact Hermitian operators $x \in \mathcal{L}_c(\mathcal{H})$ acting on a Hilbert space $\mathcal{H}$, and elements of its dual $L := L_*^*$, such as trace-class operators $y \in L$ with respect to the trace pairing $\langle x, y\rangle := \operatorname{tr}\{xy\}$. The dual space $L = L_*^*$ of all continuous linear functionals is a non-commutative generalisation of the Banach space $L(\Omega)$ of all Radon measures, and the pairing $\langle\cdot,\cdot\rangle : L_* \times L \to \mathbb{C}$ is the analog of integral (1) for functions.

We shall also consider the second dual space $L^* = L_*^{**}$ (the dual of L) with respect to a non-degenerate bilinear form $\langle\cdot,\cdot\rangle : L^* \times L \to \mathbb{C}$, naturally extended from L_* onto L^*. If L_* is the algebra of compact Hermitian operators $\mathcal{L}_c(\mathcal{H})$, then $L^* = \mathcal{L}_c^{**}(\mathcal{H}) = \mathcal{L}^h(\mathcal{H})$ is the algebra of bounded Hermitian operators in the Hilbert space. Note that algebra L^* has the unit element $1 \in L^*$. The elements of algebra L^* will be used as a non-commutative generalisation of the utility functions and expected utility.

We remind that $x \in L_*$ is positive if (and only if) $x = z^\dagger z$ for some $z \in L_*$, and set $\{x \in L_* : x \geq 0\}$ is the positive cone of L_*. Positive elements in L are the elements of the dual cone $L_+ := \{y \in L : \langle x, y\rangle \geq 0\,,\ \forall x \geq 0\}$. The cone defines an order relation $\leq$ on L: $y_1 \leq y_2$ if $y_2 - y_1 \in L_+$. This order is compatible with the linear structure of L: $y_1 \leq y_2$ implies $y_1 + z \leq y_2 + z$ and $ty_1 \leq ty_2$ for any $z \in L$ and $t \geq 0$. Positive elements $y \in L_+$ such that $\langle 1, y\rangle = 1$, $1 \in L^*$, are called *states*, and the set

$$\mathcal{P} := \{y \in L_+ : \langle 1, y\rangle = 1\}$$

of all states will be referred to as *statistical manifold* by analogy with information geometry [16,17,18]. In the classical case, $\mathcal{P}$ is the set of all probability measures on Ω, and it is a simplex in L (a Choquet simplex if L is infinite-dimensional). Indeed, observe that a probability measure on σ-algebra $\mathfrak{R}(\Omega)$ is a unique convex combination of elementary measures (the Dirac δ-measures): $\delta_\omega(\Omega') = 1$ if $\omega \in \Omega' \subseteq \Omega$; $\delta_\omega(\Omega') = 0$ otherwise. Elementary measures $\delta \in \mathcal{P}$ correspond to elements $\omega \in \Omega$ or to the 'pure' states. Thus, $\Omega \equiv \Delta$, where $\Delta \subset \partial\mathcal{P}$ is the set of all pure states. Other states $y \in \mathcal{P} \setminus \Delta$ are 'mixed'.

In non-classical (quantum) case, $\mathcal{P}$ is also a compact convex set, and therefore there also exists set $\Delta \subset \partial\mathcal{P}$ of extreme points (pure states in the Hilbert space) such that all states $y \in \mathcal{P}$ are convex combinations of elements $\delta \in \Delta$. However, these representations in the quantum case are not unique, and therefore the quantum statistical manifold is not a simplex.

Mixed states represent uncertainty (classical or quantum), and the problem of choice and optimisation under uncertainty in the classical case is usually solved using the *expected utility* $E_y\{x\} := \int_\Omega x(\omega)\,dy(\omega)$, where random variable $x : \Omega \to \mathbb{R}$ is the utility representation of $(\Omega, \lesssim)$, and $y \in \mathcal{P}$ is a probability measure. Thus, $z \lesssim y$ if and only if $E_z\{x\} \le E_y\{x\}$.

Observe now that $E_y\{x\}$ is a restriction of linear functional $x(y) = \langle x, y\rangle$ to the statistical manifold $\mathcal{P} \subset L$, and $E_\delta\{x\} = x(\omega)$ for any $\delta_\omega(\Omega') \in \mathcal{P}$. Thus, the classical utility is bounded function $x \in \mathcal{C}_c^{**}(\Omega)$ preserving pre-order $(\Omega, \lesssim)$. Following this construction, we define a *utility operator* as an element of algebra L^*, the dual of L, such that its restriction to the set of its eigenstates $\Omega \subset \Delta \subset \partial\mathcal{P}$ is an order embedding of $(\Omega, \lesssim)$ into $(\mathbb{R}, \le)$:

$$\omega_1 \lesssim \omega_2 \quad \Longleftrightarrow \quad \langle x, \delta_{\omega_1}\rangle \le \langle x, \delta_{\omega_2}\rangle$$

Thus, bounded Hermitian operators $x \in \mathcal{L}_c^{**}(\mathcal{H}) = \mathcal{L}^h(\mathcal{H})$ preserving pre-order on its eigenstates give non-commutative generalisation of the classical utility. The observed values of a utility operator are the elements of its spectrum $\sigma(x)$, which is real for the Hermitian operators.

3. Utility Pre-Ordered Dual Spaces

A utility representation $x : \Omega \to \mathbb{R}$ of pre-order $(\Omega, \lesssim)$ is not unique, because its composition $w = f \circ x$ with any strictly increasing function $f : \mathbb{R} \to \mathbb{R}$ represents the same pre-order. For example, operators x and λx represent the same pre-order if $\lambda > 0$. Thus, set $W^* \subset L^*$ of utility operators representing $(\Omega, \lesssim)$ is a cone ($\lambda W^* \subset W^*$). Moreover, if x and w have the same set of eigenstates Ω, then $wx = xw$. Thus, if x and w represent pre-order $(\Omega, \lesssim)$, then $\alpha w + \beta x$ for any α, $\beta > 0$ also represents this pre-order. Thus, set W^* of all utility representations of $(\Omega, \lesssim)$ is a convex cone ($\alpha W^* + \beta W^* \subset W^*$, α, $\beta > 0$) or a wedge if $W^* \cap (-W^*) \neq 0$. The pre-dual wedge is:

$$W := \{y \in L : \langle x, y\rangle \ge 0\,,\ \forall x \in W^*\}$$

The polar wedge of W is $W^\circ := -W^*$. The wedge W (W^*) induces pre-order on L (L^*): $z \lesssim y$ in L if $y - z \in W$ ($x \lesssim w$ in L^* if $w - x \in W^*$). The dual pre-order $\gtrsim$ on L (L^*) is induced similarly by $-W$ ($-W^*$). Each $y_0 \in L$ is the bottom element of wedge $W(y_0) := W + y_0$ (i.e. $W = W(0)$).

Pre-orders $(L, \lesssim)$ and $(L^*, \lesssim)$ are 'compatible' with pre-order $(\Omega, \lesssim)$. In particular, pre-order $(\mathcal{P}, \lesssim)$ on the statistical manifold $\mathcal{P} \subset L$, inherited from $(L, \lesssim)$, is represented by the expected utility: $z \lesssim y \in \mathcal{P}$ if and only if $E_z\{x\} \le E_y\{x\}$. Indeed, $z \lesssim y$ if (and only if) $\langle x, y - z\rangle \ge 0$. It is clear also from its definition that pre-order $(\mathcal{P}, \lesssim)$ is compatible with the linear structure of L. In fact, expected utility is the unique (up to an additive constant) representation of such pre-order on the statistical manifold[2]. Traditionally, this is shown by considering two axioms with respect to $(\mathcal{P}, \lesssim)$ extending from $(\Omega, \lesssim)$:

(1) Continuity:$y_1 < z < y_2$implies$\exists\, t \in [0,1]$such that$(1-t)y_1 + ty_2 \sim z$.
(2) Substitution:$y_1 \lesssim y_2$ implies $(1-t)y_1 + tz \lesssim (1-t)y_2 + tz$ for all $z \in \mathcal{P}$ and $t \in [0,1]$.

The first axiom implies that pre-order $(\mathcal{P}, \lesssim)$ is Archimedian, and the second ensures compatibility of $(\mathcal{P}, \lesssim)$ with the linear structure of L. Here we constructed $(\mathcal{P}, \lesssim)$ by defining pre-orders on L^* an L using the wedge W^* of utility operators for $(\Omega, \lesssim)$ and its pre-dual W. We shall refer to $(L, \lesssim)$ and $(L^*, \lesssim)$ as *utility pre-ordered spaces*, if $\lesssim$ is compatible with $(\Omega, \lesssim)$. Elements $x \in W^*$ and $y \in W$ can be referred to as *utility positive*, while $-x$ and $-y$ as *utility negative*.

4. Information Value as an Optimisation Problem

Transitions of a system between states are manifested in events that convey some information about the system. Generally, information can be seen as a 'distance' (or divergence) of state y from some original state $y_0 \in \mathcal{P}$, and there are many ways to represent information[16,17]. These information distances, however, do not reflect the 'usefulness' or the *value* of the states of the system. As discussed in the previous section, the value of states can be represented by a utility operator $x \in L^*$, and different states y_1, y_2 with the same information relative to y_0 may have different values $\langle x, y_1 \rangle \neq \langle x, y_2 \rangle$. Moreover, the value of a state is of course relative to a particular utility operator chosen for a specific task (e.g. minimisation of errors, maximisation of energy, etc). This amalgamation of the information and utility concepts underpins the information value theory, which was developed by Stratonovich [12] for classical systems with constraints on Shannon information. Here, we shall consider information value as an optimisation problem in non-commutative case and with constraints on an abstract information resource. Our aim is to single out properties of solutions to this problem that do not depend on the specific form of quantum information.

First, let us show that the problem of optimisation of states on the statistical manifold is overdetermined if the information constraints are not included in the formulation of the problem. Indeed, as discussed earlier, statistical manifold $\mathcal{P}$ is a compact convex set (a simplex in the classical

case), and therefore each state is a convex combination of its extreme points $\delta \in \Delta \subset \partial\mathcal{P}$ (the pure states). Without additional constraints, any continuous linear functional $\langle x, y\rangle$ achieves the extremum in one of these extreme points (to see this, assume $\sup\langle x, y\rangle = \langle x, \bar{y}\rangle$, where $\bar{y} = (1-t)\delta_1 + t\delta_2$, $t \in [0,1]$). Thus, without additional constraints, the optimal choice under uncertainty is always an optimal pure state $\bar{\omega} \in \Omega$. This contradicts the intuition of choosing under uncertainty where such choices are not feasible (indeed, one cannot guarantee to win a lottery unless the winning lottery ticket is known in advance or drawing all lottery tickets).

Let us represent information of state y relative to y_0 by some closed functional $F : L \to \mathbb{R} \cup \{\infty\}$ such that $\inf F = F(y_0)$ and sublevel set $\{y : F(y) \leq \lambda\}$ is bounded for each $\lambda < \sup F$. We remind that F is closed if $F(y) < \infty$ for some $y \in L$, and sublevel sets $\{y : F(y) \leq \lambda\}$ are closed for each $\lambda \in \mathbb{R}$. We now consider the following optimisation problem on a non-empty set of feasible states $M := \{y \in \mathcal{P} : F(y) \leq \lambda\} \subset \mathcal{P} \setminus \Delta$:

$$\upsilon_x(\lambda) := \sup\{\langle x, y\rangle : F(y) \leq \lambda\} \tag{2}$$

We shall refer to the restriction of real function $\upsilon_x(\lambda)$ to $\mathcal{P}$ as the *value of information* λ relative to utility x. This is a generalisation of the information value, introduced in [12], where F was Shannon mutual information between random variables $a \in A$ and $b \in B$, and $\mathcal{P}$ was the set of joint probability measures on $\Omega := A \times B$.

Problem (2) can be approached using methods of convex analysis and conjugate duality [19], and as will be shown below, the solutions are subgradients of the dual of F. Recall that the dual (conjugate) functional $F^* : L^* \to \mathbb{R} \cup \{\infty\}$ is the Legendre-Fenchel transform of F:

$$F^*(x) := \sup\{\langle x, y\rangle - F(y)\}$$

The dual is always closed and convex, and $F^{**} = F$ implies F is also closed and convex. Note that solutions to problem (2) in the case $F^{**} \neq F$ are a subset of solutions to $\sup\{\langle x, y\rangle : F^{**}(y) \leq \lambda\}$. This is because $\langle x, \cdot\rangle$ is linear and $\{y : F(y) \leq \lambda\} \subseteq \{y : F^{**}(y) \leq \lambda\}$.

Closed convex functionals are continuous on the (algebraic) interior of $\operatorname{dom} F := \{y \in L : F(y) < \infty\}$, and they have the property

$$x \in \partial F(y) \quad \Longleftrightarrow \quad \partial F^*(x) \ni y \tag{3}$$

---

*Closedness of $M \subset L$ is understood here as M being represented by intersection of closed half-spaces $\{y \in L : \langle x, y\rangle \leq \upsilon, x \in L^*\}$.

where set $\partial F(y_0) := \{x \in L^* : \langle x, y - y_0 \rangle \leq F(y) - F(y_0)\,,\ \forall\, y \in L\}$ is called *subdifferential*, and its elements are called *subgradients*. In particular, $0 \in \partial F(y_0)$ implies $F(y_0) \leq F(y)$ for all $y \in L$ (i.e. $\inf F = F(y_0)$). If F is strictly convex at y or F^* is weakly (Gâteaux) differentiable on the neighbourhood of $x \in \partial F(y)$, then $\partial F^*(x) = \{y\}$.

We now derive solutions $\bar{y}$ to problem (2) when $F = F^{**}$ (i.e. F is closed and convex). The derivation is based on the standard method of Lagrange multipliers, and it is presented here for completeness of exposition.

Proposition 4.1. (Necessary and sufficient optimality conditions) *Element $\bar{y} \in L$ solves problem (2) for closed and convex $F : L \to \mathbb{R} \cup \{\infty\}$ and some fixed $x \in L^*$ if and only if the following conditions are satisfied*

$$\bar{y} \in \partial F^*(\beta x)\,, \quad F(\bar{y}) = \lambda\,, \quad \beta^{-1} \in \partial \upsilon_x(\lambda)\,, \quad \mathbb{R} \ni \beta^{-1} > 0$$

Proof. The Lagrange function for the problem is

$$K(y, \beta^{-1}, \lambda) = \langle x, y \rangle + \beta^{-1}[\lambda - F(y)]$$

where β^{-1} is the Lagrange multiplier for the constraint $\lambda \in [\inf F, \sup F)$. Zero in the sub- or sup-differentials of the Lagrange function gives the following necessary conditions for stationary points (the Kuhn-Tucker conditions):

$$\begin{aligned} \partial_y K(\bar{y}, \beta^{-1}, \lambda) &= x - \beta^{-1} \partial F(\bar{y}) \ni 0\,, \quad &\Rightarrow \quad &\beta x \in \partial F(\bar{y}) \\ \partial_{\beta^{-1}} K(\bar{y}, \beta^{-1}, \lambda) &= \lambda - F(\bar{y}) \ni 0\,, \quad &\Rightarrow \quad &F(\bar{y}) = \lambda \end{aligned}$$

Noting that $K(\bar{y}, \beta^{-1}, \lambda) = \upsilon_x(\lambda)$, the Lagrange multiplier is obtained by

$$\partial_\lambda K(\bar{y}, \beta^{-1}, \lambda) = \partial \upsilon_x(\lambda) \ni \beta^{-1}$$

Sufficient conditions are obtained by considering convexity. Because F is convex and $\langle x, \cdot \rangle$ is linear, the Lagrange function is concave for $\beta^{-1} > 0$ and convex for $\beta^{-1} < 0$. Therefore, $\bar{y} \in \partial F^*(\beta x)$ with $\beta^{-1} > 0$ defines the least upper bound of the Lagrange function, and with $\beta^{-1} < 0$ it defines the greatest lower bound. □

Remark 4.1. If there exists $y_0 \in L$ such that $\inf F = F(y_0)$, then it is useful to consider the following function:

$$\lambda_x(\upsilon) := \inf\{F(y) : \upsilon_0 \leq \upsilon \leq \langle x, y \rangle\} \tag{4}$$

where $\upsilon_0 = \langle x, y_0 \rangle$. The Lagrange function for $\lambda_x(\upsilon)$ is $K(y, \beta, \upsilon) = F(y) + \beta[\upsilon - \langle x, y \rangle]$, and the solutions are defined by the following conditions

$$\bar{y} \in \partial F^*(\beta x)\,, \quad \langle x, \bar{y} \rangle = \upsilon\,, \quad \beta \in \partial \lambda_x(\upsilon)\,, \quad \mathbb{R} \ni \beta \geq 0$$

Function $\lambda_x(\upsilon)$ defines the information amount of value υ (or υ-information according to [12]), and it is the inverse of $\upsilon_x(\lambda)$, defined by equation (2).

5. Galois Connection and Closure

We now consider monotone operators between utility pre-ordered spaces $(L, \lesssim)$ and $(L^*, \lesssim)$. Recall that subgradients of convex functionals have the following monotonicity property [20]:

$$\langle x_1 - x_2, y_1 - y_2 \rangle \geq 0\,, \quad \forall\, x_i \in \partial F(y_i) \tag{5}$$

If the inequality is strict for any $y_1 \neq y_2$, then $y \mapsto x \in \partial F(y)$ is a strictly isotone correspondence and F is strictly convex. We now use the monotonicity property together with Proposition 4.1 to prove the following result.

Theorem 5.1. (**Galois connection**) *Let $(L, \lesssim)$ and $(L^*, \lesssim)$ be a pair of utility pre-ordered linear spaces. Let $F : L \to \mathbb{R} \cup \{\infty\}$ be a closed, convex functional, and let F^* be its dual. Then the pair $(\partial F^*, \partial F)$ of mappings $\partial F^* : L^* \to \partial F^*$ and $\partial F : L \to \partial F$ sets up an isotone Galois connection:*

$$\partial F^*(x) \lesssim y \quad \Longleftrightarrow \quad x \lesssim \partial F(y)$$

If F is also strictly convex and weakly differentiable, then $F'_G : L \to L^*$ implements a pre-order isomorphism of $(L, \lesssim)$ and $(L^*, \lesssim)$.

Proof. Let $z \in \partial F^*(x)$ and $w \in \partial F(y)$. Clearly, $z \in W$ if $x \in W^*$, because z solves problem $\langle x, z \rangle = \sup\{\langle x, y \rangle : \inf F \leq F(y) \leq F(z)\}$ by Proposition 4.1 (i.e. $z \in \partial F^*(\beta x)$ with $\beta = 1 > 0$). Condition $z \lesssim y$ implies $y \in W$, and therefore $w \in W^*$ (by $y \in \partial F^*(w)$).
($\Longrightarrow$)
Condition $\partial F^*(x) \ni z \lesssim y$ implies $y - z \in W$ or $\langle x, y - z \rangle \geq 0$. Consider $\beta x \sim w \in \partial F(y)$ (i.e. $\beta x - w \in W^*$ and $w - \beta x \in W^*$). Using monotonicity $\langle w - x, y - z \rangle \geq 0$ and $\langle w, y - z \rangle = \langle \beta x, y - z \rangle$, we have $(\beta - 1)\langle x, y - z \rangle \geq 0$ and $\beta \geq 1$. Thus, $x \lesssim \beta x \sim w \in \partial F(y)$.
($\Longleftarrow$)
Condition $x \lesssim \beta x \sim w \in \partial F(y)$ implies $\beta \geq 1$. Thus, $(\beta - 1)\langle x, y - z \rangle \geq 0$ and $\langle x, y - z \rangle \geq 0$. Therefore, $y - z \in W$ or $\partial F^*(x) \ni z \lesssim y$. If F is strictly

convex, then ∂F is monomorphism embedding $(L, \lesssim)$ into $(L^*, \lesssim)$. If F is weakly differentiable (or F^* is strictly convex), then ∂F^* is monomorphism embedding $(L^*, \lesssim)$ into $(L, \lesssim)$. □

A Galois connection is an important generalisation of isomorphism between pre-ordered sets. In particular, for an isotone Galois connection, as $(\partial F^*, \partial F)$ in Theorem 5.1, the mappings ∂F^* and ∂F are isotone (order-preserving): $x \lesssim w$ implies $\partial F^*(x) \lesssim \partial F^*(w)$ and $z \lesssim y$ implies $\partial F(z) \lesssim \partial F(y)$. This property is important in the context of optimisation problems, such as the information value problem (2). Recall that elements $x \in W^* \subset L^*$ are utility representations of preference relation $(\Omega, \lesssim)$, and therefore the composition $y(\omega) \in \partial F^* \circ x(\omega)$ is order-preserving. Isotone mappings between pre-ordered sets $(\Omega, \lesssim)$, $(L, \lesssim)$ and $(L^*, \lesssim)$ are shown on the commutative diagram below.

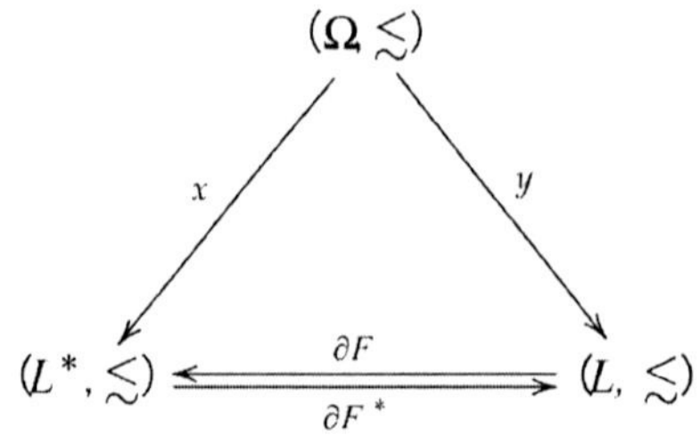

We point out also that mappings $\partial F^* : L^* \to \partial F^*$ and $\partial F : L \to \partial F$ are antitone with respect to pre-orders $\lesssim$ and its dual $\gtrsim$ on spaces L and L^*. The pair $(\partial F^* \circ -1, -1 \circ \partial F)$, where $-1 : x \mapsto -x \in L^*$, sets up an isotone Galois connection on $(L^*, \gtrsim)$ and $(L, \lesssim)$:

$$\partial F^*(-x) \lesssim y \quad \iff \quad x \gtrsim -\partial F(y)$$

This pair defines (order-theoretic) closures [21] on $(L, \lesssim)$ and $(L^*, \lesssim)$:

$$\mathrm{cl}_L := \partial F^* \circ \partial F : L \to L\,, \qquad \mathrm{cl}_{L^*} := \partial F \circ \partial F^* : L^* \to L^*$$

One can define topologies τ and τ^* on L and L^* that agree with their pre-orders, and in which the above operators agree with topological closures. Note that classical topological linear spaces in duality are defined using dual convex functions, such as the dual Young functions in Orlicz spaces (e.g. the L_p spaces). In fact, information value $v_x(\lambda)$, considered as a function of $x \in L^*$, is the support of set $\{y : F(y) \leq \lambda\}$, and it generalises norm in the dual space. Interestingly, space $L \log L$ has been considered recently in quantum theory to describe systems with bounded entropy [22,23], which is a special case of the information value problem (2).

6. Relative Information

We have discussed a generalisation of utility and value of an abstract information resource associated with classical or quantum states, as well as some general properties of the associated mappings. Let us now summarise these properties and consider a more specific definition of information.

First, the monotonic properties of subdifferentials of convex functions suggest that it is useful for information resource $F : L \to \mathbb{R} \cup \{\infty\}$ to be closed and convex. Following our construction, where L is the dual of algebra L_*, we can define F as a Legendre-Fenchel transform of some functional $F_* : L_* \to \mathbb{R} \cup \{\infty\}$. Note that if F_* is closed and convex, then $F_* = F^*$. Furthermore, if F is strictly convex and weakly differentiable, then its weak differential implements an isomorphism between pre-orders on states and utility operators.

Second, we require that sublevel sets $\{y : F(y) \leq \lambda\}$ are bounded for each $\lambda < \sup F$ (in this case $v_x(\lambda) < \infty$ for all x) and $\inf F = F(y_0)$ for some $y_0 \in L$. This is satisfied if and only if $0 \in \mathrm{Int}(\mathrm{dom}\, F^*)$ [19]. Note that y_0 is unique if F is strictly convex, and therefore y_0 is the bottom element $\bot \in \{y : F(y) \leq \lambda\}$. Information resource can also be made non-negative $F_+(y) := F(y) - \inf F \geq 0$.

Finally, one often requires information, associated with independent sources, to be additive: $F(y_1 \times \cdots \times y_n) = F(y_1) + \cdots + F(y_n)$. Observe that positive elements $y \in L_+$ (i.e. positive Radon measures) form a multiplicative semi-group $(L_+, \times)$, while utilities $x \in L^*$ form an additive semi-group $(L^*, +)$. As is well-known, the unique isomorphism between these semi-groups is given by $(L_+, \times) \ni y \mapsto \ln y \in (L^*, +)$ (and its inverse $x \mapsto \exp x$).

Taking the above into consideration, we can define *relative information* $F : L \to \mathbb{R} \cup \{\infty\}$ of $y \in L$ with respect to fixed positive element $y_0 \in L$:

$$F(y) := \left\langle \ln \frac{y}{y_0}, y \right\rangle - \langle 1, y - y_0 \rangle \tag{6}$$

with $F(0) := \langle 1, y_0 \rangle$ and $F(y) := \infty$ for negative y. It is closed, strictly convex and weakly differentiable on the interior of $\mathrm{dom}\, F$. Its subgradients are defined by the weak derivative in the following convenient form:

$$F'_G(y) = \ln \frac{y}{y_0} = x \in L^*$$

In particular, $F'_G(y_0) = 0$, and it is easy to verify that $\inf F = F(y_0) = 0$. Therefore, $F(y) \geq 0$ for all $y \in L$ (i.e. for all Radon measures, not just the probability measures).

Relative information is a generalisation of several representations of classical and quantum entropy and information. When y and y_0 belong to the statistical manifold, the term $\langle 1, y - y_0 \rangle = 0$. In the classical case with $\langle \cdot, \cdot \rangle$ given by integral (1), the relative information reduces to the classical Kullback-Leibler divergence [24]. In the non-commutative case with trace pairing $\langle \cdot, \cdot \rangle$ and $\frac{y}{y_0} := \exp\{\ln y - \ln y_0\}$, equation (6) reduces to Araki-Umegaki quantum relative information [25,26]. Note that $\frac{y}{y_0}$ can also be defined as Hermitian operator $y^{1/2} y_0^{-1} y^{1/2}$ giving a different form of quantum relative information [27].

The dual of relative information is:

$$F^*(x) = \langle 1, y_0 \, e^x \rangle$$

The solutions to problem (2) with F defined by equation (6) are obtained according to Proposition 4.1 by $F_G^{*\prime}(\beta x) = y_0 \, e^{\beta x}$ for $\beta \geq 0$. The restriction of F to the statistical manifold corresponds to transformation $F^*(x) \mapsto \ln F^*(x)$, and as is well-known, the optimal solutions are members of the exponential family:

$$\bar{y}_\beta = y_0 \exp\{\beta \, x - \Psi(\beta)\}$$

where $\Psi(\beta) = \ln\langle 1, y_0 \, e^{\beta x} \rangle$ is the cumulant generating function of measure y_0. Note that $\Psi(\beta)$ is the Legendre-Fenchel transform of convex function $\lambda_x(\upsilon)$ (υ-information): $\Psi(\beta) = \sup\{\beta\upsilon - \lambda_x(\upsilon)\}$. One can show also that free energy $\Phi(\beta^{-1}) = -\beta^{-1}\Psi(\beta)$ is the concave Legendre-Fenchel transform $\Phi(\beta^{-1}) = \inf\{\beta^{-1}\lambda - \upsilon_x(\lambda)\}$ of information value $\upsilon_x(\lambda)$. Thus, $\Psi(\beta)$ and $\Phi(\beta^{-1})$ play the role of Hamiltonians with respect to $\lambda_x(\upsilon)$ and $\upsilon_x(\lambda)$ respectively.

7. Discussion

We have described the concepts of utility and information value in the general case. Several properties of the optimal solutions were studied prior to giving a specific form of information. In particular, convexity of information gives rise to a pair of monotone operators that sets up an isotone Galois connection between dual pre-ordered spaces of states and bounded Hermitian operators. The compositions of the operators define closure operators on the spaces. A relative information functional was introduced based on these properties and satisfying also the additivity axiom. It was shown how the functional generalises several types of classical and quantum information. Moreover, the application of the relative information to the information value problem leads to several functions and relations from thermodynamics and statistical mechanics. The theory also has a close relation to the theory of quantum Orlicz spaces, used to define topological spaces of states with finite entropy [22,23]. What distinguishes the information value theory is its connection with the theory of utility and optimisation.

Apart from the classical and quantum information theories, information value has important applications in cognitive science and potentially biology. For example, it has been known in cognitive science and psychology that the decision-making strategies of human subjects, as well as some animals, are simulated better by randomised cognitive models, where actions are sampled from exponential distribution with some temperature $\beta^{-1} > 0$ [3]. Moreover, computational models reproduce the data better if the temperature parameter is controlled by the posterior entropy of success [8]. This relation is explained by solutions $\bar{y} \in \partial F^*(\beta x)$ to the information value problem (2), where $\beta^{-1} \in \partial v_x(\lambda)$ by Proposition 4.1. In particular, $\beta^{-1} \to 0$ if and only if information $\lambda \to \sup F$. The relation between β and λ helps also to optimise the exploration-exploitation balance in machine learning applications with constraints on prior information [10,11].

In biology, the diversity of genotype is directly related to randomisation of genetic sequences (mutation). The mutation operator has been known to play an important role in genetic algorithms applied in many hard optimisation problems. The author has initiated a project due to begin in 2010, which will investigate whether the relation of the mutation operator to information constraints can improve the performance of artificial as well as biological systems, and whether such a control mechanism occurs naturally in evolutionary systems.

Acknowledgements

The author acknowledges the contribution of Prof. V. P. Belavkin in the form of numerous discussions of the quantum information theory and various questions that arose during the preparation of this paper. The author also expresses his gratitude to the organisers of the QBIC'09 conference, especially to Prof. M. Ohya and Prof. N. Watanabe.

References

1. A. N. Kolmogorov, *Grundbegriffe der Wahrscheinlichkeitsrechnung* (Julius Springer, Berlin, 1933). in German.
2. J. von Neumann and O. Morgenstern, *Theory of games and economic behavior*, first edn. (Princeton University Press, Princeton, NJ, 1944).
3. J. R. Anderson and C. Lebiere, *The Atomic Components of Thought* (Lawrence Erlbaum, Mahwah, NJ, 1998).
4. M. Broom, C. Cannings and G. T. Vickers, *Journal of Mathematical Biology* **22**, 597 (1994).
5. I. M. Bomze and B. M. Potschner, *Game theoretical foundations of evolutionary stability*Lecture Notes Econom. and Math. Systems, Lecture Notes Econom. and Math. Systems (Springer, 1998).
6. A. Tversky and D. Kahneman, *Journal of Risk and Uncertainty* **5**, 297 (1992).
7. D. M. Kreps, *Notes on the Theory of Choice* (Westview Press, Colorado, USA, 1988).
8. R. V. Belavkin and F. E. Ritter, The use of entropy for analysis and control of cognitive models, in *Proceedings of the Fifth International Conference on Cognitive Modeling*, eds. F. Detje, D. Dörner and H. Schaub (Universitäts–Verlag Bamberg, Bamberg, Germany, April 2003).
9. C. T. Bergstrom and M. Lachmann, Shannon information and biological fitness, in *Proceedings of the IEEE Workshop on Information Theoery*, 2004.
10. R. V. Belavkin, Bounds of optimal learning, in *2009 IEEE International Symposium on Adaptive Dynamic Programming and Reinforcement Learning*, (IEEE, Nashville, TN, USA, 2009).
11. R. V. Belavkin, Information trajectory of optimal learning, in *Dynamics of Information Systems: Theory and Applications*, eds. M. J. Hirsch, P. M. Pardalos and R. Murphey (Springer, in press)
12. R. L. Stratonovich, *Izvestiya of USSR Academy of Sciences, Technical Cybernetics* **5**, 3 (1965), In Russian.
13. C. E. Shannon, *Bell System Technical Journal* **27**, 379(July and October 1948).
14. A. N. Kolmogorov, The theory of information transmission, in *Meeting of the USSR Academy of Sciences on Scientific Problems of Production Automatisation, 1956*, (USSR Academy of Sciences, 1957). In Russian.
15. N. Bourbaki, *Eléments de mathématiques. Intégration* (Hermann, 1963).
16. N. N. Chentsov, *Statistical Decision Rules and Optimal Inference* (Nauka, Moscow, U.S.S.R., 1972). In Russian, English translation: Providence, RI: AMS, 1982.
17. S.-I. Amari, *Differential-Geometrical Methods of Statistics*, Lecture Notes in Statistics, Vol. 25 (Springer, Berlin, Germany, 1985).
18. G. Pistone and C. Sempi, *The Annals of Statistics* **23**, 1543(October 1995).

19. R. T. Rockafellar, *Conjugate Duality and Optimization*, CBMS-NSF Regional Conference Series in Applied Mathematics, Vol. 16 (Society for Industrial and Applied Mathematics, PA, 1974).
20. R. I. Kachurovskii, *Russian Mathematical Surveys* **23**, 117 (1968).
21. B. A. Davey and H. A. Priestley, *Introduction to Lattices and Order*, second edn. (Cambridge University Press, 2002).
22. R. F. Streater, Quantum Orlicz spaces in information geometry, in *The 36th Conference on Mathematical Physics*, , Open Systems and Information Dynamics Vol. 11 (Torun, 2004).
23. S. G. Bobkov and B. Zegarlinski, *Entropy bounds and isoperimetry*, Memoirs of the American Mathematical Society, Vol. 176 (AMS, 2005).
24. S. Kullback and R. A. Leibler, *The Annals of Mathematical Statistics* **22**, 79 (1951).
25. H. Araki, *Publications of the Research Institute for Mathematical Sciences* **11**, 809 (1975).
26. H. Umegaki, *Kodai Mathematical Seminar Reports* **14**, 59 (1962).
27. V. P. Belavkin and P. Staszewski, *Reports in Mathematical Physics* **20**, 373 (1984).

Quantum Bio-Informatics III
eds. L. Accardi, W. Freudenberg & M. Ohya

SPECTRAL PROPERTIES OF ENTANGLEMENT WITNESSES AND POSITIVE MAPS

DARIUSZ CHRUŚCIŃSKI

Institute of Physics, Nicolaus Copernicus University,
Grudziądzka 5/7, 87–100 Toruń, Poland
E-mail:darch@fizyka.umk.pl

We provide partial classification of entanglement witnesses and positive maps which is based on a family of spectral conditions. This construction generalizes the entanglement witness corresponding to the celebrated Choi map which is positive but not completely positive.

1. Introduction

One of the most important problems of quantum information theory [1] is the characterization of mixed states of composed quantum systems. In particular it is of primary importance to test whether a given quantum state exhibits quantum correlation, i.e. whether it is separable or entangled. For low-dimensional systems there exists simple necessary and sufficient condition for separability. The celebrated Peres-Horodecki criterion [2,3] states that a state of a bipartite system living in $\mathbb{C}^2 \otimes \mathbb{C}^2$ or $\mathbb{C}^2 \otimes \mathbb{C}^3$ is separable iff its partial transpose is positive. Unfortunately, for higher-dimensional systems there is no single *universal* separability condition.

The most general approach to characterize quantum entanglement uses a notion of an entanglement witness (EW) [4,5]. A Hermitian operator W defined on a tensor product $\mathcal{H} = \mathcal{H}_1 \otimes \mathcal{H}_2$ is called an EW iff 1) $\mathrm{Tr}(W\sigma_{\mathrm{sep}}) \geq 0$ for all separable states σ_{sep}, and 2) there exists an entangled state ρ such that $\mathrm{Tr}(W\rho) < 0$ (one says that ρ is detected by W). It turns out that a state is entangled if and only if it is detected by some EW.

It turns out that the above problem may be reformulated in terms of positive linear maps in operator algebras: a state ρ in $\mathcal{H}_1 \otimes \mathcal{H}_2$ is separable iff $(\mathrm{id} \otimes \varphi)\rho$ is positive for any positive map φ which sends positive operators on $\mathcal{H}_2$ into positive operators on $\mathcal{H}_1$. Therefore, a classification of positive linear maps between operator algebras $\mathcal{B}(\mathcal{H}_1)$ and $\mathcal{B}(\mathcal{H}_2)$ is of primary importance. Unfortunately, in spite of the considerable effort, the structure of positive maps is rather poorly understood [6-18] (for the recent review paper see [19]).

In the present paper we perform partial classification of positive linear maps which is based on spectral conditions. Actually, the presented method enables one to construct maps with a desired degree of positivity — so called k-positive maps with $k = 1, 2, \ldots, d = \min\{\dim \mathcal{H}_1, \dim \mathcal{H}_2\}$. Completely positive (CP) maps correspond to d-positive maps, i.e. maps with the highest degree of positivity. These maps are fully classified due to Stinespring's theorem [20]. Now, any positive map which is not CP can be written as $\varphi = \varphi_+ - \varphi_-$, with $\varphi_\pm$ being CP maps. However, there is no general method to recognize the positivity of φ from $\varphi_+ - \varphi_-$. We show that suitable spectral conditions satisfied by the pair (φ_+, φ_-) guarantee the k-positivity of $\varphi_+ - \varphi_-$. This construction generalizes the celebrated Choi example of a map which is $(d-1)$-positive but not CP [8].

2. Entanglement witnesses vs. positive maps

Let M_d denote a set of $d \times d$ complex matrices and let M_d^+ be a convex set of semi-positive elements in M_d, that is, M_d^+ defines a space of (unnormalized) states of d-level quantum system. For any $\rho \in (M_{d_1} \otimes M_{d_2})^+$ denote by $\mathrm{SN}(\rho)$ the Schmidt number of ρ [22]. This notion enables one to introduce the following family of positive cones:

$$\mathrm{V}_r = \{ \rho \in (M_{d_1} \otimes M_{d_2})^+ \mid \mathrm{SN}(\rho) \le r \} . \tag{1}$$

One has the following chain of inclusions

$$\mathrm{V}_1 \subset \ldots \subset \mathrm{V}_d \equiv (M_{d_1} \otimes M_{d_2})^+ , \tag{2}$$

where $d = \min\{d_1, d_2\}$. Clearly, V_1 is a cone of separable (unnormalized) states and $V_d \smallsetminus V_1$ stands for a set of entangled states. Note, that a partial transposition $(\mathbb{1} \otimes \tau)$ gives rise to another family of cones:

$$\mathrm{V}^l = (\mathbb{1} \otimes \tau)\mathrm{V}_l , \tag{3}$$

such that $\mathrm{V}^1 \subset \ldots \subset \mathrm{V}^d$. One has $\mathrm{V}_1 = \mathrm{V}^1$, together with the following hierarchy of inclusions:

$$\mathrm{V}_1 = \mathrm{V}_1 \cap \mathrm{V}^1 \subset \mathrm{V}_2 \cap \mathrm{V}^2 \subset \ldots \subset \mathrm{V}_d \cap \mathrm{V}^d . \tag{4}$$

Note, that $\mathrm{V}_d \cap \mathrm{V}^d$ is a convex set of PPT (unnormalized) states. Finally, $\mathrm{V}_r \cap \mathrm{V}^s$ is a convex subset of PPT states ρ such that $\mathrm{SN}(\rho) \le r$ and $\mathrm{SN}[(\mathbb{1} \otimes \tau)\rho] \le s$.

Consider now a set of positive maps $\varphi : M_{d_1} \longrightarrow M_{d_2}$, i.e. maps such that $\varphi(M_{d_1}^+) \subseteq M_{d_2}^+$. Following Størmer definition [6], a positive map φ is k-positive iff

$$(\mathbb{1} \otimes \varphi)(\mathrm{V}_k) \subset (M_{d_1} \otimes M_{d_2})^+ \ , \tag{5}$$

and it is k-copositive iff

$$(\mathbb{1} \otimes \varphi)(\mathrm{V}^k) \subset (M_{d_1} \otimes M_{d_2})^+ \ . \tag{6}$$

Denoting by P_k (P^k) a convex cone of k-positive (k-copositive) maps one has the following chains of inclusions

$$\mathrm{P}_d \subset \mathrm{P}_{d-1} \subset \ldots \subset \mathrm{P}_2 \subset \mathrm{P}_1 \ , \tag{7}$$

and

$$\mathrm{P}^d \subset \mathrm{P}^{d-1} \subset \ldots \subset \mathrm{P}^2 \subset \mathrm{P}^1 \ , \tag{8}$$

where P_d (P^d) stands for a set of completely positive (copositive) maps.

A positive map $\varphi : M_{d_1} \longrightarrow M_{d_2}$ is *decomposable* iff $\varphi \in P_d + P^d$, that is, φ can be written as $\varphi = \varphi_1 + \varphi_2$, with $\varphi_1 \in P_d$ and $\varphi_2 \in P^d$. Otherwise φ is *indecomposable*. Indecomposable maps can detect entangled state from $V_d \cap V^d \equiv$ PPT, that is, bound entangled states. Finally, a positive map is *atomic* iff $\varphi \notin \mathrm{P}_2 + \mathrm{P}^2$. The importance of atomic maps follows from the fact that they may be used to detect the 'weakest' bound entanglement, that is, atomic maps can detect states from $V_2 \cap V^2$.

Using the Choi-Jamiołkowski isomorphism we may assign to any linear map $\varphi : M_{d_1} \rightarrow M_{d_2}$ the following hermitian operator $W \in M_{d_1} \otimes M_{d_2}$:

$$W = (\mathbb{1} \otimes \varphi)P^+ \in M_{d_1} \otimes M_{d_2} \ , \tag{9}$$

where P^+ stands for (unnormalized) maximally entangled state in $\mathbb{C}^{d_1} \otimes \mathbb{C}^{d_1}$. If e_i ($i = 1, \ldots, d_1$) is an orthonormal basis in $\mathbb{C}^{d_1}$, then

$$W = \sum_{i,j=1}^{d_1} e_{ij} \otimes \varphi(e_{ij}) \ , \tag{10}$$

where $e_{ij} = |i\ j|$. In the space of entanglement witnesses $\mathbf{W}$ one may introduce the following family of subsets $\mathbf{W}_r$:

$$\mathbf{W}_r = \{\, W \in M_{d_1} \otimes M_{d_2} \mid \mathrm{Tr}(W\rho) \geq 0 \ , \ \rho \in \mathrm{V}_r \,\} \ . \tag{11}$$

One has

$$(M_{d_1} \otimes M_{d_2})^+ \equiv \mathbf{W}_d \subset \ldots \subset \mathbf{W}_1 \ . \tag{12}$$

Clearly, $\mathbf{W} = \mathbf{W}_1 \setminus \mathbf{W}_d$. Moreover, for any $k > l$, entanglement witnesses from $\mathbf{W}_l \setminus \mathbf{W}_k$ can detect entangled states from $\mathrm{V}_k \setminus V_l$, i.e. states ρ with Schmidt number $l < \mathrm{SN}(\rho) \leq k$. In particular $W \in \mathbf{W}_k \setminus \mathbf{W}_{k+1}$ can detect state ρ with $\mathrm{SN}(\rho) = k$.

Consider now the following class

$$\mathbf{W}_r^s = \mathbf{W}_r + (\mathbb{1} \otimes \tau)\mathbf{W}_s \ , \tag{13}$$

that is, $W \in \mathbf{W}_r^s$ iff

$$W = P + (\mathbb{1} \otimes \tau)Q \ , \tag{14}$$

with $P \in \mathbf{W}_r$ and $Q \in \mathbf{W}_s$. Note, that $\mathrm{Tr}(W\rho) \geq 0$ for all $\rho \in \mathrm{V}_r \cap \mathrm{V}^s$. Hence such W can detect PPT states ρ such that $\mathrm{SN}(\rho) \geq r$ and $\mathrm{SN}[(\mathbb{1} \otimes \tau)\rho] \geq s$. Entanglement witnesses from $\mathbf{W}_d^d$ are called decomposable [5]. They cannot detect PPT states. One has the following chain of inclusions:

$$\mathbf{W}_d^d \subset \ldots \subset \mathbf{W}_2^2 \subset \mathbf{W}_1^1 \equiv \mathbf{W} \ . \tag{15}$$

The 'weakest' entanglement can be detected by elements from $\mathbf{W}_1^1 \setminus \mathbf{W}_2^2$. We shall call them *atomic entanglement witnesses*. It is clear that W is an atomic entanglement witness if there is an entangled state $\rho \in \mathrm{V}_2 \cap \mathrm{V}^2$ such that $\mathrm{Tr}(W\rho) < 0$. The knowledge of atomic witnesses, or equivalently atomic maps, is crucial: knowing this set we would be able to distinguish all entangled states from separable ones.

3. New family based on spectral properties

Consider the space $\mathcal{L}(\mathcal{H}_1, \mathcal{H}_2)$ of linear operators $a : \mathcal{H}_1 \longrightarrow \mathcal{H}_2$, or equivalently the space of $d_1 \times d_2$ matrices, where $d_i = \dim \mathcal{H}_i < \infty$. Let us recall that $\mathcal{L}(\mathcal{H}_1, \mathcal{H}_2)$ is equipped with a family of Ky Fan k-norms [21]: for any $a \in \mathcal{L}(\mathcal{H}_1, \mathcal{H}_2)$ one defines

$$||\, a \,||_k := \sum_{i=1}^{k} s_i(a) \ , \tag{16}$$

where $s_1(a) \geq \ldots \geq s_d(a)$ ($d = \min\{d_1, d_2\}$) are the singular values of a. Clearly, for $k = 1$ one recovers the operator norm $||\, a \,||_1 = ||\, a \,||$ and if $d_1 = d_2 = d$, then for $k = d$ one reproduces the trace norm $||\, a \,||_d = ||\, a \,||_{\mathrm{tr}}$. The family of k-norms satisfies:

(1) $||\, a \,||_k \leq ||\, a \,||_{k+1}$,
(2) $||\, a \,||_k = ||\, a \,||_{k+1}$ if and only if $\mathrm{rank}\, a = k$,

(3) if $\operatorname{rank} a \geq k+1$, then $||\, a\, ||_k < ||\, a\, ||_{k+1}$.

Consider now the Hilbert space being the tensor product $\mathcal{H}_1 \otimes \mathcal{H}_2$. It is clear that $\mathcal{H}_1 \otimes \mathcal{H}_2$ is isomorphic to the space of linear operators from $\mathcal{L}(\mathcal{H}_1, \mathcal{H}_2)$, i.e. each $\psi \in \mathcal{H}_1 \otimes \mathcal{H}_2$ corresponds to some linear operator $F : \mathcal{H}_1 \longrightarrow \mathcal{H}_2$. Indeed, taking orthonormal basis $\{e_i\}$ $(i = 1, \ldots, d_1)$ in $\mathcal{H}_1$ and $\{f_\alpha\}$ $(\alpha = 1, \ldots, d_2)$ in $\mathcal{H}_2$, one has

$$\psi = \sum_{i=1}^{d_1} \sum_{\alpha=1}^{d_2} \psi_{i\alpha}\, e_i \otimes f_\alpha \ , \tag{17}$$

where $\psi_{i\alpha}$ are complex coefficients. Defining

$$F e_i := \sum_{\alpha=1}^{d_2} \psi_{i\alpha}\, f_\alpha \ ,$$

one finds

$$\psi = \sum_{i=1}^{d_1} e_i \otimes F e_i \ , \tag{18}$$

which establishes the above-mentioned isomorphism. Let us note that the normalization condition

$$\psi|\psi = \sum_{i=1}^{d_1} \sum_{\alpha=1}^{d_2} |\psi_{i\alpha}|^2 = 1 \ , \tag{19}$$

gives rise to

$$\operatorname{tr} F F^* = 1 \ , \tag{20}$$

for the corresponding operator F. It is therefore clear that the rank-1 projector $P := |\psi\ \psi|$ may be represented in the following way

$$P = \sum_{i,j=1}^{d_1} e_{ij} \otimes F e_{ij} F^* \ . \tag{21}$$

Now, it is easy to see that

$$\mathrm{SR}(\psi) = \operatorname{rank} F \ , \tag{22}$$

where $\mathrm{SR}(\psi)$ denotes the Schmidt rank of ψ $(1 \leq \mathrm{SR}(\psi) \leq d)$, i.e. the number of non-vanishing Schmidt coefficients in the Schmidt decomposition of ψ.

It is well known that any CP map may be represented in the so-called Kraus form [23]

$$\varphi_{\rm CP}(a) = \sum_\alpha K_\alpha a K_\alpha^* \ , \tag{23}$$

where (Kraus operators) $K_\alpha \in \mathcal{L}(\mathcal{H}_1, \mathcal{H}_2)$. Any positive map is a difference of two CP maps $\varphi = \varphi_+ - \varphi_-$. However, there is no general method to recognize the positivity of φ from $\varphi_+ - \varphi_-$. Consider now a special class when W_+ and W_- are orthogonally supported and $W_- = \lambda_1 P_1$, with P_1 being a rank-1 projector (recall that $W_\pm = \sum_{i,j} e_{ij} \otimes \varphi_\pm(e_{ij})$). Let

$$\varphi(a) = \sum_{\alpha=2}^{D} \lambda_\alpha F_\alpha a F_\alpha^* - \lambda_1 F_1 a F_1^* \ , \tag{24}$$

such that

(1) all rank-1 projectors $P_\alpha = d_1^{-1} \sum_{i,j=1}^{d_1} e_{ij} \otimes F_\alpha e_{ij} F_\alpha^*$, are mutually orthogonal,
(2) $\lambda_\alpha > 0$, for $\alpha = 1, \ldots, D$, with $D := d_1 d_2$.

It turns out that the knowledge of the spectral decomposition (24) enables one to construct a map which is k-positive but not $(k+1)$-positive. One may prove the following result [24]: let $||F_1||_{k+1} < 1$. The map (24) is k-positive but not $(k+1)$-positive if

$$\frac{\lambda_1 \, ||F_1||_{k+1}^2}{1 - ||F_1||_{k+1}^2} \left(\mathbb{I}_{d_1} \otimes \mathbb{I}_{d_2} - P_1 \right) \ > \ W_+ \ \geq \ \frac{\lambda_1 \, ||F_1||_k^2}{1 - ||F_1||_k^2} \left(\mathbb{I}_{d_1} \otimes \mathbb{I}_{d_2} - P_1 \right) . \tag{25}$$

Note, that the above condition may be equivalently rewritten as the following condition for the spectrum of the corresponding entanglement witness W_+:

$$\mu_{k+1} > \lambda_\alpha \geq \mu_k \ ; \quad \alpha = 2, \ldots, D \ , \tag{26}$$

with

$$\mu_l = \frac{\lambda_1 \, ||F_1||_l^2}{1 - ||F_1||_l^2} \ , \tag{27}$$

for $l = 1, 2, \ldots, d$. It turns out [24] that if $d_1 = d_2 = d$ and P_1 is the maximally entangled state in $\mathbb{C}^d \otimes \mathbb{C}^d$, i.e. $F = U/\sqrt{d}$ with unitary U, then the above theorem reproduces 25 years old result by Takasaki and Tomiyama [25]. Moreover, for $d_1 = d_2 = d$, $k = 1$ and arbitrary P_1 the formula $\lambda_\alpha \geq \mu_1$ $(\alpha = 2, \ldots, d^2)$ was derived by Benatti et. al. [26].

4. Example: generalized Choi maps

Let us consider a family of maps

$$\varphi_\lambda \ : \ M_d \ \longrightarrow \ M_d \ ,$$

defined as follows

$$\varphi_\lambda(a) := \mathbb{I}_d \,\mathrm{tr} a - \lambda F_1 a F_1^* \ . \tag{28}$$

It generalizes the celebrated Choi map which is $(d-1)$-positive but not CP

$$\varphi_{\mathrm{Choi}}(a) := \mathbb{I}_d \,\mathrm{tr} a - \frac{1}{d-1} a \ , \tag{29}$$

which follows from (28) with $F_1 = \mathbb{I}_d/\sqrt{d}$ and $\lambda = d/(d-1)$. If $\lambda = d$, then (28) reproduces the so-called reduction map

$$\varphi_{\mathrm{red}}(a) := \mathbb{I}_d \,\mathrm{tr} a - a \ , \tag{30}$$

which is known to be completely co-positive. One easily finds

$$W_\lambda = \mathbb{I}_d \otimes \mathbb{I}_d - \lambda P_1 \ = \ W_+ - W_- \ , \tag{31}$$

where

$$W_+ = \mathbb{I}_d \otimes \mathbb{I}_d - P_1 \ , \qquad W_- = (\lambda - 1) P_1 \ ,$$

are orthogonally supported positive operators and

$$P_1 = \sum_{i,j=1}^{d} e_{ij} \otimes F_1 e_{ij} F_1^* \ . \tag{32}$$

Let $f_k := \| \, F_1 \, \|_k^2$ and assume that $f_{k+1} < 1$. The formula (25) implies that the map φ_λ is k-positive but not $(k+1)$-positive iff

$$\frac{1}{f_k} \ \geq \ \lambda \ > \ \frac{1}{f_{k+1}} \ . \tag{33}$$

Consider the family of states

$$\rho_\mu = \frac{1-\mu}{d^2-1} \left(\mathbb{I}_d \otimes \mathbb{I}_d - P_1 \right) + \mu P_1 \ . \tag{34}$$

Computing $\mathrm{tr}(W_\lambda \rho_\mu)$ one finds that $\mathrm{SN}(\rho_\mu) = k$ iff

$$f_k \ \geq \ \mu \ > \ f_{k-1} \ . \tag{35}$$

In particular ρ_μ is separable iff $\mu \geq f_1 = \| \, F_1 \, \|^2$. Note, that if P_1 is the maximally entangled state then ρ_μ defines a family of isotropic states. In this case $f_k = k/d$ and one recovers the well know result [22]: $\mathrm{SN}(\rho_\mu) = k$ iff $k/d \geq \mu > (k-1)/d$.

5. Conclusions

We provide partial classification of positive linear maps based on spectral conditions. The presented method generalizes the celebrated Choi example of a map which is positive but not CP. From the physical point of view our scheme provides a simple method for constructing entanglement witnesses. It turns out [24] that this scheme may be easily generalized to the multipartite setting. The subject deserves further studies in order to investigate which maps are indecomposable and atomic.

Acknowledgement

This work was partially supported by the Polish Ministry of Science and Higher Education Grant No 3004/B/H03/2007/33 and by the Polish Research Network *Laboratory of Physical Foundations of Information Processing*.

References

1. M. A. Nielsen and I. L. Chuang, *Quantum computation and quantum information*, Cambridge University Press, Cambridge, 2000.
2. A. Peres, Phys. Rev. Lett. **77**, 1413 (1996).
3. P. Horodecki, Phys. Lett. A **232**, 333 (1997).
4. B.M. Terhal, Phys. Lett. A **271**, 319 (2000); Linear Algebr. Appl. **323**, 61 (2000).
5. M. Lewenstein, B. Kraus, J. I. Cirac, and P. Horodecki, Phys. Rev. A **62**, 052310 (2000).
6. E. Størmer, Acta Math. **110**, 233 (1963).
7. W. Arveson, Acta Math. **123**, 141 (1969).
8. M.-D. Choi, Lin. Alg. Appl. **10**, 285 (1975); *ibid* **12**, 95 (1975).
9. S.L. Woronowicz, Rep. Math. Phys. **10**, 165 (1976).
10. S.L. Woronowicz, Comm. Math. Phys. **51**, 243 (1976).
11. A.G. Robertson, Quart. J. Math. Oxford (2), **34**, 87 (1983).
12. A.G. Robertson, Proc. Roy. Soc. Edinburh Sect. A, **94**, 71 (1983). **34**, 87 (1983).
13. W.-S. Tang, Lin. Alg. Appl. **79**, 33 (1986).
14. H. Osaka, Lin. Alg. Appl. **153**, 73 (1991); *ibid* **186**, 45 (1993).
15. K.-C. Ha, Publ. RIMS, Kyoto Univ., **34**, 591 (1998).
16. W.A. Majewski and M. Marcinek, J. Phys. A: Math. Gen. **34**, 5836 (2001).
17. D. Chruściński and A. Kossakowski, J. Phys. A: Math. Theor. **41**, 215201 (2008); *ibid* **41**, 145301 (2008).
18. E. Størmer, J. Funct. Anal. **254**, 2303 (2008).
19. D. Chruściński and A. Kossakowski, Open Systems and Inf. Dynamics, **14**, 275 (2007).

20. V. Paulsen, *Completely Bounded Maps and Operator Algebras*, Cambridge University Press, 2003.
21. R.A. Horn and C.R. Johnson, *Topics in Matrix Analysis*, (Cambridge University Press, New York, 1991).
22. B. Terhal and P. Horodecki, Phys. Rev. A **61**, 040301 (2000).
23. K. Kraus, *States, Effects and Operations: Fundamental Notions of Quantum Theory*, Springer Verlag, 1983.
24. D. Chruściński and A. Kossakowski, *Spectral conditions for positive maps*, arXiv:0809.4909 (to be published in Comm. Math. Phys.).
25. K. Takasaki and J. Tomiyama, Mathematische Zeitschrift **184**, 101 (1983).
26. F. Benatti, R. Floreanini and M. Piani, Open Systems and Inf. Dynamics, **11**, 325-338 (2004).

Quantum Bio-Informatics III
eds. L. Accardi, W. Freudenberg & M. Ohya

QUANTUM ENTANGLEMENT AND MULTIPARTITE SYMMETRIC STATES

DARIUSZ CHRUŚCIŃSKI

*Institute of Physics, Nicolaus Copernicus University,
Grudziądzka 5/7, 87–100 Toruń, Poland*

We analyze a class of multipartite states invariant under the action of unitary groups. Our class generalizes well known classes of bipartite states: Werner and isotropic states. We perform a detailed analysis of separability conditions for generalized Werner-like and isotropic-like multipartite states. It is shown that necessary and sufficient separability conditions generalize well known PPT criterion for multipartite case.

1. Introduction

One of the important problems in Quantum Information Theory [1] is the classification of multipartite quantum states. The problem is very hard and only some partial results are known [2,3]. In the bipartite case the crucial role plays so called Peres-Horodecki criterion of positive partial transposition (PPT) [4,14]: for any separable state ρ living in $\mathcal{H}_A \otimes \mathcal{H}_B$ one has $(\mathbb{1} \otimes \tau)\rho \geq 0$. Hence, a state which is not PPT is necessarily entangled. This criterion separability (it is sufficient iff $\dim\mathcal{H}_A \cdot \dim\mathcal{H}_B \leq 6$). On the other hand there are well known classes of bipartite states where the Peres-Horodecki criterion decides about separability: these are for example Werner [6] and isotropic states [7]. The characteristic feature of these states is invariance under the symmetry group, that is assuming that $\mathcal{H}_A = \mathcal{H}_B = \mathbb{C}^d$ one has for any $U \in U(d)$:

$$U \otimes U \,\mathcal{W} = \mathcal{W}\, U \otimes U \ , \tag{1}$$

in the case of Werner state, and

$$U \otimes \overline{U} \,\mathcal{I} = \mathcal{I}\, U \otimes \overline{U} \ , \tag{2}$$

in the case of isotropic state. It is well known that symmetry plays a prominent role in modern physics. In many cases it enables one to simplify the analysis of the corresponding problems and very often it leads to much deeper understanding and the most elegant mathematical formulation of the corresponding physical theory. In recent papers [8,9] we proposed a natural multipartite generalization of both Werner and isotropic states. These are $2K$-partite states living in

$$\mathcal{H}_{\rm total} = \mathcal{H}_1 \otimes \ldots \otimes \mathcal{H}_K \otimes \mathcal{H}_1 \otimes \ldots \otimes \mathcal{H}_K \ , \tag{3}$$

with $\dim\mathcal{H}_i = d_i$ which commute with all unitaries of the form

$$[U_1 \otimes \ldots \otimes U_K \otimes U_1 \otimes \ldots \otimes U_K \, , \, \mathcal{W}] = 0 \ , \tag{4}$$

for the Werner-like states, and

$$[U_1 \otimes \ldots \otimes U_K \otimes \overline{U}_1 \otimes \ldots \otimes \overline{U}_K \, , \, \mathcal{I}] = 0 \ , \tag{5}$$

for the isotropic-like states. In the above formulae $U_i \in U(d_i)$ – a group of unitary $d \times d$ matrices. In the present paper we propose a pedagogical approach to multipartite symmetric states. We perform a detailed analysis of separability conditions for generalized Werner-like and isotropic-like multipartite states. We show that necessary and sufficient separability conditions generalize well known PPT criterion for multipartite case.

2. Multipartite invariant states

Consider an N-partite quantum system living in

$$\mathcal{H}_{\rm total} = \mathcal{H}_1 \otimes \ldots \otimes \mathcal{H}_N \ , \tag{6}$$

where $\mathcal{H}_k$ denotes d_k-dimensional Hilbert space (in this paper we consider only finite dimensional Hilbert spaces). One calls a pure state represented by normalized $\psi \in \mathcal{H}_{\rm total}$ to be separable (or N-separable) if it can be represented by

$$\psi = \psi_1 \otimes \ldots \otimes \psi_N \ , \tag{7}$$

where $\psi_k \in \mathcal{H}_k$. Otherwise a state ψ is called to be entangled. In more general situation where the state is represented by a density operator ρ in $\mathcal{H}_{\rm total}$ it is called separable if it can be represented in the following form

$$\rho = \sum_\alpha p_\alpha \, \rho^{(1)}_\alpha \otimes \ldots \otimes \rho^{(1)}_\alpha \ , \tag{8}$$

where $\rho^{(k)}_\alpha$ represents density operator living in $\mathcal{H}_k$, and p_α defines a probability distribution, i.e. $p_\alpha \geq 0$ and $\sum_\alpha p_\alpha = 1$.

Consider now a special case with $N = 2K$ such that $\mathcal{H}_i = \mathcal{H}_{i+K}$, that is

$$\mathcal{H}_{\text{total}} = \mathcal{H}_1 \otimes \ldots \otimes \mathcal{H}_K \otimes \mathcal{H}_1 \otimes \ldots \otimes \mathcal{H}_K \ . \tag{9}$$

Let $G_1, \ldots, G_K$ be a set of compact Lie groups and denote by $\mathfrak{D}^{(i)}$ an irreducible unitary representation of G_i in $\mathcal{H}_i$. Then

$$\bigotimes_{i=1}^{K} \mathfrak{D}^{(1 \otimes \ldots \otimes K)}(\boldsymbol{g}) \ = \ \mathfrak{D}^{(1)}(g_1) \otimes \ldots \otimes \mathfrak{D}^{(K)}(g_K) \ , \tag{10}$$

with $\boldsymbol{g} = (g_1, \ldots, g_K)$, defines an irreducible representation of $G_1 \times \ldots \times G_K$ in $\mathcal{H}_1 \otimes \ldots \otimes \mathcal{H}_K$.

A state ρ in $\mathcal{H}_{\text{total}}$ is invariant under $\mathfrak{D}^{(1 \otimes \ldots \otimes K)} \otimes \mathfrak{D}^{(1 \otimes \ldots \otimes K)}$ iff

$$[\mathfrak{D}^{(1 \otimes \ldots \otimes K)}(\boldsymbol{g}) \otimes \mathfrak{D}^{(1 \otimes \ldots \otimes K)}(\boldsymbol{g}) \, , \, \rho] = 0 \ , \tag{11}$$

$$[\mathfrak{D}^{(1 \otimes \ldots \otimes K)}(\boldsymbol{g}) \otimes \mathfrak{D}^{(1 \otimes \ldots \otimes K)}(\boldsymbol{g}) \, , \, \rho] = 0 \tag{12}$$

for all $\boldsymbol{g} \in G_1 \times \ldots \times G_K$.

Taking as G_i a unitary group $U(d_i)$ and as $\mathfrak{D}^{(i)}$ a defining representation of $U(d_i)$, i.e. representing elements from $U(d_i)$ by $d_i \times d_i$ unitary matrices, one obtains a generalized Werner-like symmetric state $\mathcal{W}$ defined in (4). Similarly, a state ρ is invariant under $(\mathfrak{D}^{(1)} \otimes \ldots \otimes \mathfrak{D}^{(K)}) \otimes \overline{(\mathfrak{D}^{(1)} \otimes \ldots \otimes \mathfrak{D}^{(K)})}$ iff

$$[\mathfrak{D}^{(1)}(g_1) \otimes \ldots \otimes \mathfrak{D}^{(K)}(g_K) \otimes \overline{\mathfrak{D}^{(1)}(g_1) \otimes \ldots \otimes \mathfrak{D}^{(K)}(g_K)} \, , \, \rho] = 0 \ , \tag{13}$$

for all $g_i \in G_i$. Again taking as G_i a unitary group $U(d_i)$ and as $\mathfrak{D}^{(i)}$ a defining representation of $U(d_i)$ one obtains a generalized isotropic-like symmetric state $\mathcal{I}$ defined in (5).

2.1. *Werner-like states*

$2K$-partite Werner-like states satisfying (4) may be constructed a follows: define a flip operator $\mathbb{F}$ in $\mathcal{H} \otimes \mathcal{H}$ by

$$\mathbb{F} = \sum_{i,j=1}^{d} e_{ij} \otimes e_{ji} \ , \tag{14}$$

where $d = \dim\mathcal{H}$, and $e_{ij} = |e_i \, e_j|$, with $e_1, \ldots, e_d$ being on orthonormal basis in $\mathcal{H}$. Define two orthogonal projectors

$$Q^0 = \frac{1}{2}(\mathbb{I}_d \otimes \mathbb{I}_d + \mathbb{F}) \ , \qquad Q^1 = \frac{1}{2}(\mathbb{I}_d \otimes \mathbb{I}_d - \mathbb{F}) \ , \tag{15}$$

where $\mathbb{I}_d$ is an identity matrix in $\mathbb{C}^d \otimes \mathbb{C}^d$. Now, a 2-partite Werner state may be uniquely represented by

$$\mathcal{W} = q_0 \widetilde{Q}^0 + q_1 \widetilde{Q}^1 \ , \tag{16}$$

where $\widetilde{Q}^\alpha = Q^\alpha / \mathrm{Tr}\, Q^\alpha$, and $q_0, q_1 \geq 0$ together with $q_0 + q_1 = 1$. For $K > 1$ one proceeds as follows: let $\boldsymbol{\sigma} = (\sigma_1, \ldots, \sigma_K)$ be a binary K-vector, that is, $\sigma_i = 0, 1$. Let us define a $2K$-partite orthogonal projectors

$$\mathbf{Q}^{\boldsymbol{\sigma}} = Q^{\sigma_1}_{1|K+1} \otimes \ldots \otimes Q^{\sigma_K}_{K|2K} \ , \tag{17}$$

where $Q^{\sigma_i}_{i|K+i}$ is a bipartite projector defined in (18) acting on $\mathcal{H}_i \otimes \mathcal{H}_{K+i} = \mathbb{C}^{d_i} \otimes \mathbb{C}^{d_i}$. They do satisfy the following relations:

$$\mathbf{Q}^{\boldsymbol{\alpha}} \mathbf{Q}^{\boldsymbol{\beta}} = \delta_{\boldsymbol{\alpha}\boldsymbol{\beta}} \mathbf{Q}^{\boldsymbol{\alpha}} \ , \tag{18}$$

and

$$\sum_{\boldsymbol{\sigma}} \mathbf{Q}^{\boldsymbol{\sigma}} = \mathbb{I}_{d_1} \otimes \ldots \otimes \mathbb{I}_{d_K} \otimes \mathbb{I}_{d_1} \otimes \ldots \otimes \mathbb{I}_{d_K} = \mathbb{I}_{\mathrm{total}} \ , \tag{19}$$

where we use standard notation

$$\delta_{\boldsymbol{\alpha}\boldsymbol{\beta}} := \delta_{\alpha_1 \beta_1} \ldots \delta_{\alpha_K \beta_K} \ .$$

It is therefore clear[8] that any $2K$-partite Werner-like states may be uniquely written in the following form

$$\mathcal{W} = \sum_{\boldsymbol{\sigma}} q_{\boldsymbol{\sigma}} \widetilde{\mathbf{Q}}^{\boldsymbol{\sigma}} \ , \tag{20}$$

where $q_{\boldsymbol{\sigma}} \geq 0$ and $\sum_{\boldsymbol{\sigma}} q_{\boldsymbol{\sigma}} = 1$.

2.2. *Isotropic-like states*

To define $2K$-partite isotropic-like states satisfying (5) let us introduce two projectors in $\mathbb{C}^d \otimes \mathbb{C}^d$

$$P^0 = \mathbb{I}_d \otimes \mathbb{I}_d - P^+_d \ , \qquad P^1 = P^+_d \ , \tag{21}$$

where P^+_d denote a projector onto the canonical maximally entangled state

$$P^+_d = \frac{1}{d} \sum_{i,j=1}^{d} e_{ij} \otimes e_{ij} \ . \tag{22}$$

Now, a 2-partite isotropic state may be uniquely represented by

$$\mathcal{I} = p_0 \widetilde{P}^0 + p_1 \widetilde{P}^1 \ , \tag{23}$$

where $\widetilde{P}^\alpha = P^\alpha/\mathrm{Tr}\, P^\alpha$, and $p_0, p_1 \geq 0$ together with $p_0 + p_1 = 1$. For $K > 1$ one defines $2K$-partite orthogonal projectors

$$\mathbf{P}^{\boldsymbol{\sigma}} = P^{\sigma_1}_{1|K+1} \otimes \ldots \otimes P^{\sigma_K}_{K|2K} , \tag{24}$$

where $P^{\sigma_i}_{i|K+i}$ is a bipartite projector defined in (25) acting on $\mathcal{H}_i \otimes \mathcal{H}_{K+i} = \mathbb{C}^{d_i} \otimes \mathbb{C}^{d_i}$. They do satisfy the following relations:

$$\mathbf{P}^{\boldsymbol{\alpha}}\mathbf{P}^{\boldsymbol{\beta}} = \delta_{\boldsymbol{\alpha\beta}}\mathbf{P}^{\boldsymbol{\alpha}} , \tag{25}$$

and

$$\sum_{\boldsymbol{\sigma}} \mathbf{P}^{\boldsymbol{\sigma}} = \mathbb{I}_{d_1} \otimes \ldots \otimes \mathbb{I}_{d_K} \otimes \mathbb{I}_{d_1} \otimes \ldots \otimes \mathbb{I}_{d_K} = \mathbb{I}_{\mathrm{total}} . \tag{26}$$

It is therefore clear[8] that any $2K$-partite isotropic-like states may be uniquely written in the following form

$$\mathcal{I} = \sum_{\boldsymbol{\sigma}} p_{\boldsymbol{\sigma}} \widetilde{\mathbf{P}}^{\boldsymbol{\sigma}} , \tag{27}$$

where $p_{\boldsymbol{\sigma}} \geq 0$ and $\sum_{\boldsymbol{\sigma}} p_{\boldsymbol{\sigma}} = 1$.

3. Analysis of Werner-like states

Our aim now is to establish separability conditions for $2K$-partite Werner-like states (20). To do this let us define the following family of partial transpositions: for each binary K-vector $\boldsymbol{\sigma}$ let us introduce $\tau_{\boldsymbol{\sigma}}$ by

$$\tau_{\boldsymbol{\sigma}} = (\mathbb{1}_1 \otimes \ldots \otimes \mathbb{1}_K) \otimes \tau^{\sigma_1} \otimes \ldots \otimes \tau^{\sigma_K} , \tag{28}$$

where $\mathbb{1}_i$ is a identity map on $\mathcal{B}(\mathcal{H}_i)$.

3.1. $K = 1$

For $K = 1$ the partial transposition is realized by

$$(\mathbb{1} \otimes \tau)\mathcal{W}_{\mathbf{q}} = \sum_{\sigma=0}^{1} q'_\sigma \, \widetilde{P}^\sigma , \tag{29}$$

with

$$q'_\sigma = \sum_{\alpha=0}^{1} q_\alpha \, \mathbf{X}[d]_{\alpha\sigma} , \tag{30}$$

where the 2×2 matrix $\mathbf{X}[d]$ is given by

$$\mathbf{X}[d] = \frac{1}{d} \begin{pmatrix} d-1 & 1 \\ d+1 & -1 \end{pmatrix} . \tag{31}$$

Now, $\mathcal{W}_{\mathbf{q}}$ is PPT iff $q'_\sigma \geq 0$, which is equivalent to

$$q_0 \geq q_1 \ . \tag{32}$$

This condition may be rewritten as $q_1 \leq 1/2$. Therefore, the convex set of Werner PPT states is generated by 2 extremal states corresponding to $\mathbf{q}_1 = (1,0)$ and $\mathbf{q}_2 = (1/2,1/2)$. Since both extremal PPT states are separable it follows that all PPT Werner states are separable.

3.2. $K = 2$

Consider an invariant state

$$\rho = \sum_{\sigma} q_\sigma \widetilde{\mathbf{Q}}^{\sigma} \ , \tag{33}$$

living in $\mathcal{H}_1 \otimes \mathcal{H}_2 \otimes \mathcal{H}_1 \otimes \mathcal{H}_2$ with $\dim \mathcal{H}_i = d_i$. For $K = 2$ one has 3 different partial transpositions $\tau_{\boldsymbol{\sigma}}$ parameterized by 2-dimensional binary vector $\boldsymbol{\sigma} = (\sigma_1, \sigma_2) \neq 0$. A state ρ is $\boldsymbol{\sigma}$-PPT iff

$$q'_{\mu_1\mu_2} = \sum_{\nu_1,\nu_2=0}^{1} \Big(\mathbf{X}[d_1]\Big)^{\sigma_1}_{\nu_1\mu_1} \Big(\mathbf{X}[d_2]\Big)^{\sigma_2}_{\nu_2\mu_2} \geq 0 \tag{34}$$

where $\mathbf{X}^1 = \mathbf{X}$ and $\mathbf{X}^0 = \mathbf{I}$.

(1) ρ is (01)-PPT iff

$$q_{\sigma 0} \geq q_{\sigma 1} \ , \qquad \sigma = 0, 1 \ , \tag{35}$$

(2) ρ is (10)-PPT iff

$$q_{0\sigma} \geq q_{1\sigma} \ , \qquad \sigma = 0, 1 \ . \tag{36}$$

(3) ρ is (11)-PPT iff

$$\begin{aligned} (d_1 - 1)(q_{00} - q_{01}) + (d_1 + 1)(q_{10} - q_{11}) &\geq 0 \ , \\ (d_2 - 1)(q_{00} - q_{10}) + (d_2 + 1)(q_{01} - q_{11}) &\geq 0 \ , \\ (q_{00} + q_{11}) - (q_{01} + q_{10}) &\geq 0 \ . \end{aligned} \tag{37}$$

Note that (35) and (36) imply that only the last condition in (37) is nontrivial:

$$q_{00} + q_{11} \geq q_{01} + q_{10} \ . \tag{38}$$

It is clear that (35) and (36) follow from (32), i.e. they may be derived from $K = 1$ case. Condition (38) is new. Due to $q_{00} + q_{01} + q_{10} + q_{11} = 1$, it may be equivalently rewritten as follows

$$q_{01} + q_{10} \leq \frac{1}{2} \ , \tag{39}$$

which very much resembles $q_1 \leq 1/2$. Interestingly all $\boldsymbol{\sigma}$-PPT conditions (35), (36) and (39) do not depend upon dimensions d_1 and d_2! It is clear from (39) that

$$q_{01} \leq \frac{1}{2}\,, \qquad q_{10} \leq \frac{1}{2}\,, \tag{40}$$

and (39) together with (35) and (36) imply

$$q_{11} \leq \frac{1}{4}\,. \tag{41}$$

Hence, the extremal PPT states correspond to $\{\,\mathbf{q}_1, \mathbf{q}_2, \mathbf{q}_3, \mathbf{q}_4\,\}$:

$\mathbf{q}\backslash q_{\boldsymbol{\sigma}}$	q_{00}	q_{01}	q_{10}	q_{11}
$\mathbf{q}_1$	1	0	0	0
$\mathbf{q}_2$	1/2	0	1/2	0
$\mathbf{q}_3$	1/2	1/2	0	0
$\mathbf{q}_4$	1/4	1/4	1/4	1/4

(42)

It turns out that all extremal PPT states are 4-separable. Indeed, consider the following product state ρ

$$\rho = P_{\psi_1} \otimes P_{\psi_2} \otimes P_{\varphi_1} \otimes P_{\varphi_2}\,, \tag{43}$$

where $P_{\psi_k} = |\psi_k \; \psi_k|$, $P_{\varphi_k} = |\varphi_k \; \varphi_k|$, and ψ_k, φ_k are normalized vectors in $\mathbb{C}^{d_k}$; $k = 1, 2$. Upon twirling one obtains 4-separable invariant state $\rho = W_{\mathbf{q}}$ with fidelities $q_{\boldsymbol{\sigma}}$ given by:

$$\begin{aligned} q_{00} &= \frac{1}{4}(1 + a_1)(1 + a_2)\,, \\ q_{01} &= \frac{1}{4}(1 + a_1)(1 - a_2)\,, \\ q_{10} &= \frac{1}{4}(1 - a_1)(1 + a_2)\,, \\ q_{11} &= \frac{1}{4}(1 - a_1)(1 - a_2)\,, \end{aligned} \tag{44}$$

where the numbers a_1 and a_2 are defined as follows

$$a_1 = |\psi_1|\varphi_1\,|^2\,, \qquad a_2 = |\psi_2|\varphi_2\,|^2\,. \tag{45}$$

It is clear that each $\mathbf{q}_k$ may be constructed via (44) by the following choice of (a_1, a_2):

$$(1,1) \longleftrightarrow \mathbf{q}_1\,, \quad (0,1) \longleftrightarrow \mathbf{q}_2\,, \quad (1,0) \longleftrightarrow \mathbf{q}_3\,, \quad (0,0) \longleftrightarrow \mathbf{q}_4\,. \tag{46}$$

3.3. $K = 3$

Consider now a 6-partite invariant state

$$\rho = \sum_{\boldsymbol{\sigma}} q_{\boldsymbol{\sigma}} \widetilde{\mathbf{Q}}^{\boldsymbol{\sigma}} , \tag{47}$$

living in $\mathcal{H}_1 \otimes \mathcal{H}_2 \otimes \mathcal{H}_3 \otimes \mathcal{H}_1 \otimes \mathcal{H}_2 \otimes \mathcal{H}_3$. For $K = 3$ one has $8(= 2^3 - 1)$ different partial transpositions $\tau_{\boldsymbol{\sigma}}$ parameterized by 3-dimensional binary vector $\boldsymbol{\sigma} = (\sigma_1, \sigma_2, \sigma_3) \neq 0$. A state ρ is $\boldsymbol{\sigma}$-PPT iff

$$q'_{\mu_1\mu_2\mu_3} = \sum_{\nu_1,\nu_2,\nu_3=0}^{1} \Big(\mathbf{X}[d_1]\Big)^{\sigma_1}_{\nu_1\mu_1} \Big(\mathbf{X}[d_2]\Big)^{\sigma_2}_{\nu_2\mu_2} \Big(\mathbf{X}[d_3]\Big)^{\sigma_3}_{\nu_3\mu_3} \geq 0 \,. \tag{48}$$

(1) For $|\boldsymbol{\sigma}| = 1$ the corresponding $\boldsymbol{\sigma}$-PPT conditions follow from (32):

$$\boldsymbol{\sigma} = (001) \longrightarrow q_{\sigma_1\sigma_2 0} \geq q_{\sigma_1\sigma_2 1} \ , \ \sigma_1, \sigma_2 = 0, 1 \ , \tag{49}$$

$$\boldsymbol{\sigma} = (010) \longrightarrow q_{\sigma_1 0\sigma_3} \geq q_{\sigma_1 1\sigma_3} \ , \ \sigma_1, \sigma_3 = 0, 1 \ , \tag{50}$$

$$\boldsymbol{\sigma} = (100) \longrightarrow q_{0\sigma_2\sigma_3} \geq q_{1\sigma_2\sigma_3} \ , \ \sigma_2, \sigma_3 = 0, 1 \ . \tag{51}$$

(2) Additional $\boldsymbol{\sigma}$-PPT conditions for $|\boldsymbol{\sigma}| = 2$ follow from (38):

$$\boldsymbol{\sigma} = (011) \longrightarrow q_{\sigma_1 00} + q_{\sigma_1 11} \geq q_{\sigma_1 01} + q_{\sigma_1 10} \ , \ \sigma_1 = 0, 1 \ , \tag{52}$$

$$\boldsymbol{\sigma} = (101) \longrightarrow q_{0\sigma_2 0} + q_{1\sigma_2 1} \geq q_{0\sigma_2 1} + q_{1\sigma_2 1} \ , \ \sigma_2 = 0, 1 \ , \tag{53}$$

$$\boldsymbol{\sigma} = (110) \longrightarrow q_{00\sigma_3} + q_{11\sigma_3} \geq q_{01\sigma_3} + q_{10\sigma_3} \ , \ \sigma_3 = 0, 1 \ . \tag{54}$$

(3) The last (111)-PPT condition can not be derived from $K = 1$ and $K = 2$ cases. Assuming (49)–(54) the only new condition reads as follows

$$q_{000} + q_{011} + q_{101} + q_{110} \geq q_{001} + q_{010} + q_{100} + q_{111} \ , \tag{55}$$

and due to $\sum_{\boldsymbol{\sigma}} q_{\boldsymbol{\sigma}} = 1$ it may be rewritten as

$$q_{001} + q_{010} + q_{100} + q_{111} \leq \frac{1}{2} \ , \tag{56}$$

in perfect analogy to (39).

Summarizing, Werner like invariant state $W_{\mathbf{q}}$ is $\boldsymbol{\sigma}$-PPT for all binary $\boldsymbol{\sigma}$ iff (49)–(54) and (56) hold. Again, these conditions do not depend upon dimensions d_1, d_2, d_3. Now, it follows from (56) that

$$q_{001} \, , q_{010} \, , q_{100} \leq \frac{1}{2} \, . \tag{57}$$

Moreover, formula (52) implies

$$q_{001} + q_{110} + q_{010} + q_{101} \leq \frac{1}{2} \ , \tag{58}$$

and hence using (51) it leads to

$$q_{110} + q_{101} \leq \frac{1}{4} \,. \tag{59}$$

Similarly, formula (53) gives

$$q_{001} + q_{100} + q_{011} + q_{110} \leq \frac{1}{2} \,, \tag{60}$$

and upon using (50) it implies

$$q_{110} + q_{011} \leq \frac{1}{4} \,. \tag{61}$$

Finally, formula (54) implies

$$q_{010} + q_{100} + q_{011} + q_{101} \leq \frac{1}{2} \,, \tag{62}$$

which due to (49) leads to

$$q_{011} + q_{101} \leq \frac{1}{4} \,. \tag{63}$$

Note, that (59), (61) and (63) imply that

$$q_{011} \,, q_{010} \,, q_{110} \leq \frac{1}{4} \,. \tag{64}$$

Let us observe, that formulae (49)–(51) give us the following relations:

$$\begin{aligned} q_{001} &\geq q_{101} \geq q_{111} \,, \\ q_{010} &\geq q_{110} \geq q_{111} \,, \\ q_{100} &\geq q_{101} \geq q_{111} \,, \end{aligned}$$

and hence formula (56) implies

$$q_{111} \leq \frac{1}{8} \,. \tag{65}$$

Summarizing, formulae (57), (64) and (65) may be rewritten in a compact form as follows:

$$q_\sigma \leq \frac{1}{2^{|\sigma|}} \,. \tag{66}$$

The above conditions give rise to the following extremal multi-PPT states:

$$
\begin{array}{|c||c|c|c|c|c|c|c|c|}
\hline
\mathbf{q}\backslash q_\sigma & q_{000} & q_{001} & q_{010} & q_{011} & q_{100} & q_{101} & q_{110} & q_{111} \\
\hline
\mathbf{q}_1 & 1 & 0 & 0 & 0 & 0 & 0 & 0 & 0 \\
\hline
\mathbf{q}_2 & 1/2 & 0 & 0 & 0 & 1/2 & 0 & 0 & 0 \\
\hline
\mathbf{q}_3 & 1/2 & 0 & 1/2 & 0 & 0 & 0 & 0 & 0 \\
\hline
\mathbf{q}_4 & 1/2 & 1/2 & 0 & 0 & 0 & 0 & 0 & 0 \\
\hline
\mathbf{q}_5 & 1/4 & 1/4 & 0 & 0 & 1/4 & 1/4 & 0 & 0 \\
\hline
\mathbf{q}_6 & 1/4 & 1/4 & 1/4 & 1/4 & 0 & 0 & 0 & 0 \\
\hline
\mathbf{q}_7 & 1/4 & 0 & 1/4 & 0 & 1/4 & 0 & 1/4 & 0 \\
\hline
\mathbf{q}_8 & 1/8 & 1/8 & 1/8 & 1/8 & 1/8 & 1/8 & 1/8 & 1/8 \\
\hline
\end{array}
\tag{67}
$$

It is easy to show that all extremal states are 6-separable. To show this consider a product state ρ

$$
\rho = P_{\psi_1} \otimes P_{\psi_2} \otimes P_{\psi_3} \otimes P_{\varphi_1} \otimes P_{\varphi_2} \otimes P_{\varphi_3} \ , \tag{68}
$$

where $P_{\psi_k} = |\psi_k \, \psi_k|$, $P_{\varphi_k} = |\varphi_k \, \varphi_k|$, and ψ_k, φ_k are normalized vectors in $\mathbb{C}^{d_k}$; $k = 1, 2, 3$. Upon twirling one obtains 6-separable invariant state $\rho = W_{\mathbf{q}}$ with fidelities q_σ given by:

$$
q_{\sigma_1\sigma_2\sigma_3} = \frac{1}{8}(1 + (-1)^{\sigma_1} a_1)(1 + (-1)^{\sigma_2} a_2)(1 + (-1)^{\sigma_3} a_3) \ , \tag{69}
$$

with

$$
a_i = |\psi_i|\varphi_i|^2 \ , \qquad i = 1, 2, 3 \ . \tag{70}
$$

It is clear that each $\mathbf{q}_k$ may be constructed via (69) by the following choice of (a_1, a_2, a_3):

$$
\begin{array}{|c||c|c|c|c|c|c|c|c|}
\hline
 & \mathbf{q}_1 & \mathbf{q}_2 & \mathbf{q}_3 & \mathbf{q}_4 & \mathbf{q}_5 & \mathbf{q}_6 & \mathbf{q}_7 & \mathbf{q}_8 \\
\hline
a_1 & 1 & 0 & 1 & 1 & 0 & 1 & 0 & 0 \\
\hline
a_2 & 1 & 1 & 0 & 1 & 1 & 0 & 0 & 0 \\
\hline
a_3 & 1 & 1 & 1 & 0 & 0 & 0 & 1 & 0 \\
\hline
\end{array}
\tag{71}
$$

It proves, therefore, that 6-partite multi-PPT invariant states are 6-separable.

3.4. *Arbitrary K*

Consider now a $2K$-partite invariant Werner like state

$$
\rho = \sum_\sigma q_\sigma \widetilde{\mathbf{Q}}^\sigma \ , \tag{72}
$$

living in $\mathcal{H}_1 \otimes \ldots \otimes \mathcal{H}_K \otimes \mathcal{H}_1 \otimes \ldots \otimes \mathcal{H}_K$. It is clear that all partial transpositions $\tau_{\boldsymbol{\sigma}}\rho$ with $|\boldsymbol{\sigma}| < K$ may be derived from the previous cases with the number of pairs less than K. The last transposition with $|\boldsymbol{\sigma}| = K$, i.e. $\boldsymbol{\sigma} = (1, \ldots, 1)$ produces one additional condition. It is clear from (39) and (56) that this new condition reads as follows

$$\sum_{|\boldsymbol{\sigma}|=\mathrm{odd}} q_{\boldsymbol{\sigma}} \leq \frac{1}{2} \,. \tag{73}$$

Moreover, multi-PPT condition imply the following bounds upon fidelities $q_{\boldsymbol{\sigma}}$

$$q_{\boldsymbol{\sigma}} \leq \frac{1}{2^{|\boldsymbol{\sigma}|}} \,, \tag{74}$$

which generalizes (66). The above conditions give rise to 2^K extremal multi-PPT states:

- 1 extremal state corresponding to $\mathbf{q} = (1, 0, 0, \ldots, 0)$,
- K extremal states with $q_{\boldsymbol{\sigma}} = 1/2$ for each $\boldsymbol{\sigma}$ with $|\boldsymbol{\sigma}| = 1$,
- $\binom{K}{2}$ extremal states with $q_{\boldsymbol{\sigma}} = 1/4$ for each $\boldsymbol{\sigma}$ with $|\boldsymbol{\sigma}| = 2$,
- ...
- $\binom{K}{L}$ extremal states with $q_{\boldsymbol{\sigma}} = 1/2^L$ for each $\boldsymbol{\sigma}$ with $|\boldsymbol{\sigma}| = L$,
- ...
- 1 extremal state corresponding to $\mathbf{q} = (1, 1, \ldots, 1)/2^K$.

Again, it is easy to show that all extremal states are $2K$-separable. To show this consider a product state ρ

$$\rho = P_{\psi_1} \otimes \ldots \otimes P_{\psi_K} \otimes P_{\varphi_1} \otimes \ldots \otimes P_{\varphi_K} \,, \tag{75}$$

where $P_{\psi_i} = |\psi_i \, \psi_i|$, $P_{\varphi_i} = |\varphi_i \, \varphi_i|$, and ψ_i, φ_i are normalized vectors in $\mathbb{C}^{d_i}$; $i = 1, 2, \ldots, K$. Upon twirling one obtains $2K$-separable invariant state $\rho = W_{\mathbf{q}}$ with fidelities $q_{\boldsymbol{\sigma}}$ given by:

$$q_{\sigma_1\sigma_2\ldots\sigma_K} = \frac{1}{8}(1 + (-1)^{\sigma_1}a_1)(1 + (-1)^{\sigma_2}a_2) \ldots (1 + (-1)^{\sigma_K}a_K) \,, \tag{76}$$

with

$$a_i = |\psi_i|\varphi_i|^2 \,, \qquad i = 1, 2, \ldots, K \,. \tag{77}$$

It is clear that each extremal multi-PPT states may be constructed via (76) by the clever choice of $(a_1, a_2, \ldots, a_K)$.

3.5. *Biseparability*

It is interesting to ask when the $2K$-partite Wener-like state $\mathcal{W}_{\mathbf{q}}$ is $A|B$ biseparable, i.e. separable with respect $\mathcal{H}_A \otimes \mathcal{H}_B$, where $\mathcal{H}_A = \mathcal{H}_B = \mathcal{H}_1 \otimes \ldots \otimes \mathcal{H}_K$. For $K = 1$ the answer is known. In this case biseparability is nothing but the full separability which is equivalent to PPT property.

Theorem 3.1. *Let $K = 2$ and $d_1 = d_2 = d$. Then a 4-partite Werner-like state is $A|B$ biseparable iff it is $(1,1)$-PPT.*

Proof. to prove it we find extremal $(1,1)$-PPT and show that that all of them are $A|B$ biseparable. To find these one has solve (37) with $d_1 = d_2 = d$:

$$\begin{aligned} (d-1)(q_{00} - q_{01}) + (d+1)(q_{10} - q_{11}) \geq 0 \,, \\ (d-1)(q_{00} - q_{10}) + (d+1)(q_{01} - q_{11}) \geq 0 \,, \\ (q_{00} + q_{11}) - (q_{01} + q_{10}) \geq 0 \,. \end{aligned} \tag{78}$$

Actually, it turns out that the convex set of $(1,1)$-PPT states is generated by 5 extremal points: four of them are already derived in (42). Clearly, since all of them are fully 4-separable it follows that they are $A|B$ biseparable, too. The last one is given by

$$\mathbf{q}_5 = \frac{1}{2d}(d+1, 0, 0, d-1) \,. \tag{79}$$

To show that the corresponding Werner-like state

$$\mathcal{W}_5 = \frac{d+1}{2d}\widetilde{\mathbf{Q}}^{00} + \frac{d-1}{2d}\widetilde{\mathbf{Q}}^{11} \,, \tag{80}$$

is $A|B$ biseparable let us observe that $\mathcal{W}_5 =< |\Psi^{++}_{AB} < \Psi^{++}_{AB}|$, where Ψ^{++}_{AB} is a biseparable state defined by

$$\Psi^{++}_{AB} = \Psi^{+}_{1|2} \otimes \Psi^{+}_{3|4} \,, \tag{81}$$

and $\Psi^{+}_{i|j}$ is a maximally entangled state in $\mathcal{H}_i \otimes \mathcal{H}_j = \mathbb{C}^d \otimes \mathbb{C}^d$, that is

$$\Psi^{+}_{i|j} = \frac{1}{\sqrt{d}} \sum_{k=1}^{d} e_k \otimes e_k \,. \tag{82}$$

Indeed, using

$$< \Psi^{++}_{AB}| I_{1|3} \otimes \mathbf{F}^1_{2|4} |\Psi^{++}_{AB} =< \Psi^{++}_{AB}| \mathbf{F}^1_{1|3} \otimes I_{2|4} |\Psi^{++}_{AB} = \frac{1}{d} \,, \tag{83}$$

and

$$< \Psi^{++}_{AB} | \mathbf{F}^1_{1|3} \otimes \mathbf{F}^1_{2|4} | \Psi^{++}_{AB} = 1 \ , \tag{84}$$

one finds

$$q_{00} = \frac{d+1}{2d} \ , \quad q_{01} = q_{10} = 0 \ , \quad q_{11} = \frac{d-1}{2d} \ , \tag{85}$$

which finally proves the theorem. □

It should be stressed that the above theorem does not hold if $d_1 \neq d_2$ or $K > 2$ [11].

4. Analysis of isotropic-like states

In this Section we perform separability analysis for multipartite isotropic states.

4.1. $K = 1$

For $K = 1$ the partial transposition is realized by

$$(\mathbb{1} \otimes \tau)\mathcal{I}_{\mathbf{p}} = \sum_{\sigma=0}^{1} p'_\sigma \, \widetilde{Q}^\sigma \ , \tag{86}$$

with

$$p'_\sigma = \sum_{\alpha=0}^{1} p_\alpha \, \mathbf{Y}[d]_{\alpha\sigma} \ , \tag{87}$$

where the 2×2 matrix $\mathbf{Y}[d]$ is given by

$$\mathbf{Y}[d] = \mathbf{X}[d]^{-1} = \frac{1}{2} \begin{pmatrix} 1 & 1 \\ 1+d & 1-d \end{pmatrix} . \tag{88}$$

Now, $\mathcal{I}_{\mathbf{p}}$ is PPT iff $p'_\sigma \geq 0$, that is

$$p_0 + (1-d)p_1 \geq 0 \ , \tag{89}$$

which is equivalent to the well known condition

$$p_1 \leq \frac{1}{d} \ . \tag{90}$$

Therefore, the convex set of isotropic PPT states is generated by 2 extremal states corresponding to $\mathbf{p}_1 = (1,0)$ and $\mathbf{p}_2 = d^{-1}(d-1,1)$. Since both extremal PPT states are separable it follows that all PPT isotropic states are separable.

4.2. $K = 2$

Consider an isotropic-like 4-partite invariant state

$$\mathcal{I}_{\mathbf{p}} = \sum_{\boldsymbol{\sigma}} p_{\boldsymbol{\sigma}} \widetilde{\mathbf{P}}^{\boldsymbol{\sigma}} , \tag{91}$$

living in $\mathcal{H}_1 \otimes \mathcal{H}_2 \otimes \mathcal{H}_1 \otimes \mathcal{H}_2$. This state is $\boldsymbol{\sigma}$-PPT iff

$$p'_{\mu_1\mu_2} = \sum_{\nu_1,\nu_2=0}^{1} \Big(\mathbf{Y}[d_1]\Big)^{\sigma_1}_{\nu_1\mu_1} \Big(\mathbf{Y}[d_2]\Big)^{\sigma_2}_{\nu_2\mu_2} \geq 0 \, . \tag{92}$$

One has therefore

(1) $\mathcal{I}_{\mathbf{p}}$ is (01)-PPT iff

$$p_{\sigma\, 0} + (1 - d_2) p_{\sigma 1} \geq 0 \, , \qquad \sigma = 0, 1 \, , \tag{93}$$

(2) $\mathcal{I}_{\mathbf{q}}$ is (10)-PPT iff

$$p_{0\sigma} + (1 - d_1) p_{1\, \sigma} \geq 0 \, , \qquad \sigma = 0, 1 \, . \tag{94}$$

(3) $\mathcal{I}_{\mathbf{q}}$ is (11)-PPT iff

$$\begin{aligned} p_{00} + (1-d_1)p_{10} + (1+d_2)\Big(p_{01} + (1-d_1)p_{11}\Big) &\geq 0 \, , \\ p_{00} + (1-d_2)p_{01} + (1+d_1)\Big(p_{10} + (1-d_2)p_{11}\Big) &\geq 0 \, , \\ p_{00} + (1-d_2)p_{01} + (1-d_1)\Big(p_{10} + (1-d_2)p_{11}\Big) &\geq 0 \, . \end{aligned} \tag{95}$$

Note that due to (93) and (94) only the last formula in (95) is nontrivial. Using $\sum_{\boldsymbol{\sigma}} p_{\boldsymbol{\sigma}} = 1$ it gives rise to

$$d_1 p_{10} + d_2 p_{01} \leq 1 + (d_1 d_2 - d_1 - d_2) p_{11} \, . \tag{96}$$

It is analog of (39) in the Werner-like case. Note that for $d_1 = d_2 = 2$, Eq. (96) leads to

$$p_{10} + p_{01} \leq \frac{1}{2} \, , \tag{97}$$

since for $d = 2$ Werner and isotropic states have the same structure. Taking into account

$$\begin{aligned} p_{01} &\geq (d_1 - 1) p_{11} \, , \\ p_{10} &\geq (d_2 - 1) p_{11} \, , \end{aligned}$$

formula (96) implies

$$p_{11} \leq \frac{1}{d_1 d_2} \, , \tag{98}$$

which defines an analog of (41) (clearly, for $d_1 = d_2 = 2$ it reproduces (41)). Let us look for an upper bound on p_{01} and p_{01}. Suppose that one of them, say p_{01} vanishes. Formula (94) implies that $p_{11} = 0$. Hence, it follows from (96) that

$$p_{10} \leq \frac{1}{d_1} \ . \tag{99}$$

In the same way one shows that

$$p_{01} \leq \frac{1}{d_2} \ . \tag{100}$$

Summarizing formulae (98), (99) and (100) one has

$$p_{\sigma_1\sigma_2} \leq \frac{1}{d_1^{\sigma_1} d_2^{\sigma_2}} \ . \tag{101}$$

There are, therefore, 4 extremal multi-PPT states corresponding to $\{\mathbf{p}_1, \mathbf{p}_2, \mathbf{p}_3, \mathbf{p}_4\}$:

$\mathbf{p}\backslash p_\sigma$	p_{00}	p_{01}	p_{10}	p_{11}
$\mathbf{p}_1$	1	0	0	0
$\mathbf{p}_2$	λ_2	$1/d_2$	0	0
$\mathbf{p}_3$	λ_1	0	$1/d_1$	0
$\mathbf{p}_4$	$\lambda_1\lambda_2$	λ_1/d_2	λ_2/d_1	$1/d_1d_2$

(102)

where $\lambda_i = (d_i - 1)/d_i$ $(i = 1, 2)$. It turns out that all extremal multi-PPT states are 4-separable. Indeed, consider the following product state

$$\rho = P_{\psi_1} \otimes P_{\psi_2} \otimes P_{\varphi_1}^T \otimes P_{\varphi_2}^T \ , \tag{103}$$

where $P_{\psi_k} = |\psi_k\, \psi_k|$, $P_{\varphi_k} = |\varphi_k\, \varphi_k|$, and ψ_k, φ_k are normalized vectors in $\mathbb{C}^{d_k}$; $k = 1, 2$. Upon twirling one obtains 4-separable invariant state $\rho = \mathcal{I}_{\mathbf{p}}$ with fidelities p_σ given by:

$$\begin{aligned} p_{00} &= (1 - b_1)(1 - b_2) \ , \\ p_{01} &= (1 - b_1)b_2 \ , \\ p_{10} &= b_1(1 - b_2) \ , \\ p_{11} &= b_1 b_2 \ , \end{aligned} \tag{104}$$

with

$$b_i = \frac{a_i}{d_i} \ , \qquad a_i = |\psi_i|\varphi_i\,|^2 \ , \qquad i = 1, 2 \ . \tag{105}$$

It is clear that each $\mathbf{p}_k$ may be constructed via (104) by the following choice of (a_1, a_2):

$$(0,0) \longleftrightarrow \mathbf{p}_1 \ , \quad (0,1) \longleftrightarrow \mathbf{p}_2 \ , \quad (1,0) \longleftrightarrow \mathbf{p}_3 \ , \quad (1,1) \longleftrightarrow \mathbf{p}_4 \ . \tag{106}$$

4.3. $K = 3$

Consider now a 6-partite isotropic-like state

$$\mathcal{I}_{\mathbf{q}} = \sum_{\boldsymbol{\sigma}} q_{\boldsymbol{\sigma}} \widetilde{\mathbf{Q}}^{\boldsymbol{\sigma}} , \tag{107}$$

living in $\mathcal{H}_1 \otimes \mathcal{H}_2 \otimes \mathcal{H}_3 \otimes \mathcal{H}_1 \otimes \mathcal{H}_2 \otimes \mathcal{H}_3$. A state $\mathcal{I}_{\mathbf{q}}$ is $\boldsymbol{\sigma}$-PPT iff

$$p'_{\mu_1\mu_2\mu_3} = \sum_{\nu_1,\nu_2,\nu_3=0}^{1} \Big(\mathbf{Y}[d_1]\Big)^{\sigma_1}_{\nu_1\mu_1} \Big(\mathbf{Y}[d_2]\Big)^{\sigma_2}_{\nu_2\mu_2} \Big(\mathbf{Y}[d_3]\Big)^{\sigma_3}_{\nu_3\mu_3} \geq 0 \,. \tag{108}$$

(1) For $|\boldsymbol{\sigma}| = 1$ the corresponding $\boldsymbol{\sigma}$-PPT conditions follow from (89):

$$\boldsymbol{\sigma} = (001) \longrightarrow p_{\sigma_1\sigma_2 0} + (1-d_3)p_{\sigma_1\sigma_2 1} \geq 0, \ \sigma_1,\sigma_2 = 0,1, \tag{109}$$

$$\boldsymbol{\sigma} = (010) \longrightarrow p_{\sigma_1 0\sigma_3} + (1-d_2)p_{\sigma_1 1\sigma_3} \geq 0, \ \sigma_1,\sigma_3 = 0,1, \tag{110}$$

$$\boldsymbol{\sigma} = (100) \longrightarrow p_{0\sigma_2\sigma_3} + (1-d_1)p_{1\sigma_2\sigma_3} \geq 0, \ \sigma_2,\sigma_3 = 0,1. \tag{111}$$

(2) Additional $\boldsymbol{\sigma}$-PPT conditions for $|\boldsymbol{\sigma}| = 2$ follow from (96):

$$\boldsymbol{\sigma} = (011) \rightarrow d_2 p_{\sigma_1 10} + d_3 p_{\sigma_1 01} \leq 1 + (d_2d_3 - d_2 - d_3)p_{\sigma_1 11}, \tag{112}$$

$$\boldsymbol{\sigma} = (101) \rightarrow d_1 p_{1\sigma_2 0} + d_3 p_{0\sigma_2 1} \leq 1 + (d_1d_3 - d_1 - d_3)p_{1\sigma_2 1}, \tag{113}$$

$$\boldsymbol{\sigma} = (110) \rightarrow d_1 p_{10\sigma_3} + d_2 p_{01\sigma_3} \leq 1 + (d_1d_2 - d_1 - d_2)p_{11\sigma_3}, \tag{114}$$

for $\sigma_1, \sigma_2, \sigma_3 = 0,1$.

(3) The last (111)-PPT condition can not be derived from $K = 1$ and $K = 2$ cases. Assuming (109)–(114) the only new condition reads as follows

$$\begin{aligned} d_1 p_{100} + d_2 p_{010} + d_3 p_{001} \leq\ & p_{101}\Big((1-d_1)(1-d_3) - 1\Big) \\ &+ p_{011}\Big((1-d_2)(1-d_3) - 1\Big) \\ &+ p_{110}\Big((1-d_1)(1-d_2) - 1\Big) \\ &+ p_{111}\Big((1-d_1)(1-d_2)(1-d_3) - 1\Big) + 1 \,. \end{aligned} \tag{115}$$

The above formula may be rewritten in the more compact form as follows

$$\sum_{\boldsymbol{\sigma}\neq 0} \Big(1 - (1-d_1)^{\sigma_1}(1-d_2)^{\sigma_2}(1-d_3)^{\sigma_3}\Big) p_{\sigma_1\sigma_2\sigma_3} \leq 1 \,. \tag{116}$$

Fidelities p_σ satisfies

$$p_{\sigma_1\sigma_2\sigma_3} \le \frac{1}{d_1^{\sigma_1} d_2^{\sigma_2} d_3^{\sigma_3}} \ . \tag{117}$$

Proof. 1) let us take $\boldsymbol{\sigma}$ with $|\boldsymbol{\sigma}| = 2$, e.g. p_{011}. One obtains from (109) and (110):

$$p_{010} \ge (d_3 - 1)p_{011} \ ,$$
$$p_{011} \ge (d_2 - 1)p_{011} \ ,$$

and hence formula (112) implies

$$p_{011} \le \frac{1}{d_2 d_3} \ . \tag{118}$$

In the same way one shows that

$$p_{110} \le \frac{1}{d_1 d_2} \ , \qquad p_{101} \le \frac{1}{d_1 d_3} \ . \tag{119}$$

2) Let us look now for an upper bound for p_{111}. Taking into account

$$p_{010} \ge (d_3 - 1)p_{011} \ ,$$
$$p_{001} \ge (d_1 - 1)p_{101} \ ,$$
$$p_{100} \ge (d_2 - 1)p_{110} \ ,$$

one obtains from (115)

$$d_1 p_{101} + d_2 p_{110} + d_3 p_{011} \ \le \ p_{111}\Big((1-d_1)(1-d_2)(1-d_3) - 1\Big) + 1 \ . \tag{120}$$

Using

$$p_{101} \ge (d_2 - 1)p_{111} \ ,$$
$$p_{110} \ge (d_3 - 1)p_{111} \ ,$$
$$p_{011} \ge (d_1 - 1)p_{111} \ ,$$

one gets from (120)

$$p_{111} \le \frac{1}{d_1 d_2 d_3} \ . \tag{121}$$

3) Finally, consider $\boldsymbol{\sigma}$ with $|\boldsymbol{\sigma}| = 1$. Let us assume that $p_{001} = p_{010} = 0$. Then formulae (109)–(111) imply $p_{011} = p_{101} = 0$ and hence one obtains from (113)

$$p_{100} \le \frac{1}{d_1} \ . \tag{122}$$

In the same way one shows that

$$p_{010} \leq \frac{1}{d_2} \; , \qquad p_{001} \leq \frac{1}{d_3} \; , \tag{123}$$

which proves (117) □

Due to Proposition 4.3 one construct the following set of 8 extremal multi-PPT isotropic like states:

$\mathbf{p}\backslash p_\sigma$	p_{000}	p_{001}	p_{010}	p_{011}	p_{100}	p_{101}	p_{110}	p_{111}
$\mathbf{p}_1$	1	0	0	0	0	0	0	0
$\mathbf{p}_2$	λ_3	$1/d_3$	0	0	0	0	0	0
$\mathbf{p}_3$	λ_2	0	$1/d_2$	0	0	0	0	0
$\mathbf{p}_4$	λ_1	0	0	0	$1/d_1$	0	0	0
$\mathbf{p}_5$	$\lambda_2\lambda_3$	λ_2/d_3	λ_3/d_2	$1/d_2d_3$	0	0	0	0
$\mathbf{p}_6$	$\lambda_1\lambda_3$	λ_1/d_3	0	0	λ_3/d_1	$1/d_1d_3$	0	0
$\mathbf{p}_7$	$\lambda_1\lambda_2$	0	λ_1/d_2	0	λ_2/d_1	0	$1/d_1d_2$	0
$\mathbf{p}_8$	$\lambda_1\lambda_2\lambda_3$	$\lambda_1\lambda_2/d_3$	$\lambda_1\lambda_3/d_2$	λ_1/d_2d_3	$\lambda_2\lambda_3/d_1$	λ_2/d_1d_3	λ_3/d_1d_2	$1/d_1d_2d_3$

where $\lambda_i = (d_i - 1)/d_i$ $(i = 1, 2, 3)$. It is easy to show that all extremal states are 6-separable. To show this consider a product state ρ

$$\rho = P_{\psi_1} \otimes P_{\psi_2} \otimes P_{\psi_3} \otimes P^T_{\varphi_1} \otimes P^T_{\varphi_2} \otimes P^T_{\varphi_3} \; , \tag{124}$$

where $P_{\psi_k} = |\psi_k \; \psi_k|$, $P_{\varphi_k} = |\varphi_k \; \varphi_k|$, and ψ_k, φ_k are normalized vectors in $\mathbb{C}^{d_k}$; $k = 1, 2, 3$. Upon twirling one obtains 6-separable invariant state $\rho = W_{\mathbf{q}}$ with fidelities q_σ given by:

$$p_{\sigma_1\sigma_2\sigma_3} = (1-\sigma_1-(-1)^{\sigma_1}b_1)(1-\sigma_2-(-1)^{\sigma_2}b_2)(1-\sigma_3-(-1)^{\sigma_3}b_3) \; , \tag{125}$$

with

$$b_i = \frac{a_i}{d_i} \; , \qquad a_i = |\psi_i|\varphi_i\,|^2 \; , \qquad i = 1, 2, 3 \; . \tag{126}$$

It is clear that each $\mathbf{p}_k$ may be constructed via (69) by the following choice of (a_1, a_2, a_3):

	$\mathbf{p}_1$	$\mathbf{p}_2$	$\mathbf{p}_3$	$\mathbf{p}_4$	$\mathbf{p}_5$	$\mathbf{p}_6$	$\mathbf{p}_7$	$\mathbf{p}_8$
a_1	0	0	0	1	0	1	1	1
a_2	0	0	1	0	1	0	1	1
a_3	0	1	0	0	1	1	0	1

It proves, therefore, that 6-partite multi-PPT isotropic-like states are 6-separable.

4.4. *Arbitrary K*

Consider now a $2K$-partite invariant isotropic-like state

$$\mathcal{I} = \sum_{\boldsymbol{\sigma}} p_{\boldsymbol{\sigma}} \widetilde{\mathbf{P}}^{\boldsymbol{\sigma}} , \tag{127}$$

living in $\mathcal{H}_1 \otimes \ldots \otimes \mathcal{H}_K \otimes \mathcal{H}_1 \otimes \ldots \otimes \mathcal{H}_K$. It is clear that all partial transpositions $\tau_{\boldsymbol{\sigma}}\rho$ with $|\boldsymbol{\sigma}| < K$ may be derived from the previous cases with the number of pairs less than K. The last transposition with $|\boldsymbol{\sigma}| = K$, i.e. $\boldsymbol{\sigma} = (1, \ldots, 1)$ produces one additional condition. It is clear from (116) that this new condition reads as follows

$$\sum_{\boldsymbol{\sigma} \neq 0} \Big(1 - (1-d_1)^{\sigma_1} \ldots (1-d_K)^{\sigma_K}\Big) p_{\sigma_1 \ldots \sigma_K} \leq 1 . \tag{128}$$

Note that if $d_1 = \ldots = d_K = d$, then the above formula reduces to

$$\sum_{\boldsymbol{\sigma} \neq 0} \Big(1 - (1-d)^{|\boldsymbol{\sigma}|}\Big) p_{\boldsymbol{\sigma}} \leq 1 , \tag{129}$$

and hence for $d = 2$ it gives

$$\sum_{|\boldsymbol{\sigma}| = \mathrm{odd}} p_{\boldsymbol{\sigma}} \leq \frac{1}{2} . \tag{130}$$

which reproduces (73). Moreover, one may easily generalizes Proposition 4.3 as follows

Fidelities $p_{\boldsymbol{\sigma}}$ satisfy

$$p_{\sigma_1 \ldots \sigma_K} \leq \frac{1}{d_1^{\sigma_1} \ldots d_K^{\sigma_K}} . \tag{131}$$

Again, one shows that all extremal multi-PPT states states are $2K$-separable. To show this one considers a product state ρ

$$\rho = P_{\psi_1} \otimes \ldots \otimes P_{\psi_K} \otimes P_{\varphi_1}^T \otimes \ldots \otimes P_{\varphi_K}^T , \tag{132}$$

where $P_{\psi_i} = |\psi_i \, \psi_i|$, $P_{\varphi_i} = |\varphi_i \, \varphi_i|$, and ψ_i, φ_i are normalized vectors in $\mathbb{C}^{d_i}$; $i = 1, 2, \ldots, K$. Upon twirling one obtains $2K$-separable invariant state $\rho = \mathcal{I}_{\mathbf{p}}$ with fidelities $p_{\boldsymbol{\sigma}}$ given by:

$$p_{\sigma_1 \ldots \sigma_K} = (1 - \sigma_1 - (-1)^{\sigma_1} b_1) \ldots (1 - \sigma_K - (-1)^{\sigma_K} b_K) , \tag{133}$$

with

$$b_i = \frac{a_i}{d_i} , \qquad a_i = |\psi_i | \varphi_i|^2 , \qquad i = 1, 2, \ldots, K . \tag{134}$$

It turns out that each extremal multi-PPT state may be constructed via (133) by the clever choice of $(a_1, a_2, \ldots, a_K)$.

4.5. *Biseparability*

In analogy to Theorem 3.1 one has

Theorem 4.1. *Let $K = 2$ and $d_1 = d_2 = d$. Then a 4-partite isotropic-like state is $A|B$ biseparable iff it is $(1,1)$-PPT.*

Proof. to prove it we find extremal $(1,1)$-PPT and show that that all of them are $A|B$ biseparable. To find these one has solve (95) with $d_1 = d_2 = d$:

$$\begin{aligned} p_{00} + (1-d)p_{10} + (1+d)\Big(p_{01} + (1-d)p_{11}\Big) &\geq 0 \,, \\ p_{00} + (1-d)p_{01} + (1+d)\Big(p_{10} + (1-d)p_{11}\Big) &\geq 0 \,, \\ p_{00} + (1-d)p_{01} + (1-d)\Big(p_{10} + (1-d)p_{11}\Big) &\geq 0 \,. \end{aligned} \tag{135}$$

Actually, it turns out that the convex set of $(1,1)$-PPT states is generated by 5 extremal points: four of them are already derived in (102). Clearly, since all of them are fully 4-separable it follows that they are $A|B$ biseparable, too. The last one is given by

$$\mathbf{p}_5 = \frac{1}{d^2}(d^2-1, 0, 0, 1) \,. \tag{136}$$

To show that the corresponding isotropic-like state

$$\mathcal{I}_5 = \frac{d^2-1}{d^2}\,\widetilde{\mathbf{P}}^{00} + \frac{1}{d^2}\,\widetilde{\mathbf{P}}^{11} \,, \tag{137}$$

is $A|B$ biseparable let us observe that $\mathcal{I}_5 =< |\Psi^{++}_{AB} < \Psi^{++}_{AB}|$, where Ψ^{++}_{AB} is defined in (81). Using

$$< \Psi^{++}_{AB}|I_{1|3} \otimes P^1_{2|4}|\Psi^{++}_{AB} =< \Psi^{++}_{AB}|P^1_{1|3} \otimes I_{2|4}|\Psi^{++}_{AB} = \frac{1}{d^2} \,, \tag{138}$$

and

$$< \Psi^{++}_{AB}|P^1_{1|3} \otimes P^1_{2|4}|\Psi^{++}_{AB} = \frac{1}{d^2} \,, \tag{139}$$

one finds

$$p_{00} = \frac{d^2-1}{d^2} \,, \quad p_{01} = p_{10} = 0 \,, \quad p_{11} = \frac{1}{d^2} \,, \tag{140}$$

which finally proves the theorem. □

It should be stressed that the above theorem does not hold if $d_1 \neq d_2$ or $K > 2$ [11].

5. Conclusions

In the present paper we developed a pedagogical approach to multipartite symmetric states. We performed a detailed analysis of separability conditions for generalized Werner-like and isotropic-like multipartite states. We show that necessary and sufficient separability conditions generalize well known PPT criterion for multipartite case. One may easily generalize our scheme to other symmetry groups (for example an orthogonal, symplectic or rotational SO(3) group[12]). For different approach to multipartite symmetric states see Ref.[13]. There is also an interesting approach to the class of states invariant under the action of discrete groups, like for example permutation group (see Refs.[14,15,16,17]).

Acknowledgments

I thank Professors Noboru Watanabe and Takashi Matsuoka for their kind hospitality during ICQBIC 09 and the Suwa Satellite Conference. This work was partially supported by the Polish Ministry of Science and Higher Education Grant No 3004/B/H03/2007/33 and by the Polish Research Network *Laboratory of Physical Foundations of Information Processing.*

References

1. M. A. Nielsen and I. L. Chuang, *Quantum computation and quantum information*, Cambridge University Press, Cambridge, 2000.
2. R. Horodecki, P. Horodecki, M. Horodecki and K. Horodecki, Rev. Mod. Phys. **81**, 865 (2009).
3. O. Gühne and G. Tóth, Phys. Rep. **474**, 1 (2009).
4. A. Peres, Phys. Rev. Lett. **77**, 1413 (1996).
5. M. Horodecki, P. Horodecki and R. Horodecki, Phys. Lett. A **223**, 1 (1996); P. Horodecki, Phys. Lett. A **232**, 333 (1997).
6. R. F. Werner, Phys. Rev. A **40**, 4277 (1989).
7. M. Horodecki and P. Horodecki, Phys. Rev. **A** 59, 4206 (1999).
8. D. Chruściński and A. Kossakowski, Phys. Rev. A **73**, 062313 (2006).
9. D. Chruściński and A. Kossakowski, Phys. Rev. A **73**, 062314 (2006).
10. K.G.H. Vollbrecht and R.F. Werner, Phys. Rev. A **64**, 062307 (2001).
11. M. Piani and C. Mora, Phys. Rev. A **75**, 02305 (2007).
12. D. Chruściński, A. Kossakowski, Open Sys. and Inf. Dynamics, **14** (2007) 25.
13. T. Eggeling and R.F. Werner, Werner, Phys. Rev. A **63**, 042111 (2001).
14. D. Chruściński and A. Kossakowski, Phys. Rev. A **74**, 022308 (2006).
15. D. Chruściński and A. Kossakowski, Phys. Rev. A **76**, 032308 (2007).
16. D. Chruściński and A. Pittenger, J. Phys. A: Math. Theor. **41** (2008) 385301.
17. D. Chruściński and A. Kossakowski, Open Sys. Information Dyn. **15** (2008) 189.

Quantum Bio-Informatics III
eds. L. Accardi, W. Freudenberg & M. Ohya

ON A QUANTUM MODEL OF BRAIN ACTIVITIES

K.-H. FICHTNER

Unversity Jena, Institute of Applied Mathematics, 07743 Jena, Germany
E-mail: fichtner@mathematik.uni-jena.de

L. FICHTNER

Unversity Jena, Institute of Psychology, 07743 Jena, Germany
E-Mail: lars.fichtner@uni-jena.de

W. FREUDENBERG

Techn. University Cottbus, Dep. of Mathematics, 03013 Cottbus, Germany
E-mail: freudenberg@math.tu-cottbus.de

M. OHYA

Department of Information Science and Quantum Bio-Informatic Center, Tokyo University of Science, Noda City, Chiba 278-8510, Japan
E-Mail: ohya@is.noda.tus.ac.jp

One of the main activities of the brain is the recognition of signals. A first attempt to explain the process of recognition in terms of quantum statistics was given in [6]. Subsequently, details of the mathematical model were presented in a (still incomplete) series of papers (cf. [7, 2, 5, 10]). In the present note we want to give a general view of the principal ideas of this approach. We will introduce the basic spaces and justify the choice of spaces and operations. Further, we bring the model face to face with basic postulates any statistical model of the recognition process should fulfill. These postulates are in accordance with the opinion widely accepted in psychology and neurology.

Keywords: bosonic Fock space, compound Fock space, coherent states, beam splitting, recognition, brain models

1. Introduction

First we would like to point out that we do not have the intention to create a physiological model of recognition taking into account exact biological, chemical and/or physical reactions. We are searching for a *statistical* model being able to explain certain aspects of the recognition process. Specialists in modern brain research are convinced that signals in the brain should be coded by populations of excited neurons (cf. [1, 12, 15, 16]). Considering models based on classical probability theory states of signals should be identified with probability distributions of certain random point fields located inside the volume of the brain. However, researchers as for instance W. Singer [16] and H.P. Stapp [17] expressed doubts whether classical models could give satisfactory answers to fundamental problems of modern brain research, and plead for using quantum models.

As it was pointed out in [16] the procedure of recognition of signals can be described as follows: There is a set of complex signals stored in the memory. Choosing one of these signals may be interpreted as generating a hypothesis concerning an "expexted view of the world". Then the brain compares a signal arising from our senses with the signal chosen from the memory. That changes the state of both of the signals in such a manner that after the procedure the signals coincide in a certain sense. Furthermore, measurements of that procedure like EEG or MEG are based on the fact that recognition of signals causes a certain loss of excited neurons, i.e., the neurons change their state from "excited" to "nonexcited". For that reason a statistical model of the recognition process should reflect both, that change of the signals, and the loss of excited neurons.

In Section 2 we enumerate 10 postulates any statistical model of brain activities has to fulfill. In the subsequent chapters we present some aspects of the general outline of our model and we check its compliance with the postulates. To avoid elaborate formulae we confine ourselves to a simplified description of the basic spaces and operators used in this approach. Details of the sketched approach including all proofs are given in [6, 14, 7, 2, 5, 10, 11] or will be given in some forthcoming papers ([3, 4]). In this series of papers also the memory space, the procedures of creation of signals from the memory, amplification, accumulation and transformation of input signals, and measurements like EEG and MEG are treated in detail.

2. Postulates Concerning the Recognition of Signals

Wolf Singer a leading specialist in modern brain research summarizes in [16] some common experiences mainly found out in the last decade. Following the main ideas in [16] we extract some postulates any mathematical model of the procedure of recognition of signals should fulfill:

(P1) *The brain acts discrete in time.*
(P2) *Signals are represented by populations of excited neurons.*
(P3) *Signals can be decomposed into parts in compliance with the fact that there are different regions of the brain being responsible for different tasks.*
(P4) *The brain acts parallel corresponding to the different regions.*
(P5) *Signals stored in the brain are superpositions of finitely many elementary signals.*
(P6) *The brain permanently creates complex signals representing a hypothesis concerning an "expected view of the world".*
(P7) *Recognition of a signal produced by our senses is a random event which can occur as a consequence of the interaction of that signal and a signal created by the brain.*
(P8) *Recognition causes a loss of excited neurons in some regions of the brain.*
(P9) *Recognition changes the state of the signal coming from our senses. One will be aware of that changed signal.*
(P10) *Changes in some region of the brain have immediate consequences concerning the other regions.*

In what follows we will introduce some basic components of our model - the basic spaces for the signals and the memory and the basic operators describing the process of recognition. We will check whether this approach is in accordance with the above formulated postulates. Main components of our model are

- a Hilbert space $\mathcal{H}$ representing the space of *signals*,
- a Hilbert space $\overline{\mathcal{H}}$ representing the *memory*,
- an isometry $\mathcal{D} : \overline{\mathcal{H}} \to \overline{H} \otimes \mathcal{H}$ describing the *creation of signals from the memory*,
- a unitary operator $\mathcal{V}$ on $\mathcal{H} \otimes \mathcal{H}$ describing the *interaction of two signals.*

3. The Space of Signals

Our brain contains about 100 billions of neurons. Thus a lot of signals could be described in terms of a classical probabilistic model according to postulate (P2). But there are many wellknown facts being in contradiction to classical models. In a classical model it is difficult to explain that activities in one region of the brain have immediate consequences concerning the other regions (this is the so-called *binding problem*). However, because of nonlocality of quantum theory a quantum model may be in accordance with postulate (P10). H. Stapp argues in [17] for a quantum model of brain activities because *classical physics cannot explain consciousness because it cannot explain how the whole can be more than the parts.*

Starting point for our quantum model is a suitable compact subset $G \subset \mathbb{R}^d$ representing the physical volume of the brain. The Hilbert space $L^2(G)$ of square integrable complex-valued functions on G describes a quantum particle inside of G. Now, a neuron is excited if a certain amount of energy (electric potential) is stored in it. For that reason we identify *an excited neuron* with a *quantum particle localized inside the neuron*, i. e. in our model the quantum particle represents that amount of energy which makes the difference between an excited and a nonexcited neuron. In this sense the bosonic Fock space

$$\Gamma(L^2(G)) := \bigoplus_{n=0}^{\infty} \frac{1}{n!} L^2(G)^{\otimes n}_{\text{symm}} \tag{1}$$

represents *populations of excited neurons* inside the volume G of the brain. In accordance with postulate (P2) we will use (a subspace of) $\mathcal{H} := \Gamma\left(L^2(G)\right)$ as our basic *space of signals.* Incoming signals and signals taken from the memory are states on the Fock space $\mathcal{H}$. The use of the Fock space has several advantages. The most striking one is that the Fock space over G can be identified with the tensor product of the bosonic Fock spaces corresponding to the subsets $G_1, \ldots, G_n$ for an arbitrary disjoint partition of the space G:

$$\Gamma(L^2(G)) = \Gamma(L^2(G_1)) \otimes \ldots \otimes \Gamma(L^2(G_n)). \tag{2}$$

The property (2) is conformed to postulate (P3). Postulate (P5) states that there is a finite number of elementary signals generating all signals. Because of the exceptional properties of *exponential vectors* being in accordance with our postulates we restrict ourselves to signals given by *exponential vectors.* Besides the above decomposition property the Fock space has the property that *exponential vectors* from $\mathcal{H}$ decompose in the same way into *independent partial signals* from $\Gamma(L^2(G_r))$, $r \in \{1, \ldots, n\}$, and the partial signals are again given by exponential vectors. More precisely, an exponential vector $\exp(f) \in \mathcal{H}$ corresponding to $f \in L^2(G)$ is given by

$$\exp(f) = (\exp(f)_n)_{n=0}^{\infty} \in \Gamma(L^2(G)) \tag{3}$$

with

$$\exp(f)_0 = 1, \quad \exp(f)_n(x_1,\dots,x_n) = \prod_{k=1}^{n} f(x_k) \quad (n \geq 1,\ x_1,\dots,x_n \in G).$$

For $f \in L^2(G)$ and for a disjoint decomposition $G_1,\dots,G_n$ of G one has

$$\exp(f) = \exp(f_{|1}) \otimes \dots \otimes \exp(f_{|n}) \tag{4}$$

where $f_{|k}$ is the restriction of f to G_k, $k \in \{1,\dots,n\}$. This property is in accordance with postulate (P3). Moreover, this property of the exponential vectors enables us to reflect in a clear sense the fundamental experimental experience (P4) that the brain is acting parallel in the disjoint regions.

Now, for a partition $G_1,\dots,G_n$ of G we fix positive numbers $\lambda_1,\dots,\lambda_n$ representing the *intensity* of the signals in the single regions and we fix finite orthonormal systems

$$f_1^{(k)},\dots,f_{r_k}^{(k)} \in \Gamma(L^2(G_k)) \qquad (f^{(k)} \in L^2(G_k),\ n_k \in \mathbb{N},\ k \in \{1,\dots,n\}).$$

The exponential vectors $\exp(\lambda_k f_1^{(k)}),\dots,\exp(\lambda_k f_{r_k}^{(k)}) \in \Gamma(L^2(G_k))$ span a finite dimensional subspaces $\mathcal{H}_k \subset \Gamma(L^2(G_k))$ representing the *space of signals in the region* G_k. Since $|\lambda_k|^2$ is the expectation of the number operator with respect to the pure (coherent) state given by $\lambda_k f_j^{(k)}$ the intensity of the signals in the region G_k is represented by $|\lambda_k f_j^{(k)}|^2$. By choosing the above mentioned orthonormal systems we pass from the infinite-dimensional spaces $\Gamma(L^2(G_k))$ to the finite dimensional subspaces $\mathcal{H}_k$, and on G we get a finite dimensional space $\mathcal{H}_{\text{sig}} := \mathcal{H}_1 \otimes \dots \otimes \mathcal{H}_n$ of *regular* signals. The functions $f_j^{(k)}$ represent the type of signal and the number λ_k the intensity of the signals in the region G_k.

4. The Memory Space

Now we want to describe the (longterm) memory. We have used the Hilbert space $\mathcal{H} = \Gamma\left(L^2(G)\right)$ in order to describe signals (coded by sets of excited neurons). The memory contains the information concerning sets of signals. For that reason it seems to be reasonable to describe the memory using a space consisting of sets of configurations. Mathematically this can be done by constructing the (symmetric) Fock space $\overline{\mathcal{H}}$ over the Hilbert space $\mathcal{H}$, i. e.

$$\overline{\mathcal{H}} := \Gamma(\mathcal{H}) = \bigoplus_{n=0}^{\infty} \frac{1}{n!} \mathcal{H}_{\text{symm}}^{\otimes n}. \tag{5}$$

We will call this space *compound Fock space over* G. Let us remark that $\mathcal{H}$ itself can be represented as L^2-space over a space of configurations M. Consequently, the compound Fock space $\overline{\mathcal{H}}$ is a Fock space of the same kind as $\mathcal{H}$. In an analogous way to (4) we can define exponenential vectors $\mathrm{EXP}(\Psi)$ in the memory space $\overline{\mathcal{H}}$ (one only has to replace points from G by configurations of points from M:

$$\mathrm{EXP}(\Psi) = \left((\mathrm{EXP}(\Psi))_n\right)_{n=0}^{\infty}, \qquad (\mathrm{EXP}(\Psi))_0 = 1,$$

$$(\mathrm{EXP}(\Psi))_n\,(\varphi_1,\ldots,\varphi_n) = \prod_{k=1}^{n} \Psi(\varphi_k) \qquad (\Psi \in \mathcal{H},\ \varphi_k \in M,\ n \in \mathbb{N}).$$

It is well known [16] that there are finite sets of certain elementary signals corresponding to the different regions of the brain, and the only signals which can be stored in the memory are superpositions of such elementary signals. That's why we first passed over from $\mathcal{H}$ to the finite dimensional subspace $\mathcal{H}_{\mathrm{sig}}$ of regular signals. Now, the same arguments lead to a restriction of $\overline{\mathcal{H}}$ to $\overline{\mathcal{H}}_{\mathrm{mem}} \subset \overline{\mathcal{H}}$. This subspace we will call the *memory space.*

Now, according to postulate recognition is based on a comparison of incoming signals with signals coming from the memory. For that reason according to postulate (P6) we have to define an operator describing the removal of single signals from the sets of signals stored in the memory. This will be done with the aid of an isometry being a bounded variant of the *Hida-Malliavin derivative.*

Let $\mathcal{D} : \overline{\mathcal{H}}_{\mathrm{mem}} \to \overline{\mathcal{H}}_{\mathrm{mem}} \bigotimes \mathcal{H}_{\mathrm{sig}}$ be the isometry given on the exponential vectors $\mathrm{EXP}(R)$ by

$$\mathcal{D}\mathrm{EXP}(R) = \frac{1}{\sqrt{N+1}} \cdot \mathrm{EXP}(R) \otimes R \qquad (R \in \mathcal{H}_{\mathrm{reg}})$$

where N denotes the number operator in $\mathcal{H}$.

More precisely, if φ is a set of excited neurons, and Φ a family of such sets of neurons we have

$$(\mathcal{D}\mathrm{EXP}(R))\,(\Phi,\varphi) = \frac{1}{\sqrt{|\Phi|+1}} \cdot \mathrm{EXP}(R)(\Phi) \cdot R(\varphi).$$

5. The recognition operator

Now let us consider a unitary operator $\mathcal{V}$ on $\mathcal{H}\otimes\mathcal{H}$ describing the interaction of two signals one coming from our senses and one created by the memory with the aid of the operator $\mathcal{D}$. We fix a decomposition of G into subsets $G_1, \ldots, G_n$ in compliance with the fact that there are different parts of the brain being responsible for different tasks. Because of the decomposition (2) of the Fock space we can make the following identification

$$\mathcal{H}\otimes\mathcal{H} = \big(\Gamma(L^2(G_1))\otimes\Gamma(L^2(G_1))\big)\otimes\ldots\otimes\big(\Gamma(L^2(G_n))\otimes\Gamma(L^2(G_n))\big)\,. \quad (6)$$

Postulate (P4) states that the brain acts parallel corresponding to the different regions. For that reason the unitary operator $\mathcal{V}$ should decompose in the same way as the Fock space does. Experiments show that the regions can increase or decrease [13]. Further, they can change their location on the surface of the brain [18]. On the other hand the physical structure of the brain doesn't change, i. e. there are always the same neurons, nonspecialized concerning the task they are involved.

Consequently, for each possible finite disjoint decomposition $G_1,\ldots,G_n$ of G the operator $\mathcal{V}:\mathcal{H}\otimes\mathcal{H}\to\mathcal{H}\otimes\mathcal{H}$ has to have the representation

$$\mathcal{V} = \mathcal{V}_{G_1}\otimes\ldots\mathcal{V}_{G_n} \quad (7)$$

where $\mathcal{V}_{G_r}$ is a unitary operator on $\Gamma(L^2(G_r))\otimes(L^2(G_r))$, $r\in\{1,\ldots,n\}$. Condition (7) is a very strong one. It was shown in [11] that only operators $\mathcal{V}:\mathcal{H}\otimes\mathcal{H}\to\mathcal{H}\otimes\mathcal{H}$ being on exponential vectors of the type

$$\mathcal{V}\exp(f)\otimes\exp(g) = \exp\left(b_1 f + b_2 g\right)\otimes\exp\left(b_3 f + b_4 g\right) \qquad \big(f,g\in L^2(G)\big)$$

fulfill the condition (7). Hereby, $(b_1,b_2,b_3,b_4): G\to\mathbb{C}^4$ are functions satisfying

$$\begin{aligned} &|b_1(x)|^2 + |b_2(x)|^2 = 1 = |b_3(x)|^2 + |b_4(x)|^2 &&(x\in G)\\ &b_1(x)\bar{b}_3(x) + b_2(x)\bar{b}_4(x) = 0 &&(x\in G)\end{aligned}$$

Moreover, taking into account that neurons are "nonspecialized" (cf. [16]) the functions b_k should be constant. Adding a symmetry condition we finally conclude

$$b_4 = -\frac{1}{\sqrt{2}}, \quad b_1 = b_2 = b_3 = \frac{1}{\sqrt{2}}.$$

Summarizing we can conclude that the interaction of two signals should be described by the *symmetric beam splitter* $\mathcal{V}$ on $\mathcal{H}\otimes\mathcal{H}$ given on exponential vectors $f,g\in L^2(G)$ by

$$\mathcal{V}\exp(f)\otimes\exp(g) = \exp\left(\frac{1}{\sqrt{2}}(f+g)\right)\otimes\exp\left(\frac{1}{\sqrt{2}}(f-g)\right) \quad (8)$$

Observe that the operator $\mathcal{V}$ is unitary and selfadjoint what implies $\mathcal{V}\mathcal{V} = I_{\mathcal{H}\otimes\mathcal{H}}$ (I denotes the identical operator). Further $\mathcal{V}$ is a so-called *exchange operator*, i. e., $\mathcal{V}$ preserves the number of excited neurons supporting both of the signals in each part of G. For that reason the use of the unitary operator $\mathcal{V}$ reflects the hypothesis that signals permanently exchange their supports without any loss of information [16]. A detailed investigation of the class of exchange operators was given in [8].

6. Processing of the signals

Processing is described on the space

$$\boxed{\tilde{\mathcal{H}} := \mathcal{H}_{\text{sig}} \otimes \mathcal{H}_{\text{sig}} \otimes \mathcal{H}_{\text{mem}}}$$

↑	↑	↖
signal chosen from memory	incoming signal	memory before and after processing

With the aid of $\mathcal{D}$ a signal is created from memory. Then this signal and the incoming signal interact via the operator $\mathcal{V}$, and two new signals arise. The operator $\mathcal{D}^*$ adjoint to $\mathcal{D}$ reimplants the changed signal to the memory. This is of course only the processing in a very simplified way where only one step is taken into account. In a more realistic model one has to incorporate also repeated interactions, amplification processes etc. Schematically one step of the processing could be illustrated by the following picture

$$\underbrace{\mathcal{H}_{\rm sig}}_{I} \otimes \underbrace{\mathcal{H}_{\rm mem}}_{\mathcal{D}}$$

$$\Downarrow$$

$$\overbrace{\underbrace{\mathcal{H}_{\rm sig} \otimes \mathcal{H}_{\rm sig}}_{\mathcal{V}} \otimes \underbrace{\mathcal{H}_{\rm mem}}_{I}}$$

$$\Downarrow$$

$$\overbrace{\underbrace{\mathcal{H}_{\rm sig}}_{I} \otimes \mathcal{H}_{\rm sig}} \otimes \mathcal{H}_{\rm mem}, \qquad \underbrace{\mathcal{H}_{\rm sig} \otimes \mathcal{H}_{\rm mem}}_{\mathcal{D}^*}$$

$$\Downarrow$$

$$\mathcal{H}_{\rm sig} \otimes \mathcal{H}_{\rm mem}$$

7. Recognition of Signals

In order to describe the basic idea of recognition we consider the special case of coherent states of the input-signal given by the exponential vectors $\Psi^{\rm in} = \exp(g)$ and coherent signals $\Psi^{\rm mem} = \exp(f)$ chosen from the memory. The interaction with the beam splitter $\mathcal{V}$ of the signals gives again a pair of coherent states

$$\mathcal{V}(\Psi^{\rm mem} \otimes \Psi^{\rm in}) = \exp\left(\frac{1}{\sqrt{2}}(f+g)\right) \otimes \exp\left(\frac{1}{\sqrt{2}}(f-g)\right).$$

If we have $f = g$ then $\exp\left(\frac{1}{\sqrt{2}}(f-g)\right) = \exp(0)$ represents the vacuum state. Even in the case $f \neq g$ the vacuum can occur with probability $e^{-\frac{1}{2}||f-g||^2}$. In the case $f = g$ the next step gives the pair of states

$$\mathcal{V}\left(\exp\left(\frac{1}{\sqrt{2}}(f+g)\right) \otimes \exp(0)\right) = \exp\left(\frac{1}{2}(f+g)\right) \otimes \exp\left(\frac{1}{2}(f+g)\right)$$

i. e. the pair of created signal and signal arising from the senses coincide after that two steps of exchange if the above mentioned event occurs.
For that reason we will interpret this event as "full recognition" (postulate (P7)). If nothing happens the second step of exchange reconstructs the

original pair of states

$$\mathcal{V}\mathcal{V}\left(\Psi^{\text{mem}} \otimes \Psi^{\text{in}}\right) = \Psi^{\text{mem}} \otimes \Psi^{\text{in}}$$

and the procedure can start again.
Observe that the event of recognition causes a change of the signals. Furthermore, the repetitions of the procedure after the first recognition will not cause further changes of the pair of states. The recognition causes a certain loss of excites neurons, with reflects postulate (P8). In the case of our example the expectation of that loss is

$$\|f\|^2 + \|g\|^2 - \frac{1}{2}\|f+g\|^2 = \frac{1}{2}\|f-g\|^2 \ .$$

Furthermore the model reflects another wellknown experience, namely that in case of signals being "unexpected" from the point of view of the memory, the probability of recognition is small but the measured activity, i. e. the loss of excited neurons is large.

Normally full recognition doesn't occur - only some kind of partial recognition. We fix a decomposition $G_1, \ldots, G_n$ of G. In a first step in one of the regions G_k recognition can occur. All further applications of V_{G_k} will not cause any change. However, occassionally after several applications of $\mathcal{V}$ in further regions we may get additional partial recognition. Such a procedure is in accordance with postulate (P1) claiming that recognition is a process discrete in time. That means the recognition of the signal will be improved step by step up to the maximal level. Such a model agrees with experimental experiences. Following [18] one step of this process lasts $10^{-13} - 10^{-10}$ sec, and the whole procedure lasts $10^{-3} - 10^{-1}$ sec.

8. Measurements of the EEG-Typ

Let u be a function on G representing the electric potential of one excited neuron measured by an electrode placed on the surface of the scalp. If φ denotes a finite subset of G representing the positions of a configuration of excited neurons then electric potential of that configuration measured by the electrode is given by

$$U(\varphi) := \left(\sum_{x\in\varphi} u(x)\right)$$

We mentioned already that the bosonic Fock space $\mathcal{H}$ can be identified with a certain L^2-space of functions of finite point configurations, i. e. functions of finite subsets φ of G [9], and the Fock space $\overline{\mathcal{H}}$ can be identified with a certain L^2-space of functions of finite families Φ of point configurations from G. For that reason the elements of $\overline{\mathcal{H}} \otimes \mathcal{H} \otimes \mathcal{H}$ are functions $\Psi(\Phi, \varphi_1, \varphi_2)$ where φ_1 and φ_2 are finite subsets of G and Φ denotes a finite family of point configurations from G. One may interpret

$$\begin{array}{ll} \varphi_1 & -\text{"support" of the created signal ,} \\ \varphi_1 & -\text{"support" of the arriving signal ,} \\ \bigcup_{\varphi \in \Phi} \varphi & -\text{"support" of all signals in the memory} \end{array}$$

Then

$$Z\left(\Phi, \varphi_1, \varphi_2\right) := \varphi_1 \cup \varphi_2 \cup \bigcup_{\varphi \in \Phi} \varphi$$

represents the positions of all excited neurons in the brain. Consequently the operator of multiplication corresponding to the function $U\left(Z(\varphi_1, \varphi_2, \Phi)\right)$ represents the measurement of the electric potential of the brain corresponding to one electrode.
Now, in the case of an EEG-device one uses a sequence of potentials $(u_k)_{k=1}^r$ being of the type $u_k(x) := u(x - y_k) \quad (x \in G)$.
Hereby, y_k represents the position of the k-th electrode placed on the surface of the scalp. The corresponding operators of multiplication commute [7], i. e., one is able to perform the corresponding measurements simultaneously.

References

1. J.W. Philips et al. Imaging neural activity using MEG and EEG. *IEEE Engineering in Medicine and Biology*, pages 34–42, May/June 1997.
2. K.-H. Fichtner, L. Fichtner, W. Freudenberg, and M. Ohya. On a mathematical model of brain activities. In *Quantum Theory, Reconsideration of Foundations - 4*, volume 962 of *AIP Conference Proceedings*, pages 85 – 90, Melville, New York, 2007. American Institute of Physics.
3. K.-H. Fichtner, L. Fichtner, W. Freudenberg, and M. Ohya. Quantum models of the recognition process - a convergence theorem. Technical report, 2007. 32 pages.
4. K.-H. Fichtner, L. Fichtner, W. Freudenberg, and M. Ohya. Quantum models of the recognition process - mathematical prerequisites. Technical report, 2007. 44 pages.
5. K.-H. Fichtner, L. Fichtner, W. Freudenberg, and M. Ohya. On a quantum model of the recognition process. In L. Accardi, W. Freudenberg, and M.Ohya, editors, *Quantum Bio-Informatics*, volume XXI of *QP–PQ: Quantum Probability and White Noise Analysis*, pages 64 – 84, New Jersey London Singapore, 2008. World Scientifc.

6. K.-H. Fichtner and L. Fichtner. Bosons and a quantum model of the brain. Jenaer Schriften zur Mathematik und Informatik Math/Inf/08/05, FSU Jena, Faculty of Mathematics and Informatics, Jena, 2005. 27 pages.
7. K.-H. Fichtner and L. Fichtner. Quantum markov chains and the process of recognition. Jenaer Schriften zur Mathematik und Informatik Math/Inf/02/07, FSU Jena, Faculty of Mathematics and Informatics, Jena, 2007. 24 pages.
8. K.-H. Fichtner, W. Freudenberg, and V. Liebscher. On exchange mechanisms for bosons. *Random Operators and Stochastic Equations*, 12:no. 4, 331–348, 2004.
9. K.-H. Fichtner and W. Freudenberg. Characterization of states of infinite boson systems I. On the construction of states of boson systems. *Commun. Math. Phys.*, 137:315—357, 1991.
10. K.-H. Fichtner and W. Freudenberg. The compound Fock space and its application to brain models. In L. Accardi, W. Freudenberg, and M. Ohya, editors, *Quantum Bio-Informatics II*, volume XXIV of *QP–PQ: Quantum Probability and White Noise Analysis*, pages 55 – 67, New Jersey London Singapore, 2009. World Scientifc.
11. L. Fichtner and M. Gäbler. Characterisation of beam splitters. In L. Accardi, W. Freudenberg, and M. Ohya, editors, *Quantum Bio-Informatics II*, volume XXIV of *QP–PQ: Quantum Probability and White Noise Analysis*, pages 68 – 80, New Jersey London Singapore, 2009. World Scientifc.
12. R. Hari and O.V. Lounasmaa. Neuromagnetism: tracking the dynamics of the brain. *Physical World*, pages 33–38, May 2000.
13. Akinori Nakamura and et.al. Somatosensory homunculus as drawn by meg. *NeuroImage*, 7(4):377–386, 1998.
14. M. Ohya, K.-H. Fichtner, and W. Freudenberg. Recognition and teleportation. In T. Hida and K. Saitô, editors, *Quantum Information V*, pages 1–17, New Jersey London Singapore, 2006. World Scientific.
15. Wolf Singer. Consciousness and the structure of neuronal representations. *Phil. Trans. R. Soc. Lond.*, B 353:1829–1840, 1998.
16. Wolf Singer. *Der Beobachter im Gehirn. Essays zur Hirnforschung.* Suhrkamp Verlag, Frankfurt a.M., 2002.
17. Henry Pierce Stapp. *Mind, Matter and Quantum Mechanics.* Springer, 2nd edition, 2003.
18. Annette Sterr and et.al. Perceptual correlates of changes in cortical representation of fingers in blind multifinger braille readers. *The journal of neuroscience*, 18(11):4417, 1998.

Quantum Bio-Informatics III
eds. L. Accardi, W. Freudenberg & M. Ohya

SOME OF THE RECENT TOPICS IN WHITE NOISE THEORY

TAKEYUKI HIDA

Professor Emeritus Nagoya University and Meijo University

2000 AMS Subject Classification 60H40

The identity of white noise B(t) has been given, so that we proceed to the space that is spanned by the B(t)'s. The system consists of continuously many independent variables and, in fact, spans a separable space. The system is total in that space. Harmonic analysis by introducing transformation groups will be discussed briefly. Together with other notions we can prepare for the analysis of nonlinear functionals of white noise.

1. Introduction

The symbol $\dot{B}(t)$ is familiar for us in stochastic analysis and it is used to describe an elemental random element. It has been called "white noise". As the stochastic analysis, in particular white noise analysis develops, it has become necessary to give it a correct and rigorous definition. We are going to give it a suitable definition from the probabilistic viewpoint and to establish its identity, so that we can carry on advanced stochastic calculus of functionals of $\dot{B}(t)$'s.

Before we come to the definition, we present a short historical note regarding the white noise $\dot{B}(t)$, which is the time derivative of a Brownian motion $B(t)$.

1) White noise is used in communication engineering. It is stationary in time and has the flat spectrum. The probability distribution is Gaussian, although it is not an ordinary random variable, but an infinitesimal random element.

A formal understanding is that it is a stationary Gaussian process in a generalized sense. A spectral representation is of the form

$$\dot{B}(t) = \int e^{it\lambda} dZ(\lambda),$$

with $E(|dZ(\lambda)|^2) = c^2 d\lambda$, c being a constant of the form

2) The stochastic integral, or the Wiener integral

$$\int f(t)dB(t)$$

with non-random function f, uses the random measure $dB(t)$, which may be written as $\dot{B}(t)dt$.

P. Lévy, since 1940 (Amer. J. Math. vol.62), used to write $\xi_t\sqrt{dt}$ in place of $\dot{B}(t)dt$, where ξ_t is a standard Gaussian random variable. Often, t is missing!

3) I.M. Gel'fant (in [1]) was clever enough to introduce a generalized stochastic process, the sample functions of which are generalized functions. Use the technique of smearing it by a test function, say ξ in the Schwartz space S. And we have an ordinary random variable.

A good example is the white noise $\dot{B}(t)$:

$$\dot{B}(\xi) = \int \xi(t)\dot{B}(t)dt.$$

We now have a consistent family of random variables parameterized by ξ. The characteristic functional

$$C(\xi) = E[e^{i\dot{B}(\xi)}],$$

which is equal to $\exp[-\frac{1}{2}\|\xi\|^2]$, determines the probability distribution μ of $\dot{B}$.

4) It was until 1975 when $\dot{B}(t)$'s and their nonlinear functionals have been defined as generalized Brownian (in fact, white noise) functionals. See [2]. Details will be reviewed in the next section.

5) We now revisit the idea in [2] and try to let it be well understood. See [3] for details. The basic idea of understanding the $\dot{B}(t)$ is the reductionism. Namely, we wish to express an evolutional random complex system as a functional of idealized elemental random variables, where elemental means atomic and basic. White noise is the best example of such a system. Another example is the Poisson noise, which is the time derivative of a Poisson process. For some more discussions, see [3].

2. Infinite dimensional spaces spanned by $\dot{B}(t), t \in R^1$

Linear spaces generated by $\dot{B}(t)$'s are discussed in this section. They might be considered as simple spaces, but not quite. We need to give rigorous definitions and to have them correctly understood.

1) The space $H_1^{(-1)}$.

The collection of stochastic integrals of the form

$$\dot{B}(f) = \int f(t)\dot{B}(t)dt, \quad f \in L^2(R^1),$$

spans a Hilbert space H_1. It is well known that H_1 is isomorphic to $L^2(R^1)$:

$$H_1 \;\cong\; L^2(R^1).$$

This isomorphism extends to

$$H_1^{(-1)} \;\cong\; K^{-1}(R^1),$$

where $K^{-1}(R^1)$ is the Sobolev space of order -1 over R^1.
Here and hereafter we use notations such as $K^m(R^n)$ to denote the Sobolev space of order m over R^n. The subspace of $K^m(R^n)$ involving symmetric functions is denoted by $\widehat{K}^m(R^n)$.

Since δ_t is a member of $K^{-1}(R^1)$, the $\dot{B}(t) = \dot{B}(\delta_t)$ is a well-defined member of $H_1^{(-1)}$. Thus, the exact identity is given to $\dot{B}(t)$.
2) The space $H_1^{(-1)}$ is infinite dimensional.

This is an obvious assertion, since any finite collection of $\dot{B}(t_j), 1 \le j \le n$, where t_j's are different, is a system of linearly independent vectors in $H_1^{(-1)}$. This is true for any n.
3) The system $\{\dot{B}(t), t \in R^1\}$ is total in $H_1^{(-1)}$.

This means that all finite linear combinations of the system span the entire space $H_1^{(-1)}$. This assertion is proved by using the fact that $\{e^{it\lambda}, t \in R^1\}$ is total in $L^2(R^1)$.

The above total system plays some role of the system of variables of random functions, although it is not an orthonormal base in the ordinary sense.

It is noted that the random system $\{\dot{B}(t), t \in R\}$ is stationary in time, i.e. its probability distribution μ on E^* (a space of generalized functions on R^1) is invariant under the time shift. In addition, the spectrum is flat.
4) Construction of a Brownian motion (white noise).

We emphasize the significance of the Lévy construction of a Brownian motion which is done by successive interpolations. It follows the following steps. Prepare standard Gaussian i.i.d. (independent identically distributed) random variables $\{Y_n = Y_n(\omega), \omega \in \Omega, n \ge 1\}$ on a probability space (Ω, B, P). Start with $\{X_1(t)\}$ given by

$$X_1(t) = tY_1. \tag{1}$$

The sequence of processes $\{X_n(t), t \in [0,1]\}$ is formed by induction. Let T_n be the set of binary numbers $k/2^{n-1}, k = 0, 1, 2, \cdots, 2^{n-1}$, and set

$T_0 = \cup_{n\geq 1} T_n$. Assume that $X_j(t) = X_j(t,\omega), j \leq n$, are defined. Then, we set

$$X_{n+1}(t) = \begin{cases} X_n(t),\ t \in T_n, \\ \frac{X_n(t+2^{-n})+X_n(t-2^{-n})}{2} + 2^{-\frac{n}{2}} Y_k,\ t \in T_{n+1} - T_n, \\ \qquad\qquad\qquad\qquad k = k(t) = 2^{n-1} + \frac{2^n t+1}{2}, \\ (k+1-2^n t) X_{n+1}(k2^{-n}) + (2^n t - k) X_{n+1}((k+1)2^{-n}), \\ \qquad\qquad\qquad\qquad t \in [k2^{-n}, (k+1)2^{-n}], \end{cases} \tag{2}$$

where ω is omitted.

We now claim

Theorem 2.1. *i) The sequence $X_n(t), n \geq 1$, is consistent and that the uniform L^2-limit of the $X_n(t)$ exists. The limit is denoted by $\tilde{X}(t)$, which is a Brownian motion.*
ii) The time derivative $X_n'(t)$ converges to a white noise in $H_1^{(-1)}$.

Proof is easy and is omitted.

As was mentioned before, this method of approximation is significant in many ways. For instance, the approximation is uniform in t, and is consistent in the sense that it is getting finer and finer as $n \to \infty$ and the σ-field associated to $X_n(t)$ is strictly monotone increasing.

5) Some developments. The above method of approximation, applied to white noise, is used in many places in an essential manner; some typical cases are now in order.

a) We shall use Lévy's approximation to define a motion group acting on $H_1^{(-1)}$. This will be done in Section 4.

b) The method is used to define various spaces of nonlinear functionals of white noise, where the approximation works efficiently.

c) Shimizu's approach to operator calculus, in particular quadratic forms of annihilations ∂_t and their adjoints, i.e. creations ∂_t^* (which will be defined in Section 4, acting on the space of generalized white noise functionals.

d) Si Si's approach (see [4]) to the study of duality between Gaussian and Poisson noises. Her actual technique heavily depends on the method of the approximation just mentioned.

e) The Volterra form for variation of functional is a result of the idea the so-called "passage du fini à l'infini". The approximation discussed above is a realization of this idea.

Let $U(\xi)$ be a functional of ξ, where $\xi = \xi(t), t \in [0,1]$. Then, the variation δU, if it exists, is expressed in the form (Volterra form)

$$\delta U = \int_0^1 U'(\xi, t)\delta\xi(t)dt.$$

The $U'(\xi, t)$ is the Fréchet derivative. Lévy has mainly used this formula and in fact, he succeeded to Volterra's idea (see e.g. [5]). This formula is really an infinite dimensional analogue of the variation of functions on R^n:

$$du(x) = \sum_1^n u_j dx_j, \quad x = (x_1, x_2, \cdots, x_n).$$

Note that the expression is uniform in the coordinate x_j. The Volterra form expresses a continuous sum and is summed up uniformly in $\delta\xi(t)$. Such a property is most significant, if the idea is applied to the variational calculus for white noise functionals.

3. Spaces of generalized white noise functionals

Recall the Fock space

$$(L^2) = L^2(E^*, \mu) = \oplus_n H_n,$$

where H_n is the space of homogeneous chaos of degree n or it is called the space of the n-ple Wiener integrals in Itô's sense. We know that

$$H_n \cong \widehat{L}^2(R^n)$$

up to constant $\sqrt{n!}$. Remind the notation $\widehat{K}^{-(n+1)/2}(R^n)$. Then, the above isomorphism extends to an isomorphism

$$H_n^{(-n)} \cong \widehat{K}^{-(n+1)/2}(R^n).$$

Actually, this isomorphism is nothing but the definition of $H_n^{(-n)}$, which is called the space of generalized white noise functionals of degree n.

Theorem 3.1. *The collection of all Wick products of the form*

$$:\dot{B}(t_1)\dot{B}(t_2)\cdots\dot{B}(t_n):$$

with t_j's may or may not be different, is total in the space $H_n^{(-n)}$.

Proof. The Sobolev space of order $\frac{-(n+1)}{2}$ involving those functions f on R^n such that its Fourier transform $\hat{f}$ satisfies the condition that $\frac{\hat{f}(\lambda)}{(1+|\lambda|^2)^{(n+1)/4}}$ is square integrable. Also, it is known that the above Wick product corresponds to the tensor products of δ-functions via the isomorphism just established. The Fourier transform of it is expressed in the form

$\Pi e^{it_j\lambda_j}$. The collection in question is total in $\widehat{K}^{-(n+1)/2}(R^n)$. Thus the theorem is proved. □

Definition 3.1. Set

$$(L^2)^- = \oplus_n c_n H_n^{(-n)},$$

which is called the space of generalized white noise functionals.

The sequence c_n is taken to be monotone decreasing.

Having remind the definition of $(L^2)^-$ we know how to define the total family of Wick products.

4. Transformation groups

Generally, given a pair (X, G), the transformation group G and the space X on which G acts as a transformation group. Then, a harmonic analysis can start. As a realization of this idea, we present two examples.
1) Rotation group of a nuclear space E.

The time parameter space is now taken to be R^1.
Definition of $O(E)$ is as follows. A transformation $g \in O(E)$ means that g is linear isomorphism of E and is orthogonal. The compact-open topology is introduced.

The adjoint operator g^* of g is defined. Set

$$O^*(E^*) = \{g^*; g \in O(E)\}.$$

The following assertions are known,

Proposition 4.1. *i) The group $O^*(E^*)$ gives an invariance of the white noise measure μ:*

$$g^*\mu = \mu, \ g \in O(E).$$

ii) The time shift S_t is in $O(E)$ and $\{T_t = S_t^*\}$ is ergodic on (E^*, μ). (Maximality!)
iii) Lévy's projective invariance of Brownian motion is described by a subgroup of $O^*(E^*)$.

Thus, a harmonic analysis of white noise functionals arises from $O(E)$.
Note. We note that any g^* is a mapping that transforms a sample function of white noise to another.
2) Motion group $M(H_1^{(-1)})$.

First we restrict our attention to the unit time interval [0,1]. Hence, the space $H_1^{(-1)}$ is isomorphic to the Sobolev space $K^{-1}([0,1])$. The system $\{\dot{B}(t_j), t_j \in T_0\}$ is total in $H_1^{(-1)}$, where T_0 is the set determined in Section 2, 4). In principle, we have to start with the approximation of white noise by slightly modifying the method explained in Section 2, then define the motion group with the help of the $\dot{B}(t)$'s. We shall, however, omit the approximation procedure and directly show the results only.

The motion group acting on $H_1^{(-1)}$ consists of translations and rotations.

i) Translations.

Again, by regarding each $\dot{B}(t)$ a variable of white noise functionals, we first consider the translation in the direction of $\dot{B}(t)$:

$$\dot{B}(.) \;\rightarrow\; \dot{B}(.) + \alpha\dot{B}(t), \;\; \alpha \in R^1.$$

The translation defines an operator acting on functional φ of the $\dot{B}(t)$'s to have

$$\varphi(\dot{B}(.)) \;\rightarrow\; \varphi(\dot{B}(.) + \alpha\dot{B}(t)).$$

The infinitesimal generator is

$$\partial_t\varphi(\dot{B}(.)) = \frac{\partial}{\partial\dot{B}(t)}\varphi(\dot{B}(.)).$$

The ∂_t is an annihilation operator and its adjoint ∂_t^* is a creation operator. Commutation relations are

$$[\partial_t, \partial_s] = 0, \;\; [\partial_t^*, \partial_s^*] = 0.$$

$$[\partial_t, \partial_s^*] = \delta_{t-s}.$$

ii) Rotations.

We then come to the rotations of $H_1^{(-1)}$. Elementary rotation is defined by taking two directions and Euler angle so as to be rotated as that much. Many pairs of two directions can be chosen so that this method of rotation would generate all the possible rotations. We are afraid this might not be successful, since we prefer to choose the total system involving continuously many independent vectors and unfortunately the system is not a coordinate system. To give necessary interpretations we have to be back to the stage of approximation of white noise, which we have omitted, since one can know what should be done.

Thus, translations and rotations form a motion group $M(H_1^{(-1)})$ acting on the space $H_1^{(-1)}$. The topology is defined by operator norm as usual. We can further carry on harmonic analysis. This topic will be discussed in another paper.

5. Concluding remarks

In this note we have mainly discussed linear functionals of white noise $\dot{B}(t)$. We are, of course, ready to discuss nonlinear functions of white noise. Particular interest can be seen in quadratic (generalized) functionals of white noise (see [9]), in particular quadratic normal functionals. There we can see interesting duality between subspaces.

It is interesting to note that general normal functionals of higher degree are also significant. Systematic approach to those functionals will be reported in the next report.

Acknowledgement

The author wishes to express his deep thanks to the organizers of the QBIC Conference 2008 who have given him an opportunity to present this report.

References

1. I.M. Gel'fand, Generalized random processes (in Russian) Doklady Acad. Nauk. 100 (1955), 853-856.
2. T. Hida, Analysis og Brownian functionals. Carleton Math. Notes no.13. Carleton University. 1975.
3. T. Hida and Si Si, Lectures on white noise functionals. World Scientific Pub. Co. 2009.
4. Si Si, An aspect of quadratic Hida distributions in the realization of a duality between Gaussian and Poisson noises. IDAQP 11 (2008), 109-118.
5. V. Volterra, Leçons sur les équations intégrals et les équations intégro-differentielles. Gauthier-Villars. 1913.
6. P. Lévy, Problèmes concrets d'analyse fonctionnelle. Gauthier-Villars. 1951.
7. T. Hida, Si Si and T. Shimizu, The $\dot{B}(t)$'s as idealized elemental random variables. Volterra Center Notes. N.614. Feb. 2008.
8. H. Umegaki, M. Ohya and F. Hiai, Introduction to operator algebras, (in Japanese) Kyoritsu Pub. Co. 2003.
9. Si Si and T. Hida, Some aspects of quadratic generalized white noise functionals. to appear in this Proceedings.

Quantum Bio-Informatics III
eds. L. Accardi, W. Freudenberg & M. Ohya

NOTE ON GENERALIZED WHITE NOISE FUNCTIONALS

TAKEYUKI HIDA

Professor Emeritus Nagoya University and Meijo University

The purpose of this report is to give a detailed interpretations to white noise variables $\dot{B}(t), t \in R^1$ and their functionals, including the idea of renormalization. By doing so, we can see an infinite dimensional analysis and some applications follow naturally

Keywords: Generalized functional, Renormalization, Duality, Infinite, Windmill

1. Introduction

The main purpose of this talk is to present a note that illustrates the real meaning of the generalized white noise functionals and to show a beautiful structure of the space of those functionals.

So far as the research of random evolutional and complex phenomena is concerned, it would be a good idea to determine the elemental system of independent random variables first, and their functionals. It was, therefore, quite natural and inevitable to introduce a space of generalized white noise functionals (see [2]). At present, we wish

1) to reconfirm this significant fact, for which many scientists have contributed, and

2) to observe its internal beauty of the structure of the functional space of white noise and of the analysis developed there.

3) In addition to the beauty, there should be validity of defining various notions and of introducing new operators in white noise analysis.

By White Noise we mean a time derivative $\dot{B}(t)$ of a Brownian motion, or a probability measure space (E^*, μ), where E^* is a space of generalized functions and μ is the probability distribution of $\dot{B}(t), t \in R^1$ on E^*. The space of ordinary (classical) functionals of white noise is denoted by (L^2) which involves functions of smeared variables of $\dot{B}(t), t \in R^1$. The space admits a direct sum decomposition, called Fock space:

$$(L^2) = \oplus H_n.$$

A space $(L^2)^-$ of generalized white noise functionals is obtained by

$$(L^2)^- = \oplus c_n H_n^{(-n)},$$

where $H_n^{(-n)}$ is an extension of H_n.

Another method to have the space of generalized white noise functionals is defined by an analogous, but generalized, method of defining the Schwarz distributions:

$$(S) \subset\subset (L^2) \subset (S)^*.$$

The $(S)^*$ is a space of generalized white noise functionals.

In this talk, we shall discuss

i) Identity of $\dot{B}(t)$.
ii) Renormalization of white noise functionals.
iii) Continuously many coordinate vectors that span the space in question.
iv) some of new directions including the analysis that is essentially infinite dimensional.

They should be discussed in line with beauty and validity.

2. Basic ideas of generalized white noise functionals

We prepare some notations, although they are well-known.

A functional of white noise is denoted by $\varphi(\dot{B})$ or by $\varphi(x)$, where $x \in E^*$ is viewed as a sample function of $\dot{B}$.

I. The system $\{\dot{B}(t)\}$ and the space $H_n^{(-n)}$. A Review

It is a collection of idealized elemental random variables, in fact, generalized random variables..

As was emphasized in [4], we follow the method of reduction in order to analyze random complex systems. Namely, we take a system of i.e.r.v.'s (idealized elemental random variables). Presently, we take a white noise $\{\dot{B}(t), t \in R\}$. satisfying the conditions

1) a system of *coordinate vectors* (i.e. linearly independent) in $H_1^{(-1)}$ and is *total* there.

2) $\dot{B}(t), t \in R^1$, is a *generalized* stationary stochastic process with *independent* value at every point (in the sense of Gel'fand 1955 Dokl. Acad Nauk SSSR). And it is Gaussian in distribution.

Note that in the space $H_1^{(-1)}$ they are not orthogonal. In fact, we have to pay a price. For example, if $t \neq s$, then it is not orthogonal, but we have

$$\langle \dot{B}(t), \dot{B}(s) \rangle_{H_1^{(-1)}} = \pi e^{-|t-s||}.$$

We have a generalization. More details will be stated in Theorem 2.2.

Theorem 2.1. *The collection* $\{: \Pi_{j=1}^n \dot{B}(t_j)$ *forms a system of linearly independent vectors in* $H_n^{(-n)}$.

The space $H_n^{(-n)}$ is infinite dimensional. It can, however, not be called continuously many dimensional.

II. Renormalization.

In order to avoid misunderstanding on renormalization of white noise functionals, we now give another interpretation than what is familiar.

For $\dot{B}(t)^2$, which is surely not an ordinary random variable, and even not a generalized functional, we have therefore introduced a trick, that is called renormalization, to let it have some meaning. It is as follows.

In the ordinary stochastic calculus the formula, we have the equality :

$$(dB(t))^2 = dt.$$

This approximation is useful in ordinary calculus, how ever we note that it is an approximation and that the left-hand side is a random quantity although it is infinitesimal. Now we have to think of the order of being infinitesimal. More precise approximation, in fact up to the second order, is

$$(dB(t))^2 = dt + X(dt)^2,$$

where X is some random element, We are interested in the evaluation of X, so that divide both side by $\frac{1}{(dt)^2}$ to obtain the renormalized square of $\dot{B}(t)$ which is X:

$$X =: \dot{B}(t)^2 := \dot{B}(t)^2 - \frac{1}{dt}.$$

Although the computation above looks like a formal calculus, we can rigorously proceed the calculus in the space $H_2^{(-2)}$ of quadratic generalized white noise functionals.

Note that one might think that the procedure is just an orthogonalization of $\dot{B}(t)^2)$ to the space of lower degree, however we must be careful with the fact that $\dot{B}(t)^2$ is not a well-defined vector in the space $(L^2)^-$ of generalized white noise functionals, so that we can not speak of orthogonal projection. Even we have to note the topology in $(L^2)^-$.

Also, it may be helpful for our calculus to consider another simple example how to renormalize $\dot{B}(t)^3$.

Now one may ask if there is any general method of renormalization of any polynomials in $\dot{B}(t)$'s. The answer is this:

Remind the S-transform of a functional $\varphi(x)$ in the space $(L^2)^-$.

$$(S\varphi)(\xi) = C(\xi) \int_{E^*} e^{<x,\xi>} \varphi(x) d\mu(x),$$

where $C(\xi)$ is the characteristic functional of the white noise measure μ on E^*, the space of generalized functions on R^1.

We know that the S-transform of $\dot{B}(t)$ is $\xi(t)$. It is natural to think of the polynomials in $\xi(t)$'s. Namely, we take such a polynomial in ξ of degree n as

$$U(\xi) = \Pi_{j=1}^n \xi(t_j),$$

where t_j's may not be different. According to the Potthoff-Streit characterization (see [3], [4], [7]), $U(\xi)$ is a U-functional, i.e. $U(\xi)$ is the S-transform of some functional $\varphi(x)$ in $(L^2)^-$. Since $U(\xi)$ is a monomial in ξ of degree n, so that the φ must be in $H_n^{(-n)}$. We use the notation

$$(S^{-1}U)(\dot{B}) =: \Pi_{j=1}^n \dot{B}(t_j) :$$

and call it the renormalized polynomial of degree n and it is in agreement with the Wick product of $\dot{B}$ given in [3].

We then explain the theorem in a different manner.

Theorem 2.2. *The renormalized polynomials of degree n are all in the space $H_n^{(-n)}$ and they form a total system for $H_n^{(-n)}$.*

We need a proof for the second assertion. The space $H_n^{(-n)}$ is isomorphic to the symmetric Sobolev space
$\hat{K}(R^n)^{(-(n+1)/2)}$ over R^n of order $-(n+1)/2$. There the collection of the tensor products of the delta-functions form a total set. This set corresponds to the collection of the renormalized polynomials in $\dot{B}(t)$'s of degree n, This proves the theorem.

3. Infinite dimensional calculus. Windmills.

We need some note on the concept of infinite dimensional calculus. Our calculus based on the coordinate vectors $\{\dot{B}(t), t \in R^1\}$. Each member has the identity in the space $H_1^{(-1)}$; there are members as many as continuum many. However, we can not say continuously multidimensional space is with us. For one thing, every space that we are concerned with is separable. Correct understanding of this fact at present is necessary in order to develop our analysis.

We shall remind two known facts and one proposal in connection with the above note.

1) From quantum dynamics.

An example is taken from the representation theory of the Heisenberg canonical commutation relations. For the case of finite degree of freedom the irreducible unitary representation is unique up to unitary equivalence as is well-known. While, in the case of infinite degree of freedom, either countably many or continuously many degree, irreducible unitary representation is "not" unique. Actually this fact can be proved by using white noise theory.

2) Le passage du fini à l'infini (due to V. Volterra and P. Lévy).

The story goes all the way back to the classical theory of functional analysis.

It is well known that infinite dimensional space there is no measure like the Lebesgue measure in the finite dimensional euclidean space. Instead, we can introduce a white noise measure that satisfies favorable properties for our calculus of functionals. The measure is uniformly distributed in all the directions.

2.i) Quadratic forms. In the classical theory of functional analysis, they tried to introduce a Lebesgue like measure, instead they got average or mean of functionals, which we now understand integral of functionals with respect to white noise measure. For actual computation they used a method so-called Le passage du fini à l'infini.

Consider a quadratic form of n-dimensional vector $x = (x_1, x_2, \cdots, x_n)$:

$$Q(x) = \sum_{1}^{n} a_{j,k} x_j x_k.$$

When we let n tend to ∞, $Q(x)$ must be divided into two parts; $Q_1(x)$ and

$Q_2(x)$:

$$Q_1(x) = \sum_j a_{j,j} x_j^2,$$

and

$$Q_2(x) = \sum_{j \neq k} a_{j,k} x_j x_k.$$

It is quite interesting to try to give plausible reasons on why it is necessity to divide Q into two parts Q_1 and Q_2.

2.ii) The classical functional analysis asserts that we shall have functionals of the forms

$$U_1(\xi) = \int f(u)\xi(u)^2 du.$$

and

$$U_2(\xi) = \int\int F(u,v)\xi(u)\xi(v)dudv,$$

where ξ runs through a certain function space, say a nuclear space E.

The quadratic functionals of these forms are called normal functionals and play significant roles in white noise analysis.

There are significant remarks regarding the limiting procedure. Note that the indices j,k turns in to continuous u,v. This follows the idea of passage due to Volterra and Lévy.

(1) From $Q(x)$ to $Q(\xi)$, we may consider the case where x_j's are i.i.d. standard Gaussian variables. Regarding the convergence, there is a difference between Q_1 and Q_2.

(2) As for the passage from finite to infinite, we may refer the Rohlin's method to construct an abstract Lebesgue space. We may rephrase in white noise case.

(3) As for a general quadratic form of ξ, one may think of a kernel F taken to be a tempered distribution in the sense of L. Schwartz, since $\xi \otimes \xi$ is a test function. It is, however, not favorable for our purpose.

This idea can easily be generalized to the space of higher degree.

2.iii) The Volterra form for the variation of functionals.

In order to carry on differential and integral calculus for functionals $U(\xi)$, we do "not" employ operators like $\frac{\partial}{\partial \xi_n}$, n being discrete and countable. We use $\partial_t = \frac{\partial}{\partial \dot{B}(t)}$. This is motivated by the so-called Volterra form (see [9]) for variation of functional $U(\xi)$:

$$\delta U(\xi) = \int U'_\xi(\xi;u)\delta\xi(u).$$

We call $U_\xi(\xi;t)$ a functional derivative and use the notation

$$U'_\xi(\xi;t) = \frac{\delta U(\xi)}{\delta\xi(t)}$$

Needless to say, the $\delta\xi(u)$, indexed by u, formally speaking, is a continuous analogue of dx_j, the differential in the variable x_i.

The second order functional derivatives $U''_{\xi^2}(\xi,u) = U_1(\xi,u)$ and $U''_{\xi_1,\xi_2}(\xi;u,v) = U_2(\xi;u,v)$ are determined by

$$\delta U'_\xi(\xi,t) = \int U''(\xi;u,v)\delta\xi(u)\delta\xi(v)dudv.$$

We are permitted to consider the case where $U''(\xi;u,v)$ has singularity in the (u,v)-plane, where the singularities should be only on the diagonal. This restriction comes from the possible quadratic form $U_1(\xi)$ discussed in 2.i). Thus, the favorite cases of the second functional variation are

$$\int U''_1(\xi;u)(\delta\xi(u))^2du$$

and

$$\int\int U''_2(\xi;u,v)\delta\xi(u)\delta\xi(v)dudv.$$

As for the $U_1(\xi)$, we note that it is in the domain of the Laplacian

$$\Delta_L = \int (\partial_t)^2(dt)^2.$$

Once again we note that the notation ∂_t and the expression of Δ_L come from the Volterra form where continuously many coordinates appear explicitly.

4. Infinite dimensional rotation group $O(E)$. In particular, the Windmill subgroup

We know that the infinite dimensional rotation group $O(E)$ shows an invariance of white noise measure μ. We can tell some significant cases where the rotation group describes "essentially infinite dimensional" properties enjoyed by our white noise analysis.

We first consider the case where the subgroups of $O(E)$ which can be considered as the limit (in a suitable sense) of finite dimensional rotations that can be expressed by using a complete orthonormal base $\xi_n, n \in Z$. To fix the idea, the basic nuclear space E is taken to be the Schwartz space S.

Let $n(k)$ is an increasing sequence of positive integers satisfying the condition

$$(n(k+1) - n(k))\frac{n(k+1)}{n(k)} \leq K \qquad (K > 1).$$

Assume, for computational convenience, that ξ_n is the eigenfunction of the operator D:

$$D = -\frac{d^2}{du^2} + u^2 + 1$$

with eigenvalue $2(n+1)$.

Let E_k be the $(n(k+1) - n(k))$-dimensional subspace of S spanned by $\{\xi_{n(k)+1}, \xi_{n(k)+2}, \cdots, \xi_{n(k+1)}\}$. Let G_k be the rotation group acting on E_k. Then, we set $W = \otimes_k G_k$.

Theorem 4.1. *The product W forms a subgroup of $O(S)$. It contains infinitely many members that are essentially infinite dimensional.*

[Note] The term "essentially infinite dimensional rotation g" means that its average power

$$a.p.(g) = \limsup_N \frac{1}{N} \sum_1^N \langle x, g\xi_n - \xi_n \rangle^2$$

is positive almost surely. Roughly speaking, this means that the transformation g can not be approximated by finite dimensional transformations.

To come to the continuous case, we shall also keep the line of the idea "le passage du fini à l'infini" with a realization of the Volterra form of functional variation.

We therefore use the operator ∂_t, Δ_L, multiplication $\partial_t + \partial_t^*$, rotation $\partial_t^* \partial_s - \partial_s^* \partial_t$ and so forth.

There is a beautiful theorem.

Theorem 4.2. *A nontrivial quadratic form $Q(\partial_t)$ of the ∂_t's*

$$Q(\partial_t) = \int\int F(u,v)\partial_u \partial_v du dv$$

that commutes with the rotations is the Lévy Laplacian Δ_L up to constant.

Further we note that quadratic forms involving ∂_t and ∂_t^* play dominant roles in the study of operator algebra generated by them.

5. Concluding remark

One more important beautiful property is the duality. The idea of introducing the notion of duality has been motivated by Prof. I. Ojima's approach. Various kind of dualities in the space of generalized white noise functionals have been found and more are going to be discovered. They certainly can be good characteristics of white noise functionals. Some of them will be reported in this QBIC Conf. paper by Prof. Si Si.

References

1. L. Accardi et al, Selected papers of Takeyuki Hida. World Scientific Pub. Co. 2001.
2. T. Hida, Analysis of Brownian functionals. Carleton Math. Notes, no.13, Carleton Univ. 1975,
3. T. Hida et al, White noise,. An infinite dimensional calculus. Kluwer Academic Pub. Co. 1993.
4. T. Hida and Si Si, Lectures on white noise functionals. World Scientific Pub. Co. 2008.
5. P. Lévy, Leçons d'analyse fonctionnelle. Gauthier-Villars, 1922.
6. P. Lévy, Problèmes concréts d'analyse fonctionnell. Gauthier-Villars. 1951.
7. J. Potthoff and L. Streit, Characterization of Hida distributions. J. Functional Analysis,
8. Si Si, Win Win Htay and L. Accardi, T-transform of Hida distribution and factorizations. Volterra Center Notes, N. 625, February 2009.
9. V. Volterra et J. Pérés, Théorie générale des fonctionnelle. Gauthier-Villars, 1936.

Quantum Bio-Informatics III
eds. L. Accardi, W. Freudenberg & M. Ohya

ON ESTIMATION OF THE POSITION DISTRIBUTION OF THE IDEAL BOSE GAS

KARL-HEINZ FICHTNER[1], KEI INOUE[2] AND MASANORI OHYA[3]

1 *Friedrich-Schiller-Universität Jena,*
Fakultät für Mathematik und Informatik,
Institut für Angewandte Mathematik,
07737 Jena, Germany.
E-mail: karl-heinz.fichtner@uni-jena.de
2 *Department of Electrical Engineering,*
Tokyo University of Science, Yamaguchi,
Sanyo-Onoda, Yamaguchi 756-0884, Japan.
E-mail: inoue@ed.yama.tus.ac.jp
3 *Department of Information Sciences,*
Tokyo University of Science,
Noda City, Chiba 278-8510, Japan.
ohya@rs.noda.tus.ac.jp

In [2] clustering representations of the position distribution of the ideal Bose gas were considered. In principle that gives rise to possibilities concerning simulations of the system of positions of the particles. But one has to take into account that in case of low temperature the clusters are very large and their origins are far from a fixed bounded volume. For that reason we will consider some estimations of the influence of these clusters on the behaviour of the subsystem of particles located in a fixed bounded volume.

1. Introduction

In [3] it was shown that to each locally normal state ω of boson systems on R^d one can associate a distribution P^ω of a random point system in R^d that can be interpreted as the position distribution of the state. If the reduced density matrices $\rho_n^\omega(x^1, \ldots, x^n, y^1, \ldots, y^n)$ of the state ω (cf. [1]) are continuous then a density $\lambda_{P^\omega}^{(n)}$ (w.r.t. Lebesgue's measure) of the n-th factorial moment measure of P^ω is given by (cf. [4])

$$\lambda_{P^\omega}^{(n)}(x^1, \ldots, x^n) = \rho_n^\omega(x^1, \ldots, x^n, x^1, \ldots, x^n); \quad x^1, \ldots, x^n \in R^d. \tag{1}$$

Now, let us consider the (infinite-volume) ideal Bose gas (cf. [1]) corresponding to the inverse temperature $\beta > 0$ and chemical potential $\mu < 0$.

$z = \exp[\mu\beta]$ denotes the activity. The conventional quantum-mechanical Hamiltonian for free particles of mass m is $-\frac{\hbar^2}{2m}\Delta$ but we now choose units such that $\frac{\hbar^2}{2m} = 1$. Then the reduced density matrices ρ_n, $n = 1, 2, \ldots$ of the ideal Bose gas exist and (cf. [1])

$$\rho_n(x^1, \ldots .x^n, y^1, \ldots, y^n)$$
$$= \sum_{(i_1,\ldots,i_n)} \prod_{k=1}^{n} \sum_{m=1}^{+\infty} \frac{z^m}{(4\pi\beta m)^{d/2}} \exp\left[-\frac{\|x^k - y^{i_k}\|^2}{4m\beta}\right]; \tag{2}$$
$$x^1, \ldots, x^n, y^1, \ldots, y^n \in R^d$$

where $(i_1, \ldots, i_n)$ denotes a permutation of $1, \ldots, n$. Consequently, densities of the factorial moment measures of the position distribution P of this ideal Bose gas are given by

$$\lambda_P^{(n)}(x^1, \ldots, x^n) = \sum_{(i_1,\ldots,i_n)} \prod_{k=1}^{n} \sum_{m=1}^{+\infty} \frac{z^m}{(4\pi\beta m)^{d/2}} \exp\left[-\frac{\|x^k - x^{i_k}\|^2}{4m\beta}\right]; \tag{3}$$
$$n = 1, 2, \ldots, \quad x^1, \ldots, x^n \in R^d.$$

It was proved in [2] that P is an infinitely divisible point process characterized by (3). Furthermore, clustering representation of P was considered in [2]. Hence P may be interpreted as distribution of a random point field in R^d which forms independent clusters or random point-families the distribution of which is given by the clustering field (16). This corresponds to the description of the ideal Bose gas one can find in the physical literature.

In principle one can use the clustering representations for simulations of the system of positions of the quantum particle system. But one has to take into account that in case of low temperature the clusters are very large and their origins are far from a fixed bounded volume. For that reason we have to estimate the influence of these clusters on the behaviour of the subsystem of particles located in a fixed bounded volume. More precisely, we will estimate the expectation of the number of particles inside a fixed bounded volume being part of any cluster with origin outside of a certain bigger volume (for the corresponding main result cf. sect.5).

Further details are discussed in [5].

2. Random Point Fields

In this section we briefly review some basic notions and notations from point process theory. For details see the monographs [6,7]. Let A be a complete separable metric space, $\mathcal{A}$ the corresponding σ-algebra of Borel subsets of A, and $\mathcal{B}$ the ring of all bounded sets in $\mathcal{A}$. Further, let N denote the set of all measures on $\mathcal{A}$ taking only finite values on $\mathcal{B}$. A measure $\varphi \in N$ is called counting measure if it takes only integer values on $\mathcal{B}$. The set of all counting measures in N we denote by M. It is well-known (cf.[6,7]) that any $\varphi \in M$ is of the form

$$\varphi = \sum_{a \in \operatorname{supp} \varphi} \varphi(\{a\})\delta_a \tag{4}$$

where δ_a denotes Dirac's measure corresponding to $a \in A$. Hence any counting measure φ may be interpreted as an (at most countable) system of points or individuals in the phase space A. Assume that a (possibly void) population $\varphi \in M$ is partioned into non-void subpopulations or "families" $\varphi_i \in M$ via

$$\varphi = \sum_{i \in I} \varphi_i. \tag{5}$$

Then, for each $B \in \mathcal{B}$ only finitely many of the φ_i's have individuals in B. On the other hand, any collection $(\varphi_i)_{i \in I}$ with this property builds up a population $\varphi \in M$ via (5). This indicates that the "family structures" in the non-void populations on A may be described as counting measures on $\tilde{A} := M \backslash \{0\}$ equipped with a suitable metric. From 1.6.1 in [6] it follows that there exists such a complete separable metric. In particular, the corresponding σ-algebra $\tilde{\mathcal{A}}$ of Borel subsets of $\tilde{A}$ is generated by the mappings $\varphi \to \varphi(B)$, $B \in \mathcal{B}$, and a subset of $\tilde{\mathcal{A}}$ is bounded iff it is contained in

$$\{\varphi \in M \,|\varphi(B) > 0\} \tag{6}$$

for a suitable $B \in \mathcal{B}$. In the following, such a metric (cf. 1.6.1 in [6]) will be kept fixed. All objects referring to the new phase space A will be indicated by $\tilde{}$.

Now, let $\mathfrak{M}$ be the σ-algebra on M generated by the mappings $\varphi \to \varphi(B)$, $B \in \mathfrak{M}$. We define a $\mathfrak{M}$, $\mathfrak{M}$-measurable mapping sp from $\tilde{M}$ into M by setting (cf. [6])

$$\operatorname{sp} \Phi(B) := \int \Phi(d\varphi)\, \varphi(B); \quad B \in \mathcal{B} \tag{7}$$

Passing from Φ to spΦ means abandoning the family structure. Obviously sp Φ is the zero measure $o \in M$ iff the system Φ is void, i.e., coincides with zero measure $o \in \tilde{M}$.

By a random point field on A we mean a random element of the measurable space $[M, \mathfrak{M}]$. Probability measures on $[M, \mathfrak{M}]$ are called point processes on A, i.e., point processes are distributions of random point fields. According to the interpretation of elements of $\tilde{M}$ a point process on $\tilde{A}$ may be interpreted as distribution of random systems of point-families in A. For an arbitrary point process Q on $\tilde{A}$ we denote by sp$Q$ the point processes on A induced by the mapping $\Phi \to \operatorname{sp}\Phi$.

For an arbitrary $\varphi \in M$ and $k = 1, 2, \ldots$ we denote by $\varphi^{(k)}$ the measure on $\mathcal{A}^{\otimes k}$ characterized by

$$\begin{aligned}
&\varphi^{(k)}(B_1 \times \cdots \times B_k) \\
:= &\int_{B_1} \varphi(dx^1) \int_{B_2} (\varphi - \delta_{x^1})(dx^2) \cdots \int_{B_k} \left(\varphi - \sum_{i=1}^{k-1} \delta_{x^i}\right)(dx^k); \qquad (8) \\
&B_1, \ldots, B_k \in \mathcal{B}.
\end{aligned}$$

We may form, for each measure H on $\mathfrak{M}$, the measure

$$\Lambda_H^{(k)}(\cdot) := \int \varphi^{(k)}(\cdot) H(d\varphi); \quad k = 1, 2, \ldots \qquad (9)$$

on $\mathcal{A}^{\otimes k}$, which is called $k - th$ factorial moment measure of H.

$$\Lambda_H := \Lambda_H^{(1)} \qquad (10)$$

is called intensity measure of H. The Poisson point process (cf. [7]) Q_ν with intensity measure $\nu \in N$ is characterized by (cf. [8])

$$\Lambda_{Q_\nu}^{(k)} = \nu^{\otimes k}; \quad k = 1, 2, \ldots \qquad (11)$$

From 1.6.12 in [6] it follows that the mapping $Q \to$ sp$Q$ gives a one-to-one correspondence between the set of Poisson process on $\tilde{A}$ and the set of infinitely divisible point process on A.

3. Clustering Representation of the Position Distribution

We consider the special case $A = R^d$. For given $\mu < 0 < \beta$, we put

$$\begin{aligned} z &:= \exp[\mu\beta] \\ \nu(X) &:= \sum_{n=1}^{+\infty} \frac{z^n}{n(4\pi\beta)^{(nd)/2}} \int dx^1 \cdots \int dx^n \\ &\quad \times \exp\left[-\frac{1}{4\beta}\left(\|x^1 - x^n\|^2 + \sum_{i=1}^{n-1} \|x^{i+1} - x^i\|^2\right)\right] \\ &\quad \times \chi_X\left(\sum_{i=1}^{n} \delta_{x^i}\right); \\ X &\in \tilde{\mathcal{A}} \end{aligned} \tag{12}$$

where χ_X denote the indicator function of a set X. One easily checks that ν belongs to $\tilde{N}$, i.e., there holds

$$\nu\left(\{\varphi\,|\varphi(B) > 0\}\right) < +\infty; \quad B \in \mathcal{B}. \tag{13}$$

Then the following theorem has been proved in [2].

Theorem 3.1. *There exists exactly one point process P on R^d such that for all $n = 1, 2, \ldots$ a density (w.r.t. Lebesgue's measure) of $\Lambda_P^{(n)}$ is given by*

$$\lambda_P^{(n)}(x^1, \ldots, x^n) := \sum_{(i_1,\ldots,i_n)} \prod_{k=1}^{n} \sum_{m=1}^{+\infty} \frac{z^m}{(4\pi\beta m)^{d/2}} \exp\left[-\frac{\|x^k - x^{i_k}\|^2}{4m\beta}\right]; \tag{14}$$

$$n = 1, 2, \ldots, \quad x^1, \ldots, x^n \in R^d.$$

Furthermore, this point process P is given by

$$P = \operatorname{sp} Q_\nu. \tag{15}$$

Now, let $(w_1(t))_{t\geq 0}, \ldots, (w_d(t))_{t\geq 0}$ be an independent Wiener-processes with parameter 2β. For all $n = 1, 2, \ldots$ we consider the d-dimensional Brownian bridge given by

$$b_n(t) := \left[w_1(t) - \frac{t}{2\beta n} w_1(2\beta n), \ldots, w_d(t) - \frac{t}{2\beta n} w_d(2\beta n)\right]; \quad 0 \leq t \leq 2\beta n \tag{16}$$

For all $x \in R^d$, let $q_{x,n}$ denote the distribution of the following random point field in R^d

$$\sum_{k=0}^{n-1} \delta_{b_n(2\beta k)+x}.$$

Finally, we define a homogenous clustering field $K = (K_x)_{x\in R^d}$ (cf. [7] chapt.11) by setting

$$K_x(\cdot) := \frac{1}{\sum_{n=1}^{+\infty} \frac{z^n}{n(4\pi\beta n)^{d/2}}} \sum_{n=1}^{+\infty} \frac{z^n}{n(4\pi\beta n)^{d/2}}\, q_{x,n}(\cdot) \tag{17}$$

K_x may be interpreted as the distribution of a random offspring of an individual at site $x \in R^d$. The independent superposition of all individuals or points of a (possible random) configuration $\varphi \in M$ leads to be the "population's random offspring distribution" (cf.[6,7]). It can easily be shown that there holds

$$\int K_x(\{\varphi\,|\varphi(B) > 0\})\, dx < +\infty; \quad B \in \mathcal{B}. \tag{18}$$

Hence, by 11.2.3 in [7] there follows that the population's random offspring distribution $Q_{(K)}$ is again a point process on R^d if the point process Q is the distribution of a homogeneous (initial) population with finite intensity, i.e., $\Lambda_Q \in N$ and Q is stationary.

Then the following theorem was proved in [2]

Theorem 3.2. *Let Q be a stationary Poisson process on R^d with intensity*

$$\lambda_Q \equiv \sum_{n=1}^{+\infty} \frac{z^n}{n(4\pi\beta n)^{d/2}} \tag{19}$$

and let P be given by (15). Then it holds

$$P = Q_{(K)}. \tag{20}$$

4. Construction of the Clustering Representations

Theorem 3.2 means that a random point field Ψ with distribution P can be constructed as follows: Starting point is a random point field Φ with distribution Q, i.e., Φ is a stationary Poisson random point field with intensity $\lambda_Q = \lambda_Q(\beta,\mu)$ given by (19). Further let $\gamma = (\gamma_x)_{x\in R^d}$ be a family of independent random point fields such that K_x is the distribution of γ_x, i.e., γ_x represents a random offspring of an individual at site $x \in R^d$ called cluster with origin x. Finally we assume that Φ and γ are independent. Then formally we put

$$\Psi := \int \Phi(dx)\, \gamma_x, \tag{21}$$

i.e., we identify the random point field Ψ representing the system of the positions of the particles of the ideal Bose gas with the independent superpositions of the offsprings of all points of the point field Φ called to be the underlying random point field.

4.1. *On the Underlying Random Point Field*

In order to illustrate the behaviour of the ideal Bose gas in the case of low temperature, and in the case of high temperature we will consider the special case $d = 2$. Further we fix

$$\lambda_P^{(1)} = \lambda. \tag{22}$$

Using $z = \exp[\mu\beta]$, we obtain from the definition (3) of $\lambda_P^{(1)}$

$$\lambda = -\frac{1}{4\pi\beta}\ln(1-z) = -\frac{1}{4\pi\beta}\ln(1-\exp[\mu\beta]), \tag{23}$$

that means if we consider the limit $\beta \to +\infty$ then μ has to go towards 0.

From (23) we get

$$\mu = \mu(\beta, \lambda) = \frac{1}{\beta}\ln(1-\exp\{-4\pi\beta\lambda\}). \tag{24}$$

Using $z = \exp[\mu\beta]$ and (24), λ_Q will be a function $\lambda_Q(\beta, \lambda)$ of β, λ

$$\lambda_Q := \lambda_Q(\beta, \lambda) = \sum_{m\geq 1} \frac{(1-\exp\{-4\pi\beta\lambda\})^m}{4\pi\beta m^2}. \tag{25}$$

Because of

$$\sum_{m\geq 1} \frac{1}{m^2} = \frac{\pi^2}{6} < +\infty,$$

from (24) we obtain

$$\lim_{\beta\to+\infty} \lambda_Q(\beta, \lambda) = 0. \tag{26}$$

Now, let $\alpha(\beta, \lambda)$ be the expectation of the number of points in a cluster distributed by K_x (cf. (17)), i.e., we have

$$\alpha(\beta, \lambda) = \Lambda_{K_x}(R^2) \quad (x \in R^2) \tag{27}$$

where according to (10), Λ_{K_x} denotes the intensity measure of the point process K_x. Then from Theorem 3.2 it follows

$$\lambda = \alpha(\beta, \lambda)\ \lambda_Q(\beta, \lambda). \tag{28}$$

Because of (26) that implies

$$\lim_{\beta \to +\infty} \alpha(\beta, \lambda) = +\infty. \tag{29}$$

Summarizing we can state that in case of low temperature the underlying point system is very "thin" and the clusters are very large. Furthermore, all points in a fixed bounded set are parts of a cluster with origin being far from this set if the temperature is low.

Now, let us consider the case of high temperature.

Directly we obtain

$$\begin{aligned} \lambda_Q(\beta, \lambda) &= \sum_{m \geq 1} \frac{(1 - \exp\{-4\pi\beta\lambda\})^m}{(4\pi\beta)m^2} \\ &= \frac{1 - \exp\{-4\pi\beta\lambda\}}{4\pi\beta} + \sum_{m \geq 2} \frac{(1 - \exp\{-4\pi\beta\lambda\})^m}{(4\pi\beta)m^2}. \end{aligned} \tag{30}$$

We get

$$\lim_{\beta \to 0} \frac{1 - \exp\{-4\pi\beta\lambda\}}{4\pi\beta} = \lambda. \tag{31}$$

Further we get

$$\begin{aligned} 0 &\leq \sum_{m \geq 2} \frac{(1 - \exp\{-4\pi\beta\lambda\})^m}{(4\pi\beta)m^2} \\ &= (1 - \exp\{-4\pi\beta\lambda\}) \sum_{m \geq 2} \frac{(1 - \exp\{-4\pi\beta\lambda\})^{m-1}}{(4\pi\beta)m^2} \\ &\leq (1 - \exp\{-4\pi\beta\lambda\}) \frac{(1 - \exp\{-4\pi\beta\lambda\})}{4\pi\beta} \sum_{m \geq 2} \frac{1}{m^2}. \end{aligned} \tag{32}$$

We have

$$\lim_{\beta \to 0} (1 - \exp\{-4\pi\beta\lambda\}) = 0, \tag{33}$$

$$\sum_{m \geq 2} \frac{1}{m^2} < +\infty. \tag{34}$$

(32), (33), (34) and (31) imply

$$\lim_{\beta \to 0} \sum_{m \geq 2} \frac{(1 - \exp\{-4\pi\beta\lambda\})^m}{(4\pi\beta)m^2} = 0. \tag{35}$$

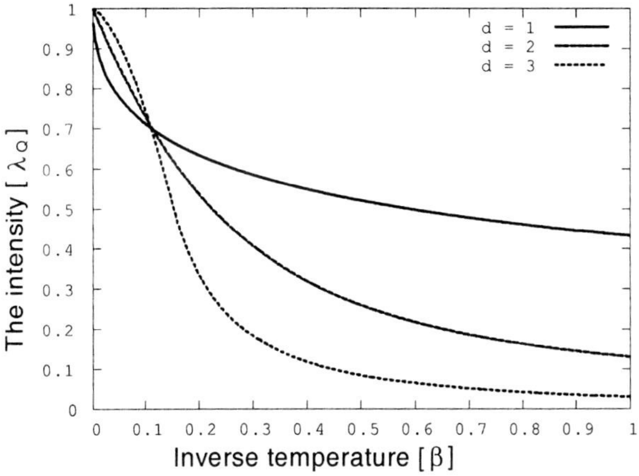

Figure 1. λ_Q versus β on [0.001,1]

(30), (31) and (35) imply

$$\lim_{\beta \to 0} \lambda_Q(\beta, \lambda) = \lambda. \tag{36}$$

Because of (28) that implies

$$\lim_{\beta \to 0} \alpha(\beta, \lambda) = 1 \tag{37}$$

i.e., the clusters become trivial(consisting of exactly one point).

Summarizing we can state that in case of high temperature the system of positions of the ideal Bose gas coincide with the underlying point system, being a stationary random Poisson field. That reflects the well-known fact that in case of high temperature the ideal Bose gas behaves like the classical ideal gas.

Finally, let us remark that one can obtain similar results in the case of general dimensions d. The following figure illustrates what happens in the case of high temperature in dimensions $d = 1, 2, 3$.

4.2. *Random cluster*

For each $x \in R^d$, let γ_x be again a random cluster (or offspring) of an individual at site x with probability distribution K_x.

Further let η_x be the number of points in the cluster γ_x, i.e., we have

$$\eta_x := \gamma_x(R^d).$$

Then from (17) we obtain the probability

$$\Pr\left(\eta_x = n\right) = \frac{1}{\lambda_Q} \frac{z^n}{n(4\pi\beta n)^{d/2}}. \tag{38}$$

Now we denote by $q_t^{(d)}(x,y)$ the densities of the transition probabilities of the d-dimensional Wiener process $(W(t))_{t\geq 0}$, i.e., we have

$$q_t^{(d)}(x,y) = \frac{1}{\sqrt{(2\pi t)^d}} \exp\left\{-\frac{\|x-y\|^2}{2t}\right\},$$

$$x = [x_1, \ldots, x_d], \;\; y = [y_1, \ldots, y_d] \in R^d, \;\; t \geq 0.$$

Using these transition probabilities we define the following functions.

$$q_n(x^1, \ldots, x^{n-1}, x) := \frac{1}{q_{2\beta n}^{(d)}(x,x)} q_{2\beta}^{(d)}(x, x^1) q_{2\beta}^{(d)}(x^{n-1}, x) \prod_{k=1}^{n-2} q_{2\beta}^{(d)}(x^k, x^{k+1}) \tag{39}$$

$$x^1, \ldots, x^{n-1}, x \in R^d.$$

Then for all measurable $Y \subseteq M$ we have the following equations.

$$\Pr(\gamma_x \in Y \,|\eta_x = n) = \int dx^1 \cdots \int dx^{n-1} q_n(x^1, \ldots, x^{n-1}, x) \times \chi_Y\left(\delta_x + \sum_{k=1}^{n-1} \delta_{x^k}\right),$$

$$n = 2, 3, \ldots, \tag{40}$$

$$\Pr(\gamma_x \in Y \,|\eta_x = 1) = \delta_{\delta_x}(Y). \tag{41}$$

Now, putting $p_n = \Pr(\eta_x = n)$, then K_x can be rewritten as

$$K_x(Y) = \sum_{n=1}^{+\infty} p_n \Pr\left(\gamma_x \in Y \,|\eta_x = n\right), \quad Y \subseteq M. \tag{42}$$

According to (10), Λ_{K_x} denotes the intensity measure of the point process K_x.

We have proven the following theorem [5].

Theorem 4.1. *Let $N_{x,r}$ be a normal distribution with expectation vector $x \in R^d$ and $d \times d$ covariance matrix r. For each measurable subset B of R^d it holds*

$$\Lambda_{K_x}(B) = \chi_B(x) + \frac{1}{\lambda_Q} \sum_{n\geq 2} \frac{z^n}{n(4\pi\beta n)^{d/2}} \sum_{k=1}^{n-1} N_{x,2\beta\frac{k(n-k)}{n}I}(B). \tag{43}$$

4.3. *System of the Positions of the Particles of the Ideal Bose Gas*

Again let $\gamma = (\gamma_x)_{x\in R^d}$ be a family of random clusters with distributions K_x. Now we assume that the random point field Φ with distribution Q and γ are independent. Remember that the system Ψ of positions of the particles of the ideal Bose gas is given by (cf. 4.1)

$$\Psi := \int \Phi(dx)\gamma_x.$$

Then the distribution P_Ψ of Ψ is $P = Q_{(K)}$ (cf. 20).

Let Φ_A be the system of points of Φ being inside of measurable subset A of R^d

$$\Phi_A := \Phi(\cdot \cap A)$$

and by Q^A we denote the point process corresponding to Φ_A, i.e., Q^A is the distribution of Φ_A. Then for the intensity measure Λ_{Q^A} we get

$$\Lambda_{Q^A} = \lambda_Q \; l(\cdot \cap A)$$

where l is Lebesgue's measure on R^d.

Using the Campbell theorem (cf.[7]) we obtain immediately the following lemma.

Lemma 4.1. *Let f be a measurable nonnegative function on R^d. Then it holds*

$$\int \Lambda_{Q^A}(dx)\; f(x) = E \int \Phi_A(dx) f(x) = \lambda_Q \int_A dx\; f(x).$$

Finally we denote by Ψ^A the system of all points of clusters γ_x with origin in A

$$\Psi^A := \int \Phi_A(dx)\gamma_x. \tag{44}$$

having the distribution $Q^A_{(K)}$.

Because of the independence of Φ and γ we obtain using lemma 4.1.

$$\Lambda_{Q^A_{(K)}} = \lambda_Q \int_A dx\, \Lambda_{K_x}. \tag{45}$$

Using theorem 4.1 and (45) we proved the following theorem [5].

Theorem 4.2. *Let A, B be measurable subsets of R^d. Then it holds*

$$\Lambda_{Q^A_{(K)}}(B) = \lambda_Q l(A\cap B) + \sum_{n\geq 2} \frac{z^n}{n(4\pi\beta n)^{d/2}} \sum_{k=1}^{n-1} \int_A dx\, N_{x,2\beta\frac{k(n-k)}{n}I}(B) \tag{46}$$

5. Main Results

Let A, B be measurable subsets of R^d. Without loss of generality we assume that

$$A = [-a, a]^d, \quad B = [-b, b]^d, \quad a \geq b \geq 0, \quad a, b \in R.$$

We proved the following theorem [5].

Theorem 5.1. *Let* $\Lambda_{Q^A_{(K)}}$ *be the intensity measure of* Ψ^A. *Then it holds*

$$\Lambda_{Q^A_{(K)}}(B) \leq f_1(a, b, d, \beta, \mu)$$

where the function $f_1(a, b, d, \beta, \mu)$ *is given by*

$$f_1(a, b, d, \beta, \mu) := (2b)^d$$
$$\times \left[\lambda_Q(\beta, \mu) + \left(1 - \exp\left\{-\frac{(a+b)^2}{\beta}\right\}\right)^{d/2} \left(\lambda_P^{(1)}(\beta, \mu) - \lambda_Q(\beta, \mu)\right)\right]. \tag{47}$$

In the followings, by A^c we denote the compliment of A.
We proved the following theorem [5].

Theorem 5.2. *Let* $\Lambda_{Q^{A^c}_{(K)}}$ *be the intensity measure of* $\Psi^{A^c} = \int \Phi_{A^c}(dx)\gamma_x$. *Then it holds*

$$\Lambda_{Q^{A^c}_{(K)}}(B) \leq f_2(a, b, d, \beta, \mu)$$

where the function $f_2(a, b, d, \beta, \mu)$ *is given by*

$$f_2(a, b, d, \beta, \mu) := (2b)^d \sum_{n \geq 2} \frac{(n-1)\left[1 - \left(1 - \exp\left\{-\frac{(a-b)^2}{n\beta}\right\}\right)^{d/2}\right] z^n}{n(4\pi\beta n)^{d/2}}. \tag{48}$$

Using theorem 5.2 we proved the following proposition [5].

Proposition 5.1. *Let* $f_2(a, b, d, \beta, \mu)$ *be the function on* a, b, d, β, μ *given by (48). Then it holds*

$$E \sup_{\tilde{B} \subseteq B} \left| \Psi^A(\tilde{B}) - \Psi(\tilde{B}) \right| \leq f_2(a, b, d, \beta, \mu). \tag{49}$$

Using the formula (48) we can consider the following problems concerning simulations of the system of positions of the particles.

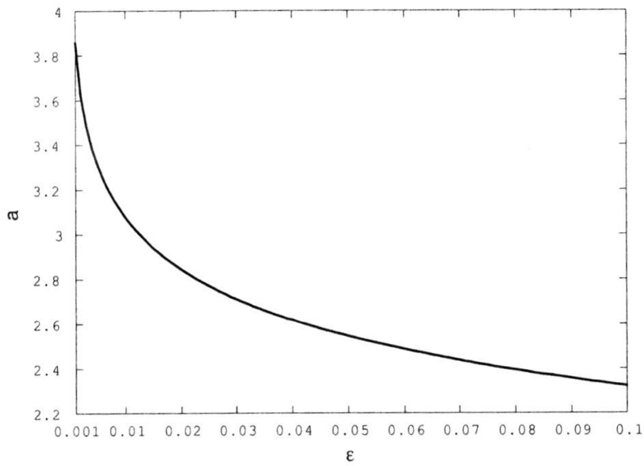

Figure 2. a versus ε for $d = 3$

(1) Fix b, β, λ and find a such that $f_2(a, b, \beta, \mu) = \varepsilon$ for different ε's,
(2) Fix b, λ, ε and find a such that $f_2(a, b, \beta, \mu) = \varepsilon$ for different β's.

Firstly let us consider the special case $d = 3$. Further we fix $\lambda_P^{(1)}(\beta, \mu) = \lambda$. Corresponding to (1) let us fix $b = 1.0, \beta = 0.1, \lambda = 1.0$. Figure 2 shows a versus ε for $d = 3$.

Secondly let us consider the special case $d = 2$ and we also fix $\lambda_P^{(1)}(\beta, \mu) = \lambda$. Then from (24) we can rewrite

$$z = \exp[\mu\beta] = 1 - \exp\{-4\pi\beta\lambda\}. \tag{50}$$

That means we replace f_2 by a function f_3 such that $f_3(a, b, \beta, \lambda) = f_2(a, b, \beta, \mu)$. That is, from (54) f_3 is given as

$$f_3(a, b, \beta, \lambda) = b^2 \sum_{n \geq 2} \frac{n-1}{n^2} \exp\left\{-\frac{(a-b)^2}{n\beta}\right\} \frac{(1 - \exp\{-4\pi\beta\lambda\})^n}{\pi\beta}. \tag{51}$$

Corresponding to (1) let us fix $b = 1.0, \beta = 0.1, \lambda = 1.0$. Figure 3 shows a versus ε for $d = 2$.

Let n be the particle density, m the mass per boson, h Planck's constant, k_B the Boltzmann constant, and ζ the Riemann zeta function $\zeta(3/2) \approx 2.614$. It is known that a transition called Bose Einstein condensate occurs for a uniform tree-dimensional gas consisting of non-interacting particle

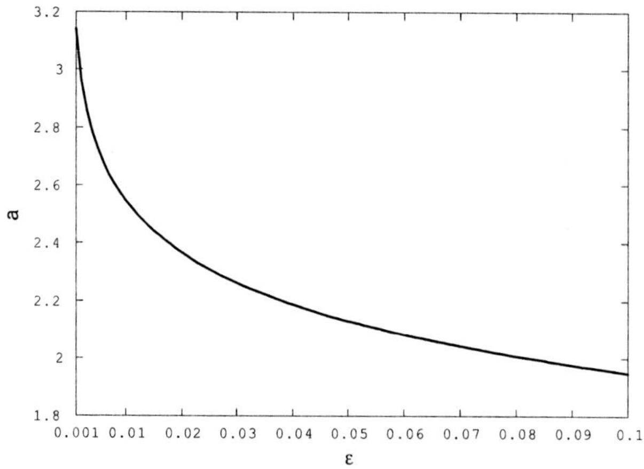

Figure 3. a versus ε for $d = 2$

with no apparent internal degrees of freedom below a critical temperature T_c, which is defined by

$$T_c = \left(\frac{n}{\zeta(3/2)}\right)^{2/3} \frac{h^2}{2\pi m k_B}. \tag{52}$$

In this paper the parameters are set as

$$n = \lambda_P^{(1)}, \quad \frac{h^2}{m} = 1, \quad d = 3. \tag{53}$$

Since the critical inverse temperature β_c is given as

$$\beta_c = \frac{1}{T_c \cdot k_B}, \tag{54}$$

it holds

$$\beta_c = 2\pi \left(\frac{\zeta(3/2)}{\lambda_P^{(1)}}\right)^{\frac{2}{3}}. \tag{55}$$

On the other hand, the intensity $\lambda_P^{(1)}$ of the position distribution P for $d = 3$ is given as

$$\lambda_P^{(1)} = \sum_{m \geq 1} \frac{z^m}{(4\pi\beta m)^{3/2}}, \quad z = \exp[\mu\beta]. \tag{56}$$

In case of $z < 1$ and $d = 3$, we have

$$\lambda_P^{(1)} < \sum_{m \geq 1} \frac{1}{(4\pi\beta m)^{3/2}} = \frac{1}{(4\pi\beta)^{3/2}} \sum_{m \geq 1} \frac{1}{m^{3/2}} = \frac{1}{(4\pi\beta)^{3/2}} \zeta\left(\frac{3}{2}\right). \quad (57)$$

From (57), we obtain

$$\beta < \frac{1}{4\pi} \left(\frac{\zeta(3/2)}{\lambda_P^{(1)}} \right)^{\frac{2}{3}}. \quad (58)$$

From (55), (58), it is clear that it holds

$$\beta_c > \frac{1}{4\pi} \left(\frac{\zeta(3/2)}{\lambda_P^{(1)}} \right)^{\frac{2}{3}}. \quad (59)$$

That means we have considered a temperature higher than the critical temperature.

References

1. O. Bratteli, D.W. Robinson, Operator algebras and quantum statistical mechanics II, Springer-Verlag, New York-Heidelberg- Berlin, 1981.
2. K.-H. Fichtner, On the position distribution of the ideal Bose gas, Math. Nachr. **151**, 59-67, 1991.
3. K.-H. Fichtner, W. Freudenberg, Point processes and the position distribution of infinite boson systems, J. Stat. Phys. **47**, 959-978, 1987.
4. K.-H. Fichtner, W. Freudenberg, Point processes and states of infinite boson systems, Preprint NTZ d.Karl-Marx-Universität Leipzig, 1986.
5. K.-H. Fichtner, K. Inoue and M. Ohya, On the low-temperature behaviour of the infinite-volume ideal Bose gas, TUS preprint.
6. A. Liemant, K. Matthes, A. Wakolbinger, Equilibrium distribution of branching processes, Akademie-Verlag, Berlin, 1988.
7. K. Matthes, J. Kerstan, J. Mecke, Infinitely divisible point processes, Wiley, Chichester-New York, 1978.
8. H. Zessin, The method of moments for random measure, Z. Wahrscheinlichkeitsth. verw. Gebiete **62**, 395-409, 1983.

Quantum Bio-Informatics III
eds. L. Accardi, W. Freudenberg & M. Ohya

ON GENERALIZATION OF QUANTUM MUTUAL ENTROPY BY USING LIFTINGS

S. IRIYAMA AND M. OHYA

Department of Information Sciences, Tokyo University of Science
2641, Yamazaki, Noda City, Chiba, Japan

A lifting is a continuous map from a system to a compound system, and it is useful to describe several physical processes. Ohya and Accardi introduced the notion of it and characterized liftings[1]. In this paper, we define a generalized quantum mutual entropy through liftings satisfying some conditions. In some cases, it satisfies the Shannon inequality and some cases not.

1. Introduction

In this section, we introduce a mathematical foundation of channels and liftings. Let $\mathcal{H}_1$ and $\mathcal{H}_2$ be two separable Hilbert spaces and $\mathcal{B}(\mathcal{H})$ the set of all bounded linear operators on $\mathcal{H}$. For the set $\mathfrak{S}(\mathcal{H})$ of all density operators on $\mathcal{H}$; $\mathfrak{S}(\mathcal{H}) = \{\rho; \rho \geq 0, \mathrm{tr}\rho = 1\}$, we call a map from $\mathfrak{S}(\mathcal{H}_1)$ to $\mathfrak{S}(\mathcal{H}_2)$ a channel. If Λ^* is affine, we call it a linear channel. We denote $\Lambda : \mathcal{B}(\mathcal{H}_2) \to \mathcal{B}(\mathcal{H}_1)$ by a dual map of Λ^*; i.e., $\mathrm{tr}\Lambda^*\rho A = \mathrm{tr}\rho\Lambda A$ for all $\rho \in \mathfrak{S}(\mathcal{H}_1)$ and $A \in \mathcal{B}(\mathcal{H}_2)$. If Λ is a complete positive map (i.e., for all $n \in \mathbb{N}, A_j \in \mathcal{B}(\mathcal{H}_2), B_k \in \mathcal{B}(\mathcal{H}_1)$ holding $\sum_{j,k=1}^{n} B_j^*\Lambda\left(A_j^*A_k\right)B_k \geq 0$), Λ^* is called a complete positive channel. Channel is a mathematical tool to describe various physical processes[12].

Lifting was introduced by Accardi and Ohya in C^*-dynamical systems[1] to integrate various channels and open system dynamics. Here let our C^*- algebras be realized on some separable Hilbert spaces $\mathcal{H}_1$ and $\mathcal{H}_2$. A continious map $\mathcal{E}^*$ from the state space $\mathfrak{S}(\mathcal{H}_1)$ to the compound state space $\mathfrak{S}(\mathcal{H}_1 \otimes \mathcal{H}_2)$ is called a lifting:

$$\mathcal{E}^* : \mathfrak{S}(\mathcal{H}_1) \to \mathfrak{S}(\mathcal{H}_1 \otimes \mathcal{H}_2).$$

If $\mathcal{E}^*$ is affine and its dual is a completely positive map, we call it a CP linear lifting. If it maps pure states into pure states, we call it pure. Remark that a pure lifting sends a mixed state to either a pure or a mixed state. A lifting from $\mathfrak{S}(\mathcal{H}_1)$ to $\mathfrak{S}(\mathcal{H}_1 \otimes \mathcal{H}_2)$ is called *non-demolition* for a state $\rho_1 \in \mathfrak{S}(\mathcal{H}_1)$ if $\mathcal{E}^*$ holds the following condition

$$\mathrm{tr}_2 \mathcal{E}^* \rho_1 = \rho_1$$

The concept of lifting can be used to understand noncommutative probability[1]. Given a state $\rho_1 \in \mathfrak{S}(\mathcal{H}_1)$ and a channel $\Lambda^* : \mathfrak{S}(\mathcal{H}_1) \to \mathfrak{S}(\mathcal{H}_2)$, the following problem is important, that is, to find a standard lifting $\mathcal{E}^* : \mathfrak{S}(\mathcal{H}_1) \to \mathfrak{S}(\mathcal{H}_1 \otimes \mathcal{H}_2)$ such that it describe the correlation between of ρ_1 and $\Lambda^* \rho_1 = \mathrm{tr}_1 \mathcal{E}^* \rho_1$. There are several solutions of this problem in the papers [1,10,12].

Example 1.1. Let $V : \mathcal{H}_1 \to \mathcal{H}_1 \otimes \mathcal{H}_2$ be an isometry(i.e. $V^*V = 1_{\mathcal{H}_1}$). Then a lifting

$$\mathcal{E}^* \rho \equiv V \rho V^*$$

for a density operator on $\mathcal{H}_1$ is called an isometric lifting.

For a normalized coherent vector

$$|\theta\rangle = \sum_k \frac{\theta^k}{\sqrt{k!}} \exp\left(-\frac{1}{2}|\theta|^2\right)|k\rangle$$

where $|k\rangle$ is a photon number state vector, an attenuation process in Figure1 is described by a lifting

$$\mathcal{E}^* \rho \equiv V_{\alpha,\beta} \rho V_{\alpha,\beta}^*$$

where $V_{\alpha,\beta} : \mathcal{H}_1 \to \mathcal{H}_1 \otimes \mathcal{H}_2$ is the map defined as

$$V_{\alpha,\beta}|\theta\rangle = |\alpha\theta\rangle \otimes |\beta\theta\rangle, \quad |\alpha|^2 + |\beta|^2 = 1.$$

and

$$\rho = |\theta\rangle\langle\theta|$$

Then $V_{\alpha,\beta}$ is an isometric in fact

$$V_{\alpha,\beta}^* V_{\alpha,\beta} = I$$

2. Quantum Mutual Entropy

The classical mutual entropy was introduced by Shannon to discuss the transmission of information from an input system to an output system[5], then Kolmogorov[6], Gelfand and Yaglom[2] gave a measure theoretic expression for the mutual entropy by means of the relative entropy defined by Kullback and Leibler. Shannon's expression for mutual entropy was generalized for the finite-dimensional quantum (matrix) case by Holevo[3,4] and

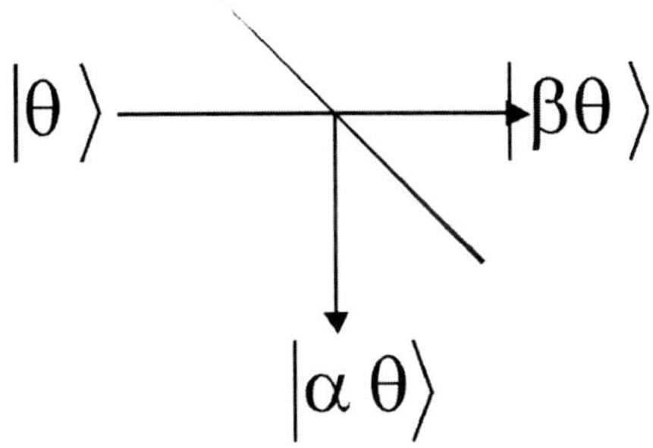

Figure 1. Attenuation process

Levitin[7]. Ohya took the measure theoretic expression of KGY to define quantum mutual entropy by means of quantum relative entropy[10,12].

Let $\mathfrak{S}$ be the set of all states in a certain C^*-algebra (or von Neumann algebra) describing a quantum system, and μ a measure decomposing the state φ into extremal orthogonal states in $\mathfrak{S}$. Ohya's definition of QM entropy is

Definition 2.1. Quantum mutual entropy w.r.t. φ and Λ^* is defined as[10,12]

$$I(\varphi;\Lambda^*) \equiv \sup\left\{\int_{\mathfrak{S}} S^{\text{Araki}}(\Lambda^*\omega,\Lambda^*\varphi)\,d\mu;\varphi=\int_{ex\mathfrak{S}}\omega d\mu\right\}$$

where S^{Araki} is Araki's relative entropy[8].

Definition 2.2. In the case that the C^*-algebra is $B(\mathcal{H})$ and $\mathfrak{S}$ is the set of all density operators, the above definition goes to

$$I(\rho;\Lambda^*) \equiv \sup\left\{\sum_n \lambda_n S^{\text{Umegaki}}(\Lambda^* E_n,\Lambda^*\rho);\rho=\sum_n \lambda_n E_n\right\}$$

where ρ is a density operator, S^{Umegaki} is Umegaki's relative entropy[9] and $\rho=\sum_n \lambda_n E_n$ is the Schatten decomposition. The Schatten decomposition is no always unique, so we take the supremum over all possible decompositions.

Both are completely quantum input and quantum output case. When the input is classical, i.e., the state is a probability distribution, the von Neumann-Schatten decomposition is unique

$$\rho=\sum_n \lambda_n \delta_n$$

and if the channel is written as $\Lambda^* = \Gamma_2^* \Gamma_1^*$ where Γ_1^* is one for quantum coding, i.e., $\Gamma_1^* \delta_n = \rho_n$, then the above mutual entropy generalizes Holevo's one

$$I(\rho; \Lambda^*) = S(\Lambda^* \rho) - \sum_n \lambda_n S(\Lambda^* \rho)$$

Moreover, let $\rho = \sum_k \lambda_k E_k$ be a Schatten decomposition of $\rho \in \mathfrak{S}(\mathcal{H})$ where E_k is one-dimensional projection expressed by $E_k = |x_k\rangle \langle x_k|$ with the eigenvector $|x_k\rangle$ for λ_k and let σ_E be a compound state of ρ and $\Lambda^* \rho$

$$\sigma_E = \sum_k \lambda_k E_k \otimes \Lambda^* E_k$$

Theorem 2.1. [11] *The quantum mutual entropy $I(\rho; \Lambda^*)$ is*

$$I(\rho; \Lambda^*) = \sup_E S(\sigma_E, \sigma_0)$$

where $\sigma_0 = \rho \otimes \Lambda^* \rho$.

Theorem 2.2. [11] *$I(\rho; \Lambda^*)$ satisfies the following property:*

(1) If a channel Λ^* is an i.d., $I(\rho; \Lambda^*)$ is equal to $S(\rho)$
(2) If the system is classical, $I(\rho; \Lambda^*)$ is equal to classical mutual entropy
(3) (Shannon's inequality) $0 \leq I(\rho; \Lambda^*) \leq \min\{S(\rho), S(\Lambda^* \rho)\}$

These entropies are discussed in [14,15].

3. Quantum Mutual Entropy defined by Lifting

In this section, we generalize the quantum mutual entropy by means of liftings. We study under which conditions this generalized quantum mutual entropy (GQM entropy for short) satisfies the Shannon inequality.

Let Λ^* be a complete positive channel from $\mathfrak{S}(\mathcal{H}_1)$ to $\mathfrak{S}(\mathcal{H}_2)$ and $\mathcal{E}^*$ be a lifting from $\mathfrak{S}(\mathcal{H}_1)$ to $\mathfrak{S}(\mathcal{H}_1 \otimes \mathcal{H}_2)$. Here, we take the following two marginal conditions:

(M1) For an input state $\rho \in \mathfrak{S}(\mathcal{H}_1)$, it holds $\mathrm{tr}_2 \mathcal{E}^* \rho = \rho$ (non-demolition property).

(M2) For a given channel Λ^*, $\mathrm{tr}_1 \mathcal{E}^* \rho = \Lambda^* \rho$.

We defined the quantum mutual entropy w.r.t. $\mathcal{E}^*$ as

$$I_L(\rho; \mathcal{E}^*) \equiv S(\mathcal{E}^* \rho, \rho \otimes \Lambda^* \rho)$$

Taking a supremum of $I_L(\rho;\mathcal{E}^*)$ over some set of the liftings $\mathcal{E}^*$, the GQM entropy for a channel Λ^* is defined as

$$I_L(\rho;\Lambda^*) \equiv \sup_{\mathcal{E}^*}\left\{I_L(\rho;\mathcal{E}^*)\,;\mathrm{tr}_2\mathcal{E}^*\rho=\rho,\ \mathrm{tr}_1\mathcal{E}^*\rho=\Lambda^*\rho\right\}$$

Let us check whether $I_L(\rho;\Lambda^*)$ satisfies the Shannon inequality

$$0 \le I_L(\rho;\Lambda^*) \le \min\left\{S(\rho), S(\Lambda^*\rho)\right\}.$$

For a channel Λ^* we consider the following three liftings $\mathcal{E}_i^*$ $(i=1,2,3)$ with M1 and M2:

Case1: $\mathcal{E}_1^*\rho=\sum_k \lambda_k E_k\otimes\Lambda^* E_k$, where $\rho=\sum_k\lambda_k E_k$ is a Schatten decomposition.
Case2: $\mathcal{E}_2^*\rho=\sum_k p_k\rho_k\otimes\Lambda^*\rho_k$ for $\rho=\sum_k p_k\rho_k$. $\sum_k p_k=1, p_k\ge 0$
Case3: $\mathcal{E}_3^*$ is a pure lifting.

Concerning the Shannon inequality, we obtain the results below.

Theorem 3.1. *$\mathcal{E}_1^*$ satisfies the marginal condition M1 and M2 and the Shannon inequality.*

Proof. Since

$$\mathrm{tr}_2\mathcal{E}_1^*\rho=\sum_k\lambda_k E_k=\rho$$

$$\mathrm{tr}_1\mathcal{E}_1^*\rho=\sum_k\lambda_k\Lambda^* E_k=\Lambda^*\rho,$$

it holds M1 and M2. We compute

$$\begin{aligned}S(\mathcal{E}_1^*\rho) &= -\mathrm{tr}\mathcal{E}_1^*\rho\log\mathcal{E}_1^*\rho\\ &= -\mathrm{tr}\sum_k\lambda_k E_k\otimes\Lambda^*E_k\log\sum_k\lambda_kE_k\otimes\Lambda^*E_k\\ &= S(\rho)+\sum_k\lambda_k S(\Lambda^*E_k)\end{aligned}$$

Then we have

$$S(\Lambda^*\rho)-I_L(\rho;\mathcal{E}^*)=S(\rho)-S(\mathcal{E}_1^*\rho)\le 0$$

and

$$
\begin{aligned}
S(\Lambda^*\rho) - S(\mathcal{E}_1^*\rho) &= -\sum_k \lambda_k S(\Lambda^* E_k) + S(\Lambda^*\rho) - S(\rho) \\
&= \sum_k \lambda_k \mathrm{tr}\Lambda^* E_k \log \Lambda^* E_k - \mathrm{tr}\sum_k \lambda_k \Lambda^* E_k \log \Lambda^*\rho - S(\rho) \\
&= \sum_k \lambda_k \mathrm{tr}\Lambda^* E_k (\log \Lambda^* E_k - \log \Lambda^*\rho) - S(\rho) \\
&= \sum_k \lambda_k S(\Lambda^* E_k, \Lambda^*\rho) - S(\rho) \\
&\le \sum_k \lambda_k S(E_k, \rho) - S(\rho) \\
&= \sum_k \lambda_k \mathrm{tr}(E_k \log E_k - E_k \log \rho) - S(\rho) \\
&= 0
\end{aligned}
$$

where our inequality above comes from the monotonicity of the relative entropy. Therefore we obtain

$$0 \le I_L(\rho; \mathcal{E}_1^*) \le \min\{S(\rho), S(\Lambda^*\rho)\} \qquad \square$$

Theorem 3.2. *$\mathcal{E}_2^*$ satisfies M1, M2, and the Shannon inequality.*

Proof. Since $\mathcal{E}_2^*\rho$ is a compound state of ρ and $\Lambda^*\rho$, it holds M1 and M2.

Let

$$\rho_k = \sum_i \lambda_{k,i} E_{k,i}^1$$

and

$$
\begin{aligned}
\Lambda^*\rho_k &= \sum_i \lambda_{k,i} \Lambda^* E_{k,i}^1 \\
&= \sum_{i,j} \lambda_{k,i} \mu_{i,j} E_{i,j}^2
\end{aligned}
$$

be Schatten decompositions of ρ_k and $\Lambda^*\rho_k$ where $E_{k,i}^1$ and $E_{i,j}^2$ are orthogonal one-dimensional projections(pure states). We compute $S(\mathcal{E}_2^*\rho)$

as

$$\begin{aligned}
S\left(\mathcal{E}_2^*\rho\right) &= -\mathrm{tr}\mathcal{E}_2^*\rho\log\mathcal{E}_2^*\rho \\
&= -\mathrm{tr}\sum_{i,j,k} p_k\lambda_{k,i}\lambda_{k,i}\mu_{i,j}\log p_k\lambda_{k,i}\lambda_{k,i}\mu_{i,j}E_{k,i}^1\otimes E_{i,j}^2 \\
&= -\mathrm{tr}\left\{\sum_{i,j,k} p_k\lambda_{k,i}\lambda_{k,i}\mu_{i,j}\log p_k\lambda_{k,i}E_{k,i}^1\otimes I\right. \\
&\quad\left. -\sum_{i,j,k} p_k\lambda_{k,i}\lambda_{k,i}\mu_{i,j}\log\lambda_{k,i}\mu_{i,j}I\otimes E_{i,j}^2\right\} \\
&= S\left(\rho\right)+\sum_k p_k S\left(\Lambda^*\rho_k\right)
\end{aligned}$$

Then we have

$$\begin{aligned}
S\left(\rho\right)-S\left(\mathcal{E}_2^*\rho\right) &= S\left(\rho\right)-S\left(\rho\right)-\sum_k p_k S\left(\Lambda^*\rho_k\right) \\
&= -\sum_k p_k S\left(\Lambda^*\rho_k\right)\leq 0
\end{aligned}$$

and

$$\begin{aligned}
S\left(\Lambda^*\rho\right)-S\left(\mathcal{E}_2^*\rho\right) &= S\left(\Lambda^*\rho\right)-S\left(\rho\right)-\sum_k p_k S\left(\Lambda^*\rho_k\right) \\
&= -\mathrm{tr}\sum_k p_k\Lambda^*\rho_k\log\Lambda^*\rho - S\left(\rho\right)+\mathrm{tr}\sum_k p_k\Lambda^*\rho_k\log\Lambda^*\rho_k \\
&= \mathrm{tr}\sum_k p_k\Lambda^*\rho_k\left(\log\Lambda^*\rho_k-\log\Lambda^*\rho\right)-S\left(\rho\right) \\
&= \sum_k p_k S\left(\Lambda^*\rho_k,\Lambda^*\rho\right)-S\left(\rho\right) \\
&\leq \sum_k p_k S\left(\rho_k,\rho\right)-S\left(\rho\right) \\
&= \sum_k p_k\mathrm{tr}\left(\rho_k\log\rho_k-\rho_k\log\rho\right)-S\left(\rho\right) \\
&= -\sum_k p_k S\left(\rho_k\right) \\
&\leq 0
\end{aligned}$$

Therefore, $\mathcal{E}_2^*$ satisfies the Shannon inequality. □

From the above two theorems, we may conclude that if the lifting is a separable type, that is, $\mathcal{E}^*\rho$ is a separable state, then the Shannon inequality is satisfied. On the contrary, there exists several entangled type pure liftings, that is, $\mathcal{E}^*\rho$ is a pure entangled state that does not satisfy the Shannon inequality. In the rest of our paper, we give two examples of pure lifting $\mathcal{E}_3^*$; one satisfying the Shannon inequality and the other is not.

Example 3.1. In the case that a channel Λ^* is written as

$$\Lambda^*\rho = V\rho V^*$$

where V is a linear operator from $\mathcal{H}_1$ to $\mathcal{H}_2$, the lifting $\mathcal{E}^*\rho = \rho \otimes V\rho V^*$ is pure. Let $\rho = \sum_k \lambda_k E_k$ be a Schatten decomposition of ρ, the lifting

$$\mathcal{E}_3^*\rho = \sum_k \lambda_k E_k \otimes V E_k V^*$$

is also pure. This is same as $\mathcal{E}_1^*$ so that it holds the Shannon's inequality.

Example 3.2. Let $\{e_k^1\}$ and $\{e_k^2\}$ be two CONSs in $\mathcal{H}_1$ and $\mathcal{H}_2$ respectively, such that $\{e_k^1\}$ gives the Schatten decomposition of ρ:

$$\rho = \sum_k \lambda_k E_k$$

$$E_k = \left| e_k^1 \right\rangle \left\langle e_k^1 \right|$$

We can give a pure lifting $\mathcal{E}_3^*$ as

$$\mathcal{E}_3^*\rho = \left(\sum_k \sqrt{\lambda_k} \left| e_k^1 \otimes e_k^2 \right\rangle \right) \left(\sum_l \sqrt{\lambda_l} \left\langle e_l^1 \otimes e_l^2 \right| \right)$$

This pure lifting $\mathcal{E}_3^*$ does not satisfy the Shannon inequality.

Proof. In this case, $\mathcal{E}_3^*\rho$ can be written as

$$\mathcal{E}_3^*\rho = |\xi\rangle \langle\xi|$$

$$|\xi\rangle = \sum_k \sqrt{\lambda_k} \left| e_k^1 \otimes e_k^2 \right\rangle.$$

Since $\mathcal{E}_3^*\rho$ is a pure state, and $S(\mathcal{E}_3^*\rho) = 0$. For a general (i.e., pure or mixed) state ρ, one has

$$S(\rho) = S(\Lambda^*\rho)$$

where

$$\rho = \mathrm{tr}_2 \mathcal{E}_3^* \rho$$

$$\Lambda^* \rho = \mathrm{tr}_1 \mathcal{E}_3^* \rho$$

Then,

$$\begin{aligned} I_L(\rho; \mathcal{E}_3^*) &= -S(\mathcal{E}_3^* \rho) + S(\rho) + S(\Lambda^* \rho) \\ &= 2S(\rho) \end{aligned}$$

which does not satisfy the Shannon inequality. □

References

1. L.Accardi and M.Ohya (1999) Compound channels, transition expectations, and liftings, Appl. Math. Optim., 39, 33-59.
2. I.M.Gelfand and A.M.Yaglom (1959) Calculation of the amount of information about a random function contained in another such function, Amer.Math.Soc.Transl., 12, pp.199-246
3. A.S.Holevo (1973) Some estimates for the amount of information transmittable by a quantum communication channel (in Russian), Problemy Peredachi Informacii, 9, pp.3-11
4. R.S.Ingarden (1976) Quantum information theory, Rep.Math.Phys., 43-73
5. R.S.Ingarden, A.Kossakowski and M.Ohya (1997) Information Dynamics and Open Systems, Kluwer
6. A.N.Kolmogorov (1963) Theory of transmission of information, Amer.Math.Soc.Translation, Ser.2, 33, pp.291–321
7. L.B.Levitin (1969) On the quantum measure of information, in Proceedings of the Fourth Conference on Information Theory, Tashkent, 111–116. English translation in Annales de la Foundation Louis de Broglie 21:3, (1996).
8. H.Araki (1976) Relative entropy for states of von Neumann algebras, Publ. RIMS Kyoto Univ., 11, 809-833
9. H.Umegaki (1962) Conditional expectations in an operator algebra IV(entropy and information), Kodai Math.Sem.Rep., Vol.14, 59-85
10. M.Ohya (1983) On compound state and mutual information in quantum information theory, IEEE Trans. Information Theory, 29, No.5, 770-774.
11. M.Ohya (1984) Entropy Transmission in C*-dynamical systems, J. Math. Anal.Appl., 100, No.1, 222-235.
12. M.Ohya (1989) Some aspects of quantum information theory and their applications to irreversible processes, Rep. Math. Phys., 27, 19-47.
13. M.Ohya and D.Petz (1993) Quantum Entropy and its Use, Springer-Verlag, TMP-series.
14. M.Ohya (2003) New quantum algorithm for studying NP-complete problems, Rep.Math.Phys.,52, No.1,25-33
15. L.Accardi, T.Matsuoka and M.Ohya (2006) Entangled Markov Chains are Indeed Entangled, Infinite Dimensional Analysis Quantum Probability and Related Topics, vol 9, no 3, 379-390.

Quantum Bio-Informatics III
eds. L. Accardi, W. Freudenberg & M. Ohya

A NEW APPROACH TO STROBOSCOPIC TOMOGRAPHY OF OPEN SYSTEMS

ANDRZEJ JAMIOŁKOWSKI
Institute of Physics, Nicholas Copernicus University, 87–100 Toruń, Poland
E-mail: jam@fizyka.umk.pl

The aim of this paper is to describe new possibilities related to some applications of frames and fusion frames in stroboscopic tomography of classical and quantum open systems. We assume that the evolutions of systems in question are given by appropriate semigroups on the sets of states and that information about system states is given by expectation values of fixed observables (physical quantities) at some time instants $t_1, \ldots, t_r$. Natural ways in which frames and fusion frames appear in stroboscopic tomography can be used to reduce the influence of noise and loss of data in determination of states. Overcompletness of frames turns out to be very useful in comparison to the methods based on orthonormal sets of vectors

Keywords: stroboscopic tomography; semigroups; frames; fusion frames.

1. Introduction

According to one of fundamental ideas of quantum mechanics we assume that quantum states can be represented by appropriate operators (density operators) which allow one to statistically predict the outcomes of measurements of all observables. The inverse problem was formulated by Pauli in 1933[1]. The question was: is it possible to determine an unknown quantum state by measuring mean values (or other functions of states) for a given fixed set of observables? One of the basic consequences of the standard quantum mechanics is the statement that determination of an unknown state can be achieved by appropriate measurements only if we have at our disposal a set of identically prepared copies of the system in question. Moreover, to set up a successful approach to the above problem of state reconstruction one has to identify a collection of observables, *a quorum*[2], such that their expectation values contain complete information about the state of the system under consideration. The problems of state determination have gained new relevance in recent years, following the realisation that quantum systems and their evolutions can perform tasks that are impossible with classical devices, such as teleportation, secure communication or dense coding (c.f. e.g.[3]). In this new situation the development of methods to control and measure quantum states becomes an issue of practical interest. It is important to realise that if we identify the quorum of observables, then we have the possibility to determine the expectation values of physical quantities (observables) for which *no* measuring apparatuses are available[4].

The idea of stroboscopic tomography for open quantum systems appears for the first time in the beginning of 1980's (although expressed in different terms[5,6,7]). In particular, using the concept of observability, in [6] and [7] the question of the minimal number of observables $Q_1, \dots, Q_\eta$ for which the quantum systems can be $(Q_1, \dots, Q_\eta)$-observable is discussed.

On the other hand theory of *frames*, which are collections of vectors that provide robust and usually non-unique representations of vectors, have been the subject of research in the last two decades and have been applied in these disciplines, where redundancy played a vital and useful role. However, in some applications it is natural to model and describe considered systems by collections of families of subspaces and to split a large frame system into a set of much smaller frame systems in these subspaces. This has led to the development of a suitable theory based on *fusion frames* (families of subspaces), which provides the framework to model these more complex applications[8,9]. In particular, a sequence of the so-called *Krylov subspaces* of the form

$$K(m, Q) := \mathrm{Span}_{\mathbb{R}^1} \left\{ Q, \mathbb{L}Q, \dots, \mathbb{L}^{m-1}Q \right\}, \tag{1}$$

where Q is a fixed observable and $\mathbb{L}$ is a generator of time evolution of the system in question, constitutes a fusion frame in the Hilbert-Schmidt space $\mathcal{B}_h(\mathcal{H})$ (cf. below Sect 2.) iff[5,6,7]

$$\boxplus_{i=1}^{r} K(\mu, Q_i) = \mathcal{B}_h(\mathcal{H}). \tag{2}$$

In the above equality μ denotes the degree of the minimal polynomial of the superoperator $\mathbb{L}$ and $Q_1, \dots, Q_r$ represent fixed observables. The symbol $\boxplus$ denotes Minkowski sum of subspaces (1) (cf. e.g.[10]).

The organisation of the paper is as follows. Secton 2 presents the main ideas of stroboscopic tomography. In Section 3, we summarise some concepts and results of the theory of frames. We conclude the paper in Section 4 by formulating some applications of the notion of frames and fusion frames to stroboscopic tomography.

2. Stroboscopic tomography of open systems

The time evolution of a quantum system of finitely many degrees of freedom (a qudit) coupled to infinite quantum systems, usually called reservoirs, can be described, under certain limiting conditions, by a one-parameter semigroup of maps (cf. e.g.[5,7]). Let $\mathcal{H}$ be the Hilbert space in question ($\dim \mathcal{H} = d$) and let $\Phi(t) : \mathcal{B}_h(\mathcal{H}) \to \mathcal{B}_h(\mathcal{H})$, $t \in \mathbb{R}^1_+$, be the dynamical semigroup, where $\mathcal{B}_h(\mathcal{H})$ denotes the real vector space of all self-adjoint operators on $\mathcal{H}$. If one introduces the scalar product of operators A, B by the formula $\langle A, B\rangle = \mathrm{Tr}(A^*B)$, then $\mathcal{B}_h(\mathcal{H})$ can be considered as another inner product space, namely so-called Hilbert-Schmidt space. States of the system are described by density operators $\rho \in S(\mathcal{H})$, where

$$S(\mathcal{H}) := \{\rho \in \mathcal{B}_h(\mathcal{H});\ \rho \geq 0, \mathrm{Tr}\,\rho = 1\}\,. \tag{3}$$

Usually one assumes that the family of superoperators $\Phi(t)$ satisfies

(1) $\Phi(t)$ is linear for all $t \in \mathbb{R}^1_+$,
(2) $\Phi(t)$ is trace preserving,
(3) the dual map $\Phi^*(t)$ is completely positive and
(4) $\Phi(t_1) \circ \Phi(t_2) = \Phi(t_1 + t_2)$

for all t_1, t_2 from $\mathbb{R}^1_+$, and if $t \to 0$, then $\lim \Phi(t) = \mathbb{I}$. Since such defined $\Phi(t)$ is a contraction[5], it follows from the Hille-Yosida theorem that there exists a linear superoperator $\mathbb{K} : \mathcal{B}_h(\mathcal{H}) \to \mathcal{B}_h(\mathcal{H})$ such that $\Phi(t) = \exp(t\mathbb{K})$ for all $t \geq 0$ and

$$\frac{\mathrm{d}\rho(t)}{\mathrm{d}t} = \mathbb{K}\rho(t), \tag{4}$$

where $\rho(t) = \Phi(t)\rho(0)$ and $\rho(0) \in S(\mathcal{H})$. The above equation (usually called *master equation*) defines the assignment $\mathbb{R}^1_+ \ni t \mapsto \rho(t) \in S(\mathcal{H})$ provided that we know the initial state of the system $\rho(0) \in S(\mathcal{H})$. The fundamental question of stroboscopic tomography reads: What can we say about the trajectories (initial state $\rho(0)$) if the only information about the system in question is given by the mean values

$$E_i(t_j) = \mathrm{Tr}\,(Q_i\rho(t_j)), \tag{5}$$

of, say r, linearly independent self-adjoint operators $Q_1, \ldots, Q_r$ at some time instants $t_1, \ldots, t_p$, where $r < d^2 - 1$ and $t_j \in [0, T]$ for $j = 1, \ldots, p$, $T > 0$. In other words, the problem of the stroboscopic tomography consists in the reconstruction of the initial state $\rho(0)$, or a current state $\rho(t)$, for any $t \in \mathbb{R}^1_+$, by known expectation values (5). To be more precise we introduce the following description. Suppose that we can prepare a quantum system repeatedly in the same state $\rho(0) \in S(\mathcal{H})$ and we make a series of experiments such that we know the expectation values $E_Q(t) = \mathrm{Tr}\,(Q\rho(t))$ for a fixed set of observables $Q_1, \ldots, Q_r$ at different time instants $t_1 < t_2 < \cdots < t_p$. The basic question is:

Can we find the expectation value of any other operator $Q \in \mathcal{B}_h(\mathcal{H})$, that is any other observable from $\mathcal{B}_h(\mathcal{H})$, knowing the set of measured outcomes of a given set $Q_1, \ldots, Q_r$ at $t_1, \ldots, t_p$, i.e.

$$\begin{array}{c} \quad Q_1 \qquad Q_2 \quad \cdots \quad Q_r \\ \begin{bmatrix} E_1(t_1) & E_2(t_1) & \cdots & E_r(t_1) \\ E_1(t_2) & E_2(t_2) & \cdots & E_r(t_2) \\ \cdots & \cdots & \cdots & \cdots \\ \cdots & \cdots & \cdots & \cdots \\ E_1(t_p) & E_2(t_p) & \cdots & E_r(t_p) \end{bmatrix} \begin{matrix} t_1 \\ t_2 \\ \vdots \\ \\ t_p \end{matrix} \end{array} \tag{6}$$

where $0 \leq t_1 < t_2 < \cdots < t_p \leq T$, for an interval $[0, T]$?

Let us observe that if our operators $Q_1, \ldots, Q_r$ commute, then the above scheme can be used to describe the reconstruction of probability distributions in classical stochastic systems [11].

If the problem under consideration is static, then the state of a d-level open quantum system (a qudit) can be uniquely determined only if $r = d^2 - 1$ expectation values of linearly independent observables are at our disposal. However, if we assume that we know the dynamics of our system i.e. we know the generator $\mathbb{K}$ or $\mathbb{L} := (\mathbb{K})^*$ (in Heisenberg picture) of the time evolution, then we can use the stroboscopic approach based on (6). In general, we use the term "state-tomography" to denote any kind of state-reconstruction method.

With reference to the terminology used in system theory, we introduce the following definition:

Definition 2.1. A d-level open quantum system $\mathbb{S}$ is said to be $(Q_1, \ldots, Q_r)$-reconstructible on the interval $[0, T]$, if for every two trajectories defined by the equation 4 there exists at least one instant $\hat{t} \in [0, T]$ and at least one operator $Q_k \in \{Q_1, \ldots, Q_r\}$ such that

$$\mathrm{Tr}\,(Q_k\rho_1(\hat{t})) \neq \mathrm{Tr}\,(Q_k\rho_2(\hat{t})). \tag{7}$$

The above definition is equivalent to the following statement[5]. A d-level open quantum system $\mathbb{S}$ is $(Q_1, \ldots, Q_r)$-reconstructible on the interval

$[0,T]$, iff there exists at least one set of time instants $0 < t_1 < \cdots < t_p \leq T$ such that the state trajectory can be uniquely determined by the correspondence

$$[0,T] \ni t_j \longmapsto E_i(t_j) = \mathrm{Tr}\left(Q_i \rho(t_j)\right) \tag{8}$$

for $i = 1, \ldots, r$ and $j = 1, \ldots, p$.

In the above definitions we assume that the time evolution of the quantum system is given in fact by a completely positive semigroup of operators. Arguments in favour of completely positive semigroups as a foundation of non-Hamiltonian dynamics as well as the discussion of properties of such semigroups can be found in papers[12,13,14]. Let us observe that in the above definition of reconstructibility we discuss the problem of verifying whether the accessible information about the system in question is sufficient to determine the state uniquely and we do not insist on determining its explicitly.

The dynamical semigroup $\{\Phi(t),\ t \in \mathbb{R}^1_+\}$ is determined by the generator

$$\mathbb{K} : \mathcal{B}_h(\mathcal{H}) \to \mathcal{B}_h(\mathcal{H}) \tag{9}$$

(the Schrödinger picture) and is related to the generator $\mathbb{L}$ of the semigroup in Heisenberg picture by the duality relation

$$\mathrm{Tr}[Q(\mathbb{K}\rho)] = \mathrm{Tr}[(\mathbb{L}Q)\rho]. \tag{10}$$

For a given set of observables $Q_1, \ldots, Q_r$, the subspace spanned on the operators $Q_i, \mathbb{L}Q_i, \ldots, (\mathbb{L})^{m-1}Q_i$, will be denoted by

$$K(m, Q_i) := \mathrm{Span}_{\mathbb{R}^1}\left\{Q_i, \mathbb{L}Q_i, \ldots, \mathbb{L}^{m-1}Q_i\right\}, \tag{11}$$

as the Krylov subspace in the Hilbert-Schmidt space $\mathcal{B}_h(\mathcal{H})$. If $m = \mu$, where μ is the degree of the minimal polynomial of the generator $\mathbb{L} : \mathcal{B}_h(\mathcal{H}) \to \mathcal{B}_h(\mathcal{H})$, then the subspace $K(\mu, Q_i)$ is an invariant subspace of the superoperator $\mathbb{L}$. It can be easily seen that the subspace $K(\mu, Q_i)$ is essentially spanned on all linearly independent operators of the form $(\mathbb{L})^k Q_i$, where $k = 0, 1, \ldots$. Furthermore, it is smallest invariant subspace of the superoperator $\mathbb{L}$ containing Q_i (i.e. the common part of all invariant subspaces of the operator $\mathbb{L}$, containing Q_i).

One can now formulate the necessary and sufficient conditions for the reconstructibility of a d-level open quantum system (c.f.[5,7]).

Let $\mathbb{S}$ be a d-level open quantum system with the evolution governed by an equation (in the Heisenberg picture) of the form $\dot{Q}(t) = \mathbb{L}Q(t)$, where $\mathbb{L}$

is the generator of the dynamical semigroup $\Phi(t) = \exp(t\mathbb{L})$. Suppose that by performing measurements, the correspondence

$$[0,T] \ni t_j \longmapsto E_i(t_j) = \mathrm{Tr}\,(\rho(0)Q_i(t_j)) \tag{12}$$

can be established for fixed observables $Q_1, \ldots, Q_r$ at selected time instants $t_1, \ldots, t_p$. The system $\mathbb{S}$ is $(Q_1, \ldots, Q_r)$-reconstructible if and only if

$$\boxplus_{i=1}^{r} K(\mu, Q_i) = \mathcal{B}_h(\mathcal{H}), \tag{13}$$

where $\boxplus$ denotes the so-called Minkowski sum of subspaces $K(\mu, Q_i)$ (cf.[10]). The above equality has been obtained by using the polynomial representation of the semigroup $\Phi(t)$[14].

The question of an obvious physical interest is to find the minimal number of observables $Q_1, \ldots, Q_\eta$ for which a d-level quantum system $\mathbb{S}$ with a fixed generator $\mathbb{L}$ can be $(Q_1, \ldots, Q_\eta)$-reconstructible. It can be shown that for an d-level generator there always exists a set of observables $Q_1, \ldots, Q_\eta$, where

$$\eta := \max_{\lambda \in \sigma(\mathbb{L})} \{\dim \mathrm{Ker}(\lambda\mathbb{I} - \mathbb{L})\}, \tag{14}$$

such that the system is $(Q_1, \ldots, Q_\eta)$-reconstructible[14]. Moreover, if we have another set of observables $\widetilde{Q}_1, \ldots, \widetilde{Q}_{\widetilde{\eta}}$ such that the system is $(\widetilde{Q}_1, \ldots, \widetilde{Q}_{\widetilde{\eta}})$-reconstructible, then $\widetilde{\eta} \geq \eta$. The number η defined by (14) is called the *index of cyclicity* of the quantum open system $\mathbb{S}$[5]. The symbol $\sigma(\mathbb{L})$ in (14) denotes the spectrum of the superoperator $\mathbb{L}$.

3. Frames and fusion frames

In this paper, we consider only finite-dimensional frames the reason being their direct link to the practical applications in stroboscopic tomography of qudits. Since the same questions can be raised for infinite-dimensional frames, we state the problems in the setting of abstract frames.

Consider a Hilbert space $\mathcal{H}$ with scalar product $(\ ,\ )$. A finite or countable collection of vectors $\mathcal{F} = \{f_i : i \in \mathcal{I}\}$, $f_i \in \mathcal{H}$, is called *a frame* if there are two positive constants $\alpha, \beta > 0$ such that for every vector $x \in \mathcal{H}$

$$\alpha \parallel x \parallel^2 \leq \sum_{i \in \mathcal{I}} |(f_i, x)|^2 \leq \beta \parallel x \parallel^2 . \tag{15}$$

The frame is *tight* when the constants α and β can be chosen equal, $\alpha = \beta$. For $\alpha = \beta = 1$, $\mathcal{F}$ is called a *Parseval frame*. The numbers (f_i, x) are called *frame coefficients*.

For a given frame $\mathcal{F}$ we can associate the *analysis* Θ and *synthesis* Θ^* operators. They are defined by the equality

$$\Theta(x) = \sum_{i \in \mathcal{I}} (f_i, x)\underline{e}_i, \tag{16}$$

where $\underline{e}_i$ stands for standard basis in $\mathbb{C}^m$. Composing Θ with its adjoint Θ^*, we obtain the frame operator

$$F : \mathcal{H} \to \mathcal{H}, \tag{17}$$

$$Fx = \Theta^*\Theta x = \sum_{i=1}^{m} (f_i, x) f_i. \tag{18}$$

It is not difficult to see that the collection of vectors $\{f_i\}_{i=1}^m$ constitutes a frame for the vector space $G := \text{span}\{f_i\}_{i=1}^m$. On the other hand a family of elements $\{f_i\}_{i=1}^m$ in $\mathcal{H}$ is a frame for $\mathcal{H}$ if and only if $\text{span}\{f_i\}_{i=1}^m = \mathcal{H}$. This means that a frame might contain more elements than necessary for it to be a basis. In particular, if $\{f_i\}_{i=1}^m$ is a frame for $\mathcal{H}$ and $\{g_i\}_{i=1}^n$ is an arbitrary finite collection of elements in $\mathcal{H}$, then the set $\{f_1, \ldots, f_m, g_1, \ldots, g_n\}$ is also a frame for $\mathcal{H}$.

Generally speaking, frame theory is the study of how $\{f_i\}_{i=1}^m$ should be chosen in order to guarantee that the frame operator $\Theta^*\Theta$ is well-conditioned. In particular, $\{f_i\}_{i=1}^m$ is a frame for $\mathcal{H}$ if there exist frame bounds α, β such that

$$\alpha\,\mathbb{I} \leq \Theta^*\Theta \leq \beta\,\mathbb{I}, \tag{19}$$

and is a tight frame iff $\Theta^*\Theta = \alpha\mathbb{I}$. It is an obvious observation that $F = \Theta^*\Theta$ is self-adjoint and invertible operator.

Fusion frame theory generalises these concepts. Let us observe that when each f_i in (18) is of unit norm, the summands of the frame operator, namely the operators $x \mapsto (f_i, x)f_i$, are rank-one projections. Fusion frame theory is the study of sums of projectors whose ranks are permitted to be greater than one. According to general definition, a collection of projections $\{\mathbb{P}_k\}_{k=1}^M$ is a fusion frame for $\mathcal{H}$ iff

$$\alpha\,\mathbb{I} \leq \sum_{k=1}^{M} \mathbb{P}_k \leq \beta\,\mathbb{I}. \tag{20}$$

Of course, the above definition can be formulated also in terms of subspaces which are images of projectors $\mathbb{P}_k$ for $k = 1, \ldots, M$.

4. Frames in stroboscopic tomography

The concepts of frames and fusion frames appear in stroboscopic tomography in natural way. The conclusion is based on the so-called polynomial representations of semigroups which describe evolutions of open systems. The possibility to represent the semigroup $\Phi(t) = \exp(t\,\mathbb{L})$ in the form

$$\Phi(t) = \sum_{k=0}^{\mu-1} \alpha_k(t)\mathbb{L}^k, \tag{21}$$

where μ stands for the degree of the minimal polynomial of the superoperator $\mathbb{L}$ and $\alpha_k(t)$, $k = 0\ldots,\mu-1$, denote some functions of the eigenvalues of $\mathbb{L}$ gives as the equality (13) as a necessary and sufficient condition for stroboscopic tomography. On the other hand, this equality means that the Krylov subspaces $K(\mu, Q_i)$, $i = 1,\ldots,r$, constitute a fusion frame in the Hilbert-Schmidt space $\mathcal{B}_h(\mathcal{H})$ of all observables. Moreover, this also means that the collection of vectors

$$f_{jk} := \mathbb{L}^k Q_j, \tag{22}$$

for $j = 1,\ldots,r$ and $k = 0,1,\ldots,\mu-1$, constitute a frame in $\mathcal{B}_h(\mathcal{H})$ iff our system in question is $(Q_1,\ldots,Q_r)$-reconstructible. In this case every element X of the space $\mathcal{B}_h(\mathcal{H})$ can be represented as

$$X = \sum_{j,k}\langle F^{-1}f_{jk}, X\rangle f_{jk} = \sum_{j,k}\langle f_{jk}, X\rangle F^{-1}f_{jk}, \tag{23}$$

where F denotes the frame operator of the family of vectors (22). One can say even more. If $X \in \mathcal{B}_h(\mathcal{H})$ also has representation $X = \sum_{j,k} c_{jk}f_{jk}$ for some scalar coefficients c_{jk}, $j = 1,\ldots,r$ and $k = 0,1,\ldots,\mu-1$, then

$$\sum_{jk}|c_{jk}|^2 = \sum_{jk}|\langle F^{-1}f_{jk}, X\rangle|^2 + \sum_{j,k}|c_{jk} - \langle F^{-1}f_{jk}, X\rangle|^2. \tag{24}$$

It is obvious that every frame in finite-dimensional space contains a subfamily that is a basis. As a conclusion we can say that if $\{f_{jk}\}$ is a frame but not a basis, then there exists a set of scalars $\{b_{jk}\}$ such that $\sum_{j,k} b_{jk}f_{jk} = 0$. Therefore, any fixed element X of $\mathcal{B}_h(\mathcal{H})$ can also be represented as

$$X = \sum_{j,k}\left(\langle F^{-1}f_{jk}, X\rangle + d_{jk}\right) f_{jk}. \tag{25}$$

The above equality means that every $X \in \mathcal{B}_h(\mathcal{H})$ has many representations as superpositions of elements from the set (22). But according to equality

(24) among all scalar coefficients $\{c_{jk}\}$ for which

$$X = \sum_{j,k} c_{jk} f_{jk}, \tag{26}$$

the sequence $\{\langle F^{-1} f_{jk}, X\rangle\}$ has minimal euclidean norm. This is the main new conclusion connected with the method of stroboscopic tomography.

References

1. W. Pauli, in *Handbook der Physik*, edited by H. Geiger and K. Scheel, (Springer, Berlin, 1933).
2. W. Band, J. L. Park, *Am. J. Phys.* **47**, 188 (1979).
3. M. A. Nielsen, I. Chuang, *Quantum Computation and Quantum Information*, (Cambridge Univ. Press., 2000).
   V. Vedral, *Introduction to Quantum Information Science*, (Oxford Univ. Press., 2006).
4. S. Weigert, in *New Insight in Quantum Mechanics*, Eds. H.-D. Doebner, et al., (World Scientific, Singapore, 2000).
5. A. Jamiołkowski, *On Observability of Classical and Quantum Stochastic Systems*, (N. Copernicus Univ. Press, Toruń, 1982), pp. 1–108, (in Polish).
6. A. Jamiołkowski, *Internat. J. Theoret. Phys.* **22**, 369 (1983).
7. A. Jamiołkowski, in *Frequency Domain and State Space Methods for Linear Systems*, Eds. C. I. Byrnes and A. Lindquist, (Elsevier, 1986), p. 347.
8. P. G. Casazza and G. Kutyniok, *Frames of subspaces*, Contemp. Math. **345**, 87, (2004).
9. P. G. Casazza et al., *Appl. Comput. Harmon. Anal.* **25**, 114 (2008).
10. R. T. Farouki et al., *Geometriae Dedicata* **85**, 283 (2001).
11. A. S. Holevo, *Statistical Structure of Quantum Theory* (Springer, Berlin, 2001).
12. K. Kraus, *Ann. Phys.* **64**, 119 (1971).
13. V. Gorini et al., *J. Math. Phys.* **17**, 149 (1976).
14. A. Jamiołkowski, *Rep.Math. Phys.* **46**, 469 (2000).

Quantum Bio-Informatics III
eds. L. Accardi, W. Freudenberg & M. Ohya

AN INTRODUCTION TO FRAMES AND THEIR APPLICATIONS TO QUANTUM OPTICS

ANDRZEJ JAMIOŁKOWSKI

Institute of Physics, Nicholas Copernicus University, 87–100 Toruń, Poland
E-mail: jam@fizyka.umk.pl

The purpose of this paper is to show how useful are the notions of frame and fusion frame in the problems of trajectory reconstruction in open quantum systems. In particular, we discuss some problems of so-called stroboscopic quantum tomography. Frames are redundant sets of vectors in a Hilbert space, which assure a natural representation of each vector in the space, but may have infinitely many different representations for any given vector. For a given quantum system represented in a Hilbert space $\mathcal{H}$, the question of minimal number η of observables $Q_1, \ldots, Q_\eta$, whose expectation values at some instants $t_1, \ldots, t_s$ determine the statistical state of the system is discussed. We assume that the time evolution of the system in question is governed by a semigroup of linear transformations with generator $\mathbb{L}$.

Keywords: frames; fusion frames; stroboscopic tomography; semigroups.

1. Introduction

It is a fundamental observation concerning quantum objects that pure states of all microscopic systems satisfy the superposition principle, i.e., if $|\psi_1\rangle$ and $|\psi_2\rangle$ are states, then their linear combination with complex coefficients $\alpha_1|\psi_1\rangle + \alpha_2|\psi_2\rangle$ describes a new state of the system. This observation led to postulate that the states of any quantum system must be represented by elements of a complex linear space. A beautiful discussion of this issue is given in [1].

If we want to calculate the probability of finding a system in a final state $|\psi\rangle$, assuming that it was originally prepared in an initial state $|\varphi\rangle$, we have to equip our linear space with a scalar product $\langle\varphi|\psi\rangle$. Therefore, one usually considers a finite- or infinite dimensional Hilbert space [1,2] as the space of states of a given quantum system. We also postulate that quantum observables are represented by linear hermitean (self-adjoint) operators on $\mathcal{H}$. The above representation of quantum systems was developed for the first time by J. von Neumann in his famous book *"Mathematische Grundlagen der Quantum Mechanik"* in 1932. This work provided a rigorous mathematical framework for the theory of quantum (microscopic) objects. According to von Neumann approach, quantum systems are described in terms of observables that act as self-adjoint operators on the Hilbert space of the system and elements of this space represent pure states of the system in question. In order to be able to describe systems which are a mixture of states rather than a pure state, one has to introduce the concept of density operators, that is, normalized, positive semidefinite operators acting on $\mathcal{H}$. The dynamics of systems are assumed to be determined by self-adjoint Hamiltonians [3].

During the two decades — 1960s and 1970s — the above theory was generalized. The so-called theory of open quantum systems (systems interacting with their environments) was established and the main sources of inspiration for it were the quantum optics and the theory of lasers. This led to the generalization of states (now density operators are considered as natural representation of quantum states) and to generalized description of time evolution of quantum states. At that time the concepts of so-called quantum master equations which preserve positive semi-definiteness of density operators and the idea of quantum communication channel were born, cf. e.g. [4,5,6,7,8,9,10] On the mathematical level, this approach initiated the study of semigroups of completely positive maps and their generators.

Summing up, as the fundamental objects in modern quantum theory one considers the set of states

$$S(\mathcal{H}) := \{\rho : \mathcal{H} \to \mathcal{H}\ ;\ \rho \geq 0, \operatorname{Tr} \rho = 1\}\,, \tag{1}$$

and the set of bounded hermitean (self-adjoint) operators

$$\mathcal{B}_h(\mathcal{H}) := \{Q : \mathcal{H} \to \mathcal{H}\ ;\ Q = Q^*\}\,. \tag{2}$$

Time evolutions of systems are governed by linear master equations of the form (in the so-called Schrödinger picture)

$$\frac{\mathrm{d}\rho(t)}{\mathrm{d}t} = \mathbb{K}\rho(t), \tag{3}$$

or in the dual form (in the so-called Heisenberg picture)

$$\frac{\mathrm{d}Q(t)}{\mathrm{d}t} = \mathbb{L}Q(t), \tag{4}$$

where superoperators $\mathbb{K}$ and $\mathbb{L}$ act on operators from the sets $S(\mathcal{H})$ and $\mathcal{B}_h(\mathcal{H})$, respectively. They represent dual forms of the same physical idea.

Both sets $S(\mathcal{H})$ and $\mathcal{B}_h(\mathcal{H})$ can be considered as subsets of the vector space $\mathcal{B}(\mathcal{H})$ of all bounded linear operators on $\mathcal{H}$ and they can be treated as scenes on which problems of quantum mechanical systems should be discussed.

In this paper we will consider finite-dimensional Hilbert spaces, so in fact $\mathcal{B}(\mathcal{H})$ denotes the set of all linear operators on $\mathcal{H}$. If we introduce in $\mathcal{B}(\mathcal{H})$ the scalar product by the equality

$$\langle A, B\rangle := \mathrm{Tr}\,(A^* B), \tag{5}$$

then $\mathcal{B}_h(\mathcal{H})$ can be regarded as yet another inner product space, namely so-called *Hilbert-Schmidt space.* It is not difficult to see that $\mathcal{B}_h(\mathcal{H})$ with scalar product defined by (5) is a *real* vector space and $\dim \mathcal{B}_h(\mathcal{H}) = d^2$.

One of the fundamental results of the modern quantum mechanics is the statement that determination of an unknown state (and in consequence the trajectory defined by (3) or (4)) can be achieved by appropriate measurements only if we have at our disposal a set of identically prepared copies of the system in question. Moreover, to be able to solve the above problem of the *state reconstruction* one has to identify a collection of observables, a *quorum* [11], such that their expectation values contain complete information about the state of the system. The problem of state determination (or equivalently identification of quantum trajectories for a given generator $\mathbb{K}$) has gained new relevance in recent years, following the observation that quantum systems can perform tasks that are impossible when we use only classical resources.

The question, how to construct a quorum of meaningful observables for a given quantum system (that is, for a given generator $\mathbb{K}$) is quite fundamental. Usually, one can identify only a small number of observables $Q_1, \ldots, Q_n$, $n \ll d^2$ with clear physical meaning, so it is natural to ask about results of the measurements of these observables at different moments $t_1, \ldots, t_s$. This idea of stroboscopic tomography for quantum systems was born in the beginning of 1980s [12].

On the other hand, theory of frames, which are redundant collections of vectors that provide robust but, in general, non-unique representations of elements of the vector space, has been the subject of research in the last two decades. It appears that the concept of frame, and its generalization to collection of subspaces (fusion frame), in the space of observables (2) is very useful in quantum stroboscopic tomography. In the next section we introduce the main objects of the frame theory and fusion frames.

2. Frames and Krylov subspaces

Frames were first introduced by Dufflin and Schaffer in 1952 as a natural notion that appeared during their research in nonharmonic Fourier analysis. After more than three decades Daubechies, Grossmann and Meyer [13] in 1986 initiated the use of frame theory in description of signal processing. Today, frame theory plays an important role in dozens of applied areas, cf. e.g. [14,15].

Let us consider a Hilbert space $\mathcal{H}$ with scalar product $\langle\ ,\ \rangle$ which is linear in the second argument. A finite collection of vectors $\mathcal{F} = \{f_i : i \in \mathcal{I}\}$, $f_i \in \mathcal{H}$, is called *a frame* if there are two positive constants $\alpha, \beta > 0$ such that for every vector $x \in \mathcal{H}$

$$\alpha \parallel x \parallel^2 \leq \sum_{i\in\mathcal{I}} |\langle f_i, x\rangle|^2 \leq \beta \parallel x \parallel^2 . \tag{6}$$

The frame is *tight* when the constants α and β are equal, $\alpha = \beta$. If $\alpha = \beta = 1$, then $\mathcal{F}$ is called a *Parseval frame.* The numbers $\langle f_i, x\rangle$ are called *frame coefficients.*

For a given frame $\mathcal{F}$ we can introduce the *analysis* Θ and *synthesis* Θ^* operators. They are defined by the equality

$$\Theta(x) = \sum_{i\in\mathcal{I}} \langle f_i, x\rangle |e_i\rangle, \tag{7}$$

where $|e_i\rangle$ stands for standard basis in $\mathbb{C}^m$ (we will consider only finite dimensional frames, so that $\mathcal{I} = \{1, \ldots, m\}$). Composing Θ with its adjoint operator Θ^*, we obtain the frame operator

$$F : \mathcal{H} \to \mathcal{H}, \qquad \text{defined by} \tag{8}$$

$$Fx := \Theta^*\Theta x = \sum_{i=1}^{m} \langle f_i, x\rangle |f_i\rangle. \tag{9}$$

It is not difficult to see that the collection of vectors $\{|f_i\rangle\}_{i=1}^m$ constitutes a frame for the vector space $G := \text{span}\{|f_i\rangle\}_{i=1}^m$. On the other hand a family of elements $\{|f_i\rangle\}_{i=1}^m$ in $\mathcal{H}$ is a frame for $\mathcal{H}$ if and only if $\text{span}\{|f_i\rangle\}_{i=1}^m = \mathcal{H}$. This means that a frame might contain more elements than necessary for it to be a basis. In particular, if $\{|f_i\rangle\}_{i=1}^m$ is a frame for $\mathcal{H}$ and $\{|g_i\rangle\}_{i=1}^n$ is an arbitrary finite collection of elements in $\mathcal{H}$, then the set $\{|f_1\rangle, \ldots, |f_m\rangle, |g_1\rangle, \ldots, |g_n\rangle\}$ is also a frame for $\mathcal{H}$.

Generally speaking, frame theory is the study of how $\{|f_i\rangle\}_{i=1}^m$ should be chosen in order to guarantee that the frame operator $\Theta^*\Theta$ is well-conditioned. In particular, $\{|f_i\rangle\}_{i=1}^m$ is a frame for $\mathcal{H}$ if there exist frame bounds α, β such that

$$\alpha\,\mathbb{I} \leq \Theta^*\Theta \leq \beta\,\mathbb{I}, \tag{10}$$

and is a tight frame iff $\Theta^*\Theta = \alpha\mathbb{I}$. It is an obvious observation that $F = \Theta^*\Theta$ is self-adjoint and invertible operator.

Fusion frame theory (theory of frames of subspaces) is an emerging mathematical theory that provides a natural framework for performing a distributed data processing in many fields [16,17]. In particular, one can apply this method in quantum state tomography. The notion of fusion frame was introduced in [16] and further developed by [17]. A fusion frame in a Hilbert space $\mathcal{H} \cong \mathbb{C}^d$ is a finite collection of subspaces $\{W_i\}_{i=1}^m$ of $\mathcal{H}$, such that there exist constants $0 < \alpha < \beta < \infty$ satisfying for any $|\varphi\rangle \in \mathcal{H}$ the two inequalities

$$\alpha \parallel |\varphi\rangle \parallel^2 \leq \sum_{i=1}^{m} \parallel \mathbb{P}_i|\varphi\rangle \parallel^2 \leq \beta \parallel |\varphi\rangle \parallel^2, \tag{11}$$

where $\mathbb{P}_i$ denotes the orthogonal projection on W_i. In other words, a collection $\{W_i\}_{i=1}^m$ is a fusion frame if and only if

$$\alpha\,\mathbb{I} \leq \sum_{i=1}^{m} \mathbb{P}_i \leq \beta\,\mathbb{I}. \tag{12}$$

The constants α and β are called fusion frame bounds. An important class of fusion frames is the class of tight fusion frames, for which $\alpha = \beta$. This equality leads to the operator relation $\sum_{i=1}^m \mathbb{P}_i = \alpha\mathbb{I}$.

Now, let us recall that for a given operator $M : \mathcal{H} \to \mathcal{H}$ and a given fixed nonzero vector $|h\rangle \in \mathcal{H}$ one introduces the kth Krylov subspace of $\mathcal{H}$ by the equality

$$\mathcal{K}_k(M, h) := \text{span}\{|h\rangle, M|h\rangle, \ldots, M^{k-1}|h\rangle\}. \tag{13}$$

The above definition can also be written as

$$\mathcal{K}_k(M, h) := \text{span}\{p(M)|h\rangle;\ \deg(p) \leq k-1\}, \tag{14}$$

where p denotes an arbitrary polynomial and $\deg(p)$ is its degree. It is an obvious observation that the size of a Krylov subspace depends on both M and $|h\rangle$. Note also that there exists such k that $\mathcal{K}_k(M, h) = \mathcal{K}_{k+1}(M, h)$ and this k is the degree of the minimal polynomial of M with respect to $|h\rangle$. If by μ_M we denote the minimal polynomial of the operator M then the minimal polynomial of M with respect to any vector $|h\rangle \in \mathcal{H}$ divides μ_M.

For a given operator $M : \mathcal{H} \to \mathcal{H}$ Krylov subspaces generated by a fixed set of vectors $|h_1\rangle, \ldots, |h_r\rangle$ constitute a fusion frame in $\mathcal{H}$ if the following equality is satisfied

$$\boxplus_{i=1}^{r} \mathcal{K}_\mu(M, h_i) = \mathcal{H}, \tag{15}$$

where $\boxplus$ denotes the so-called Minkowski sum [18] of subsets of $\mathcal{H}$ and $\mu := \deg \mu_M$. It is important, that we have to use the notion of Minkowski sum of subspaces instead of direct sum. This fact follows from the observation that even if the vectors $|h_1\rangle, \ldots, |h_r\rangle$ are linearly independent, then the Krylov subspaces $\mathcal{K}_\mu(M, h_i)$, $i = 1, \ldots, r$ can partially overlap. It can be easily seen that every Krylov subspace $\mathcal{K}_\mu(M, h)$ is essentially spanned on all vectors of the form $M^k|h\rangle$, where $k = 0, 1, \ldots,$ and it is the smallest invariant subspace of the operator M containing $|h\rangle$.

3. Fusion frames in quantum optics

The idea of Krylov subspaces as a possible fusion frame in a given Hilbert space can be immediately applied to some systems considered in quantum optics or more generally in the theory of open quantum systems. Using polynomial representations of semigroups generated by operators $\mathbb{K}$ and $\mathbb{L}$ (cf. *A New Approach to Stroboscopic Tomography of Open Systems*, A. Jamiołkowski, this volume) one notices that observables $Q_1, \ldots, Q_r$ generate for a given $\mathbb{L}$ Krylov subspaces in the space $\mathcal{B}_h(\mathcal{H})$. These subspaces constitute a frame of subspaces (fusion frame) in $\mathcal{B}_h(\mathcal{H})$ if we have the equality

$$\boxplus_{i=1}^{r} \mathcal{K}_\mu(\mathbb{L}, Q_i) = \mathcal{B}_h(\mathcal{H}). \tag{16}$$

The above relation allows us to identify states of quantum systems (i.e. to conduct the process of quantum stroboscopic tomography [19]).

The question of an obvious physical interest is to find the minimal number of observables $Q_1, \ldots, Q_\eta$ for which a d-level quantum system $\mathbb{S}$ with a fixed generator $\mathbb{L}$ can be $(Q_1, \ldots, Q_\eta)$-reconstructible. It can be shown [19] that for an d-level generator there always exists a set of observables $Q_1, \ldots, Q_\eta$, where

$$\eta := \max_{\lambda \in \sigma(\mathbb{L})} \{\dim \operatorname{Ker}(\lambda \mathbb{I} - \mathbb{L})\}, \tag{17}$$

such that the system is $(Q_1, \ldots, Q_\eta)$-reconstructible. Moreover, if we have another set of observables $\widetilde{Q}_1, \ldots, \widetilde{Q}_{\widetilde{\eta}}$ such that the system is $(\widetilde{Q}_1, \ldots, \widetilde{Q}_{\widetilde{\eta}})$-reconstructible, then $\widetilde{\eta} \geq \eta$. The number η defined by (17) is called the *index of cyclicity* of the quantum open system $\mathbb{S}$. The symbol $\sigma(\mathbb{L})$ in (17) denotes the spectrum of the superoperator $\mathbb{L}$.

In order to describe processes connected with the dissipation of energy in quantum optics, one considers the time evolution of d-level quantum systems with generator given by (cf. e.g. [20])

$$\mathbb{L}Q = \frac{1}{2}\{D[Q,D] + [D,Q]D\}, \tag{18}$$

where D denotes a hermitean operator on $\mathcal{H}$ with eigenvalues $\lambda_1, \ldots, \lambda_m$.

Now, the minimal number of observables $Q_1, \ldots, Q_\eta$ for d-level quantum system with generator (18) such that the system in question can be $(Q_1, \ldots, Q_\eta)$-reconstructible, i.e. thc value of the index of cyclicity of the generator $\mathbb{L}$, is given by the following formulae

$$\eta = \max\{\kappa, \gamma_1, \ldots, \gamma_r\} \tag{19}$$

where

$$\kappa := n_1^2 + n_2^2 + \cdots + n_m^2, \tag{20}$$

$$\gamma_k := 2\sum_{i=1}^{m-k} n_i n_{i+k}, \tag{21}$$

and r depends on the value of m, namely $r = (m-1)/2$ if m is odd and $r = (m-2)/2$ if m is even. In (20) and (21) m denotes the number of different eigenvalues of the operator D and $n_1, \ldots, n_m$ are their algebraic multiplicities (cf. [19] for the proof).

In conclusion, one can say that the Krylov subspaces $\mathcal{K}_\mu(\mathbb{L}, Q_i)$ in the space $\mathcal{B}_h(\mathcal{H})$ generated by the superoperator $\mathbb{L}$ can be used in an effective way for procedures of stroboscopic tomography if they constitute appropriate fusion frames in this space. Some details of this approach will be given elsewhere [21].

References

1. R. Feynman, *The Feynman Lectures on Physics*, Vol. 3, (Addison-Wesley, 1963).
2. M. A. Nielsen, I. Chuang, *Quantum Computation and Quantum Information*, (Cambridge Univ. Press., 2000).
3. J. von Neumann, *Mathematische Grundlagen der Quantenmechanik*, (Springer, Berlin, 1932).
4. K. Kraus, *Ann. Phys.* **64**, 119 (1971).
5. A. Kossakowski, *Rep.Math. Phys.* **3**, 247 (1972).
6. A. Jamiołkowski, *Rep.Math. Phys.* **3**, 275 (1972).

7. A. Jamiołkowski, *Rep.Math. Phys.* **5**, 415 (1975).
8. V. Gorini et al., *J. Math. Phys.* **17**, 149 (1976).
9. G. Lindblad, *Comm. Math. Phys.* **48**, 119 (1976).
10. M.Ohya, *J. Math. Anal. Appl.* **84**, 318 (1981).
11. W. Band, J. L. Park, *Am. J. Phys.* **47**, 188 (1979).
12. A. Jamiołkowski, *Internat. J. Theoret. Phys.* **22**, 369 (1983).
13. I. Daubechies, A. Grossmann and Y. Meyer, *J. Math. Phys.* **27**, 1271 (1986).
14. J. Kovačević, A. Chebira, *An Introduction to Frames*, NOW, 2008.
15. Ch. Heil, (Ed.), *Harmonic Analysis and Applications*, Birkhäuser, 2006.
16. P. G. Casazza and G. Kutyniok, *Frames of subspaces*, Contemp. Math. **345**, 87, (2004).
17. P. G. Casazza et al., *Appl. Comput. Harmon. Anal.* **25**, 114 (2008).
18. R. T. Farouki et al., *Geometriae Dedicata* **85**, 283 (2001).
19. A. Jamiołkowski, *Open Systems and Inf. Dyn.* **11**, 63 (2004).
20. A. S. Holevo, *Statistical Structure of Quantum Theory* (Springer, Berlin, 2001).
21. A. Jamiołkowski, in preparation.

Quantum Bio-Informatics III
eds. L. Accardi, W. Freudenberg & M. Ohya
 (pp. 155–166)

MEMORY IN A NONLOCALLY DAMPED OSCILLATOR

D. CHRUŚCIŃSKI AND J. JURKOWSKI
Institute of Physics, Nicolaus Copernicus University,
Grudziądzka 5/7, 87–100 Toruń, Poland

We analyze the new equation of motion for the damped oscillator. It differs from the standard one by a damping term which is nonlocal in time and hence it gives rise to a system with memory. Both classical and quantum analysis is performed. The characteristic feature of this nonlocal system is that it breaks local composition low for the classical Hamiltonian dynamics and the corresponding quantum propagator.

Keywords: damped harmonic oscillator, dissipative systems, quantum propagator

1. Introduction

The damped harmonic oscillator is one of the simplest quantum systems displaying the dissipation of energy. Moreover, it is of great physical importance and has found many applications especially in quantum optics. For example it plays a central role in the quantum theory of lasers and masers [1,2,3]. The standard Hilbert space formulation of Quantum Mechanics gives rise to the unitary evolution with no room for dissipative behavior. The usual approach to include dissipation is the quantum theory of open systems [4,5,6,7,8]. In this approach the dynamics of a quantum system is no longer unitary but it is defined by a semigroup of completely positive maps in the space of density operators [9].

Another strategy to describe dissipative quantum systems is based on the old idea of Bateman [10]. Bateman has shown that in order to apply the standard canonical formalism of classical mechanics to dissipative and non-Hamiltonian systems, one can double the number of degrees of freedom, so as to deal with an effective isolated classical Hamiltonian system. In this approach the new degrees of freedom may be interpreted as variables describing a reservoir. Applying this idea to a damped harmonic oscillator one obtains a pair of damped oscillators (so called Bateman's dual system): a primary one and its time reversed image. The Bateman dual Hamiltonian has been rediscovered by Morse and Feshbach [11] and Bopp [12] and the detailed quantum mechanical analysis was performed by Feshbach and Tikochinski [13]. The quantum Bateman system was then analyzed by many authors (see for the historical review [14]). Recently it was investigated in connection with quantum field theory and quantum groups [15] (see also [16,17]). Finally, the detailed analysis of spectral properties of the Bateman system was recently performed in [18,19] (see also [20]). For other approaches to quantization of damped oscillator see also recent papers [21,22,23].

In the present paper we propose a different approach. Instead of analyzing the standard equation of motion for the damped oscillator

$$\ddot{x} + 2\gamma\dot{x} + \omega^2 x = 0 \ , \tag{1}$$

where γ is a positive damping constant, and ω is a frequency of the undamped motion, we shall study a modified equation

$$\ddot{x} + 2\gamma \tanh(\gamma(t - t_0))\dot{x} + \omega^2 x = 0 \ , \tag{2}$$

where t_0 is an arbitrary constant. It is clear that in the limit '$\gamma t \to \infty$', formula (2) reproduces the standard equation (1). Therefore, both dynamics may differ for short times only, that is, for 't' satisfying

$$t \ll 1/\gamma \ .$$

What is the origin of (2)? It turns out that both (1) and (2) give the same frequency of oscillations of the damped motion[a]

$$\Omega = \sqrt{\omega^2 - \gamma^2} \ . \tag{3}$$

Moreover, there is no other linear equation of motion with this property. Interestingly, equation (2) introduces additional parameter t_0. It turns out that t_0 is responsible for an effective memory, that is, in addition to a 'local time t' evolution governed by (2) 'remembers' about its history at t_0. It means that the initial value problem for (2) has to be supplemented by initial data $x(t_0) = x_0$ and $\dot{x}(t_0) = v_0$. There is a crucial difference between (1) and (2): the initial data for the standard 'local' equation may be provided for arbitrary initial moment of time. It is no longer true for the modified 'nonlocal' equation: now the initial moment necessarily equals t_0. Changing initial moment t_0 one changes the corresponding equation (2) and hence the whole evolution of our damped system. This property is

[a]Throughout the paper we assume $\omega > \gamma$.

responsible for a peculiar feature of (2). The composition law for classical dynamics is no longer true, that is, evolution from t_0 to t_1, and then from t_1 to t_2 differs from the direct evolution from t_0 to t_2.

Now, the quantum system corresponding to (2) may be easily constructed. Our damped system turns out to be governed by nonlocal Hamiltonian and hence the standard composition law for quantum propagators

$$K(x_2,t_2;x_0,t_0) = \int dx_1 K(x_2,t_2;x_1,t_1)K(x_1,t_1;x_0,t_0) \ , \tag{4}$$

does not hold. We stress that the local composition law (4) is satisfied by the quantum propagator derived from the standard equation (1) via well known Caldirola-Kanai approach [24,25]. The violation of (4) defines the basic difference between two descriptions based on (1) and (2), respectively. We hope that our analysis sheds new light on this old problem.

2. Hamiltonian formulation

There were many attempts for hamiltonian formulation of Newton equation for systems with one degree of freedom (see e.g. [25,26,27,28,29]). It is well known that (1) may be derived from the so called Caldirola-Kanai Hamiltonian [24,25]

$$H(x,p,t) \ = \ \frac{1}{2}m_0\omega^2x^2\, e^{2\gamma t} + \frac{1}{2m_0}p^2\, e^{-2\gamma t} \ . \tag{5}$$

It is clear that Caldirola-Kanai Hamiltonian may be considered as the standard Hamiltonian of the harmonic oscillator with time-dependent mass given by

$$m(t) = m_0 e^{2\gamma t} \ . \tag{6}$$

Consider now a family of Hamiltonians with an arbitrary time-dependent mass $m(t)$:

$$H(x,p,t) = \frac{p^2}{2m(t)} + \frac{1}{2}m(t)\omega^2x^2 \ . \tag{7}$$

The corresponding Newton equation reads as follows

$$\ddot{x} + \frac{\dot{m}(t)}{m(t)}\,\dot{x} + \omega^2 x = 0 \ . \tag{8}$$

Let us assume that in the asymptotic regime '$\gamma t \to \infty$', equation (8) reproduces the standard equation (1), that is

$$\kappa(t) := \frac{1}{2}\frac{\dot{m}(t)}{m(t)} \longrightarrow \gamma \ . \tag{9}$$

To solve (8) one represents complexified $x(t)$ as follows

$$x(t) = \lambda(t)\, e^{i\phi(t)} , \tag{10}$$

and obtains the following equations for time-dependent real quantities λ, ϕ and κ:

$$\ddot{\lambda} - \lambda\dot{\phi}^2 + 2\kappa\dot{\lambda} + \omega^2\lambda = 0 , \tag{11}$$

$$\ddot{\phi}\lambda + 2\dot{\phi}\dot{\lambda} + 2\kappa\dot{\phi}\lambda = 0 , \tag{12}$$

where we omitted explicit time-dependence. Now, we make our basic assumption that $\dot{\phi} = \Omega$ with Ω defined in (3). Equation (12) reduces then to

$$\dot{\lambda} + \kappa\lambda = 0 ,$$

and hence equation (11) implies

$$\Omega^2 = \omega^2 - \kappa^2 - \dot{\kappa} ,$$

which together with (3) gives finally

$$\dot{\kappa} + \kappa^2 = \gamma^2 .$$

The above equation has the following solutions

$$\kappa(t) = \gamma \tanh(\gamma(t - t_0)) \quad \text{or} \quad \kappa(t) = \pm\gamma , \tag{13}$$

with t_0 being an integration constant. Note, however, that due to (9) the solution $\kappa(t) = -\gamma$ is not acceptable. It is clear that $\kappa(t) = \gamma$ corresponds to the Caldirola-Kanai model with $m(t) = e^{2\gamma t}$, whereas $\kappa(t) = \gamma \tanh(\gamma(t - t_0))$ gives rise to our new equation (2) described by a Hamiltonian with time-dependent mass

$$m(t) = m_0 \cosh^2(\gamma(t - t_0)) , \tag{14}$$

satisfying $m(t_0) = m_0$. We may therefore interpret (2) as a Newton equation for a particle with constant mass m_0 under the action of the time dependent nonlocal force

$$F(t; t_0) = -2m_0\gamma \tanh(\gamma(t - t_0))\dot{x} - m_0\omega^2 x ,$$

or as a Newton equation for a particle with time-dependent nonlocal mass (14) under the action of perfectly local force $-m_0\omega^2 x$.

3. Classical dynamics

Consider now classical Hamiltonian dynamics generated by (8) with $m(t)$ defined in (14). It is given by

$$\begin{pmatrix} x(t) \\ p(t) \end{pmatrix} = \Lambda^{\text{new}}(t;t_0) \begin{pmatrix} x_0 \\ p_0 \end{pmatrix} , \tag{15}$$

where the 2×2 matrix

$$\Lambda^{\text{new}}(t;t_0) = \begin{pmatrix} \Lambda^{\text{new}}_{xx} & \Lambda^{\text{new}}_{xp} \\ \Lambda^{\text{new}}_{px} & \Lambda^{\text{new}}_{pp} \end{pmatrix} ,$$

reads as follows

$$\Lambda^{\text{new}}_{xx} = \frac{\cos(\Omega(t-t_0))}{\cosh(\gamma(t-t_0))} , \qquad \Lambda^{\text{new}}_{xp} = \frac{1}{m_0\Omega}\frac{\sin(\Omega(t-t_0))}{\cosh(\gamma(t-t_0))} , \tag{16}$$

and

$$\begin{aligned} \Lambda^{\text{new}}_{px} &= -m_0\Big[\Omega\cosh(\gamma(t-t_0))\sin(\Omega(t-t_0)) \\ &\quad + \gamma\sinh(\gamma(t-t_0))\cos(\Omega(t-t_0))\Big] , \end{aligned} \tag{17}$$

$$\begin{aligned} \Lambda^{\text{new}}_{pp} &= \Omega\cosh(\gamma(t-t_0))\cos(\Omega(t-t_0)) \\ &\quad - \frac{\gamma}{\Omega}\sinh(\gamma(t-t_0))\sin(\Omega(t-t_0)) . \end{aligned} \tag{18}$$

The crucial feature of (15) is that it breaks the local composition law

$$\Lambda(t_2;t_0) = \Lambda(t_2;t_1) \circ \Lambda(t_1;t_0) , \tag{19}$$

for $t_0 \leq t_1 \leq t_2$, that is, formula (19) does not hold for $\Lambda = \Lambda^{\text{new}}$.

Let us compare our new evolution (15) with the classical dynamics of the Caldirola-Kanai system. Note, that we may trivially change the formula (6) for time-dependent mass

$$m(t) = m_0 e^{2\gamma(t-t_0)} , \tag{20}$$

without affecting the equation of motion. Now, these two models contain additional parameter t_0. One easily finds for the corresponding Caldirola-Kanai dynamics $\Lambda^{\text{CK}}(t;t_0)$:

$$\Lambda^{\text{CK}}_{xx} = \Big[\cos(\Omega(t-t_0)) + \frac{\gamma}{\Omega}\sin(\Omega(t-t_0))\Big]e^{-\gamma(t-t_0)}, \tag{21}$$

$$\Lambda^{\text{CK}}_{xp} = \frac{1}{m_0\Omega}\sin(\Omega(t-t_0))e^{-\gamma(t-t_0)}, \tag{22}$$

$$\Lambda^{\text{CK}}_{px} = -m_0\Omega\Big(1+\frac{\gamma^2}{\Omega^2}\Big)\sin(\Omega(t-t_0))e^{\gamma(t-t_0)}, \tag{23}$$

$$\Lambda^{\text{CK}}_{pp} = \Big[\cos(\Omega(t-t_0)) - \frac{\gamma}{\Omega}\sin(\Omega(t-t_0))\Big]e^{\gamma(t-t_0)}. \tag{24}$$

One may check that $\Lambda^{\rm CK}$ satisfies

$$\Lambda^{\rm CK}(t_2;t_0) = \Lambda^{\rm CK}(t_2;t_1) \circ \Lambda^{\rm CK}(t_1;t_0) \ , \tag{25}$$

for $t_0 \leq t_1 \leq t_2$. Note, that computing $\Lambda^{\rm CK}(t_2,t_1)$ one applies (21)–(24) with t_0 replaced by t_1, and m_0 replaced by $m_1 := m(t_1) = m_0 e^{2\gamma(t_1-t_0)}$. Then formula (25) easily follows as a simple consequence of $e^{a+b} = e^a e^b$. It is no longer true if we replace (20) by (14).

Finally, let us compare both approaches in the simplified situation corresponding to pure dissipation, that is, $\omega = 0$. In this case the corresponding formula for $\Lambda^{\rm new}(t;t_0)$ considerably simplifies

$$\Lambda_0^{\rm new}(t;t_0) \ = \ \begin{pmatrix} 1 & (m_0\gamma)^{-1}\tanh(\gamma(t-t_0)) \\ 0 & 1 \end{pmatrix} . \tag{26}$$

However, $\Lambda_0^{\rm new}$ still breaks (19). In the case of Caldirola-Kanai dynamics one finds

$$\Lambda_0^{\rm CK}(t;t_0) \ = \ \begin{pmatrix} 1 & (m_0\gamma)^{-1}\sinh(\gamma(t-t_0))\, e^{-\gamma(t-t_0)} \\ 0 & 1 \end{pmatrix} . \tag{27}$$

Note, that in a asymptotic regime $\gamma t \longrightarrow \infty$ one obtains

$$\Lambda^{\rm new}_{\rm asympt} \ = \ \begin{pmatrix} 1 & (m_0\gamma)^{-1} \\ 0 & 1 \end{pmatrix} , \qquad \Lambda^{\rm CK}_{\rm asympt} \ = \ \begin{pmatrix} 1 & (2m_0\gamma)^{-1} \\ 0 & 1 \end{pmatrix} , \tag{28}$$

that is, both dynamics have different asymptotic states: $(x_0 + p_0/(m_0\gamma), p_0)$ and $(x_0 + p_0/(2m_0\gamma), p_0)$ for the Caldirola-Kanai system.

4. Quantum propagator

Consider now a quantum propagator corresponding to Hamiltonian (7) with $m(t)$ given by (14). Since the Hamiltonian is quadratic in (x,p) one easily finds for the propagator (see e.g. [34])

$$K(x,t;x_0,t_0) = N(t-t_0)\, A(x,t;x_0,t_0) \cdot B(x,t;t_0) \ , \tag{29}$$

where

$$\begin{aligned} A(x,t;x_0,t_0) = \exp\left\{ \frac{i}{2\hbar} \frac{m_0\Omega}{\sin(\Omega(t-t_0))} \Big[(x_0^2 + x^2)\cos(\Omega(t-t_0)) \right. \\ \left. - \ 2x_0 x \cosh(\gamma(t-t_0)) \Big] \right\} , \end{aligned}$$

$$\begin{aligned} B(x,t;t_0) = \exp\left\{ \frac{i}{2\hbar}\, m_0 x^2\, \frac{\sinh(\gamma(t-t_0))}{\sin(\Omega(t-t_0))} \Big[\Omega\cos(\Omega(t-t_0))\sinh(\gamma(t-t_0)) \right. \\ \left. - \ \gamma\sin(\Omega(t-t_0))\cosh(\gamma(t-t_0)) \Big] \right\} , \end{aligned} \tag{30}$$

and the normalization factor is given by

$$N(t-t_0) = \sqrt{\frac{m_0\Omega\cosh(\gamma(t-t_0))}{2\pi i\hbar\sin(\Omega(t-t_0))}} \ . \tag{31}$$

Note, that when $\gamma = 0$ the factor B equals to 1 and one recovers standard propagator for the harmonic oscillator [30]. On the other hand when $\omega = 0$, that is, in a pure damping case, formula (29) reduces to

$$K_0(x,t;x_0,t_0) = N_0(t-t_0)\exp\left\{\frac{i\gamma}{2\hbar}\frac{m_0(x-x_0)^2}{\tanh(\gamma(t-t_0))}\right\} , \tag{32}$$

with

$$N_0(t-t_0) = \sqrt{\frac{\gamma m_0}{2\pi i\hbar\tanh(\gamma(t-t_0))}} \ . \tag{33}$$

For $\gamma = 0$ formula (32) reproduces quantum propagator for a free particle [30]. Interestingly, the formula (32) for a quantum propagator for a particle undergoing linear damping was recently derived in [33] by using purely geometrical considerations. Here we showed that it may be derived via standard methods by taking suitable nonlocal Hamiltonian with memory. Moreover, the propagator for the damped oscillator derived in [33] is given by

$$\widetilde{K}(x,t;x_0,t_0) = N(t-t_0)\,A(x,t;x_0,t_0) \ , \tag{34}$$

and therefore differs from (29) only by a factor $B(x,t;t_0)$. Both propagators K and $\widetilde{K}$ produce the same limits for $\gamma = 0$ (harmonic oscillator) and for ($\gamma = 0$ & $\omega = 0$) (free particle). Observe that $B(x,t;t_0)$ produces only an additional phase factor at the final position 'x'. Therefore both formulae give the same results for a mean position of a particle: starting from the initial wave function $\psi(x_0,t_0)$ one finds

$$\psi(x,t;t_0) = \int_{-\infty}^{\infty} K(x,t;x_0,t_0)\,\psi(x_0,t_0)dx_0 \ , \tag{35}$$

and

$$\widetilde{\psi}(x,t;t_0) = \int_{-\infty}^{\infty} \widetilde{K}(x,t;x_0,t_0)\,\psi(x_0,t_0)dx_0 \ = \ \frac{1}{B(x,t;t_0)}\,\psi(x,t;t_0) \ , \tag{36}$$

and hence

$$\langle x(t;t_0)\rangle = \int_{-\infty}^{\infty} x\,|\psi(x,t;t_0)|^2 dx = \int_{-\infty}^{\infty} x\,|\widetilde{\psi}(x,t;t_0)|^2 dx\,. \tag{37}$$

It is easy to show that $\langle x(t;t_0)\rangle$ satisfies our modified nonlocal equation (2):

$$\frac{d^2}{dt^2}\langle x(t;t_0)\rangle + 2\gamma\tanh(\gamma(t-t_0))\frac{d}{dt}\langle x(t;t_0)\rangle + \omega^2\langle x(t;t_0)\rangle = 0\,. \tag{38}$$

In the Appendix we show that this additional phase factor $B(x,t;t_0)$ may be eliminated by a suitable canonical transformation. The price we pay for a simpler formula for a propagator is the much more complex form of the Hamiltonian which apart from the standard quadratic terms x^2 and p^2 contains also a mixed term xp (see Appendix).

Let us compare the above formulae with the propagator for the Caldirola-Kanai system. One easily finds the CK propagator corresponding to $m(t)$ given by (20)

$$\begin{aligned} K_{\rm CK}(x,t;x_0,t_0) = N_{\rm CK}(t-t_0)\exp\Big\{\frac{i}{2\hbar}\,m_0\Big[\gamma\Big(x_0^2e^{2\gamma t_0} - x^2e^{2\gamma t}\Big) \\ + \Omega\cot(\Omega(t-t_0))\Big(x_0^2e^{2\gamma t_0} + x^2e^{2\gamma t}\Big) \\ - 2\frac{\Omega}{\sin(\Omega(t-t_0))}xx_0e^{\gamma(t+t_0)}\Big]\Big\}, \end{aligned} \tag{39}$$

where the normalization factor

$$N_{\rm CK}(t-t_0) = \sqrt{\frac{m_0\Omega\,e^{\gamma(t-t_0)}}{2\pi i\hbar\sin(\Omega(t-t_0))}}\,. \tag{40}$$

Again (39) correctly reproduces both the propagator for the harmonic oscillator (for $\gamma = 0$) and the propagator for the free particle ($\gamma = 0$ and $\omega = 0$). However, for pure damping both propagators are different.

There is a crucial difference between K and $K_{\rm CK}$. The Caldirola-Kanai propagator satisfies local composition law

$$K_{\rm CK}(x_2,t_2;x_0,t_0) = \int dx_1 K_{\rm CK}(x_2,t_2;x_1,t_1)K_{\rm CK}(x_1,t_1;x_0,t_0)\,. \tag{41}$$

It is no longer true for K defined in (29). Therefore, as in the classical case memory term encoded into dynamical equation is responsible for the breaking of the local composition law for the quantum evolution.

Finally, let us consider a damped (without oscillations) evolution of the initial Gaussian wave packet

$$\psi_0(x;t_0) = C\, e^{-(x-x_0)^2/\sigma + ip_0x/\hbar} \ . \tag{42}$$

One easily finds for the probability density $\rho(x,t;t_0) = |\psi(x,t;t_0)|^2$:

$$\rho(x,t;t_0) = \frac{|C|^2}{\sqrt{\sigma(t-t_0)/\sigma}} \exp\left[-2\frac{(x-x(t))^2}{\sigma(t-t_0)}\right] \ , \tag{43}$$

where $x(t)$ is given by the classical formula

$$x(t) = x_0 + p_0 \frac{\tanh(\gamma(t-t_0))}{m_0\gamma} \ , \tag{44}$$

and

$$\sigma(t-t_0) = \sigma\left[1 + \left(\frac{\hbar\tanh(\gamma(t-t_0))}{m_0\sigma\gamma}\right)^2\right] \ . \tag{45}$$

It shows that the presence of the damping term modifies the way the initial way function spreads in time. Note, that in the asymptotic regime one obtains

$$\rho_{\rm asympt}(x;t_0) = \frac{|C|^2}{\sqrt{\sigma_{\rm asympt}/\sigma}} \exp\left[-2\frac{(x-x_{\rm asympt})^2}{\sigma_{\rm asympt}}\right] \ , \tag{46}$$

with the classical formula for the asymptotic position

$$x_{\rm asympt} = x_0 + \frac{p_0}{m_0\gamma} \ , \tag{47}$$

and asymptotic dispersion

$$\sigma_{\rm asympt} = \sigma\left[1 + \left(\frac{\hbar}{m_0\gamma\sigma}\right)^2\right] \ . \tag{48}$$

It is therefore clear that the damping prevents the wave function from the total spreading. An initial Gaussian probability density $\rho_0(x;t_0)$ cannot 'spread more' than $\rho_{\rm asympt}(x;t_0)$. A similar phenomenon appears for Caldirola-Kanai dynamics with different asymptotic state

$$x^{\rm CK}_{\rm asympt} = x_0 + \frac{p_0}{2m_0\gamma} \ , \tag{49}$$

but with the same asymptotic dispersion $\sigma_{\rm asympt}$ given by (48).

5. Conclusions

We performed an analysis of the new equation of motion for the damped oscillator (2). It differs from the standard one by a damping term $-2\gamma\tanh(\gamma(t-t_0))\dot{x}$ – which is non-local in time and nonlinear in the damping constant γ. The new parameter t_0 introduces effective memory. For long time behavior $t \gg 1/\gamma$ one recovers standard equation with a damping term $-2\gamma\dot{x}$. Both classical and quantum analysis is performed. The characteristic feature of this nonlocal system is that it breaks local composition law for the classical Hamiltonian dynamics and the corresponding quantum propagator.

In principle, our modified mass (14) can be applied in other models known in the literature (see for instance [31,32]). Interestingly, the same propagator was recently derived in [33]. Without referring to Hamiltonian formulation and using purely geometric methods the author of [33] derived the corresponding propagator starting from the standard equation (1). We have shown that the corresponding classical limit leads to the new damping equation (2). Therefore, we conclude that the quantum nonlocal propagator derived in [33] does originate in nonlocal equation (2). Finally, it was shown that the purely damped behavior modifies well known property of the free quantum dynamics leading to the perfect spreading of the initial wave function. It is no longer true when the dissipation is present. Now, instead of the perfect spreading there is an asymptotic state giving rise to the asymptotic probability density (46).

Appendix

Using a general formula for a propagator corresponding to quadratic Hamiltonian (see e.g. [34,35]) one easily finds that

$$H(x,p,t) = \frac{1}{2}\mu(t)p^2 + \nu(t)xp + \frac{1}{2}\lambda(t)x^2, \tag{A.1}$$

with

$$\mu(t) = \frac{1}{m_0\cosh^2(\gamma(t-t_0))} \tag{A.2}$$

and

$$\nu(t) = \tanh(\gamma(t-t_0))\Big(\Omega\frac{\tanh(\gamma(t-t_0))}{\tan(\Omega(t-t_0))} - \gamma\Big), \tag{A.3}$$

$$\lambda(t) = m_0\Omega^2\left\{\frac{2\gamma}{\Omega}\frac{\tanh(\gamma(t-t_0))}{\tan(\Omega(t-t_0))} + \frac{1-[1+\tanh^2(\gamma(t-t_0))]\cos^2(\Omega(t-t_0))}{\sin^2(\Omega(t-t_0))}\right\}, \tag{A.4}$$

produces (34), that is, it annihilates additional phase factor $B(x,t;t_0)$ out of (29).

References

1. D.F. Walls, G.J. Millburn, *Quantum Optics*, Springer, Heildelberg, 1994.
2. M.O. Scully, M.S. Zubairy, *Quantum Optics*, Cambridge University Press, Cambridge 1997.
3. M. Sargent III, M.O. Scully, W.E. Lamb Jr., *Laser Physics*, Addison-Wesley, Reading 1974.
4. F. Haake, *Statistical treatment of open systems by generalized master equations*, Springer, Berlin, 1973.
5. E.B. Davies, *Quantum Theory of Open Systems*, Academic Press, London 1976.
6. H.-P. Breuer, F. Petruccione, *The Theory of Open Quantum Systems*, Oxford University Press, Oxford 2002.
7. H. Carmichael, *An Open System Approach to Quantum Optics*, Springer, Heidelberg 1991.
8. V.E. Tarasov, Phys. Lett. A **288**, (2001) 173–182.
9. R. Alicki, K. Lendi, *Quantum Dynamical Semigroups and Applications*, Springer, Heidelberg 1987.
10. H. Bateman, Phys. Rev. **38**, (1931) 815.
11. P.M. Morse and H. Feshbach, *Methods of Theoretical Physics*, vol. 1, McGraw-Hill, New York, 1953.
12. F. Boop, Sitz.-Ber. Bayer. Akad. Wiss.,Math.-Natur. Kl. (1973), 67.
13. H. Feshbach and Y. Tikochinsky, in: A Festschrift for I.I. Rabi, Trans. New York Ac. Sc. Ser. 2 **38**, (1977) 44.
    Y. Tikochinsky, J. Math. Phys. **19** (1978) 888.
14. H. Dekker, Phys. Rep. **80**, (1981) 1–112.
15. E. Celeghini, M. Rasetti and G. Vitiello, Ann. Phys. (N.Y.), **215** (1992) 156–170.
16. Y. N. Srivastava, G. Vitiello, Ann. Phys. (N.Y.), **238**, (1995) 200–207.
17. A. Iorio, G. Vitiello, A. Widom, Ann. Phys. (N.Y.), **241**, (1995) 496–506.
18. D. Chruściński, J. Jurkowski, Ann. Phys. **321**, 854 (2006).
19. D. Chruściński, Ann. Phys. **321**, 840 (2006).
20. M. Blasone and P. Jizba, Ann. Phys. (N.Y.), **312**, (2004) 354–397.

21. M. Montesinos, Phys. Rev. A **68**, 014101 (2003).
22. G. Dito, F. J. Turrubiates, Phys. Lett. A **352**, 309 (2006).
23. V. K. Chandrasekar, M. Senthilvelan, M. Lakshmanan, J. Math. Phys. **48**, 032701 (2007).
24. P. Caldirola, Nuovo Cim. **18**, 393 (1941).
25. E. Kanai, Prog. Theor. Phys. **3**, 440 (1948).
26. P. Havas, Nuovo Cim. (Suppl.) **5**, 363 (1957).
27. J. A. Kobussen, Acta Phys. Austriaca **51**, 293 (1979).
28. G. Lopez, Ann. Phys. **251**, 372 (1996).
29. G. Lopez, P. Lopez, quant-ph/0506179.
30. R. F. Feynman and A. R. Hibbs, *Quantum Mechanics and Path Integrals*, McGraw–Hill, 1965.
31. U. Das, S. Ghosh, P. Sarkar, B. Talukdar, Physica Scripta **71**, 235 (2005).
32. Kyu-Hwang Yeon, Suk-Sung Kim, Young-Mo Moon, Suc-Kyoung Hong, Chung-In Um, T. F. George, J. Phys. A: Math. Gen **34**, 7719, (2001).
33. D. Kochan, *Direct quantization of equations of motion: from classical dynamics to transition amplitudes via strings*, arXiv: hep-th/0703073.
34. Kyu Hwang Yeon, Kang Ku Lee, Chung In Um, T. F. George, and L. N. Pandey, Phys. Rev. A **48**, 2716 (1993).
35. Kyu-Hwang Yeon and Duk-Hyeon Kim, Phys. Rev. A **55**, 4023 (1997).

Quantum Bio-Informatics III
eds. L. Accardi, W. Freudenberg & M. Ohya

AN INTRODUCTION TO QUANTIZATION OF DISSIPATIVE SYSTEMS. THE DAMPED HARMONIC OSCILLATOR CASE

JACEK JURKOWSKI

Institute of Physics, Nicolaus Copernicus University,
Grudziądzka 5/7, 87–100 Toruń, Poland

We give an elementary introduction to the problems of quantization of dissipative systems considering the simplest but still not fully understood model of damped harmonic oscillator. We recall some old and introduce some new methods of quantization of the model starting from the classical damped harmonic oscillator equation. Our methods are based on Hamiltonian formulation of dynamical systems adopted to the case of dissipation and loss of energy. Interestingly, in our approach we can interpret the poles of the analytical continuation of the energy spectrum as resonant states, well-known in atomic physics..

1. Introduction

There is a common agreement that dissipative phenomena go beyond the standard formulation of quantum mechanics due to perfect time-reversibility of the Schrödinger equation and, in consequence, the fact that evolution operators form a group. The usual approach, which includes the interaction of the system with its environment, is the quantum theory of open systems. In this framework, the dynamics is no longer unitary but it is defined by a semigroup of completely positive maps. In principle, the dissipation phenomena can be described in such a way, but there is no simple (if any) connection with the damped harmonic oscillator equation which is the most powerful and the simplest classical realisation of dissipation [1].

Another way is to apply the standard canonical formalism to a system with the doubled number of degrees of freedom and dealing with an *effective* isolated classical Hamiltonian (or Lagrangian) system. The new degrees of freedom may be assumed to represent the reservoir. Quantization in this case seems to be straightforward using the standard methods of canonical quantization [2,4,3], but the results one obtains are still not fully understood. The main difficulty is the nonuniqueness of the *effective* Hamiltonian, one has to deal with, leading to nonequivalent quantum formulations. Some authors investigate also methods of quantization different from the canonical ones such as: deformation quantization [5,6], velocity quantization [7] and other ones [8,9].

In what follows, we recall some classical Hamiltonian functions leading to the damped harmonic oscillator equation and we show how to quantize them canonically. We mention also some relatively new results concerning the (complex) energy spectrum and its analytical continuation leading to a description of resonant states. We point out the necessity of using a broader structure than the Hilbert space formalism in order to contain dissipative phenomena.

2. Classical Damped Harmonic Oscillator Models

Due to an introductory character of this article, we focus on the simplest classical dissipative system — *one-dimensional damped harmonic oscillator* beginning its description from the second order differential equation

$$\ddot{x} + 2\gamma\dot{x} + \omega^2 x = 0\,, \tag{1}$$

where $\gamma > 0$ is the damping constant and $\omega > 0$ is the frequency of undamped oscillations. An old and well-known task is to find a *Hamiltonian representation* for (1), i.e., to establish at least one function $H(x,p)$ (not neccessarilly an energy function!) on the phase-space parametrized by (x,p) such that the set of first order (Hamiltonian) equations,

$$\dot{x} = \frac{\partial H}{\partial p}\,, \qquad \dot{p} = -\frac{\partial H}{\partial x} \tag{2}$$

leads to the equation (1). In fact, it is well-known that there is no time-independent energy function generating the set of equations due to the energy loss! Indeed, if one canonically splits (1) into two first order equations in the following way:

$$\ddot{x} + 2\gamma\dot{x} + \omega^2 x = 0 \quad \longrightarrow \quad \begin{cases} \dot{x} = p \\ \dot{p} = -2\gamma p - \omega^2 x\,, \end{cases}$$

then there is no Hamiltonian function which could generate right-hand sides in accordance with (2).

But if one looks for a time-dependent Hamiltonian, it is easy to check that the spliting

$$\ddot{x} + 2\gamma\dot{x} + \omega^2 x = 0 \quad \longrightarrow \quad \begin{cases} \dot{x} = p\,e^{-2\gamma t} \\ \dot{p} = \omega^2 x\,e^{2\gamma t} \end{cases}$$

is provided by the so-called *Caldirola-Kanai Hamiltonian*

$$H_\gamma(x,p,t) \;=\; \frac{1}{2}\omega^2 x^2\, e^{2\gamma t} + \frac{1}{2}p^2\, e^{-2\gamma t}, \tag{3}$$

which, although manifestly time-dependent, has the correct behaviour in the limit $\gamma \to 0$, i.e., $H_0(x,p,t)$ represents the energy of a harmonic oscillator. But on the other hand we can rewrite (3) as

$$H_\gamma(x,p,t) \;=\; \frac{p^2}{2m(t)} + \frac{1}{2}m(t)\omega^2 x^2$$

with $m(t) = e^{2\gamma t}$. $m(t)$ can be interpreted as the time-dependent mass term and the Hamiltonian (3) as generating *free* oscillations of the system with growing mass. It is the consequence of a rather general remark that friction linear in velocity is equivalent to a free motion with growing mass. So, we should not expect that the Hamiltonian (3) provides us with new physical phenomena for dissipative systems. There are, however, other representations.

Let us consider a more general method of looking for the Hamiltonian representation [7]. Recall that the *constant of motion* is a function $K(x,v)$ of position x and velocity $v = \dot{x}$ which is constant along the trajectory, i.e,

$$\frac{dK}{dt} = \frac{\partial K}{\partial x}\dot{x} + \frac{\partial K}{\partial v}\dot{v} = 0\,. \tag{4}$$

In our case

$$\ddot{x} + 2\gamma\dot{x} + \omega^2 x \;=\; 0 \quad\longrightarrow\quad \begin{cases} \dot{x} = v \\ \dot{v} = -2\gamma v - \omega^2 x \end{cases} \tag{5}$$

and

$$\frac{\partial K}{\partial x}\dot{x} + \frac{\partial K}{\partial v}\dot{v} = v\frac{\partial K}{\partial x} - (\omega^2 x + 2\gamma v)\frac{\partial K}{\partial v}\,. \tag{6}$$

Solving (6) one can show [7] that constants of motion have the form $K_\gamma(x,v) \equiv f(C_\gamma(x,v))$, where f is an arbitrary function and C_γ is an integral curve of (5), i.e.,

$$C_\gamma(x,v) = \ln\!\left(\frac{1}{2}(v^2 + 2\gamma v x + \omega^2 x^2)\right) - 2\gamma G_\gamma(v/x,\omega)$$

with

$$G_\gamma(v/x,\omega) \;=\; \frac{1}{\sqrt{\omega^2-\gamma^2}}\arctan\frac{v/x+\gamma}{\sqrt{\omega^2-\gamma^2}}\,.$$

In principle, $K_\gamma(x,v)$ can be used to construct the Lagrangian function $L(x,v)$ using the relation [10]

$$L_\gamma(x,v) \;=\; v\int \frac{K_\gamma(x,y)}{y^2}dy\,|_{y=v} \tag{7}$$

and it is clear that there are *infinitely* many Lagrangians depending on the choice of f.

For example, choosing

$$K_\gamma(x,v) = \exp(C_\gamma(x,v)) = \frac{1}{2}(v^2 + 2\gamma vx + \omega^2x^2)e^{-2\gamma G_\gamma(v/x,\omega)}$$

one reproduces in the limit $\gamma \to 0$ the Lagrangian for harmonic oscillator

$$L_0(x,v) = \frac{1}{2}(v^2 - \omega^2x^2)\,,$$

but for $\gamma \neq 0$ one cannot calculate analytically the integral in (7). On the other hand, choosing $K_\gamma(x,v) = C_\gamma(x,v)$ one gets the Lagrangian

$$L_\gamma(x,v) \;=\; s\frac{v/x+\gamma}{\sqrt{\omega^2-\gamma^2}}\arctan\frac{v/x+\gamma}{\sqrt{\omega^2-\gamma^2}}\,. \tag{8}$$

Calculating the canonical momentum

$$p = \frac{\partial L_\gamma}{\partial v}$$

and performing the Legendre transformation

$$H_\gamma = pv - L_\gamma$$

provides us with the following time-independent Hamiltonian function [11],

$$H_\gamma(x,p) \;=\; \frac{1}{2}\ln\frac{x^2}{\cos^2(xp\sqrt{\omega^2-\gamma^2})} - \gamma xp\,. \tag{9}$$

Unfortunately, $H_\gamma(x,p)$ fails to be the energy even in the limit $\gamma \to 0$.

Another method of dealing with the damped harmonic oscillator was proposed by Bateman [12]. The idea was to consider the model with two oscillators — the damped one (losing the energy) and the mirror-image oscillator (absorbing the energy), so the energy of the system as a whole is conserved. The supplementary degree of freedom y fulfils the mirror-image equation with $\gamma \to -\gamma$,

$$\ddot{y} - 2\gamma\dot{y} + \omega^2 y = 0\,.$$

Such a system can be generated by the Lagrangian function

$$L(x,y,\dot{x},\dot{y}) = \dot{x}\dot{y} - \omega^2xy + \gamma(x\dot{y} - \dot{x}y) \quad\longrightarrow\quad \begin{cases} \ddot{x} + 2\gamma\dot{x} + \omega^2 x &= 0 \\ \ddot{y} - 2\gamma\dot{y} + \omega^2 y &= 0\,. \end{cases}$$

Canonical momenta p_x and p_y provide the coupling between the systems

$$p_x = -\frac{\partial L}{\partial x} = \dot{y} - \gamma y\,, \qquad p_y = -\frac{\partial L}{\partial y} = \dot{x} + \gamma x$$

and after the Legendre transformation we arrive at the following Hamiltonian

$$H_B(x, y, p_x, p_y) = p_x p_y + \Omega^2 xy - \gamma(xp_x - yp_y)\,, \qquad \Omega = \sqrt{\omega^2 - \gamma^2}\,, \tag{10}$$

called the *Bateman Hamiltonian*, defined on an extended phase-space (x, y, p_x, p_y). What is promising, is that the Bateman Hamiltonian is quadratic in positions and momenta which enables its quantization, contrary to highly nonlinear dependence of x and p in (9). On the other hand, H_B is not related to energy function even in the limit $\gamma \to 0$. In fact, there is no generator of the energy-form

$$\widetilde{H}(x, y, p_x, p_y) = \frac{1}{2}(p_x^2 + \Omega^2 x^2) + \frac{1}{2}(p_y^2 + \Omega^2 y^2) + \gamma f(x, y, p_x, p_y)$$

which would yield correct evolution equations!

To sum up, even at the classical level one has no unique description of damped oscillations in both the Lagrange and the Hamilton frameworks. In consequence, one can expect that the quantum description of dissipative phenomena are even less clear.

3. Quantization procedures

In principle, any classical model presented in Sect. 2 can be *canonically quantized.* By this procedure we mean replacing the pair (x, p) of canonical variables by their operator representation in a certain Hilbert space,

$$x \to \widehat{x} = x\,, \qquad p \to \widehat{p} = -i\hbar\frac{\partial}{\partial x}\,.$$

The evolution is then governed by the Schrödinger equation

$$i\hbar\partial_t\psi(t) = \widehat{H}\psi(t) \tag{11}$$

with the Hamiltonian operator $\widehat{H} \equiv H(\widehat{x}, \widehat{p})$ and the time-dependent wave function of the system $\psi(t)$. If the Hamiltonian operator is time-independent (and fulfills some technical conditions), then the solution of (11) is generated by the group of unitary operations $\widehat{U}_t = \exp(t\widehat{H}/i\hbar)$, i.e., it has the form

$$\psi(t) = \widehat{U}_t\psi(0)\,. \tag{12}$$

The situation is much more complicated when the Hamilton operator is time-dependent, as it happens, for example, in the Caldirola-Kanai case. Other difficulties arrive when interpreting the results. It turns out that different classical models which led to the same results, at the quantum level provide different predictions.

3.1. *Quantization of the Caldirola-Kanai Hamiltonian*

The canonical quantization procedure leads in the present case to the following Schrödinger equation (see [13]):

$$i\hbar\partial_t\psi(x,t) = \widehat{H}\psi(x,t) = \left[\frac{1}{2}\omega^2 x^2\, e^{2\gamma t} - \frac{\hbar^2}{2}e^{-2\gamma t}\partial_{xx}\right]\psi(x,t)$$

which can be analytically solved yielding the wave function

$$\psi_n(x,t) = N\exp\left[\left(-\frac{i}{\hbar}\epsilon_n + \frac{\gamma}{2}\right)t - \left(\frac{\Omega + i\gamma}{2\hbar}\right)x^2 e^{2\gamma t}\right] H_n\left(\sqrt{\frac{\Omega}{\hbar}}x e^{\gamma t}\right),$$

where N is the normalization constant and

$$\epsilon_n = \hbar\Omega\left(n + \frac{1}{2}\right), \qquad n = 0, 1, 2, \ldots .$$

$H_n(\xi)$ denote Hermite polynomials. Now, one can calculate the mean energy at a time instant t

$$E_n(t) = \int_{\mathbb{R}} \psi_n^*(x,t)\widehat{H}\psi_n(x,t)dx = \hbar\omega\left(\frac{\omega}{\Omega}\right)e^{-2\gamma t}\left(n + \frac{1}{2}\right), \ n = 0, 1, 2, \ldots$$

What is astonishing in the above result, is that the mean energy tends to 0 as $t \to \infty$ (even for the ground state) decreasing below the vaccum energy $\frac{1}{2}\hbar\omega$ of the undamped oscillator. We will not go into details here [13], but there is a similar problem with uncertainty relations which are violated in the limit $t \to \infty$.

3.2. *Feshbach-Tikochinsky quantization of Bateman system*

The Bateman Hamiltonian (10)

$$H_B(x, y, p_x, p_y) = p_x p_y + \Omega^2 xy - \gamma(xp_x - yp_y), \qquad \Omega = \sqrt{\omega^2 - \gamma^2}$$

after suitable symmetrization can be rewritten at the quantum level as

$$\widehat{H}_B = \widehat{H}_0 + \widehat{H}_1,$$

where

$$\widehat{H}_0 = \widehat{p}_x\widehat{p}_y + \Omega^2\widehat{x}\widehat{y}\,, \qquad \widehat{H}_1 = -\frac{\gamma}{2}(\widehat{x}\widehat{p}_x + \widehat{p}_x\widehat{x} - (\widehat{y}\widehat{p}_y + \widehat{p}_y\widehat{y}))\,. \tag{13}$$

The standard procedure to transform (13) into the simpler form is to introduce annihilation and creation operators

$$\widehat{A} = \frac{1}{2\sqrt{\hbar\Omega}}\left[(\widehat{p}_x + \widehat{p}_y) - i\Omega(\widehat{x} + \widehat{y})\right]\,,$$
$$\widehat{B} = \frac{1}{2\sqrt{\hbar\Omega}}\left[(\widehat{p}_x - \widehat{p}_y) - i\Omega(\widehat{x} - \widehat{y})\right]\,.$$

As a consequence we arrive at the *quantum Feshbach-Tikochinsky Hamiltonian* (see [13] for details)

$$\widehat{H}_{FT} \;=\; \hbar\Omega(\widehat{A}^\dagger\widehat{A} - \widehat{B}^\dagger\widehat{B}) + i\hbar\gamma(\widehat{A}^\dagger\widehat{B}^\dagger - \widehat{A}\widehat{B}) \tag{14}$$

In order to get energy levels for $\widehat{H}_{FT}$ we introduce three generators of su(1,1) algebra

$$\widehat{J}_1 = \frac{1}{2}(\widehat{A}^\dagger\widehat{B}^\dagger + \widehat{A}\widehat{B})\,, \quad \widehat{J}_2 = \frac{i}{2}(\widehat{A}^\dagger\widehat{B}^\dagger - \widehat{A}\widehat{B})\,, \quad \widehat{J}_3 = \frac{1}{2}(\widehat{A}^\dagger\widehat{A} + \widehat{B}\widehat{B}^\dagger)$$

and the Casimir opeartor $\widehat{J}_0 = \frac{1}{2}(\widehat{A}^\dagger\widehat{A} - \widehat{B}^\dagger\widehat{B})$. The generators fulfill commutation relations

$$[\widehat{J}_1, \widehat{J}_2] = i\widehat{J}_3\,, \quad [\widehat{J}_3, \widehat{J}_2] = i\widehat{J}_1\,, \quad [\widehat{J}_1, \widehat{J}_3] = i\widehat{J}_2\,.$$

Now, we can rewrite (14) in terms of su(1,1) generators as

$$\widehat{H}_{FT} \;=\; 2\hbar\Omega\widehat{J}_0 + 2\hbar\gamma\widehat{J}_2$$

and to obtain the spectrum

$$E^{\pm}_{jm} = 2\hbar\Omega j \pm i\hbar\gamma(2m+1)\,, \tag{15}$$

$$j = 0, \pm\frac{1}{2}, \pm 1, \pm\frac{3}{2}, \ldots\,, \qquad m = |j|, |j|+1, \ldots$$

Note, that the eigenvalues are complex in spite of hermiticity of the quantum Feshbach-Tikochinsky Hamiltonian (14). In fact, one can show that the eigenvectors of $\widehat{H}$ corresponding to complex energies cannot be normalized and should be considered as generalized eigenvectors not belonging to the Hilbert space.

3.3. *Pontriagin's method*

There is a general old method going back to Pontriagin [14] of representing any dynamical system

$$\dot{x}_i = f_i(x_1, \dots, x_n)\,, \quad i = 1, \dots, n\,, \quad x = (x_1, \dots, x_n) \in Q\,, \tag{16}$$

described on a configuration space Q as a Hamiltonian one but defined on the extended space $Q \times \mathbb{R}^n$ with additional variables $(p_1, \dots, p_n)$. The Hamiltonian function defined as

$$H(x,p) = \sum_{i=1}^{n} p_i f_i(x)\,, \qquad (x,p) \in Q \times \mathbb{R}^n,$$

generates the Hamiltonian equations

$$\dot{x}_k = \{x_k, H\} \;=\; f_k(x)\,, \tag{17}$$

$$\dot{p}_k = \{p_k, H\} \;=\; -\sum_l p_l \frac{\partial f_l(x)}{\partial x_k} \tag{18}$$

with the standard Poisson bracket

$$\{F, G\} \;=\; \sum_{k=1}^{n} \left(\frac{\partial F}{\partial x_k} \frac{\partial G}{\partial p_k} - \frac{\partial G}{\partial x_k} \frac{\partial F}{\partial p_k} \right).$$

Equation (17) reproduces (16), whereas (18) provides the dynamics of variables $(p_1, \dots, p_n)$ which could be interpreted as additional degrees of freedom describing the influence of environment.

Unfortunately, such a representation is nonunique. For example, for a damped harmonic oscillator one can construct at least two Hamiltonians:

$$\mathfrak{H}(x_1, x_2, p_1, p_2) = \Omega(p_1 x_2 - p_2 x_1) - \gamma(p_1 x_1 + p_2 x_2) \tag{19}$$

producing

$$\ddot{x} + 2\gamma\dot{x} + \omega^2 x \;=\; 0 \quad \longrightarrow \quad \begin{cases} \dot{x}_1 = -\gamma x_1 + \Omega x_2 \\ \dot{x}_2 = -\gamma x_2 - \Omega x_1\,, \end{cases}$$

and another one

$$\widetilde{\mathfrak{H}}(x_1, x_2, p_1, p_2) \;=\; p_1 x_2 - \omega^2 x_1 p_2 - 2\gamma x_2 p_2$$

yielding

$$\ddot{x} + 2\gamma\dot{x} + \omega^2 x \;=\; 0 \quad \longrightarrow \quad \begin{cases} \dot{x}_1 = x_2 \\ \dot{x}_2 = -2\gamma x_2 - \omega^2 x_1\,. \end{cases}$$

In what follows we focus on a quantization procedure based on (19) due to its close relation with the Bateman model. Surprisingly, the Hamiltonians

(19) and (10) are related by the following simple canonical transformation $(x, y, p_x, p_y) \to (x_1, x_2, p_1, p_2)$

$$x_1 = \frac{p_y}{\sqrt{\omega}}, \qquad p_1 = -\sqrt{\omega}y,$$
$$x_2 = -\sqrt{\omega}x, \qquad p_2 = -\frac{p_x}{\sqrt{\omega}}.$$

As a consequence, they are equivalent, at least at the classical level.

3.4. *Quantization of the Hamiltonian* $\mathfrak{H}$

The quantum version of $\mathfrak{H}$ reads

$$\widehat{\mathfrak{H}} = \underbrace{i\hbar\Omega\left(x_1\frac{\partial}{\partial x_2} - x_2\frac{\partial}{\partial x_1}\right)}_{\hat{\mathfrak{H}}_0} + \underbrace{i\hbar\gamma\left(x_1\frac{\partial}{\partial x_1} + x_2\frac{\partial}{\partial x_2} + 1\right)}_{\hat{\mathfrak{H}}_1}.$$

Following the quantization procedure we construct the unitary operator

$$\widehat{U}_t = \exp(t\widehat{\mathfrak{H}}/i\hbar) = e^{-(i/\hbar)\hat{\mathfrak{H}}_0 t}\, e^{-(i/\hbar)\hat{\mathfrak{H}}_1 t}$$

and the wave function (12) for the Hamiltonian (19). Using the dilation generator

$$\exp\left(tx\frac{d}{dx}\right) f(x) = f(e^t x)$$

and the planar rotation generator

$$\exp\left[t\left(x\frac{\partial}{\partial y} - y\frac{\partial}{\partial x}\right)\right] = f(x\cos t - y\sin t, x\sin t + y\cos t)$$

one arrives and the following time-dependent wave function

$$\begin{aligned}\psi_t(x_1, x_2) &= U_t\psi(x_1, x_2) \qquad (20)\\ &= e^{\gamma t}\psi\left(e^{\gamma t}x_1\cos\Omega t - e^{\gamma t}x_2\sin\Omega t, e^{\gamma t}x_1\sin\Omega t + e^{\gamma t}x_2\cos\Omega t\right).\end{aligned}$$

Let us note that the same result can be obtained even simpler using polar coordinate representation (r, φ) defined as

$$x_1 + ix_2 = re^{i\varphi}.$$

Indeed, the classical Hamiltonian $\mathfrak{H}$ simplifies considerably and takes the form

$$\mathfrak{H} = -\Omega p_\varphi - \gamma r p_r,$$

where p_r and p_φ are momenta conjugate to (r, φ),

$$p_\varphi = L_3, \qquad p_r = \frac{\mathbf{x}\cdot\mathbf{p}}{r},$$

with L_3 denoting the 3rd component of angular momentum $\mathbf{L} = \mathbf{x} \times \mathbf{p}$. At the quantum level we obtain the Hamiltonian

$$\widehat{\mathfrak{H}} = \underbrace{i\hbar\omega\frac{\partial}{\partial\varphi}}_{\hat{\mathfrak{H}}_0} + \underbrace{i\hbar\gamma\left(r\frac{\partial}{\partial r} + 1\right)}_{\hat{\mathfrak{H}}_1} . \tag{21}$$

and, consequently, the unitary evolution of the initial wave function in polar coordinates takes the form

$$U_t\psi(r,\varphi) = e^{\gamma t}\exp\left(\Omega t\frac{\partial}{\partial\varphi}\right)\exp\left(\gamma t r\frac{\partial}{\partial r}\right)\psi(r,\varphi) = e^{\gamma t}\psi\left(e^{\gamma t}r, \varphi + \Omega t\right) .$$

In polar coordinates it is straightforward to calculate energy levels and complete the set of eigenfunctions of (21). One finds that [15]

$$\widehat{\mathfrak{H}}\Psi_{l\lambda} = E_{l\lambda}\Psi_{l\lambda} \tag{22}$$

is solved by

$$\Psi_{l\lambda}(r,\varphi) = \Phi_l(\phi)R_\lambda(r) ,$$

where

$$\Phi_l(\varphi) = \frac{e^{-il\varphi}}{\sqrt{2\pi}} , \quad l = 0, \pm1, \pm2, \dots ,$$

and

$$R_\lambda(r) = \frac{r^{-(i\lambda+1)}}{\sqrt{2\pi}} , \quad \lambda \in \mathbb{R} .$$

Note that R_λ does not belong to $L^2(\mathbb{R}_+, rdr)$. The energy values

$$E_{l\lambda} = \hbar(l\omega + \lambda\gamma) .$$

fulfilling (22) are real and continuous but we can relate them to discrete complex eigenvalues of the quantum Feshbach-Tikochinsky Hamiltonian (14), which describes the quantized Bateman system, via analytical continuation of $\Psi_{l\lambda}$ in λ (see [15] for details). One can show that $\Psi_{l\lambda}$, $\lambda \in \mathbb{C}$, considered as a distribution, has poles in $\lambda_{nl} = -i(|l|+2n+1)$, $n = 0, 1, 2, \dots$ The corresponding residues of $\Psi_{l\lambda}$ in λ_{nl} give raise to the set of generalized eigenvectors $|\mathfrak{f}^-_{nl}$ of $\widehat{\mathfrak{H}}$ and the residues of the complex conjugated distribution $\Psi^*_{l\lambda}$ introduce the set $|\mathfrak{f}^+_{nl}$ of generalized eigenvectors, i.e.,

$$\widehat{\mathfrak{H}}|\mathfrak{f}^\pm_{nl} = E^\pm_{nl}|\mathfrak{f}^\pm_{nl}$$

with

$$E^\pm_{nl} = \hbar\omega l \pm i\hbar\gamma(|l| + 2n + 1) . \tag{23}$$

The above formula is equivalent to the Feshbach-Tikocinsky spectrum (15) after the following identifications

$$j = \frac{l}{2}\,, \qquad m = \frac{1}{2}(|l| + 2n)\,.$$

Discrete eigenvalues $E^{\pm}_{nl}$ constitute poles of the resolvent operator of $\widehat{\mathfrak{H}}$ and it is known [16] that these poles correspond to the so-called resonant states. Hence, the complex values (23) may be interpreted as resonances of the quantum Feshbach-Tikochinsky Hamiltonian.

In consequence, our model indicates a connection between dissipation phenomena and resonances in atomic physics. It proves that dissipation and irreversibility are caused by the presence of resonances. From the mathematical point of view our model gives rise to the formalism of Gelfand triplet (or rigged Hilbert space) [17,18] which allows one to deal with generalized (non-normalised) eigenvectors.

References

1. A. Isar, A. Sandulescu, W. Scheid, Physica A **322**, 233 (2003). A. Isar, W. Scheid, Physica A **335**, 79 (2004).
2. Kyu-Hwang Yeon, Suk-Sung Kim, Young-Mo Moon, Suc-Kyoung Hong, Chung-In Um and T. F. George, J. Phys. A: Math. Gen. **34**, 7719 (2001).
3. M. Montesinos, Phys. Rev. A **68**, 014101 (2003).
4. U. Das, S. Ghosh, P. Sarkar, B. Talukdar, Physica Scripta **71**, 235 (2005).
5. V. G. Kupriyanov, S. L. Lyakhovich, and A. A. Sharapov, J. Phys. A: Math. Gen. **38**, 8039 (2005).
6. G. Dito, F. J. Turrubiates, Phys. Lett. A **352**, 309 (2006).
7. G. Lopez, Annals of Physics **251**, 372 (1996).
8. G.'t Hooft, in: *Basics and Highlights of Fundamental Physics*, Erice, 1999, arxiv: hep-th 0003005.
9. M. Blasone, P. Jizba, G. Vitiello, Phys. Lett. A **287**, 205 (2001); M. Blasone, P. Jizba, Ann. Phys. (NY) **312**, 354 (2004).
10. J. A. Kobussen, Acta Phys. Austr. **51**, 193 (1979).
11. V. K. Chandrasekar *et al.*, J. Math. Phys. 2007.
12. H. Bateman, Phys. Rev. **38**, 815 (1931).
13. H. Dekker, Phys. Rep. **80**, 1 (1981).
14. L. S. Pontriagin, V. G. Baltanskij, R. V. Gamkrelidze, E. F. Miscenko, *The Mathematical Theory of Optimal Processes*, Wiley, New York, 1962.
15. D. Chruściński, J. Jurkowski, Ann. Phys. **321**, 854 (2006).
16. S. Albeverio, L. S, Ferreira, L. Streit (eds), *Resonances — Models and Phenomena*, Lecture Notes in Physics **211**, Springer, Berlin, 1984.
17. I. M. Gelfand, N. J. Vilenkin, *Generalized Functions*, vol. IV, Academic Press, New York, 1964.
18. A. Bohm, H.-D. Doebner, P. Kielanowski, *Irreversibility and Causality, Semigroups and Rigged Hilbert Spaces*, Lecture Notes in Physics **504**, Springer, Berlin, 1998.

Quantum Bio-Informatics III
eds. L. Accardi, W. Freudenberg & M. Ohya

CLASSICAL AND QUANTUM PROBABILITY FOR BIOLOGISTS – INTRODUCTION

ANDREI. KHRENNIKOV

International Center for Mathematical Modelling
in Physics and Cognitive Sciences
University of Växjö, S-35195, Sweden
E-mail: Andrei.Khrennikov@vxu.se

The aim of this review (oriented to biologists looking for applications of QM) is to provide a detailed comparative analysis of classical (Kolmogorovian) and quantum (Dirac-von Neumann) models. We will stress differences in the definition of conditional probability and as a consequence in the structures of matrices of transition probabilities, especially the condition of double stochasticity which arises naturally in QM. One of the most fundamental differences between two models is deformation of the classical formula of total probability (FTP) which plays an important role in statistics and decision making. An additional term appears in the QM-version of FTP – so called interference term. Finally, we discuss Bell's inequality and show that the common viewpoint that its violation induces either nonlocality or "death of realism" has not been completely justified. For us it is merely a sign of *non-Kolmogorovianity of probabilistic data* collected in a few experiments with incompatible setups of measurement devices.

1. QM as statistical theory

In this review we recall fundamentals of classical and quantum probabilistic models. The main attention will be paid to stressing of similarities and differences of these models. We hope that our review will be useful for biologists who are interested in applications of QM.

The basic message of this review is that QM is a *statistical theory.* The formalism of QM cannot tell us anything about an individual quantum system, e.g., electron or photon. It predicts only probabilities for results of measurements for huge ensembles of quantum systems. Therefore the main difference between classical and quantum physics is in probabilistic models: on the one hand, the Kolmogorovian measure-theoretic and, on the other hand, the model of Dirac- von Neumann with *Born's rule for probabilities.*

2. Classical Kolmogorovian Probability

The modern axiomatics of probability theory was invented by Andrei Nikolaevich Kolmogorov – 1933, [1]. A crucial point is representation of events by subsets of some basic set Ω.

Let Ω be a set. We recall that a *σ-algebra* (=*σ-field*) $\mathbf{F}$ is a system of subsets of Ω which is closed with respect to operations of countable intersection, union and difference of sets and containing Ω and the empty set $\emptyset$. In particular, if $A, B \in \mathbf{F}$, then their intersection $C = A \cap B$ also belongs to $\mathbf{F}$. In fact, it is a very strong restriction on the model. For example, for some quantum experiments this condition is violated.

Let Ω be a set and let $\mathbf{F}$ be a σ-algebra of its subsets. A probability measure $\mathbf{P} : \mathbf{F} \to [0,1]$, $\mathbf{P}(\Omega) = 1$, and σ-additive $\mathbf{P}(\cup_n A_n) = \sum_n \mathbf{P}(A_n)$ for disjoint sets from $\mathbf{F}$. By the Kolmogorov axiomatics, the *probability space* is a triple $\tilde{P} = (\Omega, \mathbf{F}, \mathbf{P})$. Points ω of Ω are said to be *elementary events*, elements of $\mathbf{F}$ are *events*, $\mathbf{P}$ is *probability*. *Discrete random variables* on the Kolmogorov space $\tilde{P}$ are by definition functions $a : \Omega \to X_a$, where $X_a = \{\alpha_1, ..., \alpha_n, ...\}$ is a countable set (the *range of values*), such that $C^a_\alpha = \{\omega \in \Omega : a(\omega) = \alpha\} \in \mathbf{F}, \alpha \in X_a$. The probability distribution is defined by

$$p^a(\alpha) \equiv \mathbf{P}(a = \alpha) = \mathbf{P}(C^a_\alpha).$$

In the general case random variables are defined as *measurable functions.* **Connection with experiment:** In the classical framework observables are represented by random variables, i.e., by measurable functions $f : \Omega \to \mathbf{R}$. We remark that in real experimental situations both in classical and quantum framework only discrete observables are in use. Any detector induces a scale of discretization. Therefore we shall proceed with only discrete random variables. *Conditional probability* is defined by *Bayes' formula*

$$\mathbf{P}(B|C) = \mathbf{P}(B \cap C)/\mathbf{P}(C), \mathbf{P}(C) > 0. \tag{1}$$

For a (discrete) random variable a, conditional probability to get $a = \alpha$ is given by $p^a_C(\alpha) \equiv \mathbf{P}(a = \alpha|C)$, $\alpha \in X_a$. For two random variables a and b, consider conditional probabilities

$$p_{\beta|\alpha} \equiv \mathbf{P}(b = \beta|a = \alpha), \; p_{\alpha|\beta} \equiv \mathbf{P}(a = \alpha|b = \beta).$$

Following tradition, will call these probabilities *transition probabilities.* In future considerations we shall use the matrices of conditional probabilities for successive measurements – transition probabilities $\mathbf{P}^{b|a} = (p_{\beta|\alpha})$. We remark that these matrices are always *left stochastic.* A left stochastic matrix is a square matrix whose columns consist of nonnegative real numbers whose sum is 1: $\sum_\beta p_{\beta|\alpha} = \sum_\beta \mathbf{P}(b = \beta|a = \alpha) = 1$. We remind that in a *doubly stochastic matrix* all entries are nonnegative and all rows and all columns sum to 1. The following proposition characterizes random variables inducing doubly stochastic matrices.

Proposition 2.1. *Let* $a = \alpha_1, \alpha_2$ *and* $b = \beta_1, \beta_2$ *be dichotomous random variables. Then the following conditions are equivalent:* ***DS-DS*** *Both matrices* $\mathbf{P}^{a|b}$ *and* $\mathbf{P}^{b|a}$ *are doubly stochastic.* ***UD*** *Random variables are uniformly distributed:* $p^a(\alpha) = p^b(\beta) = \frac{1}{2}$. ***SC*** *Random variables are "symmetrically conditioned" in the following sense* $p_{\beta|\alpha} = p_{\alpha|\beta}$.

Thus in the Kolmogorovian model (for dichotomous random variables) one can guarantee double stochasticity for both $b|a$- and $a|b$-conditioning only for uniformly distributed random variables. It is not the case in non-Kolmogorovian models, e.g., for the quantum probabilistic model. This result can be used as a test of Kolmogorovianity. However, it seems that this test is not useful for applications outside QM, e.g., in biology: it seems that **DS** is a very special feature of QM. By Kolmogorov's axiomatics two events $A, B \in \mathbf{F}$ are *independent* if

$$\mathbf{P}(A \cap B) = \mathbf{P}(A)\mathbf{P}(B). \tag{2}$$

Consequently, two random variables a and b are independent if the joint probability distribution can be factorized:

$$p^{ab}(\alpha, \beta) = p^a(\alpha)p^b(\beta), \tag{3}$$

where $p^{ab}(\alpha, \beta) = \mathbf{P}(\omega \in \Omega : a(\omega) = \alpha, b(\omega) = \beta)$. We remark that Kolmogorov's definition of independence involves the joint probability distribution of two random variables. It is not always possible in non-Kolmogorovian considerations. Let us write condition of independence (3) by using only transition probabilities. The latter can be defined even in non-Kolmogorovian models. By using Bayes formula and (3) we obtain for two independent random variables

$$p_{\beta|\alpha} \equiv \mathbf{P}(b = \beta|a = \alpha) = p^b(\beta) \tag{4}$$

for all values $a = \alpha$. Consider two successive measurements, first of a and then of b, then the result of the latter is the same as if a was not done. In the same way

$$p_{\alpha|\beta} \equiv \mathbf{P}(a = \alpha|b = \beta) = p^a(\alpha). \tag{5}$$

Now let a and b satisfy conditions (4) or (5). Then (3) holds.

3. Formula of total probability

In our further considerations the important role will be played by the *formula of total probability* (FTP). It is a theorem of the Kolmogorov's model. Let us consider a countable family of disjoint sets A_k belonging to $\mathbf{F}$ such that their union is equal to Ω and $\mathbf{P}(A_k) > 0, k = 1, \ldots$. Such a family is called a *partition* of the space Ω.

Theorem 3.1. *Let $\{A_k\}$ be a partition. Then, for every set $B \in \mathbf{F}$, the following formula of total probability holds*

$$\mathbf{P}(B) = \mathbf{P}(A_1)\mathbf{P}(B|A_1) + \ldots + \mathbf{P}(A_k)\mathbf{P}(B|A_k) + \ldots \tag{6}$$

Especially interesting for us is the case such that a partition is induced by a discrete random variable a taking values $\{\alpha_k\}$. Let b be another random variable. It takes values $\{\beta_j\}$. For any $\beta \in X_b$, we have

$$\begin{aligned}\mathbf{P}(b = \beta) = \mathbf{P}(a = \alpha_1)\mathbf{P}(b = \beta|a = \alpha_1) + \ldots \\ +\mathbf{P}(a = \alpha_k)\mathbf{P}(b = \beta|a = \alpha_k) + \ldots\end{aligned} \tag{7}$$

Test of Kolmogorovianity N 1 – FTP: *Suppose that data obtained from measurements of two observables a and b are such that FTP (7) is violated. We can conclude that data is not classical, so it could not be described by the Kolmogorovian model.* This test has been used in psychology and cognitive science, see [2] – [7]. Our aim to stimulate biologists to perform this test in various domains of biology, including genetics.

At the end of this short introduction to classical probability we recommend two books for further reading. First of all it is Kolmogorov's monography [1] containing axiomatics of modern probability theory as well as the ideology of the (classical) probabilistic description of measurements. We think that any young researcher who plans to work in quantum probability and information theory should read Kolmogorov's book. We also mention the book [8] of one of the authors of this review.[a]

[a]It contains all known classical probabilistic models, in particular, von Mises' frequency probability model. We remind that in Kolmogorov's model connection between probability and frequencies is established by the *law of large numbers*. In von Mises' model probability is directly defined as the limit of frequencies. This book also contains an extended discussion on measurement theory in classical and quantum frameworks, in

4. Probabilistic Incompatibility: Bell-Boole inequalities

Bell's inequality is really the central point of modern QM. However, we would like to stress that this inequality is a standard subject of classical probability theory. Moreover, such type of inequalities appeared in mathematics long before not only Bell's invention, but even discovery of QM.

Our main message to biologists: attempts to associate Bell's type inequalities with *quantum nonlocality* or *death of realism* are not sufficiently justified. So, one should be very careful by borrowing from QM some of its mystical features, e.g., quantum nonlocality.

4.1. *Views of Kolmogorov: Contextuality of Kolmogorov's space*

Andrei Nikolaevich Kolmogorov emphasized [1] that each experimental arrangement (context) generates its *own probability space.* In particular, the following problem arises. A family of observables, say $\mathbf{O} = \{a_1, a_2, a_3, \ldots\}$, is given. However, it is impossible to measure all them simultaneously. Thus the joint probability distribution is not given.

Nevertheless, it is possible to measure some groups of these observables and joint probability distributions for such groups are given. Question:

Is it possible to construct a single probability space serving for the whole family $\mathbf{O}$*?* Thus we are interested in a possibility to embed the family of observables $\mathbf{O}$ into the space of random variables on a single probability space. If the answer is yes, then such observables are called *probabilistically compatible* (PC), in the opposite case, they called *probabilistically incompatible* (PI).

It seems that G. Boole (the inventor of *Boolean logic* and *Boolean algebra*) was the first who studied this problem. He formulated a necessary condition for PC of a family of three dichotomous observables, $a_1, a_2, a_3 = \pm 1$, such that they can be measured pairwise, but not all simultaneously. This condition coincides with famous Bell's inequality!

Later the most general problem of PC (i.e., for an arbitrary family of observables) was solved by Soviet mathematician Vorobj'ev [9] who applied these results to problems of optimal control and game theory.

---

particular, the problem of realism for results of measurements as well as discussion on various interpretations of the wave function. Bell's inequality is analyzed. Its violation is considered as violation of Kolmogorovianity. Generalizations of Bell's inequality are obtained in non-Kolmogorov framework.

Open mind people would agree that in general probabilistic data collected in a few incompatible experiments cannot be described by a single probability space, or in other words, observables are not of the PC-type, cf. with views of Accardi, Aerts, Fine, Fuchs, Garola and Solombrino, Hess and Philipp, Khrennikov, Kupczynski, Larsson, Pitowsky, Rastal, Sozzo, Adenier, see works [3,8] for detailed bibliography.

Why can one hope to be able put statistical data collected in incompatible experiments into one probability space? [i.e., to find the same space of elementary events Ω and a single probability measure ¶ which will serve to such collections of data]. People do not try to do this in other domains of science. For example, in sociology such an approach may imply unfair sampling: taking statistically different samples for asking of different pairs of questions (such a measurement procedure would mimic well quantum measurements violating Bell's inequality) one can get funny sociological conclusions.

We recognized that the real problem is *not statistical data by itself,* but an attempt to combine various collections of data. Hence, we should pay attention not to special (mystical) properties of systems for which data is collected (as majority of people do in physics), but to ways in which data is collected. In this way we can find other sources of probabilistic incompatibility, different from nonlocality or death of realism, see, e.g., [3], [8].

Consider three observables $a_i, i = 1, 2, 3$. Suppose that they are dichotomous: $a_i = \pm 1$. Suppose that these observables as well as their pairs can be measured and hence joint probabilities for pairs are well defined: $p^{a_i a_j}(\alpha_i, \alpha_j) \geq 0$ and $\sum_{\alpha_i, \alpha_j = \pm 1} p^{a_i a_j}(\alpha_i, \alpha_j) = 1$.

Question 4.1. *Is it possible to construct the joint probability distribution,* $p^{a_1 a_2 a_3}(\alpha_1, \alpha_2, \alpha_3)$, *for any triple of pairwise measurable observables?*

This is the simplest case of the general problem – to find conditions for existence of probability distribution with given *marginal probabilities.* First of all, it is easy to give numerous examples of nonexistence.

Example 4.1. (Vorob'ev [9]) Suppose that

$$\mathbf{P}(a_1 = +1, a_2 = +1) = \mathbf{P}(a_1 = -1, a_2 = -1) = 1/2;$$

$$\mathbf{P}(a_1 = +1, a_3 = +1) = \mathbf{P}(a_1 = -1, a_3 = -1) = 1/2;$$

$$\mathbf{P}(a_2 = +1, a_3 = -1) = \mathbf{P}(a_2 = -1, a_3 = +1) = 1/2.$$

Hence, $\mathbf{P}(a_1 = +1, a_2 = -1) = \mathbf{P}(a_1 = -1, a_2 = +1) = 0$; $\mathbf{P}(a_1 = +1, a_3 = -1) = \mathbf{P}(a_1 = -1, a_3 = +1) = 0$, $\mathbf{P}(a_2 = +1, a_3 = +1) = \mathbf{P}(a_2 = -1, a_3 = -1) = 0$. There is no probability measure with these marginals. Suppose that one can find a family of real constants $\mathbf{P}(\epsilon_1, \epsilon_2, \epsilon_3), \epsilon_j = \pm 1$, such that

$$\mathbf{P}(\epsilon_1, \epsilon_2, +1) + \mathbf{P}(\epsilon_1, \epsilon_2, -1) = \mathbf{P}(a_1 = \epsilon_1, a_2 = \epsilon_2), ...,$$

$$\mathbf{P}(+1, \epsilon_2, \epsilon_3) + \mathbf{P}(-1, \epsilon_2, \epsilon_3) = \mathbf{P}(a_2 = \epsilon_2, a_3 = \epsilon_3).$$

Then it is easy to see that some of these numbers should be negative. We remark that Bell's inequality is violated, see later.

5. Bell's and Winger's inequality

Let $\tilde{P} = (\Omega, \mathbf{F}, P)$ be a Kolmogorov probability space.

Theorem 5.1. *(Bell inequality for covariances) Let $a_1, a_2, a_3 = \pm 1$ be random variables on $\tilde{P}$. Then Bell's inequality for pairwise covariances of these random variables:*

$$|\langle a_1, a_2 \rangle - \langle a_2, a_3 \rangle| \le 1 - \langle a_3, a_1 \rangle \tag{8}$$

holds.

We now turn to Vorobj'ev Example. If a_1, a_2, a_3 can be realized on the same probability space, then (8) would hold. On the other hand, we have $\langle a_1, a_2 \rangle = 1$; $\langle a_1, a_3 \rangle = 1; \langle a_2, a_3 \rangle = -1$. Bell's inequality should imply: $1 - (-1) = 2 \le 1 - 1 = 0$. We remark that in accordance with Boole we consider Bell's inequality just as a necessary condition for probabilistic compatibility.

We also recall the following simple mathematical result:

Theorem 5.2. *(Wigner inequality) Let $a_1, a_2, a_3 = \pm 1$ be arbitrary random variables on a Kolmogorov space $\tilde{P}$. Then the following inequality holds:*

$$\mathbf{P}(a_1 = +1, a_2 = +1) + \mathbf{P}(a_2 = -1, a_3 = +1) \ge \mathbf{P}(a_1 = +1, a_3 = +1). \tag{9}$$

5.1. *Clauser-Horne-Shimony-Holt's Inequality*

Theorem 5.3. *(CHSH-inequality) Let $\tilde{P} = (\Omega, \mathbf{F}, \mathbf{P})$ be a Kolmogorov probability space and let $\xi_j(\omega)$ and $\xi_j'(\omega), j = 1, 2$, be random variables such that*

$$|\xi_j(\omega)| \le 1,\ |\xi_j'(\omega)| \le 1,\ \textit{for almost all}\, \omega \in \Omega. \tag{10}$$

Then the following inequality for correlations holds

$$< \xi_1, \xi_1' > + < \xi_1, \xi_2' > + < \xi_2, \xi_1' > - < \xi_2, \xi_2' > \leq 2. \qquad (11)$$

Proof. It is easy to prove the following elementary algebraic inequality for numbers bounded by 1 $u_1v_1 + u_1v_2 + u_2v_1 - u_2v_2 \leq 2$. Thus

$$\xi_1(\omega)\xi_1'(\omega) + \xi_1(\omega)\xi_2'(\omega) + \xi_2(\omega)\xi_1'(\omega) - \xi_2(\omega)\xi_2'(\omega) \leq 2. \qquad (12)$$

□

Finally, we integrate this inequality with respect to the measure **P**. We see that it is crucial that all covariances are based on the same probability measure. If this condition were violated, it would be impossible to get this inequality. **Test of Kolmogorovianity N 2 - Bell's type inequalities**:

Let data obtained from pair wise measurements be such that one of Bell's type inequalities (12) is violated. We can conclude that data is nonclassical. It could not be described by the Kolmogorovian model.

5.2. *Bell's type inequalities for conditional probabilities*

Original *Boole-Bell inequality* is based on the assumption that pairwise probability distributions are well defined – observables can be measured pairwise. However, even such an assumption is not always satisfied. Sometimes it is only possible to perform conditional measurements. The simplest test of PC is based on conditional probabilities.

Theorem 5.4. *(Wigner-Khrennikov inequality[3]) Let $a_1, a_2, a_3 = \pm 1$ be arbitrary random variables on a Kolmogorov space $\tilde{P}$. Then the following inequality holds:*

$$\mathbf{P}(a_1 = +1)\mathbf{P}(a_2 = +1|a_1 = +1) + \mathbf{P}(a_2 = -1)\mathbf{P}(a_3 = +1|a_2 = -1)$$

$$\geq \mathbf{P}(a_3 = +1)\mathbf{P}(a_1 = +1|a_3 = +1). \qquad (13)$$

Test of Kolmogorovianity N 3 - violation of Bell's type inequalities for conditional probability: *Suppose that data obtained from pair wise conditional measurements be such that inequality (13) is violated. We can conclude that data is nonclassical.*

6. Postulates of QM

The mathematical formalism of quantum mechanics is the theory of self-adjoint operators on complex Hilbert spaces. The symbols H and $\langle \cdot, \cdot \rangle$ denote separable complex Hilbert space and the scalar product on it, $S = \{\psi \in H : \|\psi\| = 1\}$ is the unit sphere.

Biologists who are not familiar with general theory of Hilbert spaces can consider only finite dimensional spaces which are given by Cartesian products: $H \equiv \mathbf{C} \times \ldots \times \mathbf{C}$. Self-adjoint operators are given by Hermitian matrices; spectrum of an operator is simply the set of its eigenvalues.

Postulate 1. (Quantum states). *Quantum (pure) states (wave functions) are represented by normalized vectors ψ of H. Every normalized vector $\psi \in H$ may represent a quantum state. If a vector ψ corresponding to a state is multiplied by any complex number $c, |c| = 1$, the resulting vector will correspond to the same state.*

Postulate 2. (Physical observables). *A physical observable a is represented by a self-adjoint operator $\widehat{a}$ in complex Hilbert space H. Different observables are represented by different operators.*

Postulate 3. (Spectral) *For a physical observable a which is represented by the self-adjoint operator $\widehat{a}$ we can predict (together with some probabilities) values $\lambda \in$* Spec$(\widehat{a})$.

Consider operators which are analogous to discrete random variables. A self-adjoint operator $\widehat{a}$ has *purely discrete spectrum* if

$$\widehat{a} = \alpha_1 P^a_{\alpha_1} + \ldots + \alpha_m P^a_{\alpha_m} + \ldots, \quad \alpha_m \in \mathbf{R}, \tag{14}$$

where $P^a_{\alpha_m}$ are projections on subspaces corresponding to the eigenvalues α_m.

Postulate 4. (Born's rule– in formalization of Dirac and von Neumann) *Let a physical observable a be represented by a self-adjoint operator $\widehat{a}$ with purely discrete spectrum. The probability $\mathbf{P}_\psi(a = \alpha_m)$ to obtain the eigenvalue α_m of $\widehat{a}$ for measurement of a in a state ψ is given by*

$$\mathbf{P}_\psi(a = \alpha_m) = \|P^a_m \psi\|^2. \tag{15}$$

If the operator $\widehat{a}$ has nondegenerate (purely discrete) spectrum, then each α_m is associated with one dimensional subspace. The latter can be fixed by selecting any normalized vector, say e^a_m. In this case orthogonal projectors act simply as $P^a_{\alpha_m}\psi = \langle\psi, e^a_m\rangle e^a_m$. The formula (15) takes a very simple form

$$\mathbf{P}_\psi(a = \alpha_m) = |\langle\psi, e^a_m\rangle|^2. \tag{16}$$

It is Born's rule in the Hilbert space formalism.

Postulate 5. (Time evolution). *Let $\widehat{H}$ be the Hamiltonian of a quantum system, i.e., the self-adjoint operator corresponding to the energy observable. The time evolution of the wave function $\psi \in H$ is described by the Schödinger equation $i\frac{d}{dt}\psi(t) = \widehat{H}\psi(t)$ with the initial condition $\psi(0) = \psi$.*

7. Quantum Conditional Probability

Let physical observables a and b be represented by self-adjoint operators with purely discrete (may be degenerate) spectra:

$$\widehat{a} = \sum_m \alpha_m P^a_{\alpha_m}, \ \widehat{b} = \sum_m \beta_m P^b_{\beta_m} \tag{17}$$

Let ψ be a pure state and let $P^a_{\alpha_k}\psi \neq 0$. Then the probability to get the value $b = \beta_m$ under the condition that the value $a = \alpha_k$ was observed in the preceding measurement of the observable a on the state ψ is given by

$$\mathbf{P}_\psi(b = \beta_m | a = \alpha_k) \equiv \frac{\|P^b_{\beta_m} \ P^a_{\alpha_k}\psi\|^2}{\|P^a_{\alpha_k} \ \psi\|^2} \tag{18}$$

Let both operators have nondegenerate spectra, i.e., for any eigenvalue the corresponding eigenspace is one dimensional. We have:

$$\mathbf{P}_\psi(b = \beta_m | a = \alpha_k) = \mathbf{P}_\psi(a = \alpha_k | b = \beta_m) = |\langle e^b_m, e^a_k\rangle|^2.$$

Thus in this case the matrices of transition probabilities $P^{b|a}$ and $P^{a|b}$ coincide. We remark that, classical (Kolmogorov-Bayes) conditional probability is not symmetric, besides very special situations.

Consider two nondegenerate observables. Set $p_{\beta|\alpha} = \mathbf{P}(b = \beta | a = \alpha)$. The matrix of transition probabilities $\mathbf{P}^{b|a}$, is not only *stochastic,* but it is even *doubly stochastic.* It is easy to see that

$$\sum_\alpha p_{\beta|\alpha} = \sum_\alpha |\langle e^b_\beta, e^a_\alpha\rangle|^2 = \langle e^b_\beta, e^b_\beta\rangle = 1.$$

Double stochasticity is also a very specific property of quantum probability, cf. the Kolmogorovian model. In fact, condition **DS-DS** holds: both matrices of transition probabilities $\mathbf{P}^{a|b}$ and $\mathbf{P}^{b|a}$ are doubly stochastic. Moreover, any pair of quantum obseavables (with nondegenerate spectra) satisfies to condition **SC**; they are "symmetrically conditioned." Unlike Kolmogorovian model, observables need not be uniformly distributed. **Test of QM:** *If the matrix of transition probabilities for a pair of observables (which can be assumed to be nondegenerate) is not doubly stochastic, then they can not be described by QM.*[b]

If this test demonstrates that statistical data are nonquantum and at the same time one of tests of Kolmogorovianity, N 1 – N 3, demonstrates that data are neither classical, then one should apply more general *quantum-like models*[10–12,3] (e.g., quantum Markov chains, or so called lifting, or contextual probabilistic models).

8. Interference of Probabilities in Quantum Mechanics

We will show quantum probabilistic calculus violates the conventional FTP. Thus by Test N 1 of nonclassicality quantum probability is not classical. We stress that one need not appeal to Bell's type inequality, test N 2.

Let $H_2 = \mathbf{C} \times \mathbf{C}$ be the two dimensional complex Hilbert space and let $\psi \in H_2$ be a quantum state. Let us consider two dichotomous observables $b = \beta_1, \beta_2$ and $a = \alpha_1, \alpha_2$ represented by self-adjoint operators $\widehat{b}$ and $\widehat{a}$, respectively (one may consider simply Hermitian matrices). Let $e^b = \{e^b_\beta\}$ and $e^a = \{e^a_\alpha\}$ be two orthonormal bases consisting of eigenvectors of the operators. The state ψ can be represented in the two ways

$$\psi = c_1 e^a_1 + c_2 e^a_2, \; c_\alpha = \langle \psi, e^a_\alpha \rangle; \; \psi = d_1 e^b_1 + d_2 e^b_2, \; d_\beta = \langle \psi, e^b_\beta \rangle. \tag{19}$$

By Postulate 4 we have

$$\mathbf{P}(a = \alpha) \equiv \mathbf{P}_\psi(a = \alpha) = |c_\alpha|^2, \; \mathbf{P}(b = \beta) \equiv \mathbf{P}_\psi(b = \beta) = |d_\beta|^2. \tag{20}$$

The possibility to expand one basis with respect to another basis induces connection between the probabilities $\mathbf{P}(a = \alpha)$ and $\mathbf{P}(b = \beta)$. Let us expand the vectors e^a_α with respect to the basis e^b

$$e^a_1 = u_{11} e^b_1 + u_{12} e^b_2, \; e^a_2 = u_{21} e^b_1 + u_{22} e^b_2 \tag{21}$$

where $u_{\alpha\beta} = \langle e^a_\alpha, e^b_\beta \rangle$. Thus $d_1 = c_1 u_{11} + c_2 u_{21}$, $d_2 = c_1 u_{12} + c_1 u_{22}$. We obtain the *quantum rule* for transformation of probabilities:

$$\mathbf{P}(b = \beta) = |c_1 u_{1\beta} + c_2 u_{2\beta}|^2. \tag{22}$$

[b]In this review we do not discuss generalizations of QM; in particular, those based on so called positive operator valued measures - POVM.

On the other hand, by the definition of quantum conditional probability, see (18), we obtain:

$$\mathbf{P}(b=\beta|a=\alpha) \equiv \mathbf{P}_\psi(b=\beta|a=\alpha) = |\langle e^a_\alpha, e^b_\beta\rangle|^2. \tag{23}$$

By combining (20) and (22), (23) we obtain the *quantum formula of total probability – the formula of the interference of probabilities*:

$$\mathbf{P}(b=\beta) = \sum_\alpha \mathbf{P}(a=\alpha)\mathbf{P}(b=\beta|A=\alpha) \tag{24}$$

$$+2\cos\theta\sqrt{\mathbf{P}(a=\alpha_1)\mathbf{P}(b=\beta|a=\alpha_1)\mathbf{P}(a=\alpha_2)\mathbf{P}(b=\beta|a=\alpha_2)}$$

In general $\cos\theta \neq 0$. Thus the quantum FTP does not coincide with the classical FTP which is based on the Bayes' formula.

9. Bell's Inequality in Quantum Physics

As was pointed out in Section 4, inequalities of Boole-Bell type provide necessary conditions for probabilistic compatibility (PC) of families of observables and, hence, their violations provide sufficient conditions for probabilistic incompatibility (PI). As was pointed first by J. Bell, see, e.g., book[8] for details, quantum formalism predicts existence of such quantum states[c] that inequality (8) is violated for a special choice of a family pairwise measurable observables.[d] Thus these observables are of the PI-type. In any domain of science, one should look for special roots of PI. In particular, in physics J. Bell found two possible roots: quantum nonlocality and death of realism.

Our approach is essentially more general. By considering the problem from the PI-viewpoint, we can look for other roots of PI which need not coincide with proposed by Bell. One can still keep to realism and locality. PI can arise from, e.g., taking into account *parameters of measurement devices*[3,16] (so considering values of observables as depending on internal states not only systems, but also measurement devices).

Such sort of realism differs from naive *Einsteinian realism* and it is closer to Bohr's views; cf. also with *Accardi's chameleon effect*[14] and *Ohya's adaptive dynamics*[15].

Another possible source of PI is *unfair sampling* in the EPR-Bohm experiment for photons[16,17]. Since efficiency of detectors is extremely low,

[c]These are so called EPR-type (Einstein-Podolsky-Rosen) states.

[d]For example, spin or polarization projections to specially chosen directions.

one cannot exclude that choosing of different pairs of orientations of polarization beam splitters induces unfair sampling from the population of photons produced by the source.

This review was written during author's visit to QBIC Center, Tokyo University of Science (March 2009). It is based on lectures presented at the Satellite Workshop of QBIC-2009 Conference (Suwa Tokyo University of Science). I would like to thank all participants of this workshop (and especially, N. Watanabe and I. Yamato) for important comments. I also would like to thank L. Accardi and M. Ohya for numerous discussions.

References

1. A. N. Kolmogoroff, *Grundbegriffe der Wahrscheinlichkeitsrechnung*, Springer Verlag, Berlin, 1933; English translation: A.N. Kolmogorov, *Foundations of the probability theory*, Chelsea Publishing Company, New York, 1956.
2. A. Khrennikov, *Quantum-like brain: Interference of minds,* BioSystems **84**, 225–241 (2006).
3. A. Khrennikov, *Contextual approach to quantum formalism* (Fundamental Theories of Physics), Springer, Heidelberg- Berlin-New York, 2009.
4. A. Khrennikov, *Information dynamics in cognitive, psychological and anomalous phenomena* (Fundamental Theories of Physics), Kluwer Academic, Dordrecht, 2004.
5. E. Conte, O. Todarello, A. Federici, F. Vitiello, M. Lopane, A. Khrennikov and J. P. Zbilut, *Some remarks on an experiment suggesting quantum-like behavior of cognitive entities and formulation of an abstract quantum mechanical formalism to describe cognitive entity and its dynamics,* Chaos, Solitons and Fractals **31** (5), 1076-1088 (2007).
6. E. Conte, A. Khrennikov, O. Todarello, A. Federici, L. Mendolicchio, J.P. Zbilut, *Mental state follow quantum mechanics during perception and cognition of ambiguous figures,* Open Syst. and Information Dynamics **16**, 1-17 (2009).
7. A. Khrennikov, *Quantum-like model of cognitive decision making and information processing,* Biosystems **95**, 179-187 (2009).
8. A. Khrennikov, *Interpretations of probability,* VSP International Science Publishers, Utrecht, 1999; second addition (completed) De Gruyter, Berlin, 2009.
9. N. N. Vorob'ev, Theory of Probability and its Applications **7**, 147-162 (1962).
10. L. Accardi, *On the noncommutative Markov property,* Func. Anal. Appl. **9** (1), 1-8 (1975)(in Russian: submitted 31–1–1974); English translation in: Funct. Anal. and its Appl. **9** (1), 1-8 (1975).
11. L. Accardi, *Topics in quantum probability,* Phys. Rep. **77**, 169-192 (1981).
12. L. Accardi, M. Ohya, *Compound channels, transition expectations and liftings,* Applied Mathematics Optimization, **39**, 33–59 (1999). Volterra preprint N. 75 (1991)
13. L. Accardi, *Can mathematics help solving the interpretation problem of quantum theory?* Il Nuovo Cimento **B 110**, 685–721 (1995).

14. L. Accardi, *Urne e Camaleoni: Dialogo sulla realta, le leggi del caso e la teoria quantistica,* Rome, 1997.
15. M. Ohya, Addaptive dynamics and its applications to chaos and NPC problem. *Quantum Bio-Informatics*, Edited by L. Accardi, W. Freudenberg, M. Ohya, ser. *Quantum Prob. and White Noise An.* **21**, 2007, 181-216.
16. A. Khrennikov, *EPR-Bohm experiment and Bell's inequality: Quantum physics meets probability theory*, Theoretical and Mathematical Physics **157**, 1448-1460 (2008).
17. G. Adenier and A. Yu. Khrennikov, *Is the fair sampling assumption supported by EPR experiments?* Journal of Physics B: Atomic, Molecular and Optical Physics **40**, 131-141 (2007).

Quantum Bio-Informatics III
eds. L. Accardi, W. Freudenberg & M. Ohya
 (pp. 193–203)

2-ADIC DEGENERATION OF THE GENETIC CODE AND ENERGY OF BINDING OF CODONS

A.YU. KHRENNIKOV and S. V. KOZYREV

International Center for Mathematical Modelling in Physics and Cognitive Sciences University of Växjö, S-35195, Sweden E-mail: Andrei.Khrennikov@vxu.se

Steklov Mathematical Institute, Moscow, Russia E-mail: kozyrev@mi.ras.ru

Recently systems of p-adic numbers were applied to genetics. These systems are fundamentally hierarchical. Geometrically they are represented by homogeneous trees (here p is the number of branches leaving each vertex). It was shown that p-adics provides a natural number theoretic representation of the genetic code. In our previous paper we constructed a representation of the genetic code on so called 2-adic plane (so $p = 2$ and the dimension of the 2-adic space equals to two as well). In the present paper we investigate degeneracy of the genetic code with the help of 2-adic analysis and show that there exists a linear condition on the coordinates of the 2-adic plane of codons which separates the 2-adic plane in the domains of strong and weak degeneracy of the genetic code. We discuss the relation of this result and the Rumer's symmetry of the genetic code. We extend Rumer's argument that degeneracy of the genetic code related to the energy of binding of the codons with the complementary triples of nucleotides (anticodons). In the 2-adic framework this argument became especially natural.

1. Introduction

The structure of the genetic (amino acid) code, i.e., the way of encoding of amino acids by the triples of nucleotides (codons) is one of the basic structures of molecular biology [1]. There exist several approaches to description of the structure of the genetic code. One can mention approaches based on the coding theory [2,3,4] and the investigation of the symmetry of the genetic code [5,6,7,8,9].

In the papers by the authors [10,11,12,13] and by B.Dragovich and A.Dragovich [14,15] the alternative approach was suggested. It was proposed to consider a p-adic parameterization of the genetic code (this parameterization was defined in the different ways in works of A. Khrennikov [10,11,12], A. Khrennikov and S. Kozyrev[13], and B.Dragovich and A.Dragovich[14,15]. In the first series of papers the 4-adic and 2-adic parameterizations have been used, in papers of B. Dragovich and A.Dragovich[14,15] the 5-adic parameterization has been used and in paper [13] we used 2-adic parameterization and the space of dimension two, i.e., the 2-adic plane. We also remark that in articles[10,11,12] p-adic dynamical systems were applied to explain degeneration of the genetic code: clusters of codons representing amino acids were generated as attractors of a 4-adic (or equivalently 2-adic) dynamical system (code generating system).

In this paper we follow our previous work[13].

In p-adic parameterizations the genetic code is a locally constant map of a p-adic argument. Therefore, instead of discussion of the symmetry of the code, we investigate the local constancy of the code.

This approach has some advantages. Let us compare this to the approach based on symmetry. Investigation of symmetry is the standard point of view in physics. For the genetic code it is not possible to find the single natural group of symmetries and one can consider instead a large set of symmetries. Moreover for the different dialects of the code the set of symmetries will be different. In our p-adic approach instead of a large number of symmetries we consider local constancy of the single map. Variations of the genetic code break its symmetries but do not change considerably the local constancy of the corresponding p-adic map. Based on this observation we expect that in theory of complex systems local constancy and ultrametric might play the role similar to the role of symmetry in physics. For the results of p-adic mathematical physics see review[16].

The main result of the present paper is as follows. In paper [13] it was suggested using some simple explicit parameterization to express the set of codons in the form of the 8×8 table with the 2-dimensional 2-adic metric. In the present paper we investigate the way in which this 2-adic plane of codons is separated in the domains with the strong and weak degeneracy of the genetic code. These domains are related through the Rumer's symmetry[5,6].

We show that there exists the linear with respect to the coordinates of the 2-adic plane condition which separates the 2-adic plane in the mentioned above domains. We find all possible linear conditions with this property. Thus we express the Rumer's symmetry by a linear condition in our p-adic approach.

The exposition of the present paper is as follows.

In Section 2 we expose some results of paper[13] on the 2-adic parameterization of the genetic code.

In Section 3 we find all possible linear conditions which separate the 2-adic plane of codons in the domains of strong and weak degeneracy.

In Section 4 we discuss the relation of the results of Section 3 and the Rumer's symmetry of the genetic code.

2. 2-Adic parameterization of the genetic code

In the 2–adic parameterization approach of our previous work[13] we enumerate in some special way the set of 64 codons (triples of nucleotides) by pairs of digits (x, y), $x, y = 0, 1, 2, \ldots, 7$. These digits are in one to one correspondence with the triples (x_0, x_1, x_2) and (y_0, y_1, y_2) of 0 and 1 (the expansions of x and y over degrees of 2):

$$x = x_0 + 2x_1 + 4x_2, \qquad y = y_0 + 2y_1 + 4y_2, \qquad x_i, y_i = 0, 1.$$

2–Adic norm of x is equal to 2^{-i}, where i is the number of the first non zero x_i in the above expansion (analogously for y); for example, it is equal to 1 for $x = 1, 3, 5, 7$, to 1/2 for $x = 2, 6$, to 1/4 for $x = 4$ and by definition (!) to 0 for $x = 0$. 2–Adic norm of a point (x, y) on the plane is defined as maximum of 2-adic norms of its coordinates.

Each pair of digits (x_i, y_i) is defined by a nucleotide using the rule

$$\begin{array}{|c|c|}\hline A & G \\ \hline U & C \\ \hline \end{array} = \begin{array}{|c|c|}\hline 00 & 01 \\ \hline 10 & 11 \\ \hline \end{array}.$$

Since the nucleotides $A = (0, 0)$ and $G = (0, 1)$ are purines, $U = (1, 0)$ and $C = (1, 1)$ are pyrimidines, the different first digits in the above binary representation corresponds to the different chemical types of the nucleotides. Namely, the nucleotide (x, y) with $x = 0$ is a purine, and the nucleotide (x, y) with $x = 1$ is a pyrimidine.

The second digit $y = 0, 1$ in the considered parameterization describes the H–bonding character (weak for $y = 0$ and strong for $y = 1$).

Using the above correspondence between the nucleotides and the digits 0 and 1, we introduce the correspondence between the codons (triples of nucleotides) and the triples (x_0, x_1, x_2) and (y_0, y_1, y_2) of 0 and 1 (equivalently, the corresponding $x, y = 0, 1, 2, \ldots, 7$) by the following prescription.

The second nucleotide in the codon defines the pair (x_0, y_0), the first nucleotide in the codon defines the pair (x_1, y_1), and the third nucleotide in the codon defines the pair (x_2, y_2).

This rule is related to the following hierarchy of nucleotides in the codon

$$2 > 1 > 3$$

i.e the second nucleotide in the codon is the most important (and the largest in the 2–adic norm) and the third nucleotide is the least important.

To produce a natural Euclidean image of 2-adic plane, we make the special reshuffling of the values of x and y:

$$0, 4, 2, 6, 1, 5, 3, 7 \mapsto 1, 2, 3, 4, 5, 6, 7, 8.$$

similar to made in article[17 –19] for disordered physical systems (spin glasses).

AAA	AAG	GAA	GAG	AGA	AGG	GGA	GGG
AAU	AAC	GAU	GAC	AGU	AGC	GGU	GGC
UAA	UAG	CAA	CAG	UGA	UGG	CGA	CGG
UAU	UAC	CAU	CAC	UGU	UGC	CGU	CGC
AUA	AUG	GUA	GUG	ACA	ACG	GCA	GCG
AUU	AUC	GUU	GUC	ACU	ACC	GCU	GCC
UUA	UUG	CUA	CUG	UCA	UCG	CCA	CCG
UUU	UUC	CUU	CUC	UCU	UCC	CCU	CCC

The 2×2 quadrates (marked by lines) here are 2-dimensional 2-adic balls (or clusters) of the diameter $1/4$ and the 4×4 quadrates (marked by double lines) are balls of the diameter $1/2$.

After application of the eucaryotic genetic (amino acid) code (described in Appendix 3) to the above table we get the table of amino acids on the 2–adic plane

K/N	E/D	R/S	G
Ter/Y	Q/H	Ter|W / C	R
I|M / I	V	T	A
L/F	L	S	P

where Ter is the stop codon. Each square in the above table is the 2 by 2 square in the 2-adic plane of codons. In particular, the genetic code map acts as follows on the 2–adic balls

$$\begin{array}{|cc|}\hline AAA & AAG \\ AAU & AAC \\ \hline \end{array} \longrightarrow \boxed{K/N}, \qquad \begin{array}{|cc|}\hline CCA & CCG \\ CCU & CCC \\ \hline \end{array} \longrightarrow \boxed{\mathrm{P}}$$

Here we use the standard notations A, U, G, C for the nucleotides and 1-Letter notations for the amino acids.

We see, that the degeneracy of the genetic code in the above 2-adic parameterization is described by the 2-adic proximity — the codons which encode the same amino acid are 2-adically close. Moreover, the domains with the different degeneracy are symmetric at the 2-adic plane — the lower right quadrant of the plane is occupied by amino acids with the degeneracy four, and the upper left quadrant of the plane contains the amino acids with the degeneracy mainly equal to two. We also have the five cases of additional degeneracy which is not described by the 2-adic parameterization.

Analogously, for the Vertebrate Mitochondrial Code we get the table

K/N	E/D	Ter/S	G
Ter/Y	Q/H	W/C	R
M/I	V	T	A
L/F	L	S	P

3. Linear condition which describes the degeneracy of the genetic code

Let us consider in the 2-adic plane of codons the subsets with the different degeneracy. In these subsets (inside the single 2×2 square in the 2-adic plane) the degeneracy (for the mitochondrial code) takes the values 2 and 4 correspondingly. We will denote these subsets as the domain 1 and the domain 2. The domain 1 contains the upper left quadrant (ball of the diameter 1/2), and the left halves of the upper right and the lower left quadrants.

If we consider the other dialects of the genetic code the domains 1 and 2 will have the same form and will correspond to the subsets with the strong and the weak degeneracy of the code correspondingly.

In the present paper we show that the domains 1 and 2 can be separated using the linear condition on the coordinates of the 2-adic plane. We will discuss the physical meaning of this condition.

Consider the coordinates on the 2-adic plane

$$x = x_0 + 2x_1 + 4x_2, \qquad y = y_0 + 2y_1 + 4y_2, \qquad x_i, y_i = 0, 1.$$

Let us consider the real valued function on the 2-adic plane of the form

$$E = C_{00}x_0 + C_{01}y_0 + C_{10}x_1 + C_{11}y_1.$$

Here $x_i, y_j = 0, 1$ are integers and C_{ij} are real numbers.

We will find the function E (i.e. the set of coefficients C_{ij}), which in the domain 1 will satisfy $E < 1$, and in the domain 2 will satisfy $E > 1$. Therefore we will find the linear condition on the coordinates of the 2-adic plane which will separate the 2-adic plane in the union of the domains 1 and 2.

Let us describe the system of inequalities which define the function E.

1) At the upper left quadrant we have $x_0 = y_0 = 0$. The condition $E < 1$ takes the form

$$C_{10}x_1 + C_{11}y_1 < 1, \qquad \forall x_1, y_1. \tag{1}$$

2) At the lower right quadrant we have $x_0 = y_0 = 1$. The condition $E > 1$ takes the form

$$C_{00} + C_{01} + C_{10}x_1 + C_{11}y_1 > 1. \tag{2}$$

(this and the subsequent inequalities shall be satisfied for $\forall x_1, y_1$).

3) At the upper right quadrant $x_0 = 0$, $y_0 = 1$. The conditions for E at the left and the right halves of the quadrant give the corresponding inequalities

$$C_{01} + C_{10}x_1 < 1, \tag{3}$$

$$C_{01} + C_{10}x_1 + C_{11} > 1. \tag{4}$$

4) At the lower left quadrant $x_0 = 1$, $y_0 = 0$. The conditions for E at the left and the right halves of the quadrant give the corresponding inequalities

$$C_{00} + C_{10}x_1 < 1, \tag{5}$$

$$C_{00} + C_{10}x_1 + C_{11} > 1. \tag{6}$$

Comparing (3), (4) we get $C_{11} > 0$.

Also (1), (2) imply that $C_{00} + C_{01} > 0$.

Let us consider the two cases.

1) Let $C_{10} \geq 0$.

The above inequalities are systems of inequalities for C_{ij} (we get the different inequalities in the system taking the different $x_i, y_j = 0, 1$). Inequalities in these systems are dependent, and any of the systems is equivalent to the inequality where x_i, y_j are substituted by the corresponding extremal values (i.e. by zero or one).

We obtain the following system of inequalities for C_{ij}:

$$C_{11} + C_{10} < 1, \tag{7}$$

$$C_{01} + C_{10} < 1, \tag{8}$$

$$C_{00} + C_{10} < 1, \tag{9}$$

$$C_{00} + C_{01} > 1, \tag{10}$$

$$C_{01} + C_{11} > 1, \tag{11}$$

$$C_{00} + C_{11} > 1. \tag{12}$$

Let us describe the set of solutions of the above system of inequalities. It is easy to see that $C_{10} \in [0, 1/2)$, since for $C_{10} \geq 1$ inequality (7) has no solutions and for $C_{10} \in [1/2, 1)$ the sum of (8) and (9) can not be larger than 1 which contradicts to (10).

Inequality (7) implies that $C_{11} < 1$. This and (11), (12) imply that C_{00}, C_{01} are positive.

Then, by inequalities (7), (8), (9), any of C_{00}, C_{01}, C_{11} will belong to $[0, 1-C_{10})$ with $C_{10} \in [0, 1/2)$. This set is a cube in $\mathbb{R}^3$ with the coordinates (C_{00}, C_{01}, C_{11}), with the length of the edge equal to $1-C_{10}$, and sides which are parallel to the coordinate planes, one of the vertices coincides with the zero (origin of the coordinates), and the cube does not contain the sides which do not contain the zero.

Any of the inequalities (10), (11), (12) means that we have to eliminate from the described above cube the sets (prisms) for which the sums of the pairs of the coordinates are not larger than 1.

We have found that the set of solutions of the above system of inequalities (with $C_{10} \geq 0$) contains $C_{10} \in [0, 1/2)$ and (C_{00}, C_{01}, C_{11}), lying inside the polyhedron in $\mathbb{R}^3$ with the coordinates (C_{00}, C_{01}, C_{11}). The vertices of the polyhedron are

$$\left(\frac{1}{2}, \frac{1}{2}, \frac{1}{2}\right), \quad (1 - C_{10}, 1 - C_{10}, 1 - C_{10}),$$

$$(1 - C_{10}, 1 - C_{10}, C_{10}), \quad (1 - C_{10}, C_{10}, 1 - C_{10}), \quad (C_{10}, 1 - C_{10}, 1 - C_{10}).$$

When $C_{10} = 0$ then the interior of the polyhedron is empty, therefore the nontrivial solutions of the above form correspond to $C_{10} \in (0, 1/2)$.

The example of the solution of the system of inequalities (7), (8), (9), (10), (11), (12):

$$C_{10} = \frac{1}{4}, \quad C_{00} = C_{01} = C_{11} = \frac{5}{8}.$$

2) Let us consider $C_{10} < 0$. Then, using the same arguments as above, we get the system of inequalities

$$C_{11} < 1, \tag{13}$$

$$C_{01} < 1, \tag{14}$$

$$C_{00} < 1, \tag{15}$$

$$C_{00} + C_{01} > 1 - C_{10}, \tag{16}$$

$$C_{01} + C_{11} > 1 - C_{10}, \tag{17}$$

$$C_{00} + C_{11} > 1 - C_{10}. \tag{18}$$

Since $C_{11} < 1$ and $C_{10} < 0$, inequality (18) implies that $C_{00} > 0$. Analogously, inequality (17) implies $C_{01} > 0$.

The above system of inequalities possesses solutions for negative C_{10} when $C_{10} \in (-1, 0)$. In this case for any $C_{10} \in (-1, 0)$ solutions (C_{00}, C_{01}, C_{11}) belong to the interior of the polyhedron in $\mathbb{R}^3$ with the coordinates (C_{00}, C_{01}, C_{11}). This polyhedron is defined by the vertices

$$\left(\frac{1 - C_{10}}{2}, \frac{1 - C_{10}}{2}, \frac{1 - C_{10}}{2}\right), \quad (1, 1, 1),$$

$$(1 - C_{10}, 1 - C_{10}, -C_{10}), (1 - C_{10}, -C_{10}, 1 - C_{10}), (-C_{10}, 1 - C_{10}, 1 - C_{10}).$$

4. Rumer's symmetry and interpretation of E as energy

In papers[5,6] by Yu.B. Rumer the important symmetry of the genetic code was found. The construction is as follows. Let us construct the 4×4 matrix, where the lines are enumerated by the first nucleotides in the codon, the columns are enumerated by the second nucleotides in the codon, and the matrix elements are the sets of amino acids defined by the genetic code. We have put in the matrix below only the matrix elements (amino acids) which are unambiguously defined by the first two nucleotides in the codon. One can see that this set (where the genetic code does not depend on the third nucleotide in the codon) is the half of the matrix and this set of strong degeneracy is symmetric to the set of weak degeneracy (where we put $*$ instead of sets of amino acids for the case of weak degeneracy). The obtained matrix expresses the Rumer's symmetry:

*	C	G	U	A
C	P	R	L	*
G	A	G	V	*
U	S	*	*	*
A	T	*	*	*

In mentioned papers[5,6] the degeneracy of the genetic code was discussed in relation to the energy of binding of the codons with the complementary triples of nucleotides (anticodons). Here we discuss similar arguments in the language of the 2-adic parameterization of the genetic code.

The investigated in the previous Section function E can be considered as the energy of interaction of the first two nucleotides in the codon and the complementary nucleotides (in analogue to Rumer[5,6]). We assume that this energy is linear on the 2-adic plane (i.e. this energy is a sum of contributions for the nucleotides and moreover each contribution itself is linear with respect to the 2-adic coordinates of each of the nucleotides).

If the energy E of binding of the first two nucleotides is sufficiently large (larger than 1 in our model), then, if we will change the third nucleotide in the codon, the difference of energies will be small in comparison to E. In this situation one can assume that the genetic code will be totally degenerate with respect to the third nucleotide in the codon. If the binding energy E is less than one it is natural to expect the breaking of the degeneracy with respect to the third nucleotide. This is exactly what we observe for the different dialects of the genetic code in our 2-adic parameterization.

We found that the domains with the different degeneracy of the genetic code can be separated by the linear condition with respect to the coordinates of the 2-adic plane of codons. We found all the possible linear conditions which separate the domains of strong and weak degeneracy.

Acknowledgments The authors would like to thank B.Dragovich, I. V. Volovich and D.A.Semenov for fruitful discussions. The authors were partially supported by the grant of International center for mathematical modeling in physics, engineering and cognitive science, University of

Växjö. One of the authors (A.K.) was partially supported by the QPIC grant, Tokyo University of Science. One of the authors (S.K.) gratefully acknowledges being partially supported by the grants DFG Project 436 RUS 113/809/0-1 and DFG Project 436 RUS 113/951, by the grants of The Russian Foundation for Basic Research RFFI 05-01-04002-NNIO-a and RFFI 08-01-00727-a, by the grant of the President of Russian Federation for the support of scientific schools NSh-3224.2008.1, by the Program of the Department of Mathematics of Russian Academy of Science "Modern problems of theoretical mathematics" and by the program of Ministry of Education and Science of Russia "Development of the scientific potential of High School, years of 2009–2010", project 3341.

References

1. G. Gamow, Possible relation between deoxyribonucleic acid and protein structures, *Nature* **173**, 318 (1954).
2. F. H. C. Crick, The origin of the genetic code, *Journal of Molecular Biology* **38**, 367-379 (1968).
3. J. Maddox, The genetic code by numbers, *Nature* **367**, 111 (1994).
4. R. Swanson, A unifying concept for the amino acid code, *Bulletin of Mathematical Biology* **46** 187-203 (1984).
5. Yu. B. Rumer, On systematizing codons in the genetic code, *Doklady Akademii Nauk SSSR* **167**, 1393-1394 (1966).
6. Yu. B. Rumer, Systematization of codons in the genetic code, *Doklady Akademii Nauk SSSR* **183**, 225-226 (1968).
7. G. L. Findley, A. M. Findley, S. P. McGlynn, Symmetry characteristics of the genetic code, *Proc. Natl. Acad. Sci. USA* **79**(22), 7061-7065 (1982).
8. L. Frappat, A. Sciarrino, P. Sorba, A crystal base for the genetic code, *Phys. Lett. A.* **250**, 214-221 (1998).
9. D. Duplij, S. Duplij, N. Chashnin, Symmetric properties of genetic code, *Biopolymers and Cell* **16**, 449-454 (2000).
10. A. Yu. Khrennikov, p-Adic information space and gene expression. *Integrative approaches to brain complexity*, Edited by S. Grant, N. Heintz, J. Noebels, Welcome Trust Publ., Cambridge, 2006, 14.
11. A.Yu. Khrennikov, p-Adic dynamical representation of gene expression. *Foundations of probability and physics - 4,* Edited by G. Adenier, C. Fuchs, and A. Yu. Khrennikov, AIP Conference Proceeding, **889** , 2007, 324-333.
12. A.Yu. Khrennikov, Gene expression from polynomial dynamics in the 2-adic information space, *Chaos, Solitons and Fractals*, to be published 2009, doi:10.1016/j.chaos.2008.12.012.
13. A.Yu. Khrennikov, S.V. Kozyrev, Genetic code on the diadic plane, *Physica A: Statistical Mechanics and its Applications* **381**, 265-272 (2007); arXiv:q-bio.QM/0701007.
14. B. Dragovich, A. Dragovich, A p-Adic Model of DNA Sequence and Ge-

netic Code, *p-Adic Numbers, Ultrametric Analysis and Applications* **1** 34-41 (2009); arXiv:q-bio/0607018v1.

15. B. Dragovich, A. Yu. Dragovich, *p*-Adic modeling of the genome and the genetic code, *The Computer Journal,* to be published 2009, doi:10.1093/comjnl/bxm083 (2009).
16. B. Dragovich, A.Yu. Khrennikov, S.V. Kozyrev, I. V. Volovich, On p-Adic Mathematical Physics, *p-Adic Numbers, Ultrametric Analysis and Applications* **1**, 1ĄE7 (2009); arXiv:0904.4205v1.
17. V. A. Avetisov, A. H. Bikulov, S. V. Kozyrev, Application of *p*–adic analysis to models of spontaneous breaking of replica symmetry, *J. Phys. A: Math. Gen.* **32**, 8785–8791 (1999); arXiv:cond-mat/9904360.
18. A. Yu. Khrennikov, S. V. Kozyrev, Replica symmetry breaking related to a general ultrametric space I: replica matrices and functionals, *Physica* A **359**, 222-240 (2006).
19. A. Yu. Khrennikov, S. V. Kozyrev, Replica symmetry breaking related to a general ultrametric space II: RSB solutions and the $n \to 0$ limit, *Physica* A **359**, 241-266 (2006).

Quantum Bio-Informatics III
eds. L. Accardi, W. Freudenberg & M. Ohya

ON POSITIVE MAPS; PPT STATES AND ENTANGLEMENT

WŁADYSŁAW A. MAJEWSKI
Institute of Theoretical Physics and Astrophysics
Gdańsk University
Wita Stwosza 57
80-952 Gdańsk, Poland
E-mail:fizwam@univ.gda.pl
Based on joint work with M. Ohya and T. Matsuoka

1. Introduction

In Quantum Computing a characterization of states with positive partial transposition (PPT states) is of paramount importance (see [11] and also [6]). In [19] it was pointed out that there are strong relations between large class of positive maps (decomposable maps) and the subset of all PPT states. Recently, employing relations between certain positive maps and states, the full description of PPT states was obtained in [15], see also [25]. The aim of this note, based on [15], is to review characterization of PPT states with the special emphasis put on the structure of positive maps and tensor products of certain algebras.

For sake of convenience, we provide all necessary preliminaries and set up the notation. Let $\mathcal{B}(\mathcal{H})$ be the set of all linear bounded operators on a Hilbert space $\mathcal{H}$. We denote the set of all positive elements of $\mathcal{B}(\mathcal{H})$ by $\mathcal{B}(\mathcal{H})^+$. A *state* on $\mathcal{B}(\mathcal{H})$ is a linear functional $\phi : \mathcal{B}(\mathcal{H}) \longrightarrow \mathbb{C}$ such that $\phi(A) \geq 0$ for every $A \in \mathcal{B}(\mathcal{H})^+$ and $\phi(\mathbb{I}) = 1$, where $\mathbb{I}$ is the unit of $\mathcal{B}(\mathcal{H})$. The set of all states on $\mathcal{B}(\mathcal{H})$ is denoted by $\mathcal{S}_{\mathcal{B}(\mathcal{H})}$.

A linear map $\psi : \mathcal{B}(\mathcal{H}_1) \longrightarrow \mathcal{B}(\mathcal{H}_2)$ is called *positive* if $\psi(\mathcal{B}(\mathcal{H}_1)^+) \subset \mathcal{B}(\mathcal{H}_2)^+$. For $k \in \mathbb{N}$ we consider a map $\psi_k : M_k \otimes \mathcal{B}(\mathcal{H}_1) \longrightarrow M_k \otimes \mathcal{B}(\mathcal{H}_2)$ where $M_k \equiv M_k(\mathbb{C})$ denotes the algebra of $k \times k$-matrices with complex entries and $\psi_k = \mathrm{id}_{M_k} \otimes \psi$. We say that ψ is *k-positive* if the map ψ_k is positive. The map ψ is said *completely positive*(CP for short) when ψ is k-positive for every $k \in \mathbb{N}$. Due to the Stinespring theorem any CP map

$T : \mathcal{B}(\mathcal{H}_1) \to \mathcal{B}(\mathcal{H}_2)$ is of the form

$$T(A) = W^* \pi(A) W, \tag{1}$$

where $\pi : \mathcal{B}(\mathcal{H}_1) \to \mathcal{B}(\mathcal{L})$ is a *-representation, and $W : \mathcal{H}_2 \to \mathcal{L}$ is a linear bounded operator. It is worth pointing out that when T is unital (so $T(\mathbb{I}) = \mathbb{I}$), then W is an isometry.

Following Arveson, [1], we say that a completely positive map $T : \mathcal{B}(\mathcal{H}) \to \mathcal{B}(\mathcal{L})$ is pure if, for every completely positive map $T' : \mathcal{B}(\mathcal{H}) \to \mathcal{B}(\mathcal{L})$, the property that $T - T'$ is a completely positive map implies that T' is a scalar multiple of T. It was shown by Arveson, [1], that a non-zero pure CP map T is of the form (1) with π being an irreducible representation.

In this context, the so called Kraus representation of CP map is very natural. Namely, a CP map $T : \mathcal{B}(\mathcal{H}_1) \to \mathcal{B}(\mathcal{H}_2)$ can be written in the form

$$T(A) = \sum_i W_j^* A W_j, \tag{2}$$

where W_j are bounded linear operators from $\mathcal{H}_2$ to $\mathcal{H}_1$ and the sum converges in the strong operator topology. In other words, for a CP map $T(\cdot)$ there is a possibility to represent T in terms of more elementary summands. We shall come back to such decompositions in the last Section.

Let us recall that for a Hilbert space $\mathcal{L}$, every normal (so *-weakly continuous) state $\phi(\cdot)$ on $\mathcal{B}(\mathcal{L})$ has the form of $\phi(A) = \mathrm{Tr}\,(\varrho A)$, where ϱ is a uniquely determined *density matrix*, i.e. an element of $\mathcal{B}(\mathcal{L})^+$ such that $\mathrm{Tr}\,\varrho = 1$.

We also fix orthonormal bases $\{e_i\}_i$ and $\{f_j\}_j$ of the spaces $\mathcal{H}$ and $\mathcal{K}$ respectively. For simplicity we will write $\mathcal{S}$, $\mathcal{S}_{\mathcal{H}}$, $\mathcal{S}_{\mathcal{K}}$ instead of $\mathcal{S}_{\mathcal{B}(\mathcal{H})\otimes\mathcal{B}(\mathcal{K})}$, $\mathcal{S}_{\mathcal{B}(\mathcal{H})}$, $\mathcal{S}_{\mathcal{B}(\mathcal{K})}$, respectively. By $\tau_{\mathcal{L}}$ we denote transposition maps on $\mathcal{B}(\mathcal{L})$ associated with basis $\{e_i\} \subseteq \mathcal{L}$. More precisely, such map is defined as follows. Firstly

$$Jx = J\sum_i (e_i, x) e_i = \sum_i \overline{(e_i, x)} e_i \tag{3}$$

gives definition of a conjugation on a Hilbert space $\mathcal{L}$, i.e. J is an antilinear map $J : \mathcal{L} \to \mathcal{L}$ such that for any $e, f \in \mathcal{L}$

$$(Jf, e) = (Je, f), \qquad J \cdot J = \mathbb{I}. \tag{4}$$

Then, the transposition $a \to a^t$ is defined as

$$a \mapsto a^t \equiv J a^* J. \tag{5}$$

Thus, transposition can be defined in abstract way. Let us note that transposition $\tau_{\mathcal{L}} : \mathcal{B}(\mathcal{L}) \longrightarrow \mathcal{B}(\mathcal{L})$ is a positive map but not completely positive (in fact it is not even 2-positive).

A positive map $\Psi : \mathcal{B}(\mathcal{H}) \longrightarrow \mathcal{B}(\mathcal{K})$ is called *decomposable* if there are completely positive maps $\Psi_1, \Psi_2 : \mathcal{B}(\mathcal{H}) \longrightarrow \mathcal{B}(\mathcal{K})$ such that $\Psi = \Psi_1 + \Psi_2 \circ \tau_{\mathcal{H}}$. Let $\mathcal{P}$, $\mathcal{P}_{\mathrm{C}}$ and $\mathcal{P}_{\mathrm{D}}$ denote the set of all positive, completely positive and decomposable maps from $\mathcal{B}(\mathcal{H})$ to $\mathcal{B}(\mathcal{K})$, respectively. Note that

$$\mathcal{P}_{\mathrm{C}} \subset \mathcal{P}_{\mathrm{D}} \subset \mathcal{P} \tag{6}$$

(see also [6]- [8]).

A state $\varphi \in \mathcal{S}$ is said to be *separable* if it can be approximated (in norm) by states of the form

$$\varphi_N = \sum_{n=1}^{N} \lambda_n \varphi_n^{\mathcal{H}} \otimes \varphi_n^{\mathcal{K}}$$

where $N \in \mathbb{N}$, $\varphi_n^{\mathcal{H}} \in \mathcal{S}_{\mathcal{H}}$, $\varphi_n^{\mathcal{K}} \in \mathcal{S}_{\mathcal{K}}$ for $n = 1, 2, \ldots, N$, λ_n are positive numbers such that $\sum_{n=1}^{N} \lambda_n = 1$, and the state $\varphi_n^{\mathcal{H}} \otimes \varphi_n^{\mathcal{K}}$ is defined as $\varphi_n^{\mathcal{H}} \otimes \varphi_n^{\mathcal{K}}(A \otimes B) = \varphi_n^{\mathcal{H}}(A)\varphi_n^{\mathcal{K}}(B)$ for $A \in \mathcal{B}(\mathcal{H})$, $B \in \mathcal{B}(\mathcal{K})$. The set of all separable states on the algebra $\mathcal{B}(\mathcal{H}) \otimes \mathcal{B}(\mathcal{K})$ is denoted by $\mathcal{S}_{\mathrm{sep}}$. A state which is not in $\mathcal{S}_{\mathrm{sep}}$ is called *entangled* or *non-separable.*

Finally, let us define the family of *PPT* (transposable) states on $\mathcal{B}(\mathcal{H} \otimes \mathcal{K})$

$$\mathcal{S}_{\tau} = \{\varphi \in \mathcal{S} : \varphi \circ \left(id_{\mathcal{B}(\mathcal{H})} \otimes \tau_{\mathcal{K}}\right) \in \mathcal{S}\}.$$

Note that due to the positivity of the transposition $\tau_{\mathcal{K}}$ every separable state φ is transposable, so

$$\mathcal{S}_{\mathrm{sep}} \subset \mathcal{S}_{\tau} \subset \mathcal{S}. \tag{7}$$

2. Why C^*-algebras?

As it was mentioned, PPT states are of fundamental importance for Quantum Computing. It is worth pointing out that these states are closely related to another important concept - to entanglement (so to quantum correlations; see ([16] - [18]). Speaking about such fundamental problems it is tempting to start a discussion with a brief examination of fundamentals of Quantum Theory (cf I. Volovich's lecture in this conference). It is agreed in axiomatic approach to Quantum Theory (see [14], [23], [9] and [10]) that the set of all observables forms a Jordan algebra. Namely, P. Jordan , [13] (see also

[12,14]) introduced a real algebra that is not associative, but includes hermitian matrices as a special case. The motivation was to take the product of observables A and B as

$$A \circ B = \frac{1}{2}(AB + BA) \tag{8}$$

Moreover, he noted that if A, B are hermitian matrices (bounded self-adjoint operators), then this product maps the set of hermitian matrices (bounded self-adjoint operators) into itself. Furthermore

$$((A \circ A) \circ B) \circ A = (A \circ A) \circ (B \circ A) \tag{9}$$

for all $A, B \in M_n(\mathbb{C})$.

This observation leads to

Definition 2.1. A Jordan algebra over $\mathbb{R}$ is a vector space $\mathcal{A}$ over $\mathbb{R}$ equipped with a commutative bilinear product $\circ$ that satisfies the identity(9).

Thus, the set of all observables of any quantum-mechanical system can be equipped with the Jordan algebra structure. Briefly recalling the theory of Jordan algebras we note that if $\mathfrak{A}$ is any associative algebra over $\mathbb{R}$, then $\mathfrak{A}$ becomes a Jordan algebra when equipped with the product $\circ$ given by (8). A Jordan algebra which is isomorphic to a subalgebra of $\mathfrak{A}$ is called *special.* If $\mathfrak{C}$ is a C^*-algebra then $\mathfrak{A} = \mathfrak{C}_{sa} \equiv \{A \in \mathfrak{C}; A = A^*\}$ is the basic example of a special Jordan algebra.

However, there are non-special Jordan algebras. They are called *exceptional.* A nice example is provided by the algebra $\mathfrak{M}_3^8$ of all hermitian matrices of order 3 over quasi-quaternions (or Cayley numbers). Due to the existence of exceptional Jordan algebras the method of constructing of the tensor product of general Jordan algebras is unknown. The tensor product of their underlying vector spaces can be easily done. But there is a problem how to define a Jordan product on the latter one. Note, that this obstacle for special Jordan algebras is absent.

On the other hand, we emphasize that for any theory used to describe local systems (the idea of locality is that what is needed for an analysis of correlations) one should use the concept of tensor product eg. see Rule 3 in Omnés, [20], list of basic principles of Quantum Mechanics. Concluding, we must discard general Jordan algebras as a basic structure and pass to special Jordan algebras.

The next natural question is why we insist on a general approach, i.e. we do not wish to restrict ourselves to matrix algebras. To answer this question let us review very briefly the basic ingredients of canonical quantization. The first rule of basic principles of Quantum Mechanics, [20] says that the set of all pure states is described by (normalized) vectors in a separable Hilbert space $\mathcal{H}$. Take, for simplicity, $\mathcal{H} = L^2(\mathbb{R})$, the Hilbert space of all square integrable, complex-valued functions on $\mathbb{R}$ with respect to the Lebesgue measure. For $g \in L^2(\mathbb{R})$ and a complex-valued measurable function f on $\mathbb{R}$, define $T_f g = f \cdot g$. T_f is a closable linear operator in $L^2(\mathbb{R})$ with a dense domain, [21] . Let $p = \frac{d}{dt}$, $q = T_{f_0}$, where $f_0(t) = t$. Then for $g \in L^2(\mathbb{R})$ with $g' \in L^2(\mathbb{R})$,

$$(pq - qp)g(t) = g(t) + tg'(t) - tg'(t) = g(t). \tag{10}$$

Hence $pq - qp = 1$ on a dense subset of $L^2(\mathbb{R})$. Thus, canonical quantization is realized by linear operators on $L^2(\mathbb{R})$. However, both q and p are unbounded. We can not expect any "improvement" of the above example. Namely, Wintner [27] for Hilbert spaces and Wielandt, [28], for general Banach algebras proved that the canonical quantization is impossible within the scheme of bounded operators (so also it is impossible for matrices!). More precisely, see [22], one has

Theorem 2.1. *Let $\mathfrak{A}$ be a Banach algebra. Then There are no two elements p, q in $\mathfrak{A}$ such that pq-qp=1.*

Consequently, we will use C^*-algebra approach to Quantum Theory and we will use weakly closed subalgebras of $\mathcal{B}(\mathcal{H})$ (with infinite dimensional Hilbert space $\mathcal{H}$!). Note that for such structures, functional calculus is available and therefore one can speak of bounded functions of genuine quantum variables. Thus, C^*- algebra approach is very well justified.

In the sequel, we will use tensor products of (associative) algebraic structures equipped with the properly chosen norm. However, there is another subtle point. Namely, the tensor product concept for ordered Banach spaces $\mathfrak{A}_i$, $i = 1, 2$ leads to many problems. Firstly, in general, there is no *one good* norm, i.e. there is no a unique norm $||\cdot||$ such that $||x \otimes y|| = ||x||||y||$ and $(\mathfrak{A}_1 \otimes \mathfrak{A}_2, ||\cdot||)$ is a Banach space. Again, we remind, the tensor product of underlying vector spaces is doable and does not present any problem. The difficulty is due to the fact that there are , in general, many norms. The smallest is called the projective norm, while the largest is called the injective norm. There is the class of C^*-algebras, nuclear C^*-algebras, which have uniquely determined norm on their tensor products. In particular, $M_n(\mathbb{C})$ is in this class. Hence the definition of CP maps has so simple form.

Secondly, there is the next (even more difficult) problem related to order. Namely, for two ordered Banach spaces there are various cones determining the order structure in the tensor product of algebras

$$\mathcal{C}_{inj} \equiv (\mathfrak{A}_1 \otimes \mathfrak{A}_2)^+ \supseteq, ..., \supseteq \mathcal{C}_\beta \supseteq, ..., \supseteq \mathcal{C}_{pro} \equiv conv(\mathfrak{A}_1^+ \otimes \mathfrak{A}_2^+) \qquad (11)$$

where $\mathcal{C}_\beta$ is the so called the tensor cone. It is worth pointing out that in 50' Stinespring used partial transposition (so the map used for definition of PPT states!) to show that $\mathcal{C}_{inj}$, $\mathcal{C}_\beta$, and $\mathcal{C}_{pro}$ are really different; for details see [29].

To sum up this Section: we showed that C^*-algebra approach to characterization of PPT states is well motivated. However, the concept of tensor product of underlying algebraic structures should be handled with great care and a special emphasis should be put on various orders, so on various concepts of positivity.

3. Structure of positive maps versus PPT states

To get a characterization of PPT states we will exploit certain links between tensor products and properly chosen mapping spaces. To this end we need some preparations. Let $\mathfrak{A} \equiv \mathcal{B}(\mathcal{K})$ denote the space of all bounded linear operators on a Hilbert space $\mathcal{K}$. $\mathfrak{T}$ will denote the set of trace class operators on $\mathcal{H}$. $x \to x^t$ denotes the transpose map of $\mathcal{B}(\mathcal{H})$ with respect to some orthonormal basis. The set of all linear bounded (positive) maps $\phi : \mathfrak{A} \to \mathcal{B}(\mathcal{H})$ will be denoted by $\mathcal{B}(\mathfrak{A}, \mathcal{B}(\mathcal{H}))$ ($\mathcal{B}(\mathfrak{A}, \mathcal{B}(\mathcal{H}))^+$ respectively). Denote by $\mathcal{B}_1(\mathfrak{A}, \mathcal{B}(\mathcal{H}))$ the ball of radius 1 in $\mathcal{B}(\mathfrak{A}, \mathcal{B}(\mathcal{H}))$. Recall, Arveson [2], that there is BW-topology such that $\mathcal{B}_1(\mathfrak{A}, \mathcal{B}(\mathcal{H}))$ is (BW)-compact. Finally, we denote by $\mathfrak{A} \odot \mathfrak{T}$ the algebraic tensor product of $\mathfrak{A}$ and $\mathfrak{T}$ (algebraic tensor product is defined as tensor product of underlying vector spaces equipped with *-algebraic structure induced by algebraic structures of factor spaces; so the topological questions are not considered) and denote by $\mathfrak{A} \hat{\otimes} \mathfrak{T}$ its Banach space closure under the projective norm defined by

$$||x|| = \inf\{\sum_{i=1}^{n} ||a_i|| ||b_i||_1 : x = \sum_{i=1}^{n} a_i \otimes b_i, \ a_i \in \mathfrak{A}, \ b_i \in \mathfrak{T}\}, \qquad (12)$$

where $||\cdot||_1$ stands for the trace norm while $||\cdot||$ means the operator norm of $\mathcal{B}(\mathcal{K})$. Following Størmer, [24], we will start with a relation between positive maps from $\mathfrak{A}$ to $\mathcal{B}(\mathcal{H})$ and positive functionals on the corresponding tensor product. Namely:

Lemma 3.1. *(Størmer [24]) There is an isometric isomorphism $\phi \to \tilde{\phi}$ between $\mathcal{B}(\mathfrak{A}, \mathcal{B}(\mathcal{H}))$ and $(\mathfrak{A}\hat{\otimes}\mathfrak{T})^*$ given by*

$$(\tilde{\phi})(\sum_{i=1}^{n} a_i \otimes b_i) = \sum_{i=1}^{n} Tr(\phi(a_i)b_i^t), \tag{13}$$

where $\sum_{i=1}^{n} a_i \otimes b_i \in \mathfrak{A} \odot \mathfrak{T}$.

Furthermore, $\phi \in \mathcal{B}(\mathfrak{A}, \mathcal{B}(\mathcal{H}))^+$ if and only if $\tilde{\phi}$ is positive on $\mathfrak{A}^+\hat{\otimes}\mathfrak{T}^+$.

As PPT states are "dual" to decomposable maps (see [19], and Section 1) we need an adaptation of the above lemma for CP and co-CP maps. In [15], it was showed:

Lemma 3.2. *(1) Let $\mathcal{B}[\,\mathcal{B}(\mathcal{H}), \mathcal{B}(\mathcal{K})_*]$ stand for the set of all linear, bounded, normal (so weakly *-continuous) maps from $\mathcal{B}(\mathcal{H})$ into $\mathcal{B}(\mathcal{H})_*$. There is an isomorphism $\psi \longmapsto \Psi$ between $\mathcal{B}[\,\mathcal{B}(\mathcal{H}), \mathcal{B}(\mathcal{K})_*]$ and $(\mathcal{B}(\mathcal{H}) \otimes \mathcal{B}(\mathcal{K}))_*$ given by*

$$\Psi\left(\sum_i a_i \otimes b_i\right) = \sum_i Tr_{\mathcal{K}}\psi(a_i)\, b_i^t, \quad a_i \in \mathcal{B}(\mathcal{H}),\ b_i \in \mathcal{B}(\mathcal{K}). \tag{14}$$

The isomorphism is isometric if Ψ is considered on $\mathcal{B}(\mathcal{H})\hat{\otimes}\mathcal{B}(\mathcal{K})$. Furthermore Ψ is positive on $(\mathcal{B}(\mathcal{H}) \otimes \mathcal{B}(\mathcal{K}))^+$ iff ψ is complete positive.

(2) There is an isomorphism $\phi \longmapsto \Phi$ between $\mathcal{B}[\,\mathcal{B}(\mathcal{H}), \mathcal{B}(\mathcal{K})_*]$ and $(\mathcal{B}(\mathcal{H}) \otimes \mathcal{B}(\mathcal{K}))_*$ given by

$$\Phi\left(\sum_i a_i \otimes b_i\right) = \sum_i Tr_{\mathcal{K}}\phi(a_i)\, b_i, \quad a_i \in \mathcal{B}(\mathcal{H}),\ b_i \in \mathcal{B}(\mathcal{K}). \tag{15}$$

The isomorphism is isometric if Φ is considered on $\mathcal{B}(\mathcal{H})\hat{\otimes}\mathcal{B}(\mathcal{K})$. Furthermore Φ is positive on $(\mathcal{B}(\mathcal{H}) \otimes \mathcal{B}(\mathcal{K}))^+$ iff ϕ is complete co-positive.

One of the consequences of this lemma is a possibility to define so called *entanglement mappings* [3], see also [4] and [15]. Namely, following Belavkin-Ohya scheme (B-O, for short), let us fix a normal state ω on $\mathcal{B}(\mathcal{H}) \otimes \mathcal{B}(\mathcal{K})$, i.e. a density matrix ϱ describing a composite state ω is fixed. Let $\sum_i \lambda_i |e_i><e_i|$ be its spectral decomposition. We will need the embedding $T_\zeta : \mathcal{K} \to \mathcal{H} \otimes \mathcal{K}$ defined by

$$T_\zeta \eta = \zeta \otimes \eta \tag{16}$$

where $\zeta \in \mathcal{H}$, $\eta \in \mathcal{K}$. Then, we are able to define the entanglement operator H

$$H : \mathcal{H} \to \mathcal{H} \otimes \mathcal{K} \otimes \mathcal{K} \tag{17}$$

by the formula:

$$H\zeta = \sum_i \lambda_i^{\frac{1}{2}} \left(J_{\mathcal{H}\otimes\mathcal{K}} \otimes T^*_{J_{\mathcal{H}}\zeta}\right) e_i \otimes e_i \tag{18}$$

where $J_{\mathcal{H}\otimes\mathcal{K}}$ is a complex conjugation defined by $J_{\mathcal{H}\otimes\mathcal{K}} f \equiv J_{\mathcal{H}\otimes\mathcal{K}}(\sum_i (e_i^{\cdot}, f)\, e_i^{\cdot}) = \sum_i \overline{(e_i^{\cdot}, f)} e_i^{\cdot}$ where $\{e_i^{\cdot}\}$ is any CONS (complete orthonormal system) extending (if necessary) the orthogonal system $\{e_i\}$ determined by the spectral resolution of ρ. ($J_{\mathcal{H}}$ is defined analogously with the spectral resolution given by H^*H; using the explicit form of H, easy calculations show that the spectrum of H^*H is discrete.) Having the entanglement operator H, we can define the entanglement mapping ϕ, which is the crucial ingredient of B-O approach. Namely:

$$\phi(b) = (H^*(1\otimes b) H)^t = J_{\mathcal{H}} H^* (1 \otimes b)^* H J_{\mathcal{H}}. \tag{19}$$

The properties of the entanglement mapping and its dual are collected in the next proposition which was proved in [15].

Proposition 3.1. *The entanglement mapping*
(i) $\phi^* : \mathcal{B}(\mathcal{H}) \to \mathcal{B}(\mathcal{K})_*$ *has the following explicit form*

$$\phi^*(a) = Tr_{\mathcal{H}\otimes\mathcal{K}} H a^t H^* \tag{20}$$

(ii) The state ω *on* $\mathcal{B}(\mathcal{H}\otimes\mathcal{K})$ *can be written as*

$$\omega(a \otimes b) = Tr_{\mathcal{H}} a\phi(b) = Tr_{\mathcal{K}} b \phi^*(a) \tag{21}$$

where ϕ *was defined in (19).*

Turning to the characterization of PPT states we note that, by definition, any such state composed with partial transposition is again a state. This observation and Proposition (3.1) lead to

Corollary 3.1. *PPT states are completely characterized by entanglement mappings* ϕ^* *which are both CP and co-CP.*

To appreciate the characterization given by the above Corollary we note that there is a nice criterion for checking CP property (so also co-CP). Namely, one has (see Takesaki's book [26])
Criterion 3.1 Let $\mathcal{A}$ and $\mathcal{B}$ be C^* -algebras. A linear map $\phi : \mathcal{A} \longrightarrow \mathcal{B}$ is CP if and only if

$$\sum_{i,j=1}^{n} y_i^* \phi(x_i^* x_j) y_j \geq 0 \tag{22}$$

for every $x_1, ..., x_n \in \mathcal{A}$, $y_1, ..., y_n \in \mathcal{B}$, and every $n \in \mathbb{N}$

Consequently, for an arbitrary but fixed normal state, there exists the explicit construction of entanglement map. If this mapping, for given state, is both CP and co-CP (what is verifiable due to the above criterion) the considered state is PPT.

4. Decomposition theory for entanglement maps

Both basic Lemmas 3.1 and 3.2 give the one-to-one correspondence between positive functionals and positive maps where positivity is defined by projective and injective cones respectively. Furthermore, in [15], the characterization of PPT states was illustrated by showing explicit relations between pure states and corresponding entanglement mappings. Namely, let ω be a pure state on $\mathcal{B}(\mathcal{H} \otimes \mathcal{K})$. As any pure state on the factor I is a vector state so there exists $x \in \mathcal{H} \otimes \mathcal{K}$ such that

$$\omega (a \otimes b) = (x, (a \otimes b) x) .$$

It was shown, [15], that the corresponding entanglement mapping has the form

$$\phi^*_{\omega_x}(a) = Tr_{\mathcal{H}} (a \otimes 1) P_x. \tag{23}$$

As a pure state case is the special one it would be desirable to generalize the above correspondence. It is worth noting that such generalization seems to be rather hard task if one does not restrict oneself to the set of all states and to spectral resolution. However, there is the decomposition theory and its aim is to express a complex structure as a superposition of simpler components [5] . Thus to get the mentioned generalization we will apply decomposition theory. To this end we need some preliminaries. Let $\mathcal{S} \equiv \mathcal{S}(\mathcal{B}(\mathcal{H} \otimes \mathcal{K}))$ denote the set of all states and let K be a convex closed subset of $\mathcal{S}$. Note that as $\mathcal{S}$ is compact (in the weak *-topology) then K is also compact. We wish to fix a state in K and attempt to find a measure μ on K which is supported by extreme points of K, and which decomposes ω in the form

$$\omega(A) = \int d\mu(\omega')\omega'(A) \tag{24}$$

where the integral is taken in the weak topology. For a detailed description of such decompositions and the corresponding measures see Chapter IV in [5]. Combining (24) with basic lemmas, one has

$$\omega(\cdot) = \int_K d\mu(\omega_x)\omega_x(\cdot) \quad \xrightarrow{\psi} \quad \phi^*_\omega \tag{25}$$

and

$$\omega_x(A) \quad \xrightarrow{\psi} \quad \phi^*_{\omega_x} \tag{26}$$

Denote by K_ψ the image of K, i.e. $K_\psi \equiv \psi(K) \subseteq \mathcal{B}_1(\mathfrak{A}, \mathcal{B}(\mathcal{H}))$. As $\psi : \mathcal{S} \to \mathcal{B}_1(\mathfrak{A}, \mathcal{B}(\mathcal{H}))$ is continuous (in weak topologies) then K_ψ is compact. Then, one has the following correspondence of measurable spaces

$$(K, \mu) \longrightarrow (K_\psi, \mu_\psi = \mu \circ \psi^{-1}) \tag{27}$$

This suggest the following general decomposition of entanglement mapping

$$\phi^*_\omega(A) = \int_{K_\psi} d\mu_\psi(\phi^*)\phi^*(A) \tag{28}$$

for some measure μ_ψ. If one put, as the very special case, $K = \mathcal{S}(\mathcal{B}(\mathcal{H}\otimes\mathcal{K}))$, and for μ an orthogonal measure supported on extreme points then one would get the decomposition of ϕ^* in terms of entanglement maps associated with pure states. Other cases would be a decompositions of ϕ^* in terms of entanglement mappings corresponding to ergodic or factorial states. Note that such decomposition can be considered as a generalization of Kraus decomposition for entanglement mappings.

Acknowledgments

The author would like to express a deep gratitude to Masanori Ohya for kind hospitality. He would like also to acknowledge the supports of BW grant 5400-5-0089-8 and QBIC grant which make his stay in Quantum Bio-informatics Center of Tokyo University of Science possible.

References

1. W. B. Arveson, Subalgebras of C^*-algebras, *Acta Math.* **123**, 141-224 (1969)
2. W. B. Arveson, Operator algebras and invariant subspaces, *Ann. of Math.* **100** (1974), 433-532
3. V. P. Belavkin, M. Ohya, Quantum entropy and information in discrete entangled states, *Infinite analysis, quantum probability and related topics*, **4** 137 (2001)
4. V. P. Belavkin, M. Ohya, Quantum entanglement and entangled mutual entropy, *Proc. Royal. Soc. London A*, **458** 209 (2002)
5. Bratteli O., and Robinson D.W., *Operator Algebras ans Quantum Statistical Mechanics I*, Springer Verlag, 1987
6. Choi M.-D., Positive linear maps, Proc. Sympos. Pure. Math. **38**, 583-590 (1982)
7. Choi M.-D., Completely Positive Maps on Complex Matrices *Lin. Alg. Appl.* **10** (1975), 285–290.

8. Choi M.-D., Some assorted inequalities for positive linear maps on C^*-algebras, *J. Operator Th.* **4** (1980), 271–285.
9. Emch, G., G., *Algebraic methods in Statistical Mechanics and Quantum Field Theory,* Wiley-Interscience 1972
10. Haag, R. *Local Quantum Mechanics; Fields, Particles, Algebras*, Springer Verlag, 1992.
11. Horodecki M., Horodecki P. and Horodecki R., Separability of mixed states: necessary and sufficient conditions, *Phys. Lett. A* **223**, 1-8 (1996)
12. Jordan, P. Über Verallgemeinerungsmöglichkeiten des Formalismus der Quantenmechanik, *Gött. Nachr.* 209-217, (1933)
13. Jordan, P., Über die Multiplikation quantenmechanischen Grössen, *Z. Phys.* **80** 285-291 (1933)
14. Jordan, P., von Neumann, J., and E. P. Wigner, On an algebraic generalization of the Quantum Mechanics formalism, *Ann. Math* **35** 29-64 (1934)
15. W. A. Majewski, T. Matsuoka, M. Ohya, Characterization of PPT states and measures of entanglement, *J. Math. Phys.*, in print; e-print: quant-phys/0904.1351
16. W. A. Majewski, Separable and entangled states of composite quantum systems - rigorous description, *Open Sys. & Inf. Dyn.* **6**, 79-86 (1999)
17. W. A. Majewski, "On entanglement of states and quantum correlations",in *Operator algebras and Mathematical Physics*, Eds. J.M. Combes, J. Cuntz, G.A. Elliott, G. Nenciu, H. Siedentop, S. Stratila; pp. 287-297, *Theta*, Bucharest, 2003. e-print, LANL math-ph/0202030;
18. W. A. Majewski, "On quantum correlations and positive maps", *Lett. Math. Physics,* **67**, 125-132 (2004); e-print, LANL math-ph/0403024
19. W. A. Majewski, M. Marciniak, On a characterization of positive maps, *J. Phys. A* **34** 5863-5874 (2001)
20. Omnés, R., *The interpretation of Quantum Mechanics.* Princeton Series in Physics, 1994
21. Reed, M., Simon, B., *Methods of modern mathematical physics.* vol. 1 *Functional Analysis*, Academic Press, 1972.
22. Sakai,S., *Operator algebras in dynamical systems*, Cambridge University Press, 1991
23. Segal, I., E., Postulates for general Quantm Mechanics, *Ann. Math.* **48** 930-948 (1947)
24. Størmer E., Extension of positive maps, *J. Funct. Analysis,* **66** (1986), 235–254.
25. E. Størmer, Separable states and positive maps, *J. Funct. Analysis,* **254** (2008), 2303-2312
26. M. Takesaki, Theory of operator algebras I, Springer Verlag, 1979
27. Winter, A., The unboundedness of quantum mechanical matrices, *Phys. Rev.*, **71**, 737 -9 (1947)
28. Wielandt, H., Üder der unbeschränktheit der operatoren der Quantum Mechanik, *Math. Ann.* **121**, 21 (1949)
29. G. Wittstock, Ordered normed tensor products, in *Foundation of Quantum Mechanics and Ordered Linear Spaces*, Springer Verlag,1974; pp 67–84

[illegible]

[illegible]

[illegible] Commun. [illegible] 135 [illegible]

[illegible] Lecture [illegible] Physics [illegible] 1980.
[illegible]

Quantum Bio-Informatics III
eds. L. Accardi, W. Freudenberg & M. Ohya

DETECTING ENTANGLEMENT IN SPIN LATTICE MODELS

MIŁOSZ MICHALSKI
Institute of Physics, Nicolaus Copernicus University
Grudziądzka 5, 87-100 Toruń, Poland
E-mail: milosz@fizyka.umk.pl

We describe a constructive method of transforming the Hamiltonian of a system of interacting quantum spins into an entanglement witness. The resulting observable can be used as a simple tool for the detection of entanglement. We verify the efficiency of this detection on a class of analytically solvable 1-dimensional Heisenberg models.

Keywords: Entanglement, entanglement witness, quantum spin chains, Heisenberg models

1. Introduction

Entanglement — a unique property of compound quantum systems has been in the focus of intensive theoretical research for at least two decades now[1]. The yesterday's "spooky action at a distance", considered by many as merely a troublesome and paradoxical consequence of mathematical formalism, uninteresting from the point of view of experimentalists, has been recognized to play a crucial role in various practical tasks of quantum-based telecommunication, cryptography or computation. Very promising application perspectives have led on the one hand to the redefinition of entanglement as one of the central notions of today's quantum mechanics, and on the other — to the significant progress in experimental techniques and instrumentation, encompassing entanglement as a manageable resource.

However, despite the enormous research effort worldwide, many fundamental questions, both theoretical and practical, concerning the entanglement detection, classification or measurement have not yet received satisfactory answers, and our understanding of these matters is certainly far from being complete. For example, given a density operator representing a state of a compound quantum system, how does one go about checking whether the state is entangled or not? Or how does one quantify entanglement of a quantum state? As of today, questions of this kind can be answered exhaustively only for a very narrow class of systems, with no more than 4 or 6 degrees of freedom, i.e. consisting of just a pair of qubits or a qubit and a qutrit.

In the present paper, we are concerned with the problem of entanglement detection in systems of interacting magnetic spins. Examples of such systems are Heisenberg models — spin operators distributed on a regular, discrete lattice, interacting via a simple nearest-neighbour isotropic (XXX) or nonisotropic (XXZ or XYZ) interaction. Rich symmetry structure of such systems allows one to use the interaction Hamiltonian for the construction of entanglement witnesses, i.e. special observables whose expectation values characterize the entanglement of corresponding system states.

The outline of the paper is as follows. In Section 2, we briefly recall basic definitions and facts related to the problem of entanglement detection. In Section 3, Heisenberg models are reviewed. Section 4 is devoted to the problem of construction of entanglement witnesses. Finally, Section 5 contains the results of explicit calculations for a 1-dimensional Heisenberg model: an entanglement witness is constructed and applied to a family of thermal equilibrium states.

2. Quantum entanglement and its detection

We begin by recalling the definition of entanglement of pure and mixed states. We consider a quantum system whose (finite-dimensional) Hilbert space is $\mathcal{H} = \mathcal{H}_{AB} = \mathcal{H}_A \otimes \mathcal{H}_B$. $\mathcal{B}(\mathcal{H})$ denotes, as usual, the space of bounded linear operators on $\mathcal{H}$. By $\Sigma = \Sigma(\mathcal{H})$ we denote the set of mixed states of the system, i.e. $\Sigma = \{\varrho \in \mathcal{B}(\mathcal{H}) : \ \varrho \geq 0\,,\ \mathrm{Tr}\varrho = 1\}$.

Definition 2.1. A pure state $|\psi\rangle \in \mathcal{H}_{AB}$ is called *separable* iff it can be written in the form $|\psi\rangle = |\phi\rangle \otimes |\chi\rangle$, with $|\phi\rangle \in \mathcal{H}_A$ and $|\chi\rangle \in \mathcal{H}_B$. Otherwise it is called *entangled.*

A mixed state $\varrho \in \Sigma$ is *separable* iff it can be written in the form

$$\varrho = \sum p_k |\psi_k\rangle\langle\psi_k|\,, \qquad p_k \geq 0\,, \quad \sum p_k = 1\,,$$

where each $|\psi_k\rangle$ is separable. Otherwise ϱ is *entangled.*

The multipartite generalization of the above definition is straightforward: given a multipartite system with Hilbert space $\mathcal{H} = \mathcal{H}_1 \otimes \cdots \otimes \mathcal{H}_m$, its state

$|\psi\rangle \in \mathcal{H}$ is (fully) *separable* iff it can be written as a product $|\psi\rangle = |\phi_1\rangle \otimes \cdots \otimes |\phi_m\rangle$, $|\phi_i\rangle \in \mathcal{H}_i$. It is obvious that, combinatorially, there are many different ways in which a multipartite state may fail to be separable, hence there are many different types of entangled or partially separable states in this case, differing by e.g. *robustness* against particle loss or by other properties. Separability of mixed multipartite states is defined analogously to the bipartite ones.

The fundamental tool to test separability of a bipartite state is the celebrated Peres-Horodecki criterion[2,3].

Theorem 2.1. *If a state $\varrho \in \Sigma(\mathcal{H}_{AB})$ is separable, then its partial transpose*[a] ϱ^{Γ} *is a positive definite matrix. This condition is both necessary and sufficient in the case of a $2 \otimes 2$ or $2 \otimes 3$ system.*

The class of states with positive partial transpose is denoted by PPT. We have the following inclusions:

$$\mathrm{Sep} \subset \mathrm{PPT} \subset \Sigma ,$$

with Sep = PPT in the low-dimensional cases mentioned above.

The criterion 2.1 is actually a fragment of broader picture. Namely, any linear map Λ on $\mathcal{B}(\mathcal{H}_B)$ which is positive but not completely positive[b] gives rise to an analogous separability criterion: the image of a separable $\varrho \in \Sigma(\mathcal{H}_{AB})$ under $\mathrm{Id}_A \otimes \Lambda$ is a positive definite matrix. Partial transpose is obviously of such form: $\Gamma = \mathrm{Id}_A \otimes T_B$. The transposition T_B is positive but not completely positive. The property of Λ not being completely positive is crucial here if it is to reveal nonseparability of some states by mapping them to nonpositive matrices. Collectively, the maps of such type give rise to a necessary and sufficient separability condition.[2]

Theorem 2.2. *$\varrho \in \Sigma(\mathcal{H}_{AB})$ is separable iff $(\mathrm{Id}_A \otimes \Lambda)\varrho \geq 0$ for all positive but not completely positive maps $\Lambda\colon \mathcal{B}(\mathcal{H}_B) \to \mathcal{B}(\mathcal{H}_A)$.*

The sufficiency of the PPT criterion in the two special low-dimensional cases is a consequence of the fact that all positive maps are then *decomposable*[4,5], i.e. expressible in the form $\Lambda = \Lambda_1 + \Lambda_2 \circ T$, where Λ_i, $i = 1, 2$, are completely positive and T stands for the transposition. In other words, in

---

[a]This is a transpose in just one of the subsystems, denoted by "Γ" which resembles "half of T".

[b]Positive maps on $\mathcal{B}(\mathcal{H})$ preserve positive definiteness of matrices. Complete positivity of Λ means that all its finite-dimensional amplifications $\mathrm{Id}_n \otimes \Lambda$ are positive.

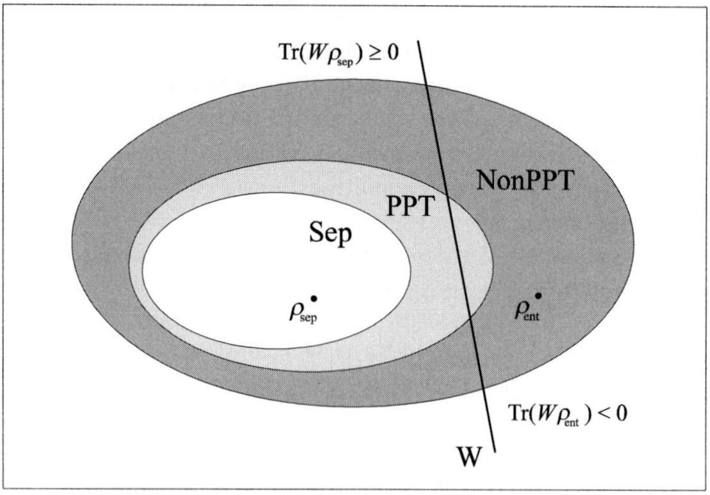

Figure 1.: Graphical representation of an entanglement witness

low dimensions transposition is the only agent among positive maps which breaks complete positivity.

An alternative way of checking the separability of states is the use of so-called *entanglement witnesses*.[2,6] These are observables $W \in \mathcal{B}(\mathcal{H})$ with the property that $\mathrm{Tr}(W\varrho) \geq 0$ for *any* separable ϱ, yet W possesses a negative eigenvalue. It means that $\mathrm{Tr}(W\varrho) < 0$ for some nonseparable ϱ. W is then said to detect the entanglement of ϱ. Again, separable states are fully characterized by a collective witness condition.[2,6]

Theorem 2.3. $\varrho \in \mathrm{Sep}$ *iff* $\mathrm{Tr}(W\varrho) \geq 0$ *for all witnesses* W.

Witness features are often depicted by a diagram represented in Fig. 1. W can be viewed as giving rise to a linear functional on $\mathcal{B}(\mathcal{H})$, $\ell(\cdot) = \mathrm{Tr}(W\,\cdot\,)$, whose kernel is a hyperplane in $\mathcal{B}(\mathcal{H})$. This hyperplane is depicted as the line splitting the set of states Σ.

An important question concerns witness optimality,[7,8] which, in simple intuitive terms, amounts to judging whether the line in Fig. 1 can be moved any closer to the set of separable states, so as to enhance the detecting capabilities of the witness.

There exists an interesting duality between positive maps and entanglement witnesses, given explicitly by the so-called *Jamiołkowski isomorphism*.[9] Assuming for simplicity that $\mathcal{H}_A = \mathcal{H}_B$, we form first the projection

Π^+ onto the maximally entangled state,

$$\Pi^+ = |\Psi^+\rangle\langle\Psi^+| , \qquad |\Psi^+\rangle = \frac{1}{\sqrt{d}} \sum_{i=0}^{d-1} |i\rangle \otimes |i\rangle ,$$

where $d = \dim \mathcal{H}_A$, and define the entanglement witness W_Λ corresponding to a positive map Λ by

$$W_\Lambda \;=\; \mathcal{J}(\Lambda) \;=\; (\mathrm{Id} \otimes \Lambda)\Pi^+ . \tag{1}$$

One should note that the general problem of systematic construction or classification of positive maps as well as entanglement witnesses is very hard.

There are analogous although operationally much more involved methods for the detection of nonseparability in the multipartite case.[10] The most straightforward generalization is the one dealing with entanglement witnesses: a witness as before is an observable $W \in \mathcal{B}(\mathcal{H})$ with positive expectation on separable states (equivalently, on pure product states), but having at least one negative eigenvalue, so that it is capable of detecting some nonseparable states ϱ via $\mathrm{Tr}(W\varrho) < 0$.

We shall also mention briefly yet another approach to entanglement detection, or more exactly — its quantification. This is achieved by the construction of various *entanglement measures*, i.e. nonnegative functions of state which vanish on separable states and are nonincreasing under local operations on subsystems supplemented by possible classical communication, the so-called *LOCC* operations[1]. Some exemplary measures are: *concurrence*[11], *entanglement of formation*[12], *negativity*[13,14] and the *degree of entanglement*[15,16] based on the concept of entanglement mapping[17].

3. Heisenberg Hamiltonians

In lattice models of magnetism one is concerned with configurations consisting of systems of spins placed in the nodes of a regular spatial lattice. Particular models are obtained by specifying how groups of spins interact among themselves. For simplicity we consider here only configurations on finite lattices with *nearest neighbour* interactions, represented by bonds between interacting pairs of lattice sites. Some typical examples are depicted in Fig. 2.

Let N be the lattice size (number of its nodes). $\boldsymbol{\sigma}_k = (\sigma_k^x, \sigma_k^y, \sigma_k^z)$ is the triple of Pauli spin operators at site k on the lattice. More formally,

$$\sigma_k^x = \mathbb{1} \otimes \cdots \otimes \sigma^x \otimes \cdots \otimes \mathbb{1},$$

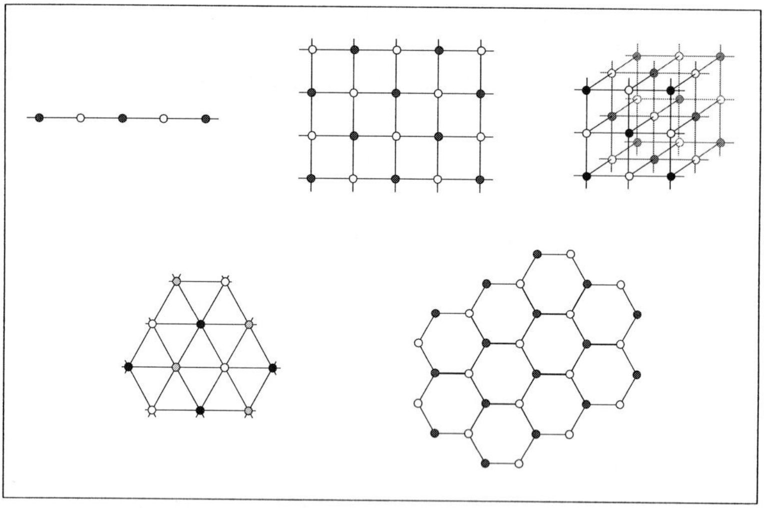

Figure 2.: Lattice models: rectangular 1-, 2- and 3-dimensional, triangular and hexagonal.

and likewise for y and z spin components, as the Hilbert space of the system is $\mathcal{H} = (\mathbb{C}^2)^{\otimes N}$.

The interaction Hamiltonian is usually of the form

$$H = \sum_{(k,l)} P(\boldsymbol{\sigma}_k, \boldsymbol{\sigma}_l)\,,$$

where the sum extends over all pairs of indexes of neighbouring lattice sites and P is a polynomial in variables $\sigma_k^x \ldots, \sigma_l^z$. The Heisenberg model[18] is an instance of such a general schema,

$$H = \sum_{(k,l)} J_x \sigma_k^x \sigma_l^x + J_y \sigma_k^y \sigma_l^y + J_z \sigma_k^z \sigma_l^z + B\sigma_k^z\,. \tag{2}$$

J_x, J_y, J_z are coupling constants and B represents external magnetic field. Positive constants J yield an antiferromagnetic interaction, while the negative — a ferromagnetic one. If $J_x = J_y = J_z$, the model is called *isotropic* or briefly an XXX model. XXZ and XYZ correspond to different *nonisotropic* models.

The richer the symmetry structure of a model, the easier it is to be solved. Isotropic 1-dimensional models can be diagonalised analytically for any N, nonisotropic ones — only for a few smaller values of N.[18]

Analytically solvable Heisenberg models are a rich source of examples of quantum states with multipartite entanglement of various kinds, and as such provide a very nice testing ground for numerous entanglement detection and quantification tools.[19,20,21,22]

4. Construction of entanglement witnesses

The Jamiołkowski isomorphism (1) can be used as a tool to construct an entanglement witness out of a positive map. For example, the operator

$$V = \sum_{i,j=0}^{d-1} |i\rangle\langle j| \otimes |j\rangle\langle i|$$

is the image under $\mathcal{J}$ of the transposition. This is, however, not a very convenient way of obtaining entanglement witnesses, because one has to have a source of suitable positive maps first. As it has been mentioned, our current knowledge of the latter is scarce. Moreover, positive maps seldom lead to witnesses with clear physical meaning, the partial transpose being a distinctive example of this. Therefore, the corresponding witnesses are useless from the experimentalist's point of view as they are *not* directly measurable physical quantities.

There is the following alternative way. Given a physical observable $A \in \mathcal{B}(\mathcal{H})$ we construct

$$W_A = A - \inf_{\varrho \in \mathrm{Sep}} \mathrm{Tr}(A\varrho)\,.$$

Remembering that a separable state ϱ can be written as a convex linear combination of 1-dimensional projections on pure *product* states, we can write the above formula in an equivalent, simpler form, namely

$$W_A = A - \inf_{|\psi\rangle \in \mathrm{Sep}} \langle\psi|A|\psi\rangle\,. \tag{3}$$

It is obvious that $\mathrm{Tr}(W_A\varrho) \geq 0$ for separable ϱ. The operation (3) amounts to shifting the spectrum of A as a whole so that separable states become immersed in the subspace spanned by eigenvectors corresponding to (new) positive eigenvalues. If there is also at least one negative eigenvalue, then W_A is a nontrivial entanglement witness.

One good choice for the operator A is the Hamiltonian of the system one deals with.[23,24] In this case, testing states for entanglement is realized via measurements of the system energy.

The infimum in (3) may be hard, if at all possible, to compute. Then symmetries present in the system may be of great help. We will see an example of such a situation in the next section.

5. Exactly solvable case: a class of Heisenberg models

We will show how to construct an entanglement witness starting from the Hamiltonian (2). According to (3), we have to compute the infimum of $\langle\psi|H|\psi\rangle$ over all product states $|\psi\rangle = |\psi_1\rangle \otimes \cdots \otimes |\psi_N\rangle$. Thus, each individual term of H, say $J_x\sigma_k^x\sigma_l^x$, yields a real scalar value, here $J_x\langle\psi_k|\sigma_k^x|\psi_k\rangle\,\langle\psi_l|\sigma_l^x|\psi_l\rangle$. We shall denote the term $\langle\psi_k|\sigma_k^x|\psi_k\rangle$ by x_k and, respectively, by y_k, z_k for other spin components. We obviously have

$$-1 \le x_k, y_k, z_k \le 1\,, \qquad \text{and} \qquad x_k^2 + y_k^2 + z_k^2 = 1 \tag{4}$$

for all k. This way the minimization of $\langle\psi|H|\psi\rangle$ over product states with independently varying factors $|\psi_k\rangle$ is equivalent to the minimization of

$$P_H = \sum_{(k,l)} J_x x_k x_l + J_y y_k y_l + J_z z_k z_l : + : B z_k, \tag{5}$$

a real polynomial in $3N$ variables subject to the constraints (4). This can be solved by using standard multivariable calculus techniques, like e.g. the Lagrange multiplier method.

Following the ideas of Tóth[24], we consider a class of Heisenberg Hamiltonians for which the minimization procedure simplifies still further. Namely, using the symmetries of interaction we aim at performing the optimization termwise.

For this to be possible, we need a few technical assumptions, which actually are fulfilled by typical Heisenberg models. The first one concerns the topology of the lattice and it is easily expressible in the language of graph theory.

Assumption 5.1. *The lattice is a k-partite graph, i.e. its sites can be partitioned into k disjoint subsets, so that no edge has both its endpoints in one of these sets.*

All lattices of Fig. 2 are 2-partite graphs except for the triangular one, which is a 3-partite one.

The second assumption concerns the form of the Hamiltonian. Let $\mathbf{r}_i = (x_i, y_i, z_i)$.

Assumption 5.2. *There exists a function of 3k variables, $h(\mathbf{r}_{i_1}, \ldots, \mathbf{r}_{i_k})$, symmetric with respect to permutations of $\mathbf{r}_{i_j}$, such that*

$$P_H = \sum h(\mathbf{r}_{i_1}, \ldots, \mathbf{r}_{i_k})\,, \tag{6}$$

where the k-tuple of indexes of every term present in the sum is such that each i_j belongs to a different set of the lattice partition as in Assumption 5.1.

For example, a periodically closed 1-dimensional Heisenberg chain with $N = 2n$ sites is a 2-partite graph (split into odd- and even-numbered sites) and taking

$$h(\mathbf{r}_i, \mathbf{r}_j) = J_x x_i x_j + J_y y_i y_j + J_z z_i z_j + \frac{B}{2}(z_i + z_j)\,,$$

we have

$$P_H = \sum_{i=1}^{N} h(\mathbf{r}_i, \mathbf{r}_{i+1}) \tag{7}$$

with identification $N+1 \equiv 1$. If the number of sites is odd, the periodically closed chain is no longer bipartite and at least one term in P_H cannot have the property postulated in Assumption 5.2. As it will follow below, our minimization strategy may not be exact then.

It is easy to observe that the 2- and 3-dimensional periodically closed rectangular lattices can be treated identically if only all their sizes are even.

Similarly, for a periodic triangular lattice with sizes in all directions divisible by 3, partitioned into sets M, P, Q, we have

$$h(\mathbf{r}_m, \mathbf{r}_p, \mathbf{r}_q) = \frac{J_x}{2}(x_m x_p + x_p x_q + x_q x_m) + \frac{J_y}{2}(\cdots) + \frac{J_z}{2}(\cdots) + \frac{B}{6}(z_m + z_p + z_q)$$

with $m \in M$, $p \in P$ and $q \in Q$.

The minimization procedure is described by the following lemma.

Lemma 5.1. *Suppose that Assumptions 5.1 and 5.2 are fulfilled. Then*

$$\min P_H = n \cdot \min h\,,$$

where n is the number of terms in (6) and the minima are conditioned by the constraints (4).

Proof. Obviously, $\min P_H \geq n{\cdot}\min h$. Let the minimum of h be attained at $(\mathbf{s}_1, \ldots, \mathbf{s}_k)$, with $\mathbf{s}_i = (x_i^0, y_i^0, z_i^0)$, $\|\mathbf{s}_i\| = 1$, $i = 1, \ldots k$. Associating one $\mathbf{s}_i$ with exactly one set V_i of the lattice partition, we perform the substitution in (6) $\mathbf{s}_{i_j} := \mathbf{s}_i$ if $i_j \in V_i$. This way, each term in (6) is precisely h evaluated at a permutation of $(\mathbf{s}_1, \ldots, \mathbf{s}_k)$. Using now the symmetry of h, we obtain the thesis. □

1-dimensional antiferromagnetic XXX chain. Setting $J_x = J_y = J_z = 1$ in (2) we have

$$H = \sum_{i=1}^{N} \sigma_i^x \sigma_{i+1}^x + \sigma_i^y \sigma_{i+1}^y + \sigma_i^z \sigma_{i+1}^z + B\sigma_i^z \,, \tag{8}$$

assuming in addition N to be even. We thus have to minimize

$$h(\mathbf{r}_p, \mathbf{r}_q) \;=\; x_p x_q + y_p y_q + z_p z_q + \frac{B}{2}(z_p + z_q)$$

subject to

$$x_p^2 + y_p^2 + z_p^2 = 1 \,, \qquad x_q^2 + y_q^2 + z_q^2 = 1 \,,$$

which yields

$$\min h = \begin{cases} -1 - \frac{B^2}{8} & \text{for } |B| \le 4 \\ 1 - |B| & \text{for } |B| > 4 \end{cases} \tag{9}$$

and consequently

$$\inf_{\psi \in \mathrm{Prod}} \langle \psi | H | \psi \rangle = N \min h.$$

It is easy to see that in the case of d-dimensional periodically closed cubic lattice we would simply have $\inf\langle\psi|H|\psi\rangle = dN \min h$.

1-dimensional antiferromagnetic XXY chain. We consider now a model with partial nonisotropy of interaction in the Z direction,

$$H = \sum_{i=1}^{N} \sigma_i^x \sigma_{i+1}^x + \sigma_i^y \sigma_{i+1}^y + (1+\Delta)\sigma_i^z \sigma_{i+1}^z + B\sigma_i^z \,.$$

Minimization of the respective $h(\mathbf{r}_p, \mathbf{r}_q)$ with normalization constraints yields

$$\min h = \begin{cases} -1 - \Delta & \text{(I) for } \;\Delta > \sqrt{1 + \left(\frac{B}{2}\right)^2} - 1 \\ -1 - \frac{B^2}{4(2+\Delta)} & \text{(II) for } \;\frac{|B|}{2} - 2 \le \Delta \le \sqrt{1 + \left(\frac{B}{2}\right)^2} - 1 \\ 1 - |B| - \Delta & \text{(III) for } \;\Delta < \frac{|B|}{2} - 2 \end{cases}$$

Fig. 3 shows the corresponding "phase" diagram in the (B, Δ) plane. Similarly as in the XXX case, $\inf\langle\psi|H|\psi\rangle = N \min h$ for even N.

If the chain length N is odd then, as we have observed, one of the terms of (7), say $h(\mathbf{r}_N, \mathbf{r}_1)$, will not conform with the requirements of Assumption

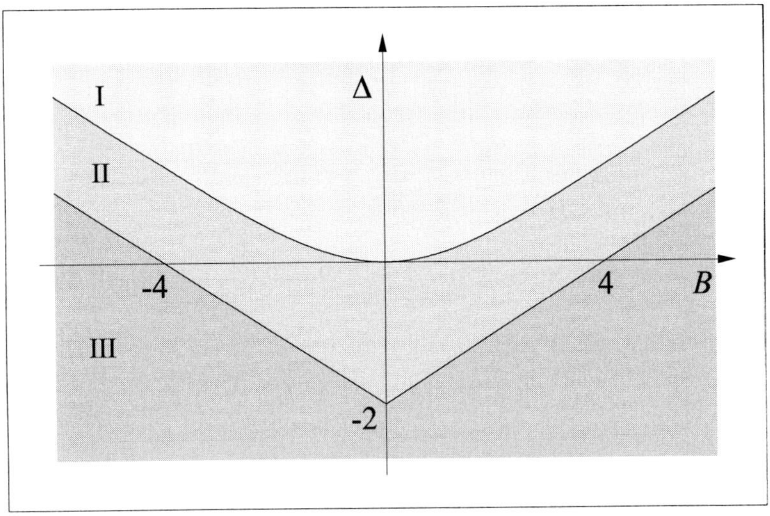

Figure 3.: Regions I-III in the (B, Δ)-plane corresponding to different values of $\min h$ for an XXY chain.

5.2, since the sites N and 1 belong to the same partition set. Applying the minimization strategy as in the proof of the lemma we get $h(\mathbf{s}_1, \mathbf{s}_2) = \min$ and $P_H = (N-1) \cdot \min h + h(\mathbf{s}, \mathbf{s})$ with $\mathbf{s} = \mathbf{s}_1$ or $\mathbf{s}_2$ depending which of the two yields a smaller value of h. Since the true minimum of P_H is certainly no greater than this, we have $N \cdot \min h \leq \min P_H \leq (N-1) \cdot \min h + h(\mathbf{s}, \mathbf{s})$. As a result, the corresponding witness $W_H = H - (N \min h)\mathbb{I}$ may not be optimal.

Exemplary computation for a 4-particle isotropic chain. We consider the Hamiltonian of the form (8) with $N = 4$. Interesting part of its spectrum and the corresponding eigenvectors are given below:

$$E_0 = -8 \qquad |\psi_0\rangle :=: \tfrac{1}{2\sqrt{3}}(\,|0011\rangle - 2|0101\rangle + |0110\rangle + |1001\rangle - 2|1010\rangle + |1100\rangle)$$

$$E_1 = -4 - 2B \qquad |\psi_1\rangle :=: \tfrac{1}{2}(-|0111\rangle + |1011\rangle - |1101\rangle + |1110\rangle)$$

$$E_2 = 4 - 4B \qquad |\psi_2\rangle \;=\; |1111\rangle$$

Depending on the value of $B \geq 0$, the ground energy is E_0, E_1 or E_2. Fig. 4 shows the corresponding level crossing diagram. So, for $0 \leq B < 4$ the ground state is entangled, with level crossing point at $B = 2$. For $B > 4$

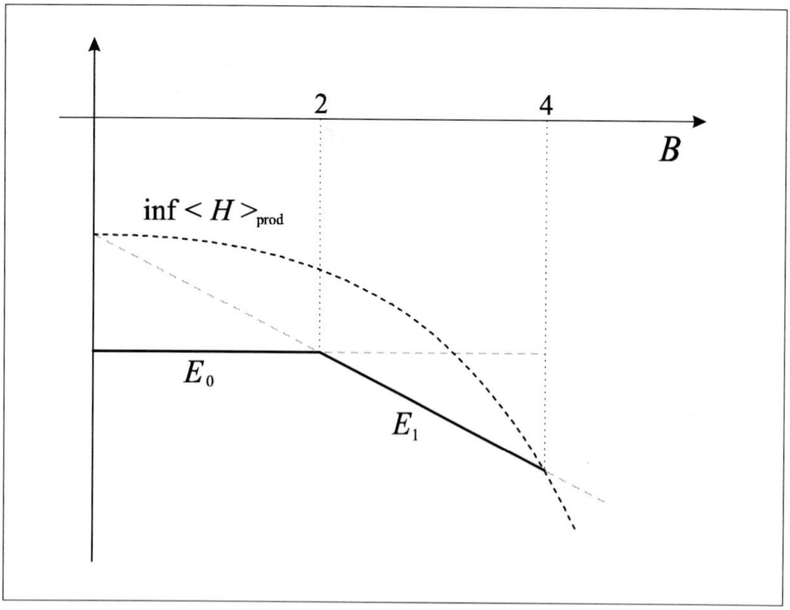

Figure 4.: Groundstate level crossing for an XXX chain with 4 spins.

the ground state becomes separable.

Fig. 4 shows also the value of $\inf\langle\psi|H|\psi\rangle$ over product states (the dashed parabolic line). When $B > 4$, this infimum becomes smaller than the ground energy so that the constructed witness W_H is positive definite, and hence trivial.

We finish this paper with the results of a test of the witness W_H effectiveness in detecting entanglement. The test has been performed on a family of thermal equilibrium states of our system,

$$\varrho_\beta = \frac{e^{-\beta H}}{Z}, \qquad Z = \mathrm{Tr}(e^{-\beta H}), \quad \beta = \frac{1}{k_B T}.$$

Fig. 5 displays the plot of $-\mathrm{Tr}(W_H \varrho_\beta)$ for a range of (β, B) values (left) in comparison with the plot of the concurrence — an exact measure of entanglement between any pair of neighbouring spins in the chain.

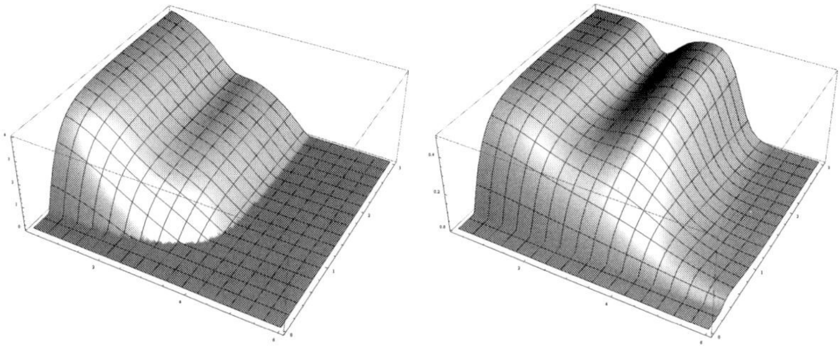

Figure 5.: $-\operatorname{Tr}(W_{HO_\beta})$ versus $C(O_{12})$ — the concurrence of the reduced density matrix with spins 3 and 4 traced out (Z-axes are not in scale)

The plots illustrate the deficiency of the witness in detecting the so-called thermally induced entanglement. This is visible to the right of the $B = 4$ line. As we have said, the ground state $|\psi_2\rangle$ is then separable which makes the witness construction ineffective — W_H is a positive definite operator then. On the other hand, for a range of positive temperatures, the equilibrium state shows traces of entanglement due to an admixture of higher energy eigenstates $|\psi_0\rangle$ and $|\psi_1\rangle$ present in it. This is correctly detected by the concurrence.

Finally, it may also be interesting to observe the trace of critical level crossing at $B = 2$ — the characteristic fold along this line in the plots.

Acknowledgment

The author is greatly indebted to Professor Masanori Ohya and the organizers of the QBIC'09 International Conference for their kind invitation and hospitality during his stay in Noda, Japan.

References

1. Horodecki R., P. Horodecki, M. Horodecki, K. Horodecki, Rev. Mod. Phys. **81**, 865 (2009).
2. Horodecki M., P. Horodecki, R. Horodecki, Phys. Lett. A **223**, 1 (1996).
3. Peres A., Phys. Rev. Lett. **77**, 1413 (1996).
4. Størmer E., Acta Math. **110**, 233 (1963).

5. Woronowicz S. L., Rep. Math. Phys. **10**, 165 (1976).
6. Terhal B., Phys. Lett. A **271**, 319 (2000).
7. Lewenstein M., B. Kraus, J. I. Cirac, P. Horodecki, Phys. Rev. A **62**, 052310 (2000).
8. Sarbicki G., arXiv:0905.0778v1 [quant-ph] (2009).
9. Jamiołkowski A., Rep. Math. Phys. **3**, 275 (1972).
10. Horodecki M., P. Horodecki, R. Horodrecki, Phys. Lett. A **283**,1 (2001).
11. Wootters W. K., Phys. Rev. Lett. **80**, 2245 (1998).
12. Bennett C. H., D. P. DiVincenzo, J. Smolin, W. K. Wootters, Phys. Rev. A **54**, 3824 (1996).
13. Życzkowski K., P. Horodecki, A. Sanpera, M. Lewenstein, Phys. Rev. A **58**, 883 (1998)
14. Vidal G., R. F. Werner, Phys. Rev. A **65**, 032314 (2002).
15. Matsuoka T., [in:] *Quantum Bio-Informatics II*, L. Accadri, W. Freudenberg. M. Ohya, eds., Quantum Probability and White Noise Analysis **XXIV**, 139 (2009).
16. Majewski W. A., T. Matsuoka, M. Ohya, arXiv:0904.1351v1 [quant-ph] (2009).
17. Belavkin, V. P., M. Ohya, Proc. Royal Soc. Lond. **A 458**, 209 (2002).
18. Baxter R. J., *Exactly solved models in statistical mechanics*, London, Academic Press, 1982.
19. D. Bruss, N. Datta, A. Ekert, L.C. Kwek, C. Macchiavello, Phys. Rev. A **72**, 014301 (2005).
20. Wang X., Phys. Rev. A **64**, 012313 (2001).
21. Wang X., P. Zanardi, Phys. Rev. A **66**, 044303 (2002).
22. Michalski M., Int. J. Quant. Inf. **6**, 727 (2008).
23. Brukner C., V. Vedral, arXiv:quant-ph/0406040v1 (2004).
24. Tóth G., Phys. Rev. A **71**, 010301(R) (2005).

Quantum Bio-Informatics III
eds. L. Accardi, W. Freudenberg & M. Ohya

HOW TO DETECT ENTANGLEMENT IN QUANTUM SYSTEMS

MIŁOSZ MICHALSKII
Institute of Physics, Nicolaus Copernicus University
Grudziądzka 5, 87-100 Toruń, Poland
E-mail: milosz@fizyka.umk.pl

The present paper is a brief, elementary introduction to the theory of entangled states in quantum systems, based on the lecture given by the author at the Suwa Satellite Conference of QBIC09, held at Tokyo University of Science in Suwa, March 6–8, 2009. We focus here on the mathematical problem of entanglement detection. The duality between positive maps and entanglement witnesses is emphasized. One particular class of experimentally relevant entanglement witnesses and the method of their construction is further discussed in another paper appearing in this volume.[1] An interested reader may find much more information about recent developments in the entanglement theory and applications in a review paper [2].

Keywords: Entanglement, positive maps, entanglement witnesses

1. Introduction

Fundamental postulates of quantum theory associate with every quantum system a Hilbert space $\mathcal{H}$ whose normalized vectors represent pure states of the system. Moreover, if A and B are two systems with Hilbert spaces $\mathcal{H}_A$ and $\mathcal{H}_B$, respectively, then the space associated with the composite (two-partite) system AB is $\mathcal{H}_{AB} = \mathcal{H}_A \otimes \mathcal{H}_B$. Multipartite generalization is straightforward, $\mathcal{H}_{\text{total}} = \mathcal{H}_1 \otimes \cdots \otimes \mathcal{H}_m$.

It is then the mere structure of tensor product space that brings in the abundance of vectors of special form: they cannot be represented as simple products $\psi_A \otimes \psi_B$ of some $\psi_A \in \mathcal{H}_A$ and $\psi_B \in \mathcal{H}_B$, but instead are linear combinations of two or more such terms.[a] Although there seems to be

[a]An example of such a vector is $|\phi_A\rangle \otimes |\phi_B\rangle + |\psi_A\rangle \otimes |\psi_B\rangle$, where $|\phi_A\rangle$ and $|\psi_A\rangle$ are arbitrary linearly independent vectors in $\mathcal{H}_A$ and likewise for $|\phi_B\rangle$ and $|\psi_B\rangle$ in $\mathcal{H}_B$.

nothing special about it from the point of view of mathematics, physical interpretation within the framework of quantum mechanics is far from being simple and in the past was a matter of serious, fundamental debates among such prominent adversaries as Albert Einstein, Erwin Schrödinger and Niels Bohr.[3,4,5,6] It was during this debate that the notion of *entanglement* was first coined by E. Schrödinger.[5]

The discrimination between separable (product) and entangled pure states is mathematically very simple, in fact it is no more troublesome than computing the rank of a matrix. However, when the notion of entanglement is properly generalized to *mixed* states which are represented by density operators, the same problem immediately becomes more complex, beyond any comparison. Suffice it to say, that despite the joint research effort of numerous physicists and mathematicians over the past two decades, only partial solutions have been obtained so far, if not to count the simplest low-dimensional cases.

Recent developments in experimental techniques and laboratory instrumentation has brought entanglement into the focus of quantum mechanical research, both theoretical and experimental. The former interpretational burden, attracting attention of a relatively narrow group of theoretical physicists interested in the logical foundations of quantum mechanics, turned out to be a manageable resource, indispensable in performing various tasks of quantum communication, cryptography and computing, opening up extremely interesting technological perspectives. The problem of determination which system states are entangled and which are not is thus of fundamental theoretical and experimental importance. This exposition is intended to introduce the reader into our current state of knowledge concerning this kind of problems.

The structure of the present paper is as follows. Section 2 contains a brief review of basic notions — quantum states, tensor products and positivity. Section 3 is devoted to the presentation of positive maps and their role in entanglement detection. Section 4 discusses entanglement witnesses, an alternative tool in the detection of entanglement, and finally, Section 5 — their duality relation with positive maps.

2. Review of basic notions

In this paper, we consider only finite-dimensional Hilbert spaces. Given such space $\mathcal{H}$, *pure* states of the corresponding quantum system are represented by normalized vectors $|\psi\rangle$ of $\mathcal{H}$.[b] In other words,

$$\mathcal{H} \simeq \mathbb{C}^d \ni |\psi\rangle = \sum_{k=1}^{d} c_k \, |k\rangle \,, \qquad \langle\psi|\psi\rangle = \sum_k |c_k|^2 = 1 \,,$$

where by the widespread convention we write $|1\rangle, \ldots, |d\rangle$ for the standard basis of $\mathcal{H}$. $\mathcal{B}(\mathcal{H})$ denotes the set of all bounded linear operators on $\mathcal{H}$, which, of course, in the finite-dimensional case discussed here is just the set of *all* linear operators on $\mathcal{H}$.

Of special interest are those linear operators which have real eigenvalues and complete systems of eigenvectors — they correspond to physical quantities, *observables*, whose numerical values are obtained in measurement processes. Mathematically, observables are selfadjoint (hermitean) operators in $\mathcal{B}(\mathcal{H})$, $A = A^*$.

In order to introduce *mixed* states into the formalism[7], we represent first pure states as the corresponding 1-dimensional projection operators,

$$|\psi\rangle \quad \longrightarrow \quad |\psi\rangle\langle\psi| \in \mathcal{B}(\mathcal{H}) \,,$$

and then define mixed states ϱ as convex linear combinations of such pure projections,

$$\varrho = \sum_k \mu_k |\psi_k\rangle\langle\psi_k| \,, \qquad \mu_k \geq 0 \,, \quad \sum_k \mu_k = 1 \,.$$

They are interpreted as statistical mixtures of the corresponding pure states, the coefficients μ_k representing our uncertainties about the current state of our system.

Equivalently, mixed states of a system with Hilbert space $\mathcal{H}$ are density matrices on $\mathcal{H}$, i.e. such $\varrho \in \mathcal{B}(\mathcal{H})$ which are selfadjoint $\varrho = \varrho^*$, nonnegative definite $\rho \geq 0$ (have all eigenvalues nonnegative), and of unit trace $\mathrm{Tr}(\varrho) = 1$ (the diagonal elements, or equivalently — the eigenvalues, sum up to one). The set of mixes states is denoted by $\Sigma = \Sigma(\mathcal{H})$. It is a convex set in $\mathcal{B}(\mathcal{H})$. Sometimes, in geometric arguments, it is convenient to disregard the normalization condition $\mathrm{Tr}(\varrho) = 1$ and view Σ as a convex cone in $\mathcal{B}(\mathcal{H})$, [8] itself a well-studied mathematical notion.

[b]Actually, the situation is mathematically a bit more involved. Namely, the Hilbert space distinguishes vectors differing by a scalar phase constant $e^{i\alpha}$, while from the point of view of quantum mechanics they represent equivalently the same state. Their identification leads to the notion of a complex projective space $\mathbb{C}\mathbf{P}^n$ which, strictly speaking, is the proper mathematical model of state space. However, we will stick with conventional Hilbert space notation in this exposition.

Density matrices are the quantum counterpart of classical (finite) probability distributions, i.e.

$$\{p_1, \dots, p_n\}, \qquad p_i \geq 0, \ \sum_i p_i = 1.$$

To localize such *classical* states in Σ we can represent them as diagonal martices

$$\begin{bmatrix} p_1 & 0 & \cdots & 0 \\ 0 & p_2 & \cdots & 0 \\ \vdots & \vdots & \ddots & \vdots \\ 0 & 0 & \cdots & p_n \end{bmatrix},$$

while for nonclassical $\varrho \in \Sigma$ we have

$$\varrho = \begin{bmatrix} \varrho_{11} & \varrho_{12} & \cdots & \varrho_{1n} \\ \varrho_{21} & \varrho_{22} & \cdots & \varrho_{2n} \\ \vdots & \vdots & \ddots & \vdots \\ \varrho_{n1} & \varrho_{n2} & \cdots & \varrho_{nn} \end{bmatrix} \xrightarrow{\text{diagonalization}} \begin{bmatrix} \mu_1 & 0 & \cdots & 0 \\ 0 & \mu_2 & \cdots & 0 \\ \vdots & \vdots & \ddots & \vdots \\ 0 & 0 & \cdots & \mu_n \end{bmatrix}.$$

The reader should note, however, that unlike the classical states, two or more "true" density matrices are typically not simultaneously diagonalizable.

As we have mentioned above in Section 1, the composition of quantum systems into multipartite ones corresponds to tensor multiplication of the respective Hilbert spaces. We intend to give here a handy operational description of tensor products in vector spaces.

Let $\mathcal{H}_A$ and $\mathcal{H}_B$ be two Hilbert spaces, with $\dim \mathcal{H}_A = m$, $\dim \mathcal{H}_B = n$. Suppose we have fixed orthonormal bases in $\mathcal{H}_A$ and $\mathcal{H}_B$. It is convenient to think of the tensor product of two vectors $|\mathcal{H}\rangle \otimes |\mathcal{H}\rangle$ in terms of blocks of size n replacing each of the m components of $|\mathcal{H}\rangle$. These blocks contain copies of the vector $|\mathcal{H}\rangle$ multiplied respectively, by the components a_i of $|\mathcal{H}\rangle$, cf. Fig. 1.

$\mathcal{H}_A \otimes \mathcal{H}_B$ contains all such product vectors and all linear combinations of them. In particular, if $|i\rangle$, $i = 1, \dots, m$, $|j\rangle$, $j = 1, \dots, n$, are bases in $\mathcal{H}_A$ and $\mathcal{H}_B$, then $|ij\rangle = |i\rangle \otimes |j\rangle$ is a basis in $\mathcal{H}_A \otimes \mathcal{H}_B$ and $\dim : \mathcal{H}_A \otimes \mathcal{H}_B = m \cdot n$. As we have already mentioned, not all $|\mathbf{c}\rangle \in \mathcal{H}_A \otimes \mathcal{H}_B$ can be written as a simple product $|\mathbf{a}\rangle \otimes |\mathbf{b}\rangle$ for some $|\mathbf{a}\rangle$, $|\mathbf{b}\rangle$.

Our construction of $|\mathbf{a}\rangle \otimes |\mathbf{b}\rangle$ given above suggests an easy way of checking whether a given $|\mathbf{c}\rangle$ is a separable (product) pure state or an entangled one. Namely, one has to decompose the coordinates of $|\mathbf{c}\rangle$ into m blocks of

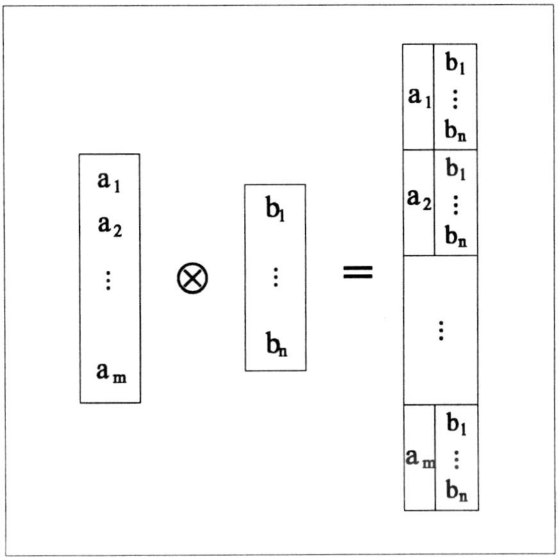

Figure 1. Forming the tensor product of two vectors $|\mathbf{a}\rangle$ and $|\mathbf{b}\rangle$

n components each and verify whether these blocks are all mutually proportional. Equivalently, we can arrange the blocks as columns of an $m \times n$ matrix C and check whether its rank is one or more. If it is one, the state $|\mathbf{c}\rangle$ is separable.

We are one step away from a systematic computational procedure of checking whether a given vector is separable or not. It is called the Schmidt decomposition and yields the minimal number of product terms needed to represent $|\mathbf{c}\rangle \in \mathcal{H}_A \otimes \mathcal{H}_B$ as their linear combination.[9] This number is called the Schmidt length of $|\mathbf{c}\rangle$.

We begin by expressing $|\mathbf{c}\rangle$ in the $|i\rangle \otimes |j\rangle$ basis,

$$|\mathbf{c}\rangle = \sum_{i=1}^{m} \sum_{j=1}^{n} c_{ij} |i\rangle \otimes |j\rangle .$$

Next we compute the square matrix $C^T C$, where $C = [c_{ij}]$, and diagonalize it. It has nonnegative eigenvalues

$$\lambda_1, \ldots, \lambda_s, 0, \ldots, 0, \qquad \sum_i \lambda_i = 1,$$

where $1 \leq s \leq \min\{m, n\}$. Then $|\mathbf{c}\rangle$ can be written as

$$|\mathbf{c}\rangle = \sum_{i=1}^{s} \sqrt{\lambda_i}\, |\phi_i\rangle \otimes |\psi_i\rangle$$

for some vectors $|\phi_i\rangle \in \mathcal{H}_A$ and $|\psi_i\rangle \in \mathcal{H}_B$, the number s being the Schmidt length of $|\mathbf{c}\rangle$. In what follows we will denote this number by $\ell_S(|\mathbf{c}\rangle)$ or simply by ℓ_S if the pure state to which it refers will be clear from the context. If $\ell_S = 1$ the state is separable, if $\ell_S > 1$ it is entangled. In the latter case ℓ_S can serve as a rough measure allowing one to compare vectors with respect to their degree of entanglement: of two vectors the one with larger Schmidt length is the more entangled one. A more refined measure is the Shannon entropy of the sequence λ_i, $S(|\mathbf{c}\rangle) = -\sum_i \lambda_i \log \lambda_i$. There exists a vast literature concerning quantitative characterization of entanglement, e.g.[2,26,27,28,29,30,31,32] but we will not discuss it in this paper any further.

The Schmidt decomposition of $|\mathbf{c}\rangle$ amounts to a careful choice of bases $|\phi_i\rangle$ and $|\psi_i\rangle$ in $\mathcal{H}_A$ and $\mathcal{H}_B$ respectively, so that the representation of $|\mathbf{c}\rangle$ is as short as possible. Note, however, that such a choice of bases, in general, does the trick only for $|\mathbf{c}\rangle$ and not necessarily for any other vector $|\mathbf{c}'\rangle$, i.e. an instance of Schmidt decomposition has no global meaning.

Similarly, block operations can be used to describe tensor products of matrices. Suppose $A = [a_{ij}]_{m\times n}$ and $B = [b_{kl}]_{p\times q}$ are two matrices. Then the tensor product $A \otimes B$ is the $mp \times nq$ matrix which is obtained from A by replacing each of its entries a_{ij} with a block of size $p \times q$ holding a copy of B multiplied respectively by a_{ij}, cf. Fig. 2.

If $\mathcal{A}$ and $\mathcal{B}$ are two matrix algebras then $\mathcal{A} \otimes \mathcal{B}$ contains all linear combinations of simple product matrices $A \otimes B$ as above. It is convenient to look at $\mathcal{A}\otimes\mathcal{B}$ as an algebra of matrices of the size like that of the elements of $\mathcal{A}$ whose entries are elements B_{ij} of $\mathcal{B}$ rather than complex numbers,

$$\left[\begin{array}{c|c|c} B_{11} & \cdots & B_{1n} \\ \hline \vdots & \ddots & \vdots \\ \hline B_{m1} & \cdots & B_{mn} \end{array}\right].$$

Suppose now that we have a density matrix $\varrho \in \mathcal{B}(\mathcal{H}_A \otimes \mathcal{H}_B)$, namely a mixed state of a bipartite quantum system, and we want to determine whether it is entangled or not. The definition of separability for mixed states, due to R.F. Werner[10], is the following.

Definition 2.1. A mixed state ϱ of a bipartite system with Hilbert space

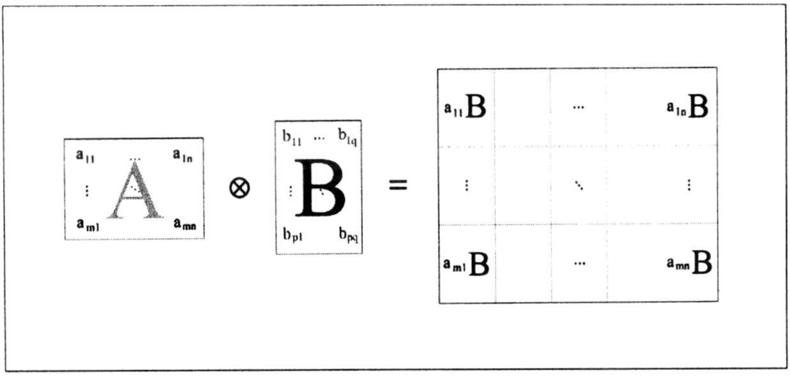

Figure 2. Forming the tensor product of two matrices A and B.

$\mathcal{H}_A \otimes \mathcal{H}_B$ is separable if it can be represented as a convex combination of product states,

$$\varrho = \sum_i p_i \, \varrho_i^A \otimes \varrho_i^B \,,$$

where $\sum_i p_i = 1$, $\varrho_i^A \in \mathcal{B}(\mathcal{H}_A)$, $\varrho_i^B \in \mathcal{B}(\mathcal{H}_B)$ or, equivalently, as a convex combination of projections onto separable states

$$\varrho = \sum_i p_i \, |\phi_i \otimes \psi_i\rangle\langle\phi_i \otimes \psi_i| \,,$$

where $|\phi_i\rangle \in \mathcal{H}_A$ and $|\psi_i\rangle \in \mathcal{H}_B$.

We should observe that, unlike in the case of pure states, there is no immediate effective way of testing a mixed state for separability. In particular, there is no simple linear dependence-based criterion referring to blocks into which the matrix ρ can be decomposed according to the underlying tensor product structure. This is due to the presence of convex combination in the definition which, in general, destroys simple proportionality relations among the blocks existing in individual product terms. As we will see below, the separability problem for mixed states has not been given a satisfactory, general solution so far.

There is a simple generalization of the Schmidt length applicable to mixed states. However, its computation is by no means a simple task. To find ℓ_S of a given density matrix ϱ we have to effectively find *all possible* convex decompositions of ϱ into pure projectors, $\varrho = \sum_i p_i \, |\psi_i\rangle\langle\psi_i|$, more or less the the same effort that one has to make when checking separability of ϱ straight from Definition 2.1. We have

$$\ell_S(\varrho) \stackrel{\text{df}}{=} \min_{\{p_i,\psi_i\}} \left\{ \max_i : \ell_S(|\psi_i\rangle) \right\} ,$$

where the minimum extends over all such decompositions of ϱ. In practice, it is hardly ever possible to find the Schmidt length of a mixed state exactly.

The notion of Schmidt length for density matrices gives rise to a filtration of the state space $\Sigma = \Sigma(\mathcal{H}_{AB})$. Writing $\Sigma_k = \{\varrho : \ell_S(\varrho) \leq k\}$, we have

$$\Sigma_{\mathsf{Sep}} = \Sigma_1 \subset \Sigma_2 \subset \cdots \subset \Sigma_n = \Sigma ,$$

n being the smaller of the dimensions of $\mathcal{H}_A$ and $\mathcal{H}_B$. We will see the role of this filtration in subsequent sections.

In order to proceed with the discussion of various separability criteria for mixed states, we need one more notion — *positivity*. It is a common practice in mathematics to define positivity of certain class of objects, in in analogy to the ordinary positivity of numbers. Then it is interesting to characterize a class of maps on these objects which preserve this positivity.

Suppose, for example, that considering some natural phenomenon we work with finite probability schemes $\mathbf{p} = \{p_1, \ldots, p_n\}$, n being fixed. Such distributions form a subset Σ in the vector space $\mathbb{R}^n$. We can conveniently characterize this subset by saying that $\mathbf{p} \in \Sigma$ iff $\mathbf{p}$ is positive, $\mathbf{p} \geq 0$, meaning that all $p_i \geq 0$, and normalized, $\sum p_i = 1$.[c] Evolution processes are then modeled by actions of some class of maps on Σ, e.g. in the simplest case — linear maps $V : \mathbf{p} \mapsto \mathbf{p}'$. Suitable maps are then, of course, the ones which take probabilities into probabilities or, somewhat more generally, the ones which preserve positivity. One can easily see that for linear maps $V = [v_{ij}]$ to preserve vector positivity it is necessary and sufficient that $v_{ij} \geq 0$ for all i and j. In addition, the property $\sum_i v_{ij} = 1$ for all j ensures the preservation of normalization. So with these two conditions satisfied we have the desired property $V : \Sigma \to \Sigma$.

Similarly, when considering classes of admissible operations on quantum states, we have to ensure that maps implementing them will not destroy the property of being a mixed state, i.e. a density matrix. Therefore, this time the objects are *matrices* and the maps are linear operations on matrices $\Lambda : \mathcal{B}(\mathcal{H}) \to \mathcal{B}(\mathcal{H})$. We are going now to define the relevant notions.

---

[c]Actually, the normalization condition is often of secondary importance, since given an arbitrary positive $\mathbf{p}$ the normalization constant $\sum p_i$ is automatically known.

Definition 2.2. A selfadjoint matrix $V \in \mathcal{B}(\mathcal{H})$ is positive definite, $V > 0$ (resp. positive semidefinite or non-negative definite, $V \geq 0$) iff all its eigenvalues are positive (resp. nonnegative) or, equivalently, iff for all $|\psi\rangle \in \mathcal{H}$ we have $\langle\psi|V\psi\rangle > 0$ (resp. $\ldots \geq 0$).

Definition 2.3. A linear map $\Lambda : \mathcal{B}(\mathcal{H}) \to \mathcal{B}(\mathcal{H})$ is called positive iff it preserves positive definiteness of matrices.

In other words, positive maps preserve the positivity of spectra of matrices they act on. It is much more complicated property than the preservation of vector positivity by matrices we have discussed above. In particular, no effective conditions are known so far to check maps for their being positive.

A simple but important example of a positive map is matrix transposition. It is linear, $(\alpha U + \beta V)^T = \alpha U^T + \beta V^T$, and does not affect the spectrum of its argument, hence it is positive.

It turns out that, from the physical point of view, positivity is still too general property to qualify a map as a valid representation of a quantum operation. Namely, it is natural to consider the evolution of a quantum system as an open one, taking into account its contact and interaction with the surrounding environment or the *reservoir* in thermodynamically motivated terminology. Formally, if the system S has the associated Hilbert space $\mathcal{H}$ and the reservoir, being a quantum system itself, has the space $\mathcal{H}_R$, then the space of the entire system is $\mathcal{H}_{\text{total}} = \mathcal{H} \otimes \mathcal{H}_R$. States of such an open system are density matrices in $\mathcal{B}(\mathcal{H} \otimes \mathcal{H}_R) \simeq \mathcal{B}(\mathcal{H}) \otimes \mathcal{B}(\mathcal{H}_R)$. Consequently, the action of an operation Λ on S not affecting the reservoir in any way is described by the operation $\Lambda \otimes \mathbb{1}_R$, where $\mathbb{1}_R$ denotes the identity mapping on $\mathcal{B}(\mathcal{H}_R)$. If the dimension of the space $\mathcal{H}_R$ is r, then the map $\Lambda \otimes \mathbb{1}_R$ is said to be an r-dimensional extension (or amplification) of Λ.[d]

Now, positivity of Λ does not guarantee, in general, positivity of its extensions. If however an r-dimensional extension happens to be positive then Λ is called r-positive. Moreover,[11,12,14]

Definition 2.4. A map $\Lambda : \mathcal{B}(\mathcal{H}) \to \mathcal{B}(\mathcal{H})$ is called completely positive iff it is r-positive for all r.

Going back to physical context, valid quantum operations are implemented by completely positive maps and their action on density matrices,

[d] $\mathbb{1}_R \otimes \Lambda$ is similarly an amplification of Λ if we change the order of spaces in $\mathcal{H}_{\text{total}}$ which is a matter of convention.

so that their extensions remain positive regardless of the dimensionality of the reservoir space. One can moreover show the following sequence of inclusions

$$\mathcal{P} = \mathcal{P}_1 \supset \mathcal{P}_2 \supset \cdots \supset \mathcal{P}_n = \mathcal{P}_c\,,$$

where $\mathcal{P}$ denotes the class of positive maps, $\mathcal{P}_r$ — of r-positive and $\mathcal{P}_c$ — of completely positive ones. n is the dimension of the initial system space $\mathcal{H}$. So if Λ is n-positive, it is automatically r-positive for any $r > n$ and, consequently, completely positive.

Let us also mention that often r-positivity for certain $r < n$ already suffices to guarantee complete positivity of Λ. This is related to the structure of Λ characterized by its so-called minimal length.[13]

As we will see shortly, there is a connection between the above system of inclusions and the filtration of the state space Σ induced by the Schmidt length, as described earlier.

We conclude this section by referring again to the very important example of transposition map. We have seen it is positive. However, it is not 2-positive, and hence certainly not completely positive, as demonstrated by the following computation. Suppose for simplicity that $\mathcal{H}$ and $\mathcal{H}_R$ are 2-dimensional. Therefore $\mathcal{H} \otimes \mathcal{H}_R \simeq \mathbb{C}^4$. A mixed state of the system + reservoir is therefore a 4×4 density matrix ϱ. The 2-amplification of the 2×2 transposition, $T \otimes \mathbb{1}_2$, exchanges the off-diagonal 2×2 blocks of ϱ, leaving them internally intact,

$$\varrho = \begin{bmatrix} A & B \\ C & D \end{bmatrix} \xrightarrow{T\otimes \mathbb{1}_2} \begin{bmatrix} A & C \\ B & D \end{bmatrix},$$

with A, B, C, D being 2×2 matrix blocks. For example

$$\varrho = \left[\begin{array}{cc|cc} \frac{1}{2} & 0 & 0 & \frac{1}{2} \\ 0 & 0 & 0 & 0 \\ \hline 0 & 0 & 0 & 0 \\ \frac{1}{2} & 0 & 0 & \frac{1}{2} \end{array}\right] \xrightarrow{T\otimes \mathbb{1}_2} \left[\begin{array}{cc|cc} \frac{1}{2} & 0 & 0 & 0 \\ 0 & 0 & \frac{1}{2} & 0 \\ \hline 0 & \frac{1}{2} & 0 & 0 \\ 0 & 0 & 0 & \frac{1}{2} \end{array}\right].$$

The spectrum of ϱ is here $\lambda_1 = 1, \lambda_2 = \lambda_3 = \lambda_4 = 0$, while after the amplified transposition application it becomes $\lambda_1^{'} = -\frac{1}{2}, \lambda_2 = \lambda_3 = \lambda_4 = \frac{1}{2}$. Hence, the transposition is not 2-positive.

The importance of transposition stems from the fact that in low-dimensional cases, $\Lambda : \mathcal{B}(\mathbb{C}^p) \to \mathcal{B}(\mathbb{C}^q)$ with $p = 2$ and $q = 2$ or 3, it is *the only* nontrivial positive but not completely positive map. It turns out that *every* positive Λ in these cases can be represented in the form

$$\Lambda = \Pi_1 + \Pi_2 \circ T\,,$$

where Π_j are completely positive.[15,16] In higher dimensional spaces, maps Λ admitting representation as above are called *decomposable* and they form a relatively simple subclass among positive ones. Very few examples of indecomposable maps are known so far and the construction of such maps proves to be difficult, see e.g. [17] for a recent example.

3. Separability criteria and positive maps

One of the most well known tests for separability of mixed states is the following so-called *positive partial transpose*, in short PPT, or Peres-Horodecki criterion[18,19]. In what follows we consider bipartite systems with Hilbert space $\mathcal{H}_{AB} = \mathcal{H}_A \otimes \mathcal{H}_B$. $\Sigma = \Sigma_{AB}$ is again the set of mixed states of the compound system. Γ_A, Γ_B stand for *partial transpositions* in systems A and B, respectively, that is $T_A \otimes \mathbb{1}_B$ and $1_A \otimes T_B$.[e]

Theorem 3.1. *If the state $\varrho \in \Sigma_{AB}$ is separable, then its partial transpose (Γ_A or equivalently Γ_B) is positive semidefinite (i.e. has no negative eigenvalues).*

It is important to observe that the PPT condition is only necessary for separability and in general not sufficient. If, however, $\dim \mathcal{H}_{AB} \leq 6$ then it becomes also a sufficient one. The latter is a consequence of decomposability of all low-dimensional positive maps mentioned just a few paragraphs above.[19]

The image resulting from the PPT condition is the following, cf. Fig. 3,

$$\Sigma_{\text{Sep}} \subseteq \Sigma_{\text{PPT}} \subset \Sigma\,,$$

(Σ_{PPT} denotes the set of states with positive partial transpose) with equality replacing the first inclusion in low-dimensional systems. There are, however, known examples of nonseparable PPT states,[20] meaning that this inclusion is proper, in general. The PPT entangled states turn out to have qualitatively different properties from non-PPT entangled ones. These differences become crucial when entangled states are used for quantum communication tasks.

The reader may become aware now that the PPT criterion relies, in fact, on the non-complete positivity of the transposition. Partial transposition is nothing but an amplification of T treated as a map on $\mathcal{B}(\mathcal{H}_A)$. If T were

---

[e] Note that symbol Γ resembles "half of" the letter T, hence its use to denote partial transposition.

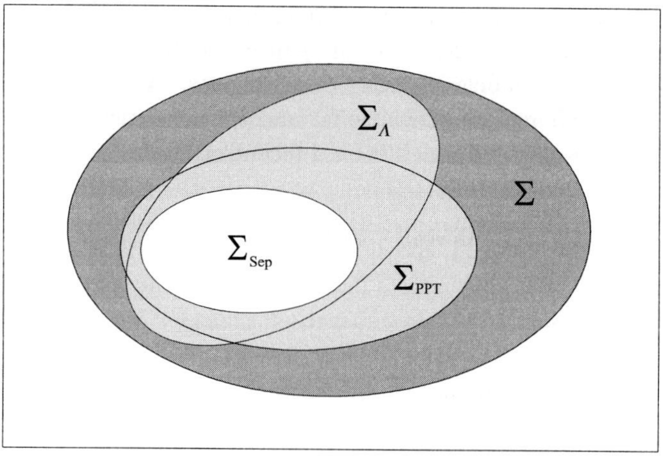

Figure 3. The set Σ of mixed states of a bipartite system and some of its subsets, cf. text.

completely positive, then all its amplifications would likewise be positive, and would never produce any negative eigenvalues when applied to states in Σ, be it separable or not. So the Peres-Horodecki criterion would remain true but completely useless as it would not see any difference between separable and entangled states. It can be guessed that PPT test is actually only an element of a broader image.

Namely, let the map $\Lambda : \mathcal{B}(\mathcal{H}_A) \to \mathcal{B}(\mathcal{H}_A)$ be positive but not completely positive. One can then use its amplification $\Lambda \otimes \mathbb{1}_B$ and attempt to design a criterion analogous to PPT: if $\varrho \in \Sigma_{AB}$ is separable then $(\Lambda \otimes \mathbb{1}_B)(\varrho)$ has no negative part in its spectrum. Such an approach is in fact successful. Collectively, all such "Λ criteria" provide a necessary and sufficient condition for separability:[19]

Theorem 3.2. *$\varrho \in \Sigma_{AB}$ is separable iff for all positive but not completely positive maps Λ on $\mathcal{B}(\mathcal{H}_A)$ $(\Lambda \otimes \mathbb{1}_B)(\varrho)$ (or equivalently $(\mathbb{1}_A \otimes \Lambda)(\varrho)$ if Λ is taken to act on $\mathcal{B}(\mathcal{H}_B)$) remains positive semidefinite.*

So for every suitable map Λ we have in analogy with the PPT-related image

$$\Sigma_{\text{Sep}} \subset \Sigma_\Lambda \subset \Sigma\,,$$

where by Σ_Λ we mean the set of states on which the amplification $\Lambda \otimes \mathbb{1}_B$ produces positive semidefinite outcomes. Different sets Σ_Λ overlap (one of them is of course $\Sigma_{\rm PPT}$) and their intersection equals $\Sigma_{\rm Sep}$, Fig. 3.

We should add that if Λ is decomposable, then $\Sigma_\Lambda \subseteq \Sigma_{\rm PPT}$, so that it provides no new information about state separability above that produced by the PPT criterion. This explains how the sufficiency of the PPT test relates with decomposability of all positive maps in low dimensions. Also note, that the immanent difficulty in designing indecomposable maps makes the problem of detection of PPT entangled states quite involved. There are few examples of such states[19], also in the generalized multipartite setting[21,22].

The above theorem, despite its beauty, is unfortunately yet another impractical tool when it comes to testing states for entanglement. Its direct application requires the systematic knowledge of all positive maps, or optimally — the indecomposable ones. We have already said that testing whether a particular map preserves the positivity of spectra is a very difficult task of its own. What then can be expected of the problem of systematic construction of positive maps?

We can still refine our image of the structure of Σ a bit further. Suppose that we have a map Λ which is r-positive but not $(r+1)$-positive, $\Lambda \in \mathcal{P}_r - \mathcal{P}_{r+1}$. This gives rise to the following criterion relating the filtration of Σ according to the Schmidt length with r-positivity: if $\varrho \in \Sigma_r$, that is $\ell_S(\varrho) \leq r$, then $\Lambda \otimes \mathbb{1}_B$ leaves ϱ positive semidefinite. Similarly as before, Σ_r coincides with the intersection of all sets Σ_Λ for different maps $\Lambda \in \mathcal{P}_r - \mathcal{P}_{r+1}$.

To sum up the current section, the problem of complete characterization of separability turns out to be intimately related with classification of positive maps, r-positivity and complete positivity. Each positive but not completely positive map gives rise to a different necessary condition of state separability, but collectively, all these criteria completely characterize $\Sigma_{\rm Sep}$.

4. Entanglement witnesses

While the separability criteria discussed in the previous section have a sound theoretical foundation, their practical value is, however, not so immediately obvious. One reason is that in general we do not know much about indecomposable positive maps — in particular, there exist very few examples of them in the literature. So our arsenal of available separability tests is by and large limited to the PPT one.

Secondly, it is not obvious how to design experimental set-ups to perform such tests on real quantum systems in laboratory. The maps we want to apply to states are, by definition, *not* completely positive and hence nonphysical: they do not correspond to admissible quantum operations, partial transpose being a primary example. However, there are ways to go around this difficulty.

The approach we are going to discuss now is an instance of the general problem of state estimation whose objective is to gain as much information as possible about an unknown mixed state of a system. This is achieved by performing measurements of various physical quantities while the system is in the state ϱ. Each measurement returns the expectation value $\mathrm{Tr}(V\varrho)$ of the corresponding observable $V \in \mathcal{B}(\mathcal{H})$. Collecting such outcomes for different noncommuting observables (or for just one observable but measured at different instants of time, assuming that the form of dynamical evolution the system is subjected to is fully known[23]) one is then able to localize the state ϱ in Σ with higher and higher precision. As ϱ is an $n \times n$ matrix, the complete information allowing one to reconstruct it consists in general of $O(n^2)$ numbers.[f]

Our goal is obviously more modest here: we only want to be able to decide, on the basis of measurement outcome, whether ϱ lies in Σ_{Sep} or not.[19,24]

Definition 4.1. An observable $W \in \mathcal{B}(\mathcal{H}_{AB})$ is an *entanglement witness* iff

(i) W has at least one negative eigenvalue;
(ii) $\mathrm{Tr}(W\varrho) \geq 0$ for all separable ϱ.

Any (entangled) state ϱ such that $\mathrm{Tr}(W\varrho) < 0$ is said to be *detected by* W.

The sense of the above definition is illustrated by Fig. 4.

An observable W gives rise a linear functional on $\mathcal{B}(\mathcal{H})$ by $\omega(\cdot) = \mathrm{Tr}(W\,\cdot)$. Then, by Hahn-Banach theorem, the set $\{V \in \mathcal{B}(\mathcal{H}) : \mathrm{Re}\,\omega(V) = 0\}$ is a hyperplane splitting $\mathcal{B}(\mathcal{H})$ into two halfspaces corresponding to $\mathrm{Re}\,\omega(\cdot) > 0$ and $\mathrm{Re}\,\omega(\cdot) < 0$. For W satisfying conditions (i) and (ii) of Def. 4.1 this hyperplane passes through the set of sates Σ without intersecting Σ_{Sep} — the white oval (note that ω assumes *real* values on Σ). This

[f]The procedure just described is fully analogous to the classical estimation problem of statistics: one attempts to reconstruct an unknown probability distribution on the basis of knowledge of expectation values of some random variables. The difference is that currently we work with noncommutative objects.

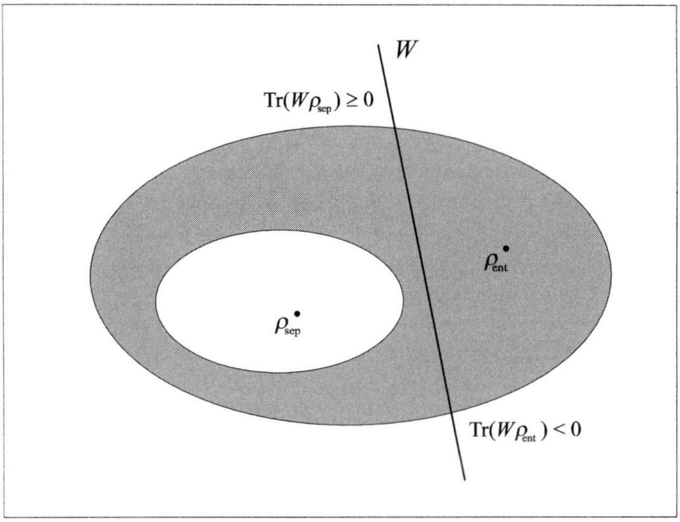

Figure 4. Graphical representation of an entanglement witness W (cf. text).

situation is presented in Fig. 4, where with a slight abuse of notation we have used the letter W to indicate the splitting hyperplane. Thus, when such a W is measured in a given state ϱ, the negative outcome indicates that ϱ is entangled, while the positive one does not (though ϱ may actually be entangled!).

As different choices of W correspond to different sections of Σ, there is a valid question of witness optimality. Intuitively, this corresponds to asking whether the line representing W can be moved any closer to Σ_{Sep} without intersecting it. Optimal witnesses correspond roughly to the lines tangent to Σ_{Sep}. The image of Fig. 4 using ellipsoidal smooth shapes is, however, somewhat deceiving. Σ and Σ_{Sep} are certainly convex but they are rather complex, multidimensional polyhedra, with faces, edges, corners etc., so tangency as such still may leave some room for finer witness optimization.

Formally, optimality is defined as follows. Let $D(W) = \{\varrho : \operatorname{Tr}(W\varrho) < 0\}$. We say that W is *finer* than W' if $D(W') \subseteq D(W)$. W is then called *optimal* if no W' can be finer.

There are known necessary[25] and sufficient[8] conditions characterizing optimal entanglement witnesses, the respective theorem being

Theorem 4.1. *W is optimal iff the set*

$$\mathcal{P}_W = \{|\omega\rangle = |\phi\rangle \otimes |\psi\rangle : \langle\omega|W|\omega\rangle = 0\}$$

spans the whole space $\mathcal{H}_{AB}$.

Similarly to characterization of separability by positive maps, Σ_{Sep} is collectively described by all entanglement witnesses.[19,24]

Theorem 4.2. $\varrho \in \Sigma_{\text{Sep}}$ *iff* $\text{Tr}(W\varrho) \geq 0$ *for all entanglement witnesses* $W \in \mathcal{B}(\mathcal{H})$.

As we shall see in the next section, there is a much deeper connection between witnesses and positive maps.

An example of entanglement witness on $\mathcal{H}_{AB} = \mathcal{H} \otimes \mathcal{H}$ is provided by the so called swap operator,[10]

$$S = \sum_{i,j} |i\rangle\langle j| \otimes |j\rangle\langle i| ,$$

where $|1\rangle, \ldots, |n\rangle$ is a basis in $\mathcal{H}$. Taking arbitrary $|\Phi\rangle = |\phi\rangle \otimes |\psi\rangle \in \Sigma_{\text{Sep}}$ we get

$$\text{Tr}(S|\Phi\rangle\langle\Phi|) = \langle\phi| \otimes \langle\psi|S|\phi\rangle \otimes |\psi\rangle = |\langle\phi|\psi\rangle|^2 \geq 0 ,$$

so that the witness condition (ii) is satisfied. In addition, it can be immediately seen that e.g. $|\Psi\rangle = \frac{1}{\sqrt{2}}(|1\rangle \otimes |2\rangle - |2\rangle \otimes |1\rangle)$ satisfies $S|\Psi\rangle = -|\Psi\rangle$, showing that S has eigenvalue -1. So S is indeed a witness. In particular, $\varrho = |\Psi\rangle\langle\Psi|$ is detected by S: $\text{Tr}(S|\Psi\rangle\langle\Psi|) = -1$.

A class of entanglement witnesses can be constructed by the following general scheme:[33,34] given an observable $Q \in \mathcal{B}(\mathcal{H}_{AB})$ we compute

$$W_Q \stackrel{\text{df}}{=} Q - \inf_{|\psi\rangle \in \text{Sep}} \langle\psi|Q|\psi\rangle\, \mathbb{I} ,$$

$\mathbb{I}$ being the identity matrix. Obviously, $\text{Tr}(W_Q\,\varrho) \geq 0$ for separable ϱ. Then W_Q is a witness if it has a negative eigenvalue. This occurs if the eigenvector corresponding to the smallest eigenvalue of Q is not separable. One should be aware that the infimum in the definition of W_Q may be very hard to compute. The most suitable observables Q for this construction are the directly measurable ones, like e.g. the Hamiltonian of a system we work with. As we have just indicated, it is then essential that the ground state be entangled. An explicit construction of this type is presented in [1].

Finally, let us mention that there is a notion of Schmidt rank k witness detecting states with Schmidt length not exceeding k. In this generalization,

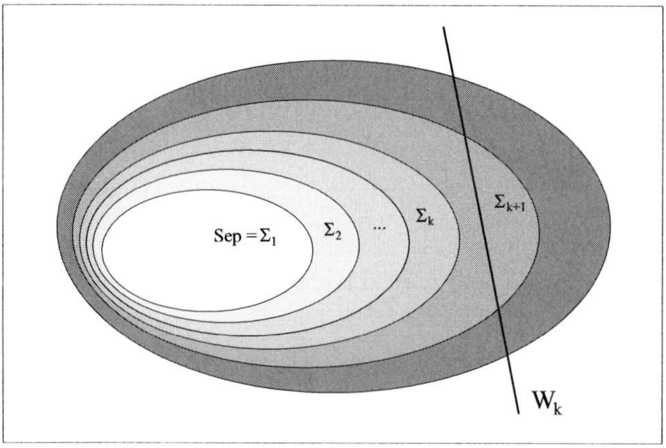

Figure 5. Graphical representation of an entanglement witness W_k of Schmidt rank k.

ordinary entanglement witnesses become the special case corresponding to $k = 1$. The difference with Definition 4.1 is that now condition (ii) reads

(ii) $\mathrm{Tr}(W\varrho) \geq 0$ for all $\varrho \in \Sigma_k$

or, equivalently, $\langle\Psi|W|\Psi\rangle \geq 0$ for all vectors Ψ with $\ell_S \leq k$. The intuitive image is presented in Fig. 5.

5. Relating positive maps and entanglement witnesses

Throughout this section we suppose for simplicity of exposition that $\mathcal{H}_{AB} = \mathcal{H} \otimes \mathcal{H}$, with $\dim \mathcal{H} = n$. An entanglement witness W as a linear operator on $\mathcal{H}_{AB}$, namely an $n^2 \times n^2$ matrix, is an element of n^4-dimensional linear space $\mathcal{B}(\mathcal{H}_{AB})$. On the other hand, any linear map $\Lambda : \mathcal{B}(\mathcal{H}) \to \mathcal{B}(\mathcal{H})$ is naturally an element of likewise n^4-dimensional, but different linear space $\mathcal{B}(\mathcal{B}(\mathcal{H}))$. Having the same dimensionality, the two spaces are naturally isomorphic, $\mathcal{B}(\mathcal{H}_{AB}) \simeq \mathcal{B}(\mathcal{B}(\mathcal{H}))$.

Of course, there are very many ways to set up a concrete example of an isomorphism between these spaces, but it is not so obvious how to design one which will take completely positive maps to positive definite matrices or will have some other desirable properties with respect to positivity. A realization of such mapping is the so-called *Jamiołkowski isomorphism*,[35] a very useful tool in the theory of entanglement detection. It is defined as the map $\mathcal{J} : \mathcal{B}(\mathcal{B}(\mathcal{H})) \to \mathcal{B}(\mathcal{H}_{AB})$ in the following way. Let

$$|\Psi^+\rangle = \frac{1}{\sqrt{n}} \sum_{i=1}^{n} |i\rangle \otimes |i\rangle \qquad \text{and} \qquad P^+ = |\Psi^+\rangle\langle\Psi^+| .$$

We then have

$$W_\Lambda = \mathcal{J}(\Lambda) \stackrel{\text{df}}{=} (\mathbb{I} \otimes \Lambda)P^+ .$$

It can be demonstrated that indeed W_Λ is positive definite if Λ is completely positive. Moreover, the image of Λ positive but not completely positive turns out to have properties of an entanglement witness: it is positive definite when restricted to separable vectors, but not such on the entire space $\mathcal{H}_{AB}$.

So making use of isomorphism $\mathcal{J}$ we establish a correspondence between positive but not completely positive maps which, as we know, give rise to various separability criteria, and entanglement witnesses. It is not difficult to observe, for example, that the swap operator S described in the previous section is the $\mathcal{J}$ image of the transposition map and thus, as a witness, it is related to the PPT test.

We have said in Section 3 that positive decomposable maps give rise to criteria which are of no greater sensitivity than the PPT test. As it can be expected, the same is true for the so-called *decomposable* witnesses which are $\mathcal{J}$ images of decomposable maps. In other words, only witnesses corresponding to indecomposable maps are capable of detecting some PPT entangled states,[17,36] cf. Fig. 6

Let us look closer at the structure of the Jamiołkowski isomorphism. It turns out that it is easy do give a simple operational description of$\mathcal{J}$ using block matrices, in the similar spirit as we have done it in Section 2 when discussing tensor products. This operational definition will, moreover, tell us an easy way to find the inverse of $\mathcal{J}$ by means of which one can find maps corresponding to particular entanglement witnesses.

First of all, note that a linear map $\Lambda : \mathcal{B}(\mathcal{H}) \to \mathcal{B}(\mathcal{H})$ is fully specified by listing the values it assumes on basis vectors $E_{ij} = |i\rangle\langle j|$ in $\mathcal{B}(\mathcal{H})$, say $\Lambda E_{ij} = \Lambda_{ij} \in \mathcal{B}(\mathcal{H})$, $i, j = 1, \ldots, n$. Now, we form a matrix consisting of $n \times n$ blocks of size $n \times n$ each, recording Λ_{ij} in the block (i, j),

$$\left[\begin{array}{c|c|c} \Lambda_{11} & \cdots & \Lambda_{1n} \\ \hline \vdots & \ddots & \vdots \\ \hline \Lambda_{n1} & \cdots & \Lambda \end{array}\right] .$$

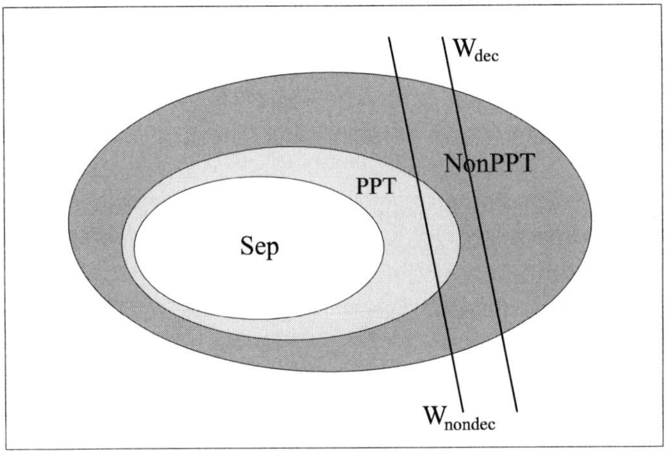

Figure 6. Decomposable and indecomposable witnesses.

A moment of thought convinces us that it is precisely the matrix $(\mathbb{I} \otimes \Lambda)P^+$, that is W_Λ. Conversely, writing the matrix of an entanglement witness W and dividing it into $n \times n$ blocks we can read-off the corresponding map Λ from it: the block (i,j) contains the matrix $\Lambda_{ij} \doteq \Lambda E_{ij}$.

Suppose finally that Λ is k-positive but not $(k+1)$-positive, $\Lambda \in \mathcal{P}_k - \mathcal{P}_{(k+1)}$. Then its image $\mathcal{J}(\Lambda) = W$ turns out to be a witness of Schmidt rank k. It shows that the Jamiołkowski isomorphism respects also the classification of positive maps and witnesses related to the filtration of the state space Σ with respect to the Schmidt number ℓ_S.

Acknowledgment

The author is greatly indebted to Professor Masanori Ohya and the organizers of the QBIC'09 International Conference for their kind invitation and hospitality during his stay in Suwa and Noda, Japan.

References

1. Michalski M., *Detecting Entanglement in Spin Lattice Models*, this volume.
2. Horodecki R., P. Horodecki, M. Horodecki, K. Horodecki, Rev. Mod. Phys. **81**, 865 (2009).
3. Einstein A., B. Podolsky, and N. Rosen, Phys. Rev. **47**, 777 (1935).
4. Schrödinger, E., Naturwissenschaften **23**, 807, 823, 844 (1935); English translation in: J.D. Trimmer, Proc. American Phil. Soc. **124**, 323 (1980).

5. Schrödinger, E., Proc. Cambridge Phil. Soc. **31**, 555 (1935); **32**, 446 (1936).
6. Bohr, N., Phys. Rev. **48**, 696 (1935).
7. Von Neumann J., *Mathematische Grundlagen der Quantenmechanik*, Springer, Berlin, 1932.
8. Sarbicki G., arXiv:0905.0778v1 [quant-ph] (2009).
9. Bengtsson I., K. Życzkowski, *Geometry of Quantum States*, Cambridge Univ. Press, Cambridge, 2006.
10. Werner R. F., Phys. Rev. A **40**, 4277 (1989).
11. Stinespring W.F., Proc. Am. Math. Soc. **6**, 211 (1955).
12. Kraus K., Ann. Phys. **64**, 311 (1971).
13. Jamiołkowski A., Open Sys. Information Dyn. **11**, 385 (2004).
14. Gorini V., A. Kossakowski, E.C.G. Sudarshan, J. Math. Phys **17**, 821 (1976).
15. Størmer E., Acta Mth. **110**, 233 (1963).
16. Woronowicz S. L., Rep. Math. Phys. **10**, 165 (1976).
17. Chruściński D., A. Kossakowski, Open Sys. Information Dyn. **14**, 275 (2007).
18. Peres A., Phys. Rev. Lett. **77**, 1413 (1996).
19. Horodecki M., P. Horodecki, R. Horodecki, Phys. Lett. A **223**, 1 (1996).
20. Horodecki M., P. Horodecki, R. Horodecki, Phys. Rev. Lett. **78**, 574 (1997).
21. Chruściński D., A. Kossakowski, Open Sys. Information Dyn. **15**, 189 (2008).
22. Jurkowski J., D. Chruściński, A. Rutkowski, Open Sys. Information Dyn. **16**, 235 (2009).
23. Jamiołkowski A., Open Sys. Information Dyn. **11**, 63 (2004).
24. Terhal B., Phys. Lett. A **271**, 319 (2000).
25. Lewenstein M., B. Kraus, J. I. Cirac, P. Horodecki, Phys. Rev. A **62**, 052310 (2000).
26. Wootters W. K., Phys. Rev. Lett. **80**, 2245 (1998).
27. Bennett C. H., D. P. DiVincenzo, J. Smolin, W. K. Wootters, Phys. Rev. A **54**, 3824 (1996).
28. Życzkowski K., P. Horodecki, A. Sanpera, M. Lewenstein, Phys. Rev. A **58**, 883 (1998)
29. Vidal G., R. F. Werner, Phys. Rev. A **65**, 032314 (2002).
30. Matsuoka T., [in:] *Quantum Bio-Informatics II*, L. Accadri, W. Freudenberg. M. Ohya, eds., Quantum Probability and White Noise Analysis **XXIV**, 139 (2009).
31. Majewski W. A., T. Matsuoka, M. Ohya, arXiv:0904.1351v1 [quant-ph] (2009).
32. Belavkin, V. P., M. Ohya, Proc. Royal Soc. Lond. **A 458**, 209 (2002).
33. Brukner C., V. Vedral, arXiv:quant-ph/0406040v1 (2004).
34. Tóth G., Phys. Rev. A **71**, 010301(R) (2005).
35. Jamiołkowski A., Rep. Math. Phys. **3**, 275 (1972).
36. Chruściński D., A. Kossakowski, J. Phys. A: Math. Theor. **41** 145301, (2008).

Quantum Bio-Informatics III
eds. L. Accardi, W. Freudenberg & M. Ohya

TUNNELING STUDY ON HIGH-T_C SUPERCONDUCTORS

M. MINEMATSU, S. KAWASHIMA AND N. MIYAKAWA*

Department of Applied Physics, Tokyo University of Science, 1-3 Kagurazaka, Shinjuku-ku, Tokyo 162-8601, Japan

We report some of our recent tunneling results on various high-T_C superconductors, including cuprate superconductors and a newly discovered iron-based pnictide superconductor. After the introduction of Section 1, in Section 2 we explain the experimental details, and in Section 3, we show the characteristic features of the superconducting gap on electron-doped cuprate superconductor, subsequently we report the two kinds of superconducting gap on an iron-based pnictide superconductor $NdFeAsO_{1-y}$ with T_C~51K.

1. Introduction

Since the discovery of Cu-based high-T_C superconductors (HTSs) in 1986[1] and of Fe-based superconductors in 2008[2], a great deal of effort has been made to understand the mechanism of pairing because HTSs are very attractive materials due to their scientific significance as well as potential applications. However the pairing mechanism for both families is still an open question. Because the Cu-based HTSs have complexities associated with unusual features that have been found such as a *d*-wave order parameter[3,4], pseudogap[5] and stripe structure[6,7] *etc*.. These unusual features were not observed for conventional superconductors where mean-field BCS theory[8] was successful. Based on the mean-field theory, the superconducting gap magnitude should be proportional to the superconducting critical temperature T_C. However, our previous tunneling results of $Bi_2Sr_2CaCu_2O_{8+\delta}$ (Bi2212) as well as $Bi_2Sr_{2-x}La_xCuO_{6+\delta}$ showed that the magnitude of superconducting gaps Δs increases monotonically with decreasing carrier concentration although T_C decreases in the underdoped regime[9-11]. These unusual features are observed on hole-doped cuprates, but it is unclear yet for electron-doped cuprates. Furthermore, a pseudo-gap structure at $T>T_C$ was observed in the underdoped regime of many hole-doped cuprates, but this pseudo-gap issue is also controversial for electron-doped cuprates.

* Corresponding author. (E-mail address:miyakawa@rs.kagu.tus.ac.jp)

For conventional superconductors, it is well known that tunneling studies have played a crucial role in determining the magnitude of superconducting gap and the glue of pairing which is the "phonon"[12]. Since tunneling conductance provides a direct measure of the quasiparticle density of states (DOS) near the Fermi level, one can determine the precise magnitude of superconducting gap. In addition, it can also display spectral features beyond the superconducting gap peak, which appear at a voltage corresponding to the energy of the glue for pairing. The spectral features appear on tunneling conductance as a result of modulation of the gap function, because a bosonic mode contributing to pairing (e.g. phonon for conventional superconductors) must modify the energy-dependent gap function as well as renormalization function at the eigen energy[12]. The modified gap function leads to spectral features on the quasiparticle DOS, *i.e.* tunneling conductance. Therefore if one can identify the origin of the spectral feature observed on tunneling spectrum, one can possibly understand the glue for pairing. However HTS cuprates (at least hole-doped cuprates) are most probably *d*-wave superconductor, and at the present time there is no theoretical framework to quantitatively determine the spectral function $\alpha^2F(\omega)$ which would reproduce the measured Δ and T_C. However, the analysis of tunneling dip structures for Bi2212 as a function of doping suggested that the magnetic resonance mode would contribute to form the Cooper pair[13].

In this article, we show our recent tunneling results on an electron-doped HTS cuprate in Section 3.1 and we will discuss the symmetry of energy gap, and the tunneling dip structure.

Up to now, the origin of superconductivity for high-T_C cuprates has been an unresolved question, but the discovery of a series of new Fe-based superconductors (SCs)[2,14-16] has led to intense research activity, and the contrast between high-T_C cuprates and the Fe-based pnictides has been extensively done to get an insight for the still unresolved problems of high-T_C cuprate superconductivity[17,18]. Since some of those pnictides showed the high superconducting critical temperature with T_C >50K[14,19,20], it is intriguing to explore the mechanism of the high-T_C superconductivity for the iron-based pnictides.

Concerning the electronic structure in Fe-based pnictides, we now know that all five Fe3d orbitals contribute to the electronic structure near the Fermi level, where they form an electronic band with various electron- and hole-Fermi sheets (pockets). That is, the Fe-based pnictides belong to a multiband system although the cuprates are a single band system. In addition, it is known that the superconductivity of these Fe-based pnictides occurs when an antiferromagnetically ordered phase has been suppressed by charge doping[21], therefore the emergence of superconductivity in Fe-based pnictide is similar with that in cuprate, although the parent compounds in Fe-based pnictides would be semi-metal (or insulator) and those in cuprates are Mott-insulator.

However the nature of superconductivity for Fe-based SCs has been still unclear and controversial. A most essential physical quantity to investigate the nature of a SC is the superconducting gap Δ, since the magnitude and the symmetry of Δ are intimately related to the pairing mechanism. However in this early stage, the reported values of Δ show a large spread from 4meV to over 20meV by various measurements[22-27]. In some cases, two-gaps[23] were asserted; on the other hand the single gap[22] feature also was reported.

In addition, concerning the symmetry of the order parameter, it is also controversial. Some models are proposed to explain the experimental results on pnictide SCs. One of them is the d-wave model, and another is the extended s-wave ($s_\pm$) model[28,29] in which a sign-changing s-wave gap opens on the hole- and electron-pocket. In the $s_\pm$ model, they reported that the multiple spin-fluctuation modes arising from the nesting across the disconnected Fermi surface realize such unusual symmetry.

In this paper, we report the tunneling conductance of a polycrystalline $NdFeAsO_{1-y}$ (Nd1111) with T_C~51K by a mechanical point contact technique using a Au tip or a GaAs tip in Section 3.2. In our junctions, we observed two-gap features, where the clear conductance peaks were observed at ~6.0$\pm$1.0mV and/or ~14.0$\pm$1.0mV.

2. Experimental details

In this paper, we will show our tunneling results of an electron-doped high-T_C cuprate, $Pr_{1-x}LaCe_xCuO_4$ (PLCCO) and an iron-based pnictide superconductor, $NdFeAsO_{1-y}$ (Nd1111) with T_C ~51K.

PLCCO single crystals were grown by traveling-solvent floating zone (TSFZ) methods, where crystal growth was performed in a double focusing mirror furnace. After the growth, the single crystals were carefully annealed under argon gas flow at 900-950C for ~10hrs and they were subsequently annealed under O_2 gas-flow at 500C for 10hrs[30].

Our fluorine-free polycrystalline Nd1111 was prepared by a high-pressure technique using a cubic-anvil-type apparatus.[19] The starting materials of NdAs, Fe, Fe_2O_3 were mixed at nominal composition of Nd:Fe:As:O=1:1:1:0.7 and ground by an agate mortar in a glove box filled with dry nitrogen gas. The samples were synthesized by heating the mixtures in BN crucibles under a pressure of about 2GPa at 1200C for 2h.

T_C was determined by dc resistivity measurement and magnetic susceptibility.

Tunneling measurement is a powerful tool to investigate the quasiparticle DOS in the pairing states, therefore we can directly determine the magnitude of the superconducting gap Δ, with high energy resolution (≤meV). In addition we can investigate the symmetry of order parameter from the shape of sub-gap region. SIN(S= superconductor, I= insulator, N= normal metal) junctions were prepared by a point contact technique using a Au-tip. The tip was mechanically and chemically cleaned prior to each run. After the sample and a Au tip are placed in the measurement system and cooled down to 4.2 K, the distance and contact force between the tip and sample are adjusted using a differential micrometer. In the ordinary way, a tip pushes onto the surface of the crystal, and a stable mechanical junction is formed between the tip and crystal and SIN junction is formed. Here the insulating tunnel barrier is the native surface layer of the crystal.

Dynamical tunneling conductance, dI/dV, was measured by standard ac lock-in technique, where for SIN spectra, the bias voltage is that of the sample with respect to the tip so that negative bias corresponds to the occupied electron states below E_F.

3. Results

3.1. $Pr_{1-x}LaCe_xCuO_4$

Hole-doped high-T_C cuprates has the order parameter with d_{x2-y2} symmetry, but that for electron-doped cuprates is still controversial. In addition, the magnitude of superconducting gap for hole-doped cuprates is large therefore high-T_C is realized. It is believed that the large Δ is due to the large energy scale for lattice (oxygen phonon energy, Ω~0.08eV), spin (superexchange coupling, J~0.15eV), and charge (hopping integral, t_{pd}~0.4eV). Therefore it is important to investigate the precise value of Δ for electron-doped cuprates.

In our tunneling study of PLCCO, the out of plane point contact junctions (PCJs) were prepared by approaching a Au tip toward the (001) surface of PLCCO single crystals, where I-V and dI/dV-V curves were measured. For each of PLCCO single crystals, we repeated measurements many times at different locations on the (001) surface of some PLCCO single crystals, and obtained the statistical distribution of superconducting gap, Δ.

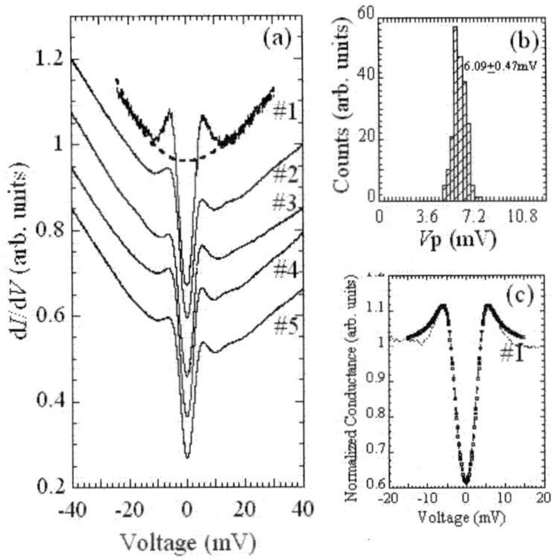

Figure 1. (a) Typical tunneling conductances on PLCCO with x=0.11 with T_C~26K measured at 4.2K. The dashed line for junction #1 shows an inferred normal state conductance $(dI/dV)n$ by the fifth-order polynominal fit of the observed dI/dV. (b) Statistical distribution of V_P values for many junctions on PLCCO. (c) Normalized tunneling DOS which is $(dI/dV)s/(dI/dV)n$ together with the fitted curve by nonmonotonic *d*-wave DOS (open circles).

Figure 1(a) shows typical tunneling conductance on PLCCO that were measured at 4.2K. In this figure, the each tunneling curves are normalized by the conductance at high-bias, and all curves except the lowest one are shifted upward for clarify. As seen in Figure 1(a) we reproducibly observed the clear superconducting gap peaks at V_p~±6.0meV. Furthermore, we fitted tunneling conductance by assuming nonmonotonic *d*-wave (NMD), and it is shown in Figure 1(c). As clearly seen Figure 1(c), the best fit is realized by NMD, but we inferred the normal conductance as shown by dot-dashed line for junction #1 in Figure 1(a). This result is consistent with ARPES result for PLCCO[31]. By assuming the superconducting gap magnitude as the peak position V_p, we plotted the statistical distribution of V_p in Figure 1(b), where the averaged V_p is ~6.09±0.47 meV. By using this value as $\Delta(0)$, we obtain the gap-T_C ratio, $2\Delta/k_BT_C$~5.43, suggesting that PLCCO with x=0.11 is in strong coupling regime. Concerning to doping dependence of Δ, we found that V_p value follows T_C but the fitted gap value using NMD increases monotonically with decreasing carrier concentration. This behavior is same as that of Bi2212[9,10]. This is due to the increase of nonmonotonic parameter η as the carrier concentration decreases. These results will be published elsewhere. In addition, we notice that all curves show V-shaped background (VB). The origin of VB is unclear, but it seems to be one of characteristic common features for all electron-doped cuprates[32-34], although the shape of background conductance for hole-doped cuprates is quite different from the VB[9-11,35-37]. Furthermore we noticed a fine structure exists at ~16mV for some junctions. Then we focus on the fine structures on the tunneling conductance that are observed at voltage beyond V_P. For conventional superconductors, phonon structures for contributing to the pairing interaction are observed on tunneling conductance at $\Delta+\Omega$, where Ω is phonon energy. The structure can be observed as dips on d^2I/dV^2. However if the structures originate from barrier phonon (namely inelastic tunneling process), they appear as peaks on d^2I/dV^2 curves. We notice a weak structure at ~10meV from V_P in Figure 1. Then due to be easy to see the fine structures, d^2I/dV^2 is shown in Figure 2 where d^2I/dV^2 is calculated as numerical differential of dI/dV. As clearly seen in Figure 2 the structures are stronger at positive bias than that at negative bias, although for hole-doped cuprates, it was stronger at negative bias than that at positive bias when we measured by PCJs[11]. It is unclear if this superficial asymmetry for the strength of fine structures is intrinsic. Now if the structure at ~16mV, reflect the bosonic mode, its energy must be Ω=eV-Δ/e~9.89$\pm$0.96meV. Therefore, the resonance mode observed by inelastic neutron study[38] may be one of candidates. Another candidate is low-energy phonon mode. However it must be need further investigation to conclude for this issue such as doping dependence of fine structures and isotope effect.

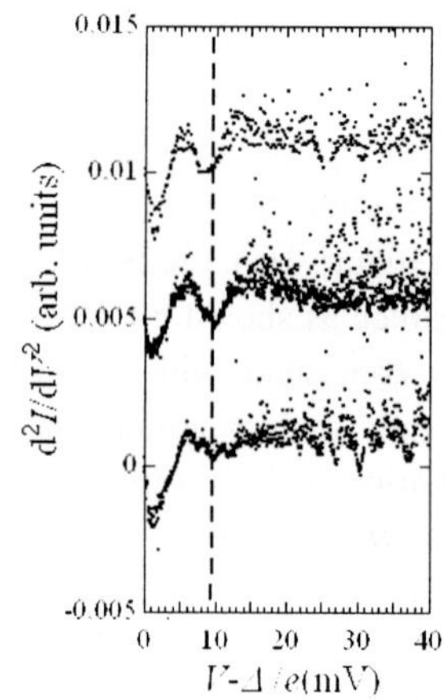

Figure 2. d^2I/dV^2 vs. (V – Δ/e) on PLCCO with x=0.11. The broken line stands at 9.8mV, which is guide for eyes.

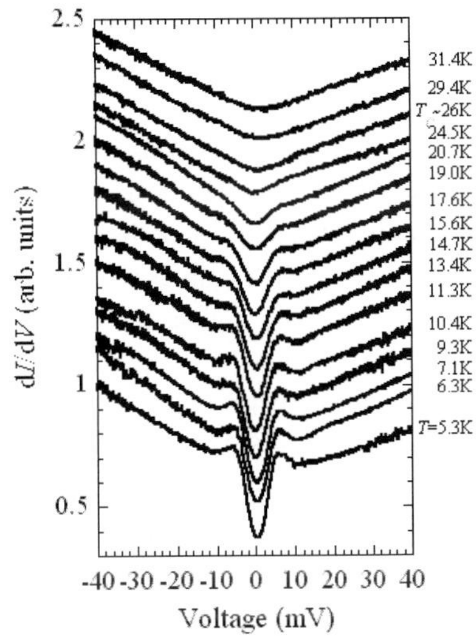

Figure 3. Temperature dependence of dI/dV on PLCCO with x=0.11.

Finally we report temperature dependence of tunneling conductance on PLCCO with x~0.11 as shown in Figure 3. From Figure 3 we found that superconducting gaps dwindle with increasing temperature. The gap structure almost disappeared at T_C, suggesting that small pseudogap temperature is $\leq T_C$ if it exists.

3.2. *NdFeAsO$_{1-y}$*

Although the extensive research for high-T_C superconductivity has been done on cuprate superconductors for more than two decades, there are still many unresolved questions for cuprate superconductors. In order to give an insight for the unresolved issues of high-T_C cuprate superconductors, it is important to compare the physical properties with those of other high-T_C superconductors. However, up to last year, there was no high-T_C superconductor with T_C>50K except for cuprate superconductors. However a series of Fe-based superconductors were discovered last year, and one of Fe-based SCs, REFeAsO (RE=rare earth) showed high-T_C superconductivity with T_C>50K, and this family is called as (RE)1111. By intensive research for Fe-based superconductors, the basic characteristic features are getting clear, but the features of superconducting state are still open questions. That is, one of the most important physical quantities is the superconducting gap. Therefore we have done tunneling study of Nd1111 in order to determine the magnitude of

Δ and give an insight for the symmetry of Δ. Through our tunneling study on a $NdFeAsO_{1-y}$ (Nd1111), mainly two types of dI/dVs were observed, where these spectra display the clear conductance peak at different energy position. One is spectra displaying a clear coherence peak at Δ_S/e~5-7mV (Type-I), and another is spectra showing a dominantly larger gap ($V_p \sim \Delta_L/e$~13-15mV) (Type-II), the representative tunneling conductance curves of Nd1111 measured at 4.2K for Type I and II are shown in Figure 4 and Figure 5 respectively.

In Type-I spectra (see Figure 4), we notice that the clear peak structures are observed reproducibly at lower voltage (V_P ~6mV), suggesting that these are intrinsic superconducting gap (SC-gap) structures. The SC-gap magnitude is estimated from V_P values, as $eV_P(\sim\Delta_S)\sim 6.0\pm 1.0$meV. As seen from #1 in Figure 4, the broader peak makes the peak position be higher, and the conductance curves with the broad peak are sometimes observed. In tunneling conductance, it is well-known that this peak position V_p (if the quasiparticle damping is not too large) roughly corresponds to a SC-gap magnitude, Δ/e. In addition, our determined gap value as Δ_S~6meV is consistent with the previously reported SC-gap magnitude of $SmFeAsO_{1-x}F_x$ by Andreev spectroscopy[22-23] and STM/STS[24,25] . Furthermore we noticed that the spectra of this Type-I display the hump structure at 20-35mV (see #1, #2 in Figure 4).

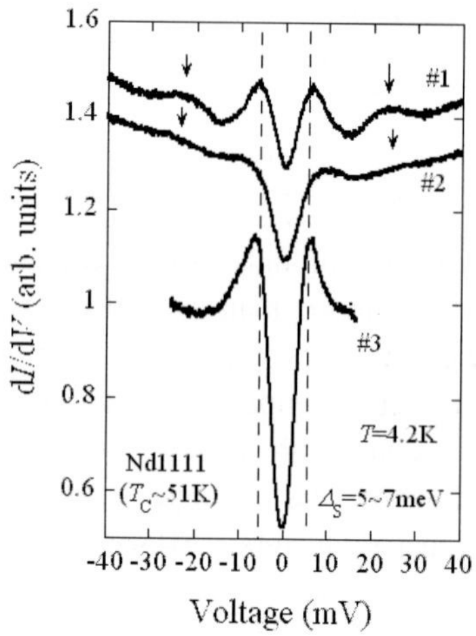

Figure 4. Representative tunneling conductances on $NdFeAsO_{0.7}$ using a Au tip for Type-I Spectra. The vertical dashed lines are drawn at $\pm$6.6mV.The hump structures are indicated by arrows.

Similar structures at higher energy have been reported[23], and the authors of Ref.23 identified it as a second SC-gap. On the other hand, T.Y. Chen *et al.* also observed a similar hump structure in the same energy range[22], but they assert that the hump structure do not represent an intrinsic SC-gap because the observed hump position changed with contact resistance. In our case, the Δ_S structure is quite clear and reproducible, but the hump structure often appears dimly, then it is difficult to determine the precise position. However the similar structures in Type-I spectra on our Nd1111-GaAs mechanical junctions were also observed at the range of 20-35mV. Concerning the origin of the hump, there are four possibilities that are (i) an intrinsic SC-gap which opens at different Fermi surface (FS), or (ii) a pseudogap, or (iii) a strong coupling structure for paring and (iv) extrinsic effects, but we need further investigation to conclude the origin.

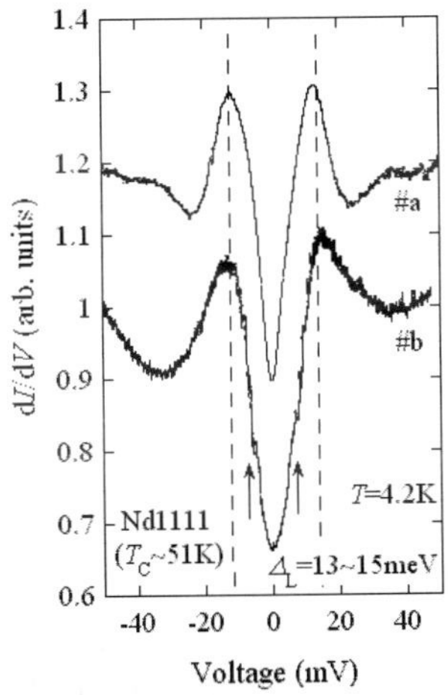

Figure 5. Representative tunneling conductances on $NdFeAsO_{0.7}$ using a Au tip for Type- II Spectra. The vertical dashed lines are drawn at ±14mV. The trace of smaller gap is simultaneously observed, and they are indicated by arrows.

On the other hand, we sometimes observed the spectra showing a clear peak structure at 13-15mV which is much larger than Δ_S/e. The representative tunneling conductances of Type-II are shown in Figure 5. As seen in Figures. 4 and 5, the shape of both type of spectra is similar each other. Namely, the spectra of Type-II also display the clear coherence peaks suggesting that the peaks are due to superconducting gaps. Therefore we identify the peak structure as a second superconducting gap, whose magnitude Δ_L is 13-15meV, and the Δ_L value is consistent with those reported by ARPES[26] and far-infrared ellipsometric measurements[27]. Kondo *et al.*[26] reported that Γ hole pocket of Nd1111 is fully gapped and its magnitude is 15±1.5meV. In addition, Ding *et al.* showed two kinds of SC-gaps by ARPES studies on $Ba_{1-x}K_xFe_2As_2$, where a small SC-gap was observed on the large hole-like Fermi surface (FS), and a large SC-gap was done on the small hole-like and electron-like Fermi sheets[39]. By linking with these ARPES results, we assign the peak at eV_p~13-15meV as a larger SC-gap Δ_L which may open at the small hole Fermi-sheet centered at the Γ point or the electron sheets at the M points. On the other hand, we assign the peak at eV_p~ 7meVas a smaller SC-gapΔ_S/e which may open at the large outer hole FS centered at the Γ point.

We noticed the difference of spectra between Type-I and –II that is a kink structure within the gap peaks. As shown by arrows in Figure5 in Type-II spectra, the kink can be simultaneously observed at the position of Δ_S/e~5-7mV suggesting the existence of multiple gaps (at least two kinds of gaps). This kink structure has been also observed on Nd1111-GaAs junctions, suggesting it is an intrinsic structure of Nd1111. In addition, the similar multiple gaps by tunneling study were observed on MgB_2 which is a two-band superconductor[40,41], and on multilayered cuprates such as $(Cu,C)Ba_2Ca_3Cu_4O_y$ [35], $TlBa_2Ca_2Cu_3O_y$[36] *etc.* that are multiband superconductors. Our tunneling results of Nd1111 give direct evidence for the existence of multiple gaps in the quasiparticle excitation spectrum of this multiband system. Furthermore, we could fit both spectra (Type-I and -II) using an *s*-wave Blonder-Tinkham-Klapwijk theory[42], in which the quasi-particle energy E is replaced by $E+i\Gamma$, (where Γ is the broadening parameter), suggesting that both gaps would be of s-wave symmetry which will be published elsewhere. However, our technique is not sensitive for the phase difference between Δ_S and Δ_L. therefore we need further study to conclude the symmetry of Δ.

Now we will compare these tunneling results with those on cuprates. In cuprates, we know the magnitude of Δ is very large[9, 10] and it gives rise to high-T_C superconductivity. On the other hand, our tunneling results on Nd1111 show two kinds of superconducting gaps. One is a small superconducting gap Δ_S and it gives $2\Delta_S/k_BT_C$~2.8 which is considerably smaller than the BCS value of 3.52. Another larger SC-gap gives $2\Delta_L/k_BT_C$~6.4 which shows that Nd1111 is a strong coupling SC. These values give a strong constraint on the consideration for the superconducting mechanism of the Fe-based SCs, and they may indicate that the small FS has an important role for high-T_C superconductivity of the pnictide. A similar smaller gap value was observed on MgB_2 with T_C~40K[40]. Since the Fe-based superconductor is a multi-band superconductors, it is plausible that the material has two (or multi) kinds of SC-gaps like MgB_2 and multilayered cuprates with n≥3.

4. Summary

In Sec. 3.1, we reported tunnelling conductance on nearly optimally doped PLCCO and we found superconducting gap magnitude Δ is about 6meV, and $2\Delta/k_BT_C$~5.3 when Vp is defined as $\Delta(0)$. Furthermore, we found that the symmetry of order parameter for PLCCO would be nonmonotonic *d*-wave. In addition, in some junctions, we observed a fine structure at~16mV which would correspond to bosonic mode with Ω~10meV although it has been observed Ω~40meV for optimally-doped Bi2212. However, we need further investigation to conclude the origin of a fine structure beyond V_P.

In Sec. 3.2, we reported tunneling results on $NdFeAsO_{0.7}$ with T_C~51K and showed the direct evidence for the existence of two kinds of s-wave SC-gaps in Nd1111. The magnitude of the two-kinds of clear SC-gaps is Δ_S/e~ 6.0±1.0 and Δ_L/e~14.0±1.0 mV, where the value of Δ_S is considerably smaller than the BCS value, but the large value of Δ_L shows that Nd1111 is a strong coupling superconductor.

Acknowledgments

The authors would like to thank M. Fujita, K. Yamada, K. Ogata, K. Miyazawa, H. Kito, P. M. Shirage, H. Eisaki, A. Iyo for cooperation with the experiment and for useful discussions. This work was partially supported by Grants-in-Aid for Scientific Research (C) from the Ministry of Education, Culture, Sports, Science and Technology, Japan.

References

1. J. G. Bednorz and K. A. Muller, *Z. Phys B* **64**, 189 (1986).
2. Y. Kamihara, T. Watanabe, M. Hirano and H. Hosono, *J. Am. Chem. Soc.* **130**, 3290 (2008)
3. C. C. Tsuei and J. Kirtly, *Rev. Mod. Phys.* **72**, 969 (2000); and references therein.
4. D. J. Van Harlingen, 1995, *Rev. Mod. Phys.* **67**, 515 (1995); and references therein.
5. T. Timusk, and B. Statt, *Rep. Prog. Phys.* **62**, 61 (1999); and references therein.
6. J. Tranquada J. D. Axe, N. Ichikawa, A. R. Moodenbaugh, Y. Nakamura, and S. Uchida, *Phys. Rev. Lett.* **78**, 338 (1997).
7. H. A. Mook Pengcheng Dai, S. M. Hayden, G. Aeppli, T. G. Perring, F. Doan, *Nature (London)* **395**, 580 (1998).
8. J. Bardeen, L. N. Cooper, and J. R. Schrieffer, *Phys. Rev. B* **106**, 162 (1957).
9. N. Miyakawa P. Guptasarma, J. F. Zasadzinski, D. G. Hinks, and K. E. Gray, *Phys. Rev. Lett.* **80**, 157 (1998).

10. N. Miyakawa, J. F. Zasadzinski, L. Ozyuzer, P. Guptasarma, D. G. Hinks, C. Kendziora, and K. E. Gray, *Phys. Rev. Lett.* **83**, 1018 (1999).
11. N. Miyakawa, J. F. Zasadzinski, M. Asano, D. Henmi, S. Oonuki, K. Sasaki, T. Kaneko, L. Ozyuzer, K. E. Gray., *Physica C* **357-360**, 126 (2001).
12. W. L. McMillan and J. M. Rowell, *Superconductivity*, edited by R.D. Parks, Marcel Dekker Inc., New York (1969).
13. J. F. Zasadzinski, L. Ozyuzer, N. Miyakawa, K. E. Gray, D. G. Hinks, and C. Kendziora, *Phys. Rev. Lett.* **87**, 067005-1 (2001).
14. Z.-A. Ren, G.-C. Che, X.-L. Dong, J. Yang, W. Lu, W. Yi, X.-L. Shen, Z.-C. Li,L.-L. Sun, F. Zhou and Z.-X. Zhao. *Europhys. Lett.* **83**, 17002 (2008)
15. M. Rotter, M. Tegel, and D. Johrendt, *Phys. Rev. Lett.* **101**, 107006 (2008) ; K. Sasmal, B. Lv, B. Lorenz, A. M. Guloy, F. Chen, Y.-Y. Xue, and C.-W. Chu., Phys. Rev. Lett. **101**, 107007 (2008)
16. F.-C. Hsu, J.-Y. Luo, K.-W. Yeh, T.-K. Chen, T.-W. Huang, P. M. Wu, Y.-C. Lee, Y.-L. Huang, Y.-Y. Chu, D.-C. Yan, M.-K. Wu, *arXiv*:0807.2369.
17. M.R. Norman, *Physics* **1**, 21 (2008)
18. S. Uchida, *J. Phys. Soc. Jpn* **77**, Suppl. C 9 (2008)
19. H. Kito, H. Eisaki and A. Iyo, *J. Phys. Soc. Jpn* **77**, 063707 (2008)
20. X. H. Chen, T. Wu, G. Wu, R. H. Liu, H. Chen, D. F. Fang, *Nature* **453**, 761 (2008)
21. C. de la Cruz Clarina de la Cruz, Q. Huang, J. W. Lynn, Jiying Li, W. Ratcliff II, J. L. Zarestky, H. A. Mook, G. F. Chen, J. L. Luo, N. L. Wang, *et al.*, *Nature* **453**, 899 (2008)
22. T. Y. Chen, Z. Tesanovic, R. H. Liu, X. H. Chen, C. L. Chien, *Nature* **453**, 1224 (2008) ; T. Y. Chen, S. X. Huang, Z. Tesanovic, R. H. Liu, X. H. Chen, C. L. Chien., *Physica C* **469**, 521 (2009)
23. D. Daghero, M. Tortello, R. S. Gonnelli, V. A. Stepanov, N. D. Zhigadlo, J. Karpi, *arXiv*:0812.1141.
24. O. Millo, I. Asulin, O Yuli, I. Felner, Z.-A. Ren, X.-L. Shen, G.-C. Che, and Z.-X. Zhao, *Phys. Rev. B* **78**, 092505 (2008)
25. J. Karpinski *et al.*, *Physica C* (2009), doi:10.1016/j.physc.2009.03.048
26. T. Kondo *et al.*, *Phys. Rev. Lett.* **101**, 147003 (2008)
27. A. Dubroka *et al.*, *Phys. Rev. Lett.* **101**, 097011 (2008)
28. I. I. Mazin, D. J. Singh, M. D. Johannes and M. H. Du, *Phys. Rev. Lett.* **101**, 057003 (2008)
29. K. Kuroki *et al.*, *Phys. Rev. Lett.* **101**, 087004 (2008)
30. M. Fujita, T Kubo, S. Kuroshima, T. Uefuji, K. Kawashima and K. Yamada, *Phys. Rev. B* **67** 014514 (2003)
31. H. Matsui, K, Terashima, T. Sato, T. Takahashi, M. Fujita and K. Yamada, *Phys. Rev. Lett.* **95** 017003 (2005)
32. S. Kashiwaya and Y. Tanaka, *Superlattices and Microstructures* **27** 271 (2000)
33. C. C. Tsuei and J. R. Kirtley *Phys. Rev. Lett.* **85** 182 (2000)

34. F. C. Niestemski, S. Kuwar, S. Zhou, S. Li, H. Ding, Z. Wang , P. Dai and V. Madhavan, *Nature* **450** 1058 (2007)
35. N. Miyakawa, K. Tokiwa, S. Mikusu, J. F. Zasadzinski, L. Ozyuzer, T. Ishihara, T. Kaneko, T. Watanabe, and K. E. Gray, *Int. J. Mod. Phys. B* **17** 3612 (2003)
36. N. Miyakawa, K. Tokiwa, S. Mikusu, T. Watanabe, A. Iyo, J. F. Zasadzinski, T. Kaneko, *Int. J. Mod. Phys. B* **19** 225 (2005)
37. N. Miyakawa, H. Hoya, A. Iyo, Y. Tanaka, K. Tokiwa, T. Watanabe, and T. Kaneko, *Int. J. Mod. Phys. B* **21** 3233 (2007)
38. S. D. Wilson, P Dai, S Li, S Chi, H. J. Kang and J. W. Lynn *Nature* **442** 59 (2006)
39. H. Ding, P. Richard, K. Nakayama, K. Sugawara, T. Arakane, Y. Sekiba, A. Takayama,S. Souma, T. Sato, T. Takahashi,, Z. Wang, X. Dai, Z. Fang, G. F. Chen, J. L. Luo and N. L. Wang, *Europhys. Lett.* **83**, 47001 (2008) ; K. Nakayama, T. Sato, P. Richard, Y.-M. Xu, Y. Sekiba, S. Souma, G. F. Chen,J. L. Luo, N. L. Wang, H. Ding and T. Takahashi, Europhys. Lett. **85**,67002 (2009)
40. H. Schmidt, J. F. Zasadzinski, K. E. Gray, and D. G. Hinks, *Phys. Rev. Lett.* **88**, 127002 (2002)
41. T. Ekino, T. Takasaki, T. Muranaka, J. Akimitsu, and H. Fujii, *Phys. Rev. B* **67**, 094504 (2003)
42. G. E. Blonder, M. Tinkham, and T. M. Klapwijk, *Phys. Rev. B* **25** 4515 (1982)

Quantum Bio-Informatics III
eds. L. Accardi, W. Freudenberg & M. Ohya

QUANTUM DYNAMICS OF SUPERCONDUCTING QUBIT READOUT WITH A DRIVEN NONLINEAR JOSEPHSON OSCILLATOR

HAYATO NAKANO*

NTT Basic Research Laboratories, NTT Corporation,
Atsugi-shi, Kanagawa 243-0198, Japan
**E-mail: nakano@will.brl.ntt.co.jp*

The Josephson bifurcation amplifier (JBA) readout process of a superconducting qubit is analyzed quantum mechanically. We examined the dynamics of the density operator of a probe (a driven nonlinear oscillator) and a qubit coupled system during the measurement process. We have observed the qubit-probe (JBA) entangled state and it is divided into two separable states at the moment of the JBA transition begins. To readout the measurement result, however, we must wait until the two JBA states are macroscopically well separated.

Keywords: Superconducting qubit; Readout; Josephson Bifurcation Amplifier; Entanglement; Projection.

1. Introduction

A qubit is a most elementary unit of quantum information processing[1]. Qubit readout is one of simple quantum measurements. According to a simple explanation of quantum measurements, it is described as if we can directly obtain an observable eigenvalue of the quantum system by a measurement. In a real measurement, however, we always have to uses a probe that interacts with the quantum system. A probe is often a macroscopic object. We actually measure a macroscopic quantity in the probe, and merely *postulate* the physical quantity of the quantum system we want to measure. In order to discuss what we really measure when we measure a qubit, we have to take into account the probe system. A measurement process is a quantum dynamics of the total system consists of a qubit and a probe.

A good qubit readout gives a probabilistic projection of a qubit state $|\psi_{\rm q}\rangle = a|\uparrow\rangle + b|\downarrow\rangle$ into $|\uparrow\rangle$ or $|\downarrow\rangle$ with a probability $|a|^2$ or $|b|^2$, respectively. Here, $|\uparrow\rangle$ and $|\downarrow\rangle$ are two of eigenstates of the *observable* in the qubit. We design the qubit-probe interaction so that the observable is coincident with the physical quantity we want to measure. Mainly, the observable is determined by the operator of the qubit appearing in the interaction Hamiltonian between the qubit and the probe. However, the observable is sometimes different from the operator because it is determined by the whole dynamics of the total system during the readout process. The process reflects not only on the interaction Hamiltonian but also on the probe Hamiltonian, the qubit Hamiltonian, and decoherence.

Below, we discuss a readout process for a superconducting qubit. We use a Josephson Bifurcation Amplifier (JBA) as the probe of the measurement[2]. The JBA is an AC-driven SQUID that behaves as a nonlinear oscillator. A JBA is a simple enough to analyze although it is a macroscopic system with many degrees of freedom. We find many important quantum characters from the investigate for the dynamics during the readout process[3].

Usually, the total system is expected to evolve as follows, for a case in which the qubit is in the superposition of two of observable eigenstates $|\psi(0)_{\mathrm{q}}\rangle = a|\uparrow\rangle + b|\downarrow\rangle$. When the interaction becomes on, entanglement starts to be formed.

$$|\psi(0)_{\mathrm{q}}\rangle \otimes |\mathrm{C}_0\rangle \to a|\uparrow\rangle \otimes |\mathrm{A}_1\rangle + b|\downarrow\rangle \otimes |\mathrm{B}_1\rangle$$

Since the probe is a macroscopic system where a finite decoherence is inevitable[4], a dephasing occurs. Then, the total system becomes a classical mixture like

$$|a|^2|\uparrow\rangle|\mathrm{A}_1\rangle\langle\uparrow|\langle\mathrm{A}_1| + |b|^2|\downarrow\rangle|\mathrm{B}_1\rangle\langle\downarrow|\langle\mathrm{B}_1|.$$

Here, $|\mathrm{C}_0\rangle$, $|\mathrm{A}_1\rangle$, $|\mathrm{B}_1\rangle$ are macroscopic states of the probe, and we can distinguish easily $|\mathrm{A}_1\rangle$ or $|\mathrm{B}_1\rangle$. Therefore, when we measure the probe state, we can postulate whether the qubit state is $|\uparrow\rangle$ or $|\downarrow\rangle$.

We neglected the dynamics of the probe in the above picture. However, in the case of the qubit readout with a JBA, as shown below, the JBA dynamics is important in order to understand the readout process.

2. Superconducting Qubit

There are three types of superconducting qubits; charge, phase, and flux ones[5,6,7,8,9,10,11]. (Recently, some intermediate types of them have appeared. ex., Quantronium, Transmon.) A superconducting flux qubit is a superconducting ring interrupted by a few Josephson junctions[11,12,13,14,15,16]. When the external magnetic field piercing the ring is in the vicinity of the half flux quantum $\Phi_0/2$, the ring becomes a superposition of two macroscopically distinctive states. Supercurrent flows along the ring in clockwise in one state $|\mathrm{R}\rangle$, and it flows in the opposite direction in the other state $|\mathrm{L}\rangle$. The qubit Hamiltonian is given by[17,18,19]

$$H_{\mathrm{q}} = \frac{1}{2}(\varepsilon\sigma_z + \Delta\sigma_x), \quad (1)$$

where σ_z and σ_x are two of Pauli matrices, and

$$\sigma_z = |\mathrm{L}\rangle\langle\mathrm{L}| - |\mathrm{R}\rangle\langle\mathrm{R}|.$$

We can control ε by changing the externally applied magnetic flux, but Δ is fixed. Energy-eigenstates of the qubit is given by

$$|g\rangle = \cos(\theta/2)|\mathrm{L}\rangle + \sin(\theta/2)|\mathrm{R}\rangle, \quad |e\rangle = \sin(\theta/2)|\mathrm{L}\rangle - \cos(\theta/2)|\mathrm{R}\rangle,$$

where $\tan\theta = \Delta/\varepsilon$. Therefore, when $|\varepsilon/\Delta| \gg 1$, energyeigenstates are eigenstates of σ_z, that is, $|\mathrm{L}\rangle$ and $|\mathrm{R}\rangle$, and contrary, when $|\Delta/\varepsilon| \ll 1$, energy-eigenstates become the bonding and anti-bonding states of $|\mathrm{L}\rangle$ and $|\mathrm{R}\rangle$.

3. Qubit Measurement with A DC-SQUID

In order to readout superconducting flux qubits, a SUperconducting Quantum Interference Device (SQUID) is used as the probe. The SQUID detects a very small magnetic flux induced by supercurrent flowing along the qubit ring, which is only one observable in a flux qubit we can measure from the outside.

In our experiments, we used a DC-SQUID to measure the small magnetic flux induced by the supercurrent along the loop. The output of the SQUID (a switching current) has a continuous value although the qubit measured by the SQUID is a two-level system. (see Fig. 1)

A SQUID is a very sensitive magnetic flux detector. For realistic experimental situations, a SQUID can be approximated as a single Josephson junction whose Josephson energy depends on the magnetic flux Φ_{SQ} piercing the SQUID ring. The Hamiltonian for the SQUID phase γ(=the Josephson phase of the effective single junction) is given by[20], $(\hbar/2e)^2\dot{\gamma}^2/(2C) - E_{\mathrm{J}}(\Phi_{\mathrm{SQ}})\cos\gamma - (\hbar/2e)I_{\mathrm{b}}\gamma = (\hbar/2e)^2\dot{\gamma}^2/(2C) - E_{\mathrm{J}}(\Phi_{\mathrm{SQ}})(1 - \gamma^2/2 + \gamma^4/24 - \cdots) - (\hbar/2e)I_{\mathrm{b}}\gamma$, where C is the effective capacitance of the junction and I_{b} is the bias current applied to the SQUID.

The qubit state slightly changes Φ_{SQ} depending on its state. So, the effective parameters of the SQUID varies with the qubit state. In early

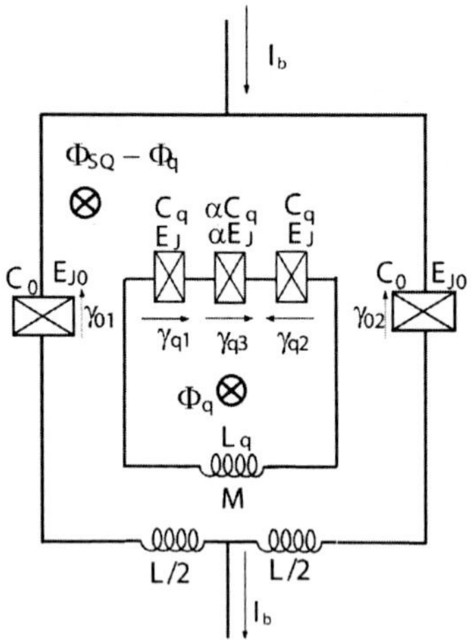

Figure 1. Readout of a flux-qubit system with a probe SQUID that has a continuous variable $\gamma = (\gamma_{01} + \gamma_{02})/2$. The inner loop is the flux qubit, and the outer is the SQUID.

days of flux qubit experiments, we measured the variation of the switching current and postulated the qubit state with the same way as a usual small magnetic flux measurement[21]. In this case, the interaction Hamiltonian is given by $H_{\rm int} \simeq k(I_{\rm b})\sigma_z \sin\gamma$, with a coupling strength of a function of the bias current. Therefore, if the measurement is a projection measurement of σ_z, the quantum average of the measured value behaves as shown in Fig. 1. as the function of ε. Suppose that the qubit state is the superposition $|\psi\rangle = a|{\rm L}\rangle + e^{i\varphi}\sqrt{1-|a|^2}|{\rm R}\rangle$. If the measurement corresponds to the projection measurement of σ_z, we obtain an eigenvalue 1 or -1, probabilistically. In order to obtain the value a, we have to prepare large numbers of the same state qubits and average the readout results and then, obtain $2|a|^2 - 1$.

However, a real measurement on a qubit is often not an ideal projection measurement, as it is not in the case of flux-qubit readout that we discuss in this paper. A half decade ago, our experimental results where raw data before averaging behave as in the right panel of Fig. 2 clearly showed that our flux-qubit measurement with a DC-SQUID is not a σ_z projection measurement[21,22]. Moreover, we theoretically clarified that it happens in the qubit measurement with a SQUID when we increase the SQUID bias current slowly, by a detailed calculation on the dynamics of the qubit-SQUID coupled system[20].

The measurement projects the coupled system to one of its energy-

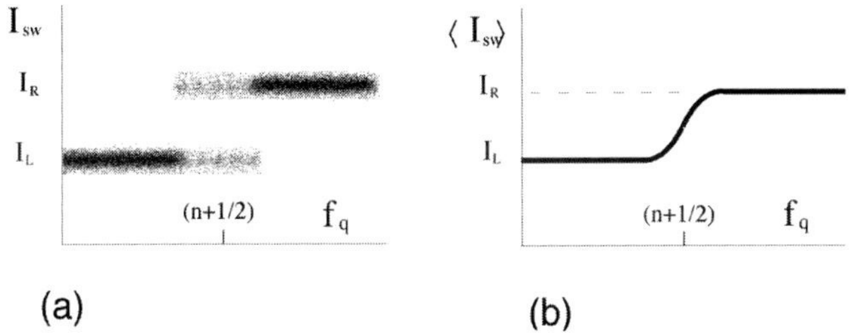

Figure 2. Expected behavior of readout data as a function of ε if the measurement is σ_z projection (schematic). (a) Bare data before averaging. (b) Averaged values over huge number of detections.

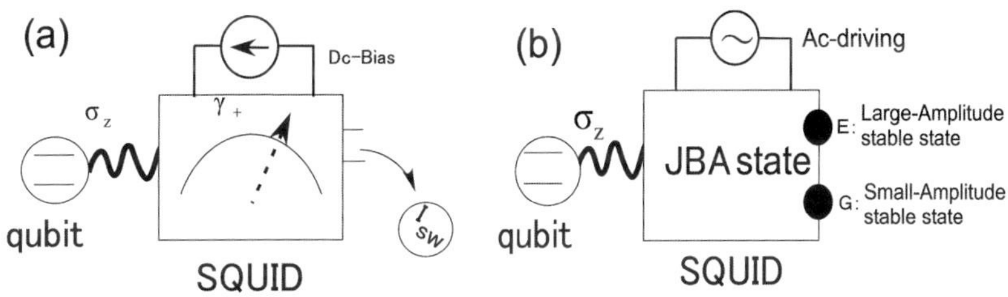

Figure 3. (a) Reading by a conventional switching electric current: We measure the switching electric current which is quantity consecutively, estimate state ($|g\rangle$ of qubit or $|e\rangle$). (2) The reading that JBA depends on: I measure a resonance state of SQUID using that I become two states (from a nonlinearity), and a resonance state of SUQUID estimates state ($|g\rangle$ of qubit or $|e\rangle$). Speedup of improvement (the amplification) and the measurement of the S/N ratio is expected hereby.

eigenstates. In the state, the qubit is approximately in an energy-eigenstate of H_q, and obtained value of a single measurement approximately corresponds to $\langle g|\sigma_z|g\rangle$ or $\langle e|\sigma_z|e\rangle$.

4. Quantum Dynamics of A Josephson Bifurcation Amplifier

When the amplitude of γ is much small, we can approximate the junction as a harmonic oscillator. However, when the AC bias current is large, that is, the driving for the oscillator is strong, the oscillator shows a nonlinearity. Mathematically, it is a driven Duffing oscillator, and becomes bistable under appropriate operation parameters [23]. One stable state has a

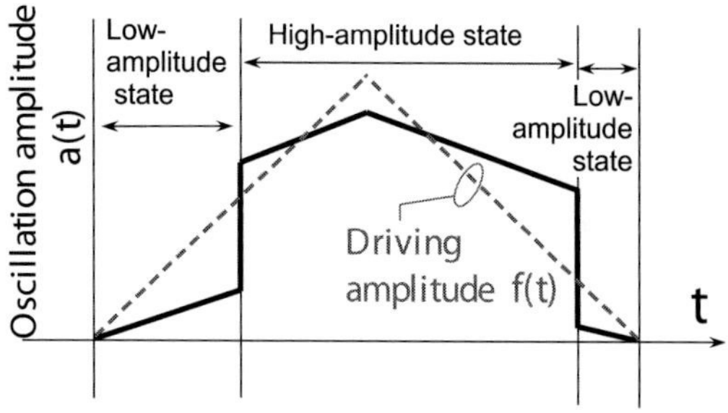

Figure 4. Hysteretic behavior of the oscillation amplitude $a(t)$ of a classical JBA.

small amplitude (low-amplitude state), and the other has a larger amplitude (high-amplitude state). Utilizing this bistability, Josephson Bifurcation Amplifier reads out the small difference depending on the qubit state[2].

The critical driving force f_c or the critical detuning δ_c for the transition between these two states is very sensitive to small changes in the operational parameters of the oscillator. For example, when we increase or decrease the driving force continuously, the amplitude of the oscillation behaves hysteretically as shown in Fig. 4. When using a JBA as a qubit state readout probe, the JBA detects a small change depending on the qubit state. However, the quantum-mechanical behavior of the JBA readout process has not been established theoretically. This is because the bifurcation phenomenon can be discussed only for classical oscillators, and is impossible from the view point of pure quantum mechanics for an isolated system [25].

A classical analysis gives no information on entanglement between the qubit and the probe (JBA) or the decoherence in the composite system, although all the quantum properties (projection, measurement backaction, etc.) in the readout are contained in such information[24]. A quantum mechanical analysis is indispensable if we are to understand the readout process[3].

A JBA can be modeled as an anharmonic oscillator in a rotating frame approximation with a Hamiltonian;

$$H_{\mathrm{J}} = (\Omega - \omega)n_a - \alpha {n_a}^2 - \frac{1}{2}f(a^\dagger + a) \tag{2}$$

where, $a^\dagger(a)$ is the creation (annihilation) operator of the Josephson plasma oscillation. Normalized amplitude and momentum of an oscillation are expressed by $x = (a^\dagger + a)/2$ and $p = (a - a^\dagger)/(2i)$, respectively. $n_a = a^\dagger a$, and

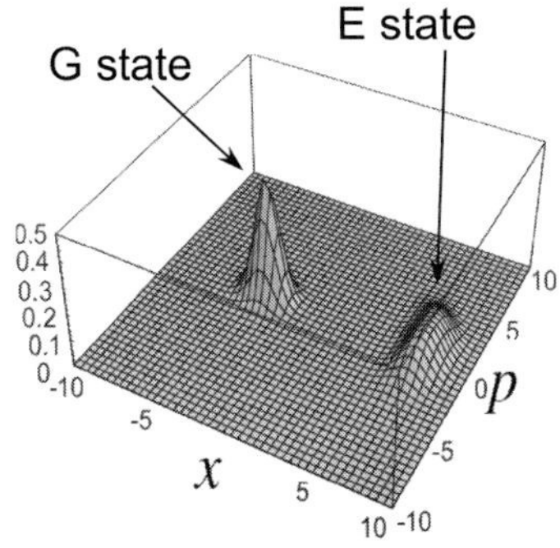

Figure 5. Two macroscopic states of JBA (quasi-distribution representation): p, x are momentum and amplitude of the oscillation, respectively. (in this figure, vibration consists of about 10-20 quanta)

Ω is the linear resonant frequency of the JBA oscillator. ω is the driving frequency, which is slightly smaller than Ω by the detuning $\delta \equiv \Omega - \omega$. f is the driving strength, and $\alpha(> 0)$ is the nonlinearity. In a classical approximation, this model shows the bifurcation in an appropriate parameter region. However, for a quantum-mechanical junction, the transition from $|G\rangle_{\mathrm{J}}$ (low-amplitude state) to $|E\rangle_{\mathrm{J}}$ (high-amplitude state) or, from $|E\rangle_{\mathrm{J}}$ to $|G\rangle_{\mathrm{J}}$ is impossible as mentioned below. [a]

Here, $|G\rangle$and $|E\rangle$ are schematic expression for the states corresponding to the classically stable two states. These two states are shown in Fig. 5 with the quasi-distribution function representation.

Without any dissipation, the JBA state only moves along the initial quasienergy-state (eigenstates of the Hamiltonian Eq. (2) in the rotating approximation) and never jumps to another quasienergy-state which the high-amplitude resonant state belongs to [23].

As an example, here, we introduce linear loss in the oscillator (JBA). The time evolution of the JBA is governed by a Liouville equation[27]:

$$\frac{\mathrm{d}\rho}{\mathrm{d}t} = \frac{1}{i}[H_{\mathrm{J}}, \rho] + \frac{\Gamma}{2}(2a\rho a^{\dagger} - a^{\dagger}a\rho - \rho a^{\dagger}a), \tag{3}$$

where ρ is the density operator of the system, and Γ is the relaxation rate due to the linear loss in the JBA.

Figure 6 shows a numerical example of JBA dynamics without a qubit. Here, we set $\delta = 0.007\Omega$, nonlinearity $\alpha = 8 \times 10^{-5}\Omega$, quality factor $Q =$

[a]Nonadiabatic change in f causes inter-eigenstate transitions. However, this does not result in a systematic transition between $|G\rangle_{\mathrm{J}}$ and $|E\rangle_{\mathrm{J}}$.

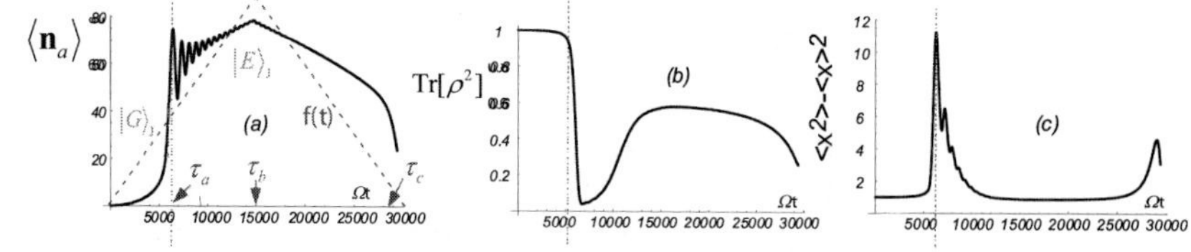

Figure 6. Time variations of some quantum expectation values when the driving force f is operated as shown in Fig. 4. (a) Number of bosons excited in JBA. (b) Purity of the JBA. (c) Fluctuation in JBA amplitude.

2500, and the driving force f is operated as $0 \to 0.025\Omega \to 0$. These parameters are similar to those used in actual experiments [2]. In order to emphasize quantumness, however, δ and Γ are a factor of 10^{-2} times smaller than real cases. More than one hundred basis states are prepared in order to avoid numerical artifacts.

Figure 6(a) approximately reproduces the hysteretic behavoir of the JBA amplitude shown in Fig. 4. The critical driving f_c is approximately 0.011Ω. The critical f_c is almost the same in a classical JBA with the same parameters. In Fig. 6(b), we found that the purity decreases abruptly ($t = \tau_a$). This corresponds to the beginning of the transition from $|G\rangle_J$ to $|E\rangle_J$ of the JBA. After the rapid decrease, the purity recovers to some extent and the JBA approaches to the classically stable state $|E\rangle_J$ ($t = \tau_b$). Since $|E\rangle_J$ is a meta-stable state (stationary point of the classical Hamiltonian), dragging JBA into the state by decoherence (linear loss) leads to the recovery of the purity. However, the purity does not reach unity because it is not a true ground state. In Fig. 6(c), we can see a divergence of the fluctuation at the moment of the rapid decrease in the purity ($t = \tau_a$).

The qubit-JBA composite system is approximately expressed by the Hamiltonian

$$H = H_J + k\sigma_z n_a + H_q, \qquad H_q = \frac{1}{2}(\varepsilon\sigma_z + \Delta\sigma_x) \tag{4}$$

k is the interaction constant between the qubit and the JBA.

5. Qubit Readout with A JBA

The qubit readout process is well understood by employing knowledge of the quantum behavior in the time evolution of the JBA discussed above.

In Fig. 7, we show a dynamics during the qubit readout process (H_J in Eq. (3) replaced with the Hamiltonian H of the qubit-JBA composite system). JBA parameters are the same as for the above example. The initial state is a separable state; $(\frac{1}{\sqrt{2}}|g\rangle_{\mathrm{q}} + \frac{1}{\sqrt{2}}|e\rangle_{\mathrm{q}}) \otimes |G\rangle_{\mathrm{J}}$. Here, $|g\rangle_{\mathrm{q}}$ and $|e\rangle_{\mathrm{q}}$ are the ground and excited states of the qubit, respectively. Qubit parameters are $\varepsilon = 0.2\Omega$, $\Delta/\varepsilon = 1/2$. The coupling between the qubit and the JBA is set at $k = 0.001\Omega$. The driving force f is increased from 0 to 0.012Ω (slightly larger than f_c of the JBA) and maintained. This parameter set gives a typical behavior of successful qubit readout.

The quasi-distribution representations of the JBA state $\mathrm{Tr_q}[\rho]$ are shown in Fig. 7, where ρ is the density operator of the qubit-JBA coupled system, and $\mathrm{Tr_q}[\cdots]$ denotes taking partial trace about qubit degrees of freedom. These peaks constitute an incoherent mixture, so they correspond to two possibilities in the measurement result.

As a measure of the entanglement between the JBA and the qubit, we adopt $E \equiv \mathrm{Tr}\left[\rho^2\right] - \mathrm{Tr}\left[(\mathrm{Tr_q}[\rho])^2\right]$. The reduction in $\mathrm{Tr}\left[(\mathrm{Tr_q}[\rho])^2\right]$ is the purity decrease in the reduced density operator of the JBA, that contains the decrease due to both decoherence and the entanglement formation.

Figire 8 shows the time variation of the entanglement measure E. This process can be schematically expressed as

$$\begin{aligned}
\rho(0) &= |G\rangle_{\mathrm{JJ}}\langle G| \otimes \left(\frac{1}{\sqrt{2}}|g\rangle_{\mathrm{q}} + \frac{1}{\sqrt{2}}|e\rangle_{\mathrm{q}}\right)\left(\frac{1}{\sqrt{2}}{}_{\mathrm{q}}\langle g| + \frac{1}{\sqrt{2}}{}_{\mathrm{q}}\langle e|\right) \\
\to \rho(\tau_1) &= \frac{1}{2}\left(|G\rangle_{\mathrm{J}}|e\rangle_{\mathrm{q}} + |G'\rangle_{\mathrm{J}}|g\rangle_{\mathrm{q}}\right)\left({}_{\mathrm{J}}\langle G|_{\mathrm{q}}\langle e| +_{\mathrm{J}} \langle G'|_{\mathrm{q}}\langle g|\right) \\
\to \rho(\tau_2) &= \frac{1}{2}|G\rangle_{\mathrm{J}}|e\rangle_{\mathrm{qJ}}\langle G|_{\mathrm{q}}\langle e| + \frac{1}{2}|G'\rangle_{\mathrm{J}}|g\rangle_{\mathrm{qJ}}\langle G'|_{\mathrm{q}}\langle g| \\
\to \rho(\tau_3) &= \frac{1}{2}|G\rangle_{\mathrm{J}}|e\rangle_{\mathrm{qJ}}\langle G|_{\mathrm{q}}\langle e| + \frac{1}{2}|E\rangle_{\mathrm{J}}|g\rangle_{\mathrm{qJ}}\langle E|_{\mathrm{q}}\langle g|. \qquad (5)
\end{aligned}$$

When we remind ourselves the scenario of the measurement process described in ß1, in the readout process with a JBA we can find another time evolution is added after the destruction of the entanglement (formation of a classical mixture caused by decoherence). At the moment (τ_2), it is impossible to obtain any information about the qubit from the observed probe (JBA) state because $|G'\rangle$ closely resembles $|G\rangle$ in a classical mechanical sense (Fig. 7(b)), and both belong to the low-amplitude oscillation state of the JBA, although ${}_{\mathrm{J}}\langle G'|G\rangle_{\mathrm{J}} \sim 0$, namely, these two states are orthogonal. When we increase the driving force, one JBA state $|G'\rangle_{\mathrm{J}}$ moves to $|E\rangle_{\mathrm{J}}$. The other $|G\rangle_{\mathrm{J}}$ stays. (Figs. 7(c),(d)). At τ_3, we can easily distinguish $|E\rangle_{\mathrm{J}}$ or $|G\rangle_{\mathrm{J}}$. This tells which qubit state ($|g\rangle_{\mathrm{q}}$ or $|e\rangle_{\mathrm{q}}$) is realized. The state of the total system is already divided into separable states $|e\rangle_{\mathrm{q}}|G\rangle_{\mathrm{J}}$

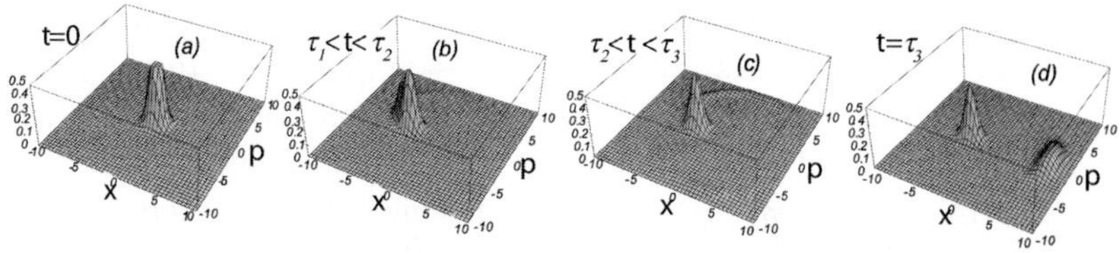

Figure 7. Time-evolution of JBA during readout. The figures show quasi-distribution representations ($\langle\alpha|\mathrm{Tr_q}[\rho]|\alpha\rangle$) of the JBA oscillator states, where $|\alpha\rangle$ is the coherent state of a complex amplitude $\alpha = x + ip$ (in the rotating frame). (a) Beginning of the readout. The state of the total system is (schematically) $\rho = |\psi_0\rangle\langle\psi_0|$, with $|\psi_0\rangle = 1/\sqrt{2}(|g\rangle_q + |e\rangle_q) \otimes |G\rangle_J$. (b) Starting the transition. Entanglement formation and projection are carried out during this period. (c) During the transition. Entanglement has already been destroyed. (d) The entire system has become a mixture of classically correlated states.

and $|g\rangle_q|G'\rangle_J$ just after the transition $|G'\rangle_J \to |E\rangle_J$ starts. Therefore, "projection" itself is successful. However, we have to wait until the time when we can distinguish $|G\rangle_J$ and $|E\rangle_J$ until the transition finishes.

The entanglement formation (τ_1) takes a time of the inverse of k^2/ω_q, where k is the coupling constant in Eq. (3), with the qubit energy $\omega = \sqrt{\varepsilon^2 + \Delta^2}$. The time $\tau_2 - \tau_1$ for entanglement destruction is mainly determined by only JBA parameters as discussed above. And the time scale is much shorter than other time scales involved. The time $\tau_3 - \tau_2$ for displacement of two macroscopic states of JBA is determined by M/Γ, where M is the number of quasienergy-states the JBA has to pass during the $|G\rangle \to |E\rangle$ transition.

When the qubit gap Δ is much smaller than other energies, the JBA readout causes only *pure dephasing* on the qubit. This does not pollute the measurement result. This corresponds to the "non-demolition measurement" condition. When Δ is not negligible compared with ε, the measurement causes *qubit relaxation* in the qubit.

6. Conclusion

We analyzed the dynamics of a composite system consisting of a superconducting qubit and a JBA in order to understand the quantum characters in qubit readout process with a JBA and clarified what happens during the process and how we obtain the qubit information by measuring the probe, JBA.

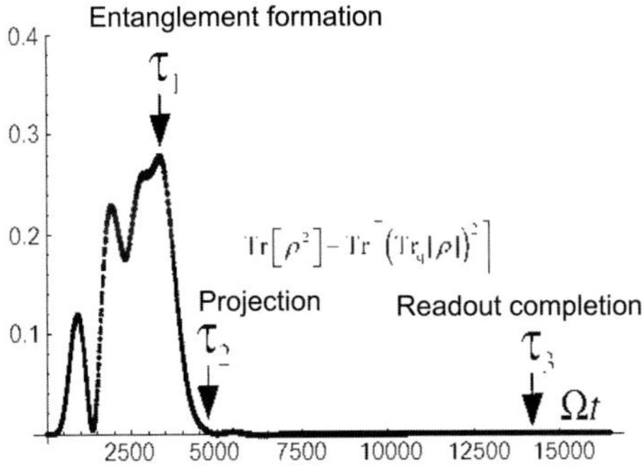

Figure 8. Of qubit and JBA A time change of entanglement strength. Time variation of the entanglement between the JBA and the qubit. $\tau_i (i = 1, 2, 3)$ corresponds to those in Eq. (5), that is, τ_1: entanglement formation, τ_2: projection, τ_3: end of the readout.

Acknowledgements

We would like to thank D. Haviland, C.J.P.M. Harmans, and H. Takayanagi for valuable comments on the early stage of this work. We also thank M. Devoret and H. Mabuchi for fruitful discussions. We should acknowledge to coworkers in Superconducting Quantum Physics Group of NTT Basic Research Laboratories. This work received support from KAKENHI (Sci. Research of Specially Promoted #18001002, Sci. Research (A) #18201018) by JSPS.

References

1. M.A. Nielsen and I.L. Chuang, '*Quantum Computation and Quantum Information*', (Cambridge Univ. Press, 2000).
2. I. Siddiqi *et al.*, Phys. Rev. Lett. **93**, 207002 (2004); I. Siddiqi *et al.*, *ibid.* **94**, 027005 (2005), A. Lupascu *et al.*, *ibid.* **96**, 127003 (2006); I. Siddiqi *et al.*, Phys. Rev. **B 73**, 054510 (2006); A. Lupascu *et al.*, Nature Physics **3**, 119 (2007); N. Boulant *et al.*, Phys. Rev. **B 76**, 014525 (2007).
3. Hayato Nakano, Shiro Saito, Kouichi Semba, Hideaki Takayanagi, Phys. Rev. Lett. **102**, 257003(2009).
4. U. Weiss, '*Quantum Dissipative Systems*' 2nd edition, (World Scientific, Singapore, 1999).
5. M. Devoret and J. Martinis, Quantum Information Processing **3**, 163 (2004), and references therein.
6. J.R. Friedman, V. Patel, W. Chen, S.K. Tolpygo, and J.E. Lukens, Nature **406**, 43 (2000).
7. D. Vion, *et al.*, Science **296**, 886(2002).
8. Y. Yu, S. Han, X. Chu, S. Chu, and Z. Wang, Science **296**, 889(2002).

9. Y. Nakamura, Yu. A. Pashkin, and J. S. Tsai, Nature **398**, 786(1999).
10. J. M. Martinis, S. Nam J. Aumentado, and C. Urbina, Phys. Rev. Lett. **89**, 117901(2002).
11. J. E. Mooij, *et al.*, Science **285**, 1036 (1999).
12. Lin Tian, S. Lloyd and T. P. Orlando, Phys. Rev. **B 65**, 144516 (2002).
13. C. van der Wal *et al.*, Science **290**, 773(2000); C. van der Wal, "Quantum superpositions of persistent Josephson currents," Ph.D. thesis, Delft Univ. Press (2001).
14. I. Chiorescu, Y. Nakamura, C.J.P.M. Harmans, and J.E. Mooij, Nature **431**, 159(2004).
15. I. Chiorescu, P. Bertet, K. Semba, Y. Nakamura, C.J.P.M. Harmans, and J.E. Mooij, Science **299**, 1869(2003).
16. P. Bertet, *et al.*, cond-mat/0412485(2004).
17. D. S. Crankshaw and T. P. Orlando, IEEE Trans. Appl. Superconductivity, **11**, 1006(2001); J. Q. You, Y. Nakamura, and Franco Nori, cond-mat/0309491; A. M. van Brink, cond- mat/0310425.
18. C. H. van der Wal, F. K. Wilhelm, C. J. P. M. Harmans, J. E. Mooij, Eur. Phys. J. **B 31**, 111 (2003).
19. T. P. Orlando, *et al.*, Phys. Rev. **B 60**, 15398(1999).
20. H. Nakano and H. Takayanagi, J. Phys Soc. Jpn. **72**, Suppl. A, 1 (2003); H. Nakano *et al.*, arXiv:cond-mat/0406622 (2004); H. Nakano and H. Takayanagi, p.359 in *Toward the Controllable Quantum States*, edited by H. Takayanagi and J. Nitta (World Scientific Pub., Singapore, 2003).
21. For review, K. Semba, J. Johansson, K. Kakuyanagi, H. Nakano, S. Saito, H. Tanaka and H. Takayanagi,
"Quantum state control, entanglement, and readout of the Josephson persistent-current qubit",
Quantum Information Processing: Volume 8, Issue 2 (2009), Page 199.
22. H. Tanaka, Y. Sekine, S. Saito, H. Takayanagi, Physica **C 368**, 300(2002); H. Takayanagi, H. Tanaka, S. Saito, and H. Nakano, Physica Scripta **T102**, 95-102 (2002); H. Tanaka, S. Saito and H. Takayanagi, p. 366 in '*Toward the Controllable Quantum States*', edited by H. Takayanagi and J. Nitta (World Scientific Pub., Singapore, 2003).
23. W. Jordan and P. Smith, 'Nonlinear Ordinary Differential Equations', third ed. (Oxford Univ. Press, 1999).
24. Y. Makhlin, A. Shnirman and G. Schön, Rev. Mod. Phys., **73**, 357 (2001).
25. M. I. Dykman and M. V. Fistul, Phys. Rev. **B 71**, 140508 (R) (2005); M. I. Dykman, Phys. Rev. **E 75**, 011101 (2007); V. Peano and M. Thorwart, Chem. Phys. **322**, 135 (2006); V. Peano and M. Thorwart, New J. Phys. **8**, 21 (2006).
26. M. Rigo *et al.*, Phys. Rev. **A 55**, 1665 (1997).
27. G. Lindblad, Commun. Math. Phys. **48**, 119(1976).

Quantum Bio-Informatics III
eds. L. Accardi, W. Freudenberg & M. Ohya

ROLES OF ASYMPTOTIC CONDITION AND S-MATRIX AS MICRO-MACRO DUALITY IN QFT

IZUMI OJIMA

Research Institute for Mathematical Sciences,
Kyoto University, Kyoto 606-8502, Japan

Various versions of "independence" are actively inverstigated in quantum probability. In the context of relativistic QFT, we show here that the physical origin of "independence" can be sought in the asymptotic condition through which asymptotic fields and states exhibiting the independence emerge from the non-independent interacting Heisenberg fields in a kind of "central limit". From the algebraic viewpoint, this condition is equivalent to the on-shell condition to pick up free one-particle modes, which also reduces to Einstein's famous formula $E = mc^2$. A scenario to reconstruct interacting Heisenberg fields as Micro-objects from these "independent"=free Macro-objects intertwined by an S-matrix as a measurable quantity is formulated according to the Micro-Macro Duality associated with a new notion of a *cocycle of K-T operators.*

1. Two questions: What do independence and $E = mc^2$ mean?

In the theory of quantum probability, several versions of *independence* [1] have been formulated with interesting results as generalizations of the bosonic tensor type. My naïve questions here are *on which physical grounds* they have appeared and *what physical meanings* they have. At least for the familiar Gaussian cases (=quasi-free states on bosonic CCR or fermionic CAR), my partial answers in the context of relativistic QFT will be given as follows:

1) ***Emergence of independence*** through the ***asymptotic condition*** $\varphi_H(x) \underset{x^0=t\to\mp\infty}{\to} \phi^{in/out}(x)$:

From *non-independent* interacting Heisenberg fields φ_H, the *asymptotic fields* $\phi^{as} = \phi^{in/out}$ and *asymptotic states* generated by ϕ^{as} from the vacuum *satisfying the* ***independence*** arise through the *asymptotic condition* [2], $\varphi_H(x) \underset{x^0=t\to\mp\infty}{\to} \phi^{in/out}(x)$, which can be interpreted as a sort of "central limit" theorem. The conceptual meaning of this "central limit" theorem can be placed in the "***Micro-Macro Duality***" [3] in QFT as follows:

Micro		Macro
	asymp.cond.	
φ_H : generic	$\rightleftarrows$	ϕ^{as}: universal
	GLZ expansion of φ_H in ϕ^{as}	
$(\Box + m^2)\varphi_H = J_H$		$p^2 = m^2 \Longleftrightarrow$
$\Longleftrightarrow \varphi_H = \Delta_{ret} * J_H + \phi^{in}$		$(\Box + m^2)\phi^{as} = 0$

(NB: There exists another local-net version of "statistical independence" based on the so-called *nuclearity condition* [4] in Algebraic QFT, which should be confused with the present version.)

2) "Units" of *independence* identified with *particles* specified by Einstein's famous formula $E = mc^2$ ($\Longrightarrow$ Sec.2):

It is worth noting that this famous formula $E = mc^2$ is meaningful only for asymptotic fields/ states as the *on-shell condition* $p^2 = p_\mu p^\mu = m^2$ to extract particles as *independent* = *free* = *non-interacting* entities from the interacting Heisenberg fields; in contrast, the latter do not satisfy this formula because of the interactions. The former, asymptotic fields and states, serve as the *vocabulary for describing state changes* taking place in the scattering processes in such a form as the state transitions described by the S-matrix $\langle \beta, out|\alpha, in\rangle = \langle \beta|S|\alpha\rangle$ from an *asymptotic incoming state* (*in-state*, for short) $|\alpha, in\rangle$ to an *outgoing state* (*out-state*, for short) $|\beta, out\rangle$. In the absence of interactions, however, the *on-shell* asymptotic fields ϕ^{as} can***not*** by themselves ignite scattering processes, which necessitates the use of *off-shell interacting Heisenberg fields* φ_H.

3) The logical basis of the *asymptotic condition* is to be found in the *cluster property* [5]:

$$\langle\Omega|A\alpha_{\vec{x}}(B)\Omega\rangle \underset{\vec{x}\to\infty}{\longrightarrow} \langle\Omega|A\Omega\rangle\langle\Omega|B\Omega\rangle,$$

following from the ergodicity valid for a unique vacuum vector Ω invariant under spacetime translations $U(x)$ and from the assumption of the local commutativity ($\Longrightarrow$ Sec.3). In this context, asymptotic fields ϕ^{as} can be understood as quantities to materialize *kinematically* this factorization (= independence) of correlations without taking limit and they can be decomposed into creation and annihilation operators $a(\vec{p}).a^*(\vec{q})$ which represent *infinite number of conserved quantities.*

4) ***Universality*** of "central limit" due to Haag-GLZ expansion ($\Longrightarrow$ Sec.5):

From the asymptotic condition, the Yang-Feldman equation $\varphi_H = \Delta_{ret} * J_H + \phi^{in}$ can be derived with the Heisenberg source current J_H formally defined by $J_H = (\Box + m^2)\varphi_H$ [6]. The analogue of the "Fock expansion" [7] in WNA can be found in the Haag-GLZ expansion [8,9],

$$SA =: (\omega_0 \otimes id)(T(A \otimes 1)\exp(iJ_H \otimes \phi^{in})) :,$$

where $\omega_0 = \langle\Omega|\cdots\Omega\rangle$ is the vacuum state and $S =: (\omega_0 \otimes id)(T(\exp(iJ_H \otimes \phi^{as})) :$ is the *S-matrix*. By this formula, the Heisenberg observables A depending on φ_H can be expressed in terms of the asymptotic fields ϕ^{as}.

2. What does $E = mc^2$ means?

While *Einstein's famous equality* $E = mc^2$ between energy and mass has been regarded as one of the most fundamental consequences of the relativity theory, however, its actual content is simply the "***on-shell condition***" to pick up *1-particle modes*, meaningful only for the *independent = free = non-interacting* asymptotic fields/states. In fact, taking m as "moving mass" $m = \dfrac{m_0}{\sqrt{1-v^2/c^2}}$, we have

$$\begin{aligned}
E = mc^2 &= \frac{m_0}{\sqrt{1-v^2/c^2}}c^2 \\
\Longrightarrow\ (m_0c)^2 &= (\frac{E}{c})^2(1-v^2/c^2) \\
&= (\frac{E}{c})^2 - \left(\frac{m_0}{\sqrt{1-v^2/c^2}}\vec{v}\right)^2 = (\frac{E}{c})^2 - (\vec{p})^2 \\
\Longrightarrow p^2 &= p_\mu p^\mu = (m_0c)^2
\end{aligned}$$

where $\dfrac{m_0}{\sqrt{1-v^2/c^2}}\vec{v} =: \vec{p}$ is the relativistic 3-momentum and $p^\mu = (\dfrac{E}{c}, \vec{p})$ is the 4-mementum. The meaning of the above equality $p^2 = p_\mu p^\mu = (\dfrac{E}{c})^2 - (\vec{p})^2 = (m_0c)^2$ can be understood as follows:

i) It is just the *mass-shell* (or, *on-shell*) *condition* to characterize a mass hyperboloid in the p-space of 4-momenta $p_\mu \in \hat{\mathbb{R}}^4$ carried by the free 1-particle states with a rest mass m_0. The spacetime geometry inherent to the special relativity is controlled by the Poincaré group $\mathcal{P}_+^\uparrow = \mathbb{R}^4 \rtimes L_+^\uparrow$ (or, its universal covering $\widetilde{\mathcal{P}_+^\uparrow} = \mathbb{R}^4 \rtimes SL(2,\mathbb{C})$) defined by the semi-direct product of spacetime translation group $\mathbb{R}^4$ and the orthochronous proper Lorentz group $L_+^\uparrow := \{\Lambda = (\Lambda^\mu_\nu);\ \Lambda x \cdot \Lambda y = x \cdot y, \Lambda^0_0 > 0, \det(\Lambda) = +1\}$

consisting of homogeneous Lorentz transformations $\Lambda = (\Lambda^{\mu}_{\nu}) \in SO(1,3)$ leaving the Minkowski metric $\eta(x,y) := x \cdot y = x^0 y^0 - \vec{x} \cdot \vec{y}$ invariant, $\Lambda^T \eta \Lambda = \eta$, without changing the time direction $\Lambda^0_0 > 0$. In Wigner's construction [6] of irreducible unitary representations of $\mathcal{P}^{\uparrow}_{+}$ or $\widetilde{\mathcal{P}^{\uparrow}_{+}}$, four orbit families, $p^2 \gtreqless 0$ and $p_\mu = 0$, appear: $p^2 = m_0^2 > 0$ corresponds to a massive particle with a rest mass m_0, $p^2 = 0$ to massless particles, $p^2 < 0$ to (unphysical) "tachyons" (with an imaginary mass) and the last one to the vacuum, each of which is induced from one of the corresponding "little groups" ($SU(2), E(2), SU(1,1)$ and $L^{\uparrow}_{+}$).

ii) Through the "first quantization" $p_\mu \to i\hbar\partial_\mu = i\hbar(\frac{1}{c}\frac{\partial}{\partial t}, \vec{\nabla})$, the *Klein-Gordon equation* $[\hbar^2 \partial_\mu \partial^\mu + (m_0 c)^2]\phi(x) = 0$ describes a *free scalar field* $\phi(x)$ with rest mass m_0.

iii) The existence of *positive/ negative energy* solutions $E = \pm\sqrt{(\vec{p}c)^2 + (m_0 c^2)^2}$ of $(\frac{E}{c})^2 - (\vec{p})^2 = m_0^2 c^2$ is related with creation and annihilation operators, particle-antiparticle pairs, time reversal T and PCT *invariance.*

Thus, the famous equivalence $E = mc^2$ between energy E and mass m gives only *partial* information for dynamical descriptions of relativistic quantum fields, with *off-shell* apects being neglected in spite of their vital importance for non-trivial scattering processes, particle decays and productions, etc., etc.

3. Free= *independent* vs. interacting= *non-independent*

A free quantum field $\phi(x)$ as the quantized solution of Klein-Gordon equation $(\Box + m^2)\phi = 0$ describes "particle pictures" in terms of creation and annihilation operators $a(f), a^*(f)$ defined as follows:

$$\phi(x) = \int \frac{d^3p}{\sqrt{(2\pi)^3 2\omega_{\vec{p}}}}(a(\vec{p})\exp(-ip_\mu x^\mu) + h.c.),$$

$$a(f) := i\int \overline{f(x)}\overleftrightarrow{\partial_0}\phi(x)d^3x = \int \overline{\tilde{f}(\vec{p})}a(\vec{p})d^3p,$$

$$a^*(f) := i\int \phi(x)\overleftrightarrow{\partial_0}f(x)d^3x = \int a^*(\vec{p})\tilde{f}(\vec{p})d^3p = [a(f)]^*$$

$$[a(f), a^*(g)] = \int \overline{\tilde{f}(\vec{p})}\tilde{g}(\vec{p})d^3p = \langle \tilde{f}, \tilde{g} \rangle,$$

$$[\phi(x), \phi(y)] = \int \frac{d^4p}{(2\pi)^3}\varepsilon(p^0)\delta(p^2 - m^2)\exp(-ip(x-y)) =: i\Delta(x-y; m^2),$$

with $\omega_{\vec{p}} := \sqrt{\vec{p}^2 + m^2}$ in the "natural unit system" with $\hbar = c = 1$ (rest mass m_0 is denoted by m, henceforth).

Although free quantum fields $\phi(x)$ with $a^*(\vec{p}), a(\vec{p})$ are customarily believed to be sufficient entities for describing ***wave-particle dualism*** inherent in elementary particles, the perpetual creation and annihilation processes of particles require ***interactions*** among elementary particles, which is not consistent with the linearity of free field equations. Concerning this point, the famous theorem of Haag [5,6] has been taken as a kind of "no-go theorem" for the theoretical description of the interactions: *Haag theorem*: Poincaré (or even, Galilei)-covariant quantum fields related to free fields by a unitary transformation are only free fields. Owing to this theorem, it is meaningless to formulate the interacting Heisenberg fields by means of a unitary transformation of free fields as is commonly done in perturbative approaches. This is in sharp contrast to the qunatum systems with *finite* degrees of freedom.

On the other hand, to describe relativistic scattering processes of elementary particles in a satisfactory way, we need inevitably the following ingredients: Poincaré-covariant quantum fields/ their interactions/ free asymptotic fields and states. Here, free fields are indispensable as the vocabulary for the description of scattering processes, where an initial state with incoming free particles is changed into a final one with outgoing particles.

Giving up the idea to connect directly Heisenberg and asymptotic fields (as is forbidden by the Haag theorem), we consider the mutual relations between *two free asymptotic fields*, $\phi^{in}(x)$ and $\phi^{out}(x)$ in terms of the *unitary S-matrix* S to control the *basis change* between the in-state basis $|\alpha, in\rangle$ and the out-state basis $|\beta, out\rangle$:

$$|\alpha, in\rangle = \sum_{\beta} |\beta, out\rangle S_{\beta,\alpha} \quad \text{with } S_{\beta,\alpha} := \langle\beta, out|\alpha, in\rangle = \langle\beta|S|\alpha\rangle,$$

The schematic picture can be summarized as follows:

	in		out
asymptotic fields	$Ad\Theta^{in} \curvearrowright$ $\phi^{in}(x)$	AdS^{-1} & $Ad\Theta$ $\rightleftarrows$ AdS & $Ad\Theta$	$\curvearrowleft Ad\Theta^{out}$ $\phi^{out}(x)$
	GLZ $\searrow\nwarrow$ $t \to -\infty$	asymp.cond. $\uparrow$ $\downarrow$ GLZ formula	$t \to +\infty$ $\nearrow\swarrow$ GLZ
Heisenberg fields		$Ad\Theta \curvearrowright$ $\varphi_H(x)$	

To treat Heisenberg fields $\varphi_H(x)$, we recapitulate briefly the essence of

Wightman axioms for relativistic quantum fields [5,6] (in the vacuum representation $(\mathcal{P}, \mathfrak{H}, U, \Omega)$) in the form of relativistic covariance, local commutativity, cyclicity or ergodicity of vacuum state and spectral condition:

a) [Heisenberg fields] as operator-valued distributions $\mathcal{D}(\mathbb{R}^4) \ni f \longmapsto \varphi^i_H(f)$ with values being (unbounded) closable operators acting on a Hilbert space $\mathfrak{H}$ are defined on the 4-dimensional Minkowski spacetime $(\mathbb{R}^4, \eta)$, where η is the Minkowski metric: $\eta(x,y) := x \cdot y = x^0 y^0 - \vec{x} \cdot \vec{y}$.

b) [Relativistic covariance]: a local net $\mathcal{P} : \mathcal{K} \ni \mathcal{O} \longmapsto \mathcal{P}(\mathcal{O})$ of *-algebras $\mathcal{P}(\mathcal{O})$ generated by local fields $\varphi^i_H(f) = \int \varphi^i_H(x) f(x) d^4x$ with $f \in \mathcal{D}(\mathcal{O})$ and their polynomials defined on the net $\mathcal{K}$ of double cones $\mathcal{O} = \mathcal{O}_{a,b} = (a + V_+) \cap (b - V_+)$ (with the forward lightcone V_+) constitute a non-commutative covariant dynamical system,

$$\begin{aligned}
\alpha_{a,\Lambda}(\varphi^i_H(x)) &= U(a,\Lambda)\varphi^i_H(x)U(a,\Lambda)^{-1} \\
&= s(\Lambda)^i_j \varphi^i_H(\Lambda^{-1}(x-a)), \\
\alpha_{a,\Lambda}(\mathcal{P}(\mathcal{O})) &= \mathcal{P}(\Lambda\mathcal{O} + a),
\end{aligned}$$

under the action α, $\mathcal{P}^\uparrow_+ \ni (a,\Lambda) \longmapsto \alpha_{a,\Lambda} \in Aut(\mathcal{P}(\mathbb{R}^4))$, of Poincaré group $\mathcal{P}^\uparrow_+$ (or its covering $\widetilde{\mathcal{P}^\uparrow_+}$) and $\mathcal{P}^\uparrow_+ \ni (a,\Lambda) \longmapsto U(a,\Lambda) \in \mathcal{U}(\mathfrak{H})$ is its unitary representation on $\mathfrak{H}$, and $s(\Lambda)^i_j$ is a finite-dimensional representation of Lorentz group $L^\uparrow_+$ associated with each field multiplet $(\varphi^i_H(x))_i$.

c) [Local commutaitivity]: the absence of propagation of physical effects exceeding the light velocity due to Einstein causality, implies the local commutativity of Heisenberg fields $\varphi^i_H(f)$:

$$[\varphi^i_H(f_1), \varphi^j_H(f_2)] = 0 \quad \text{if } (suppf_1) \times (suppf_2)$$

where $\mathcal{O}_1 \times \mathcal{O}_2$ means that any pair of points $x \in \mathcal{O}_1$,$y \in \mathcal{O}_2$ are spacelike separated: $(x-y)^2 < 0$.

Remark: By this condition, the Fourier transform of Wightman functions $\omega_0(\varphi^{i_1}_H(x_1) \cdots \varphi^{i_r}_H(x_r))$ as correlation functions of φ^i_H in the vacuum state $\omega_0(\cdot) = \langle \Omega | (\cdot) \Omega \rangle$ defined in the next d) admits an analytic continuation into a holomorphic function in the complex energy-momentum space. According to it, dispersion relations are valid.

d) [Vacuum state and spectrum condition]:

d-i) Energy-momentum spectrum $Sp(U(\mathbb{R}^4))$ of spacetime translations $\mathbb{R}^4$ realized on $\mathfrak{H}$ is within the forward light cone, $Sp(U(\mathbb{R}^4)) \subset \overline{V_+}$ in p-space $\widehat{\mathbb{R}^4}$, and the lowest energy is realized by eigenvalue 0 of the vacuum vector Ω: $U(x) := U(x,1) = \int_{p \in \overline{V_+}} \exp(ipx) dE(p)$; $U(x)\Omega = \Omega$

Remark: Similarly to p-analyticity due to local commutativity, x-space analyticity of Wightman functions $\omega_0(\varphi_H^{i_1}(x_1)\cdots\varphi_H^{i_r}(x_r))$ follows from spectrum condition, which provides powerful tools for structural analysis.

d-ii) Cyclicity $\overline{\mathcal{P}(\mathbb{R}^4)\Omega} = \mathfrak{H}$ of $\Omega \Longleftrightarrow$ irreducibility of $\mathcal{P}(\mathbb{R}^4) \Longleftrightarrow$ uniqueness of vacuum (: $U(x)\Psi = \Psi \Longrightarrow \Psi \propto \Omega$) $\Longleftrightarrow$ cluster property:

$$|\omega_0(A(x)B(y)) - \omega_0(A)\omega_0(B)| \to 0 \text{ as } (\vec{x}-\vec{y})^2 \to \infty,$$

where $A(x) := \alpha_x(A) = U(x)AU(x)^*$, $B(y) := \alpha_y(B)$ are the spacetime translates of local observables $A, B \in \mathcal{P}(\mathcal{O})$ by $x, y \in \mathbb{R}^4$, respectively. This follows from partition of unity due to spectral resolution of spacetime translations $U(x)$:

$$1 = |\Omega\rangle\langle\Omega| + \sum_i (\textit{1-particle singularities on mass-shell } p^2 = m_i^2)$$
$$+(\text{absolutely continuous } p\text{-spectra})$$

4. Independence of asymptotic fields due to on-shell asymptotic condition as "central limit" theorem

From the *cluster property* and the local commutativity, follows the *asymptotic condition* $\varphi_H(x) \underset{x^0=t\to\mp\infty}{\to} \phi^{in/out}(x)$ (as weak convergence), according to which asymptotic fields ϕ^{as} materialize *kinematically* the factorization (= independence) of correlations *without taking limits*: any n-point functions of ϕ^{as} are factorized into the products of two-point functions, $\omega_0(\phi^{as}\phi^{as}\cdots\phi^{as}) = \sum\omega_0(\phi^{as}\phi^{as})\cdots\omega_0(\phi^{as}\phi^{as})$, which is known as the "quasi-freeness" of the vacuum state ω_0 with respect to ϕ^{as}, familiar in the form of "Wick theorem" or the *independence* of Gaussian type. The creation and annihilation operators a_k, a_k^* contained in ϕ^{as} constitute an *infinite number of conserved quantities*, as they are given by the spatial integral of current densities $i\phi^{as}(x)\overleftrightarrow{\partial_\mu}f(x)$ which are conserved, $\partial^\mu[\phi^{as}(x)\overleftrightarrow{\partial_\mu}f(x)] = \phi^{as}(x)\overleftrightarrow{\Box}f(x) = \phi^{as}(x)\overleftrightarrow{(\Box+m^2)}f(x) = 0$ by the on-shell conditions: $(\Box+m^2)\phi^{as}(x) = 0 = (\Box+m^2)f(x)$.

Thus, the *independence* embodied in the asymptotic fields ϕ^{as} is seen to emerge from interacting Heisenberg fields φ_H via the asymptotic condition as a kind of "central limit" theorem. In this context, what corresponds to "Langevin equation" can be found in the *Yang-Feldman equation* [6] to connect the Heisenberg field $\varphi_H(x)$ and the asymptotic field $\phi^{as}(x)$:

$$\varphi_H(x) = \int \Delta_{ret}(x-y;m^2)J_H(y)d^4y + \phi^{in}(x) = [\Delta_{ret} * J_H + \phi^{in}](x)$$
$$= \int \Delta_{adv}(x-y;m^2)J_H(y)d^4y + \phi^{out}(x) = [\Delta_{adv} * J_H + \phi^{out}](x).$$

where $J_H = (\Box + m^2)\varphi_H$ is the Heisenberg source current and $\Delta_{ret/adv}(x-y;m^2)$: retarded/ advanced Green's functions (i.e., principal solutions) of the Klein-Gordon equation defined by

$$(\Box_x + m^2)\Delta_{ret/adv}(x-y;m^2) = \delta(x-y),$$
$$\Delta_{ret/adv}(x-y;m^2) = 0 \quad \text{for } x_0 \lessgtr y_0.$$

In the Yang-Feldman equation, the asymptotic fields $\phi^{in/out}$ and Heisenberg source current J_H appear, respectively, as the *residue* and the *quotient* in the division of φ_H by $\Delta_{ret/adv}$. What is more important is that J_H gives the *residues at the on-shell pole* $\dfrac{1}{p^2-m^2}$ to determine matrix elements of scattering amplitudes, as will be seen in the next formula.

5. Mutual relations between Heisenberg fields and asymptotic fields controlled by Micro-Macro Duality

From the asymptotic condition and the LSZ reduction formulae [2], one can derive the Haag-GLZ formulae [8] to express Heisenberg operators A in terms of the Wick products $:\phi^{as}\cdots\phi^{as}:$ of asymptotic fields:

$$SA =: \exp(\phi^{in}(\Box + m^2)\frac{\delta}{\delta J}) : \omega_0(T(A\exp(i\varphi_H J)) \restriction_{J=0}$$
$$= \sum_{k=0}^{\infty} \frac{i^k}{k!} \int dx_1 \cdots \int dx_k (\Box_{x_1} + m^2) \cdots (\Box_{x_1} + m^2)$$
$$\times \omega_0(T(A\varphi_H(x_1)\cdots\varphi_H(x_k)) : \phi^{in}(x_1)\cdots\phi^{in}(x_k) :,$$
$$S =: \exp(\phi^{as}(\Box + m^2)\frac{\delta}{\delta J}) : \omega_0(T(\exp(i\varphi_H J)) \restriction_{J=0},$$

which is similar to the Fock expansion formula [7] known in WNA. While the formulae of this type have long been known simply as those to expand the Heisenberg operators in the Wick products of ϕ^{as}, what should be emphasized here are the following novel points:

1) We can reformulate these equalities into the following form [9]:

$$\begin{aligned}
SA &= :(\omega_0 \otimes id)(T(A\otimes 1)\exp(iJ_H \otimes \phi^{in})):, \\
A &= S^{-1} :(\omega_0 \otimes id)(T[A\otimes 1]\exp(iJ_H \otimes \phi^{in})): \\
&= :(\omega_0 \otimes id)(T[A\otimes 1]\exp(iJ_H \otimes \phi^{out}): S^{-1}, \\
S &= :(\omega_0 \otimes id)(T\exp(iJ_H \otimes \phi^{in})) :=: (\omega_0 \otimes id)(T\exp(iJ_H \otimes \phi^{out}): .
\end{aligned}$$

The roles played by the system consisting of Heisenberg operators φ_H and those of asymptotic fields ϕ^{in} and ϕ^{out} are clearly separated here in the ***operational context*** in such a way that the former is an *unknown* target system to be detected and analyzed by means of the latter ones functioning as ***probe systems***. The relevant ***coupling terms*** among them are specified by $\exp(iJ_H \otimes \phi^{as})$, with the first tensor factor to be time-ordered and the second one Wick ordered, which are mutually in duality, as will be seen by the relation between $\dfrac{\delta}{i\delta\phi^{in}(x)}$ and $\phi^{in}(x)$ in 3). In this connection, the comparison to the notion of instruments in quantum measurements will be instructive: while a basic scheme for instruments can be seen in

$$\begin{array}{ccccc}
\text{neutral state of} & & & & \text{measured values} \\
\text{probe system} & & & & \text{recorded in probe} \\
\searrow & & \text{coupling between} & & \nearrow \\
 & \rightarrow & & \rightarrow & \\
\nearrow & & \text{system \& probe} & & \searrow \\
\text{initial state} & & \longrightarrow & & \text{final state} \\
\text{of system} & & \text{state changes} & & \text{of system}
\end{array},$$

the corresponding one for scattering processes in QFT is given in sharp contrast by

$$\begin{array}{ccccc}
\text{initial state} & & \text{state changes in probe } \phi^{as} & & \text{final state} \\
|\alpha, in\rangle \text{ of } \phi^{in} & & \longrightarrow & & |\beta, out\rangle \text{ of } \phi^{out} \\
\searrow & & \text{coupling } \exp(iJ_H \otimes \phi^{as}) & & \nearrow \\
 & \rightarrow & & \rightarrow & \\
\nearrow & & \text{between } \varphi_H \text{ \& } \phi^{as} & & \searrow \\
\text{vacuum} & & & & \text{vacuum} \\
|\Omega\rangle \text{ of } \varphi_H & & \text{vacuum } |\Omega\rangle \text{ unchanged} & & |\Omega\rangle \text{ of } \varphi_H
\end{array}.$$

Perhaps, they can be unified by combining the level of the object system in the instrument with that of the asymptotic fields ϕ^{as}, which results in the successive measurement processes where the state changes taking place at the level of asymptotic fields in scattering processes of quantum fields are monitored by measuring such observables as the particle momenta or spins through the instruments.

2) The mathematical and conceptual meanings of the coupling term $\exp(iJ_H \otimes \phi^{in})$: We first note by simple computation that the on-shell condition $(\Box + m^2)\phi^{as} = 0$ for the asymptotic fields implies the following equalities:

$$\begin{aligned} J_H \otimes \phi^{as} &= (\Box + m^2)\varphi_H \otimes \phi^{as} = (\Box + m^2)\varphi_H \otimes \phi^{as} - \varphi_H \otimes (\Box + m^2)\phi^{as} \\ &= \Box\varphi_H \otimes \phi^{as} - \varphi_H \otimes \Box\phi^{as} = \partial^\mu\partial_\mu\varphi_H \otimes \phi^{as} - \varphi_H \otimes \partial^\mu\partial_\mu\phi^{as} \\ &= \partial^\mu[\partial_\mu\varphi_H \otimes \phi^{as} - \varphi_H \otimes \partial_\mu\phi^{as}] = -\partial^\mu[\varphi_H \otimes \overleftrightarrow{\partial_\mu}\, \phi^{as}]. \end{aligned}$$

Combining this with the asymptotic condition we can further rewrite this quantity $iJ_H \otimes \phi^{in}$:

$$\begin{aligned} i\int_{\mathbb{R}^4} d^4x J_H(x) \otimes \phi^{in}(x) &= -i\int_{\mathbb{R}^4} d^4x \partial^\mu[\varphi_H \otimes \overleftrightarrow{\partial_\mu}\, \phi^{in}] \\ = -i\int_{\partial\mathbb{R}^4} & dS^\mu[\varphi_H \otimes \overleftrightarrow{\partial_\mu}\, \phi^{in}] \\ = -i\int_{x^0=+\infty} & d^3x[\varphi_H \otimes \overleftrightarrow{\partial_0}\, \phi^{in}] + i\int_{x^0=-\infty} d^3x[\varphi_H \otimes \overleftrightarrow{\partial_0}\, \phi^{in}] \\ = -i\int_{x^0=+\infty} & d^3x[\phi^{out} \otimes \overleftrightarrow{\partial_\mu}\, \phi^{in}] + i\int_{x^0=-\infty} d^3x[\phi^{in} \otimes \overleftrightarrow{\partial_\mu}\, \phi^{in}] \\ = -(S^{-1} & \otimes 1)iQ(S \otimes 1) + iQ = -(S^{-1} \otimes 1)[iQ, S \otimes 1] \\ = -(S^{-1} & \otimes 1)ad(iQ)(S \otimes 1), \end{aligned}$$

in terms of a *conserved charge* Q defined by

$$iQ := i\int d^3x[\phi^{in} \otimes \overleftrightarrow{\partial_\mu}\, \phi^{in}] = \sum_k [(a_k^{in})^* \otimes a_k^{in} - a_k^{in} \otimes (a_k^{in})^*].$$

It is remarkable that $:\exp(iQ):$ (with the Wick product $:\cdots:$ to be applied to the second tensor factor) is the operator to create a *coherent state* (or, *exponential vector*) $:\exp(iQ)|\Omega\rangle:$ from the vacuum $|\Omega\rangle$ with the Wick-ordered *commutative* parameters a_k^{in}. In the context of WNA, these quantities will correspond to the *U-functionals* [10] constituting a commutative algebra with respect to the Wick product.

3) Applying the above formula $A = S^{-1} : (\omega_0 \otimes id)(T[A \otimes 1]\exp(iJ_H \otimes \phi^{in})):$ to the Heisenberg source current $J_H = A$, we reproduce the expression

$$\begin{aligned} (\Box + m^2)\varphi_H &= J_H = S^{-1} : (\omega_0 \otimes id)(T[J_H \otimes 1]\exp(iJ_H \otimes \phi^{in})) \\ &: = S^{-1}\frac{\delta}{i\delta\phi^{in}(x)}S, \end{aligned}$$

once obtained by Bogoliubov, Medvedev and Polivanov [6]. Through the above relations, the asymptotic condition $\varphi_H \underset{t\to\mp\infty}{\to} \phi^{in/out}$ to be identified with the *on-shell condtion* $\Box + m^2 = 0 = m^2 - p^2$ is seen to play the essential role from the algebraic viewpoint: it extracts the asymptotic fields $\phi^{in/out}$ from the algebra of interacting Heisenberg fields φ_H as the ***fixed points*** *under the Lie subgroup* Γ generated by ϕ^{in} or Q in the infinite-dimensional Heisenberg Lie group generated by ϕ^{in} and $\dfrac{\delta}{i\delta\phi^{in}(x)}$ with ϕ^{in} taken as an element belonging to the above commutative algebra with the Wick product. Under the action of Γ, a_k^{in} and $(a_k^{in})^*$ are infinitely many conserved charges (to characterize the "integrability" of the probe system consisting of ϕ^{as}). Thus the group Γ generated by (the Wick-ordered) a_k^{as} and $(a_k^{as})^*$ characterizes the aspects of the symmetry associated with the macroscopic on-shell situation described by ϕ^{as}. The essential features here show the strong similarity to that of WNA if the functional derivatives $\dfrac{\delta}{i\delta\phi^{in}(x)}$ are put in parallel with the Hida derivatives [10].

4) ***Breakdown of invariance*** under Γ due to the interactions: The symmetry Γ preserved in the probe systems consisting of ϕ^{as} is broken in the total system containing the interacting Heisenberg fields φ_H, essentially due to the presence of $\dfrac{\delta}{i\delta\phi^{in}(x)}$ or the Heisenberg source current $J_H(x) = S^{-1}\dfrac{\delta}{i\delta\phi^{in}(x)}S = (\Box + m^2)\varphi_H(x) \neq 0$, which introduces the effects coming from the ***off-shell*** aspects of the theory. The non-trivial existence of the coupling term $iJ_H \otimes \phi^{in} = -(S^{-1} \otimes 1)[iQ, S \otimes 1]$ is equivalent to the non-triviality $S \neq 1$ of the S-matrix which implies the difference between ϕ^{in} and ϕ^{out} due to $[iQ, S \otimes 1] \neq 0$. It is interesting to note the parallelism of this situation with the mutual relation between the first and second laws in Newtonian mechanics: since the interaction term due to a(n external) force is switched off in the stage of the first law, the system enjoys a symmetry to conserve the momentum (or velocity), similarly to the above constancy of ϕ^{in} or Q. At the stage of the second law, the symmetry inherent to the first law is ***broken*** by introducing the (external) force F, which causes the ***state changes*** described by the *changes* of the momentum p according to the second law, $dp/dt = F$, in the essential use of the vocabulary provided by the first law. The state changes in QFT due to $J_H = (\Box + m^2)\varphi_H \neq 0$ is similarly described by the very breaking term $iJ_H \otimes \phi^{in} = -(S^{-1} \otimes 1)[iQ, S \otimes 1]$ in terms of the non-trivial S-matrix $S \neq 1$. At first sight,

what is conserved, momentum p and ϕ^{as} (contrasted with the "momentum variables" $\frac{\delta}{i\delta\phi^{as}}$), seems to be opposite between the Newtonian and the QFT cases, but actually it is not the case, because a particle picture with conserved $\vec{p}$ in the Newtonian first law is embedded also in ϕ^{as} through the creation and annihilation operators, $a(\vec{p})^*$ and $a(\vec{p})$.

5) ***Reconstruction*** of φ_H from $\phi^{in/out}$ intertwined by S: The essence of the GLZ-Fock expansion of the Heisenberg fields φ_H or A in terms of the asymptotic fields ϕ^{as} can be seen in the "inverse problem" to *reconstruct* the former from its fixed-point subalgebra(s) $\{\varphi_H\}^\Gamma = \{\phi^{in}\}$ or $\{\phi^{out}\}$ through the ***co-action*** of Γ (which is in parallel to the so-called "inverse scattering method" in quantum mechanics to determine a potential term responsible for scattering processes from the scattering data). This kind of *duality* relation between φ_H and ϕ^{as} ensures the *universality* of the asymptotic fields ϕ^{as} in spite of its speciality as statistically independent free objects. What is conceptually more important is such a possibility to re-construct an interacting theory of relativistic quantum fields φ_H from the knowledge of an S-matrix S intertwining the asymptotic in- and out-fields $\phi^{in/out}$: $\phi^{in} \underset{AdS}{\overset{AdS^{-1}}{\rightleftarrows}} \phi^{out}$. The crucial ingredient, $iJ_H \otimes \phi^{in} = -(S^{-1} \otimes 1)[iQ, S \otimes 1]$, as a coupling term in $:(\omega_0 \otimes id)(T((A \otimes 1)\exp(iJ_H \otimes \phi^{in})):= SA$ can be determined, at least in its integrated form, by the knowledge of ϕ^{as} and S, the former of which is easily constructible as free objects. The latter one is the highly non-trivial object to be determined phenomenologically from the experimental data of particle scattering processes to within certain limits of exactitude. Once the functional dependence of S on ϕ^{in} is specified, the formula $J_H(x) = S^{-1}\frac{\delta}{i\delta\phi^{in}(x)}S$ allows a local quantity $J_H(x)$ to be determined. Therefore, the whole scheme to control the mutual relations between φ_H (: Micro) and ϕ^{as} (: Macro) can be understood in the context of "***Micro-Macro duality***" [3] via the Fourier-Galois duality between the fixed-point subalgebra and the recovery of the total algebra as the Galois extension. A *new feature* found here is the problem related with the appearance of *two* fixed point subalgebras, $\{\phi^{in}\}$ and $\{\phi^{out}\}$, which are mutually equivalent by the intertwining actions of the S-matrix, $Ad(S)$ and $Ad(S^{-1})$:

$$S\phi^{out}S^{-1} = \phi^{in}, \qquad S^{-1}\phi^{in}S = \phi^{out}.$$

In this connection, the coupling terms $: T\exp(iJ_H \otimes \phi^{as}) :$ cannot be directly regarded as a Kac-Takesaki operator but should be interpreted as a kind of a ***cocycle*** between two such. This aspects will further be elaborated.

6) PCT invariance & Borchers classes: While the (weak) local commutativity is inevitably violated between φ_H and $\phi^{in/out}$ and between ϕ^{in} and ϕ^{out}, the fields φ_H, ϕ^{in} and ϕ^{out} enjoy the local commutativity within each system. Then the vacuum ω_0 is invariant, $\omega_0 \circ \theta = \omega_0 = \omega_0 \circ \theta^{as}$, under the PCT transformations θ, θ^{as} given by $\theta(\varphi_H(x)) = \gamma\varphi_H(-x)^*$, $\theta^{as}(\phi^{as}(x)) = \gamma\phi^{as}(-x)^*$ (with $\gamma \in \mathbb{T}$), and hence, θ, θ^{as} are implemented, respectively, by anti-unitary PCT operators Θ and Θ^{as} s.t. $\theta(\varphi_H(x)) = \Theta\varphi_H(x)\Theta$, $\theta^{as}(\phi^{as}(x)) = \Theta^{as}\phi^{as}(x)\Theta^{as}$ and $\Theta\Omega = \Theta^{as}\Omega = \Omega$ [5,6]. Then $S\phi^{out}(x)S^{-1} = \phi^{in}(x) = \Theta\gamma^{-1}\phi^{out}(-x)^*\Theta = \Theta\Theta^{out}\phi^{out}(x)\Theta^{out}\Theta$ implies

$$S = \Theta^{in}\Theta = \Theta\Theta^{out}, \quad S\Theta^{out} = \Theta = \Theta^{in}S,$$
$$S\Theta = \Theta^{in} = \Theta\Theta^{out}\Theta = \Theta S,$$

under the assumption of asymptotic completeness. These relations exhibit more detailed structures of S in terms of PCT operators. Thus, quantum fields with *the same PCT operator* Θ have the same S-matrix $S = \Theta^{in}\Theta = \Theta\Theta^{out}$: this explains the "ambiguities" in the choice of Heisenberg fields interpolating asymptotic fields $\phi^{in/out}$ connected by a given S-matrix S in such a form as the Borchers classes [6] characterized by the relative local commutativity to share the same PCT operator Θ. The consideratoins on this aspect will be crucial for discussing the above points of 2) – 5).

7) For physical applicationsin the scheme of the above 1), it would be interesting to utilze the above formula $SA =: (\omega_0 \otimes id)(T(A \otimes 1)\exp(iJ_H \otimes \phi^{in}))$: with a specific choice of A such as the electromagnetic current $A = J_\mu(x)$ to analyze the processes to measure the form factor $\langle\beta, out|J_\mu(x)|\alpha, in\rangle = \langle\beta|SJ_\mu(x)|\alpha\rangle$, or, in more general contexts of "weak values" [11] $\langle\beta, out|A|\alpha, in\rangle = \langle\beta|SA|\alpha\rangle$, as suggested by Hosoya and Shikano.

The author would like to thank Profs. M. Ohya and N. Watanabe for their invitation to QBIC2009. He is very grateful to Profs. T. Hida and Si Si and to Prof. M. Ozawa for navigating him to the important aspects, respectively, of WNA and of the instrument. In this connection, he has been so much encouraged by Mr. T. Shimizu, without whose deep insights the bridge between WNA and QFT has not been properly appreciated. He thanks also Prof. A. Hosoya and Mr. Y. Shikano very much for their encouragements, discussions on his attempts to reformulate the traditional QFT along the above line and their suggestion to examine "weak values". Last but not least, he is very grateful to all the members of the regular seminar on Math. Phys. at RIMS, Prof. S. Tanimura, Messrs. H. Andou, R. Harada, T. Hasebe, K. Okamura and H. Saigo, for their valuable comments and questions.

References

1. Barndorff-Nielsen, O.E., Franz, U., Gohm, R. Kümmerer, B. and Thorbjørnsen, S., Quantum Independent Increment Processes II, Lecture Notes in Math., Vol. 1866, Springer-Verlag, 2006.
2. Lehmann, H., Symanzik, K. and Zimmermann, W., Zur Formulierung quantisierter Feldtheorien, Nuovo Cim. **1**, 425 (1955).
3. Ojima, I., A unified scheme for generalized sectors based on selection criteria –Order parameters of symmetries and of thermal situations and physical meanings of classifying categorical adjunctions–, Open Sys. Info. Dyn. **10**, 235-279 (2003); Micro-macro duality in quantum physics, 143-161, Proc. Intern. Conf. "Stochastic Analysis: Classical and Quantum", World Sci., 2005; Ojima, I. and Takeori, M., How to observe and recover quantum fields from observational data? –Takesaki duality as a Micro-macro duality–, Open Sys. Info. Dyn. **14**, 307–318 (2007); Ojima, I. and Harada, R., A unified scheme of measurement and amplification processes based on Micro-Macro Duality – Stern-Gerlach experiment as a typical example –, Open Sys. Info. Dyn. **16**, 55–74 (2009).
4. Buchholz, D. and Wichmann, E.H.: Causal independence and the energy-level density of states in local quantum field theory, Comm. Math. Phys. **106**, 321–344 (1986).
5. Streater, R.F. and Wightman, A.S., "PCT, Spin and All That", Benjamin, 1964; Jost, R., "The General Theory of Quantized Fields", Amer. Math. Soc. Publ., Providence, 1963.
6. Bogoliubov, N.N., Logunov, A.A. and Todorov, I.T., "Introduction to Axiomatic Quantum Field Theory", Benjamin/ Cummings, 1975.
7. Obata, N., "White Noise Calculus and Fock Space", Lect. Notes in Math. Vol. 1577, Springer–Verlag, 1994.
8. Haag, R., On quantum field theories, Kgl. Danske Videnskab. Selskab. Mat.-fys. Medd., **29**, no.12 (1955); Glaser, V., Lehmann, H. and Zimmermann, W., Field operators and retarded functions, Nuovo Cim., **6**, 1122 (1957); Kugo, T. and Ojima, I., Suppl. Prog. Theor. Phys. no.66 (1979), Appendix.
9. Ojima, I., Notions of generalized observables and instruments in relation to quantum systems with infinite degrees of freedom (in Japanese), Proceedings of the 4th workshop "Quantum-Field Theoretical Approaches to Evolution Dynamics" at Yukawa Institute, Jan. 1989, "Soryushiron Kenkyu", **80** No.4 and in "Bussei Kenkyu", **52** No.5.
10. Hida, T., Kuo, H.-H., Potthoff, J. and Streit, L., "White noise: An Infinite Dimensional Calculus", Kluwer Academic Publishers, 1993; Hasebe, T., Ojima, I. and Saigo, H., No zero divisor for Wick product in $(\mathcal{S})^*$, Inf. Dim. Anal. Quant. Prob. **11**, No. 2, 307 - 311 (2008).
11. Aharonov, Y., Albert, D.Z. and Vaidman, L., Phys. Rev. Lett. **60** (1988) 1351; Aharonov, Y. and Rohrlich, D., "Quantum Paradoxes" Wiley-VCH, Weibheim, 2005; Shikano, Y. and Hosoya, A., Weak values with decoherence, arXiv:0812.4502.

Quantum Bio-Informatics III
eds. L. Accardi, W. Freudenberg & M. Ohya

GAUSSIAN MARKOV TRIPLETS

DÉNES PETZ AND JÓZSEF PITRIK
Department for Mathematical Analysis,
Budapest University of Technology and Economics
H-1521 Budapest XI., Hungary

The Markov property of states of algebras of the canonical commutation relation is studied and in the case of Gaussian states several equivalent properties are obtained. The detailed description is given in terms of a block matrix. The relation to classical multivariate Gaussian Markov triplets is also described. The minimizer of relative entropy with respect to a Gaussian markov state has the Markov property.

Keywords: Weyl unitaries, von Neumann entropy, CCR algebra, Markov triplet, Gaussian state, relative entropy.

1. Introduction

A simple, but important example of a stochastic process with dependence is one in which each random variable depends on the one preceding it and is conditionally independent of all the other preceding random variables. Such a process is said to be Markov. A stochastic process is Markov if and only if the Shannon entropy increase is constant. Multivariate normal distributions are described by a positive definite matrix and if their joint distribution is Gaussian as well then it can be represented by a block matrix. The aim of this note is to study Markov triplets (Markov process with three random variables) by using the block matrix technique. A Markov triplet is characterized by the form of its block covariance matrix and by the form of the inverse of this matrix. We also study some non-commutative analogues, namely the CCR algebraic case. The notion of Gaussian state was developed in the framework of the C*-algebraic approach to the canonical commutation relation (CCR) [17,12,5,19]. The CCR-algebra is generated by the Weyl unitaries (satisfying a commutation relation, therefore Weyl algebra is an alternative terminology). The Gaussian states on CCR algebras can be regarded as analogues of Gaussian distributions in classical probability: The n-point functions can be computed from the 2-point functions and in a kind of central limit theorem the limiting state is Gaussian and it maximizes the von Neumann entropy when the 2-point function is fixed [20]. The Gaussian states are quite tractable, for example the von Neumann entropy has an explicit expression [7,5].

The Markov property was invented by Accardi in the non-commutative (or quantum probabilistic) setting [1,2]. (Another approach is in the paper [9].) This Markov property is based on a completely positive, identity preserving map, so-called quasi-conditional expectation and it was formulated in the tensor product of matrix algebras. A state of a tensor product system is Markovian if and only if the von Neumann entropy increase is constant. This property and a possible definition of the Markov condition was suggested in [21]. A remarkable property of the von Neumann entropy is the strong subadditivity [16,11,18,23] which plays an important role in the investigations of quantum system's correlations. The above mentioned constant increase of the von Neumann entropy is the same as the equality for the strong subadditivity of von Neumann entropy.

A CCR (or Weyl) algebra is parametrized by a Hilbert space, we use the notation $CCR(\mathcal{H})$ when $\mathcal{H}$ is the Hilbert space. Assume that φ_{123} is a state on the composite system $CCR(\mathcal{H}_1)\otimes CCR(\mathcal{H}_2)\otimes CCR(\mathcal{H}_3)$. Denote by $\varphi_{12}, \varphi_{23}$ the restriction to the first two and to the second and third factors, similarly φ_2 is the restriction to the second factor. The Markov property is defined as

$$S(\varphi_{123}) - S(\varphi_{12}) = S(\varphi_{23}) - S(\varphi_2),$$

where S denotes the von Neumann entropy [18]. When φ_{123} is Gaussian, it is given by a positive operator (corresponding to the 2-point function) and the main goal of the present paper is to describe the Markov property in terms of this operator. The paper [25] studies a similar question for the CAR algebra. Although the multivariate Gaussian case (in classical probability) is rather different from the present non-commutative setting, we used the same block matrix formalism. A Gaussian state is described by a block matrix and the Markov property is formulated by the entries. A Markovian Gaussian state induces multivariate Gaussian restrictions, but they are very special in that framework.

2. Classical Markov triplets

Let $\mathcal{X} = \mathcal{X}_1 \times \mathcal{X}_2 \times \mathcal{X}_3$ be a finite set with probability distribution $p(x_1, x_2, x_3)$ $(x_i \in \mathcal{X}_i, 1 \le i \le 3)$. The Markov property is defined by conditional probabilities in the stochastic setting:

$$p(x_3|x_1,x_2) = p(x_3|x_2)$$

which means

$$\frac{p(x_1,x_2,x_3)}{p(x_1,x_2)} = \frac{p(x_2,x_3)}{p(x_2)}, \tag{1}$$

and we say that the random variables X_1,X_2 and X_3 form a Markov chain, denoted by $X_1 \to X_2 \to X_3$. We remark that $X_1 \to X_2 \to X_3$ implies $X_3 \to X_2 \to X_1$. The computation

$$p(x_1,x_3|x_2) = \frac{p(x_1,x_2,x_3)}{p(x_2)} = \frac{p(x_1,x_2)p(x_3|x_2)}{p(x_2)} = p(x_1|x_2)p(x_3|x_2)$$

shows that the Markov property holds if and only if X_1 and X_3 are conditionally independent. If X_3 has the interpretation as "future", X_2 is the "present" and X_1 is the "past", then having the Markov property means that, given the present state, future state is independent of the past state. In other words, the description of the present state fully captures all the information that could influence the future evolution of the process. In order to move to the C*-algebra setting, we denote by $\mathcal{A}_{123}$ the algebra of functions on $\mathcal{X}$. The subalgebra $\mathcal{A}_{12}$ consists of those functions of the variables x_1, x_2, x_3 whose values are actually do not depend on x_3. The subalgebras $\mathcal{A}_{23}$ and $\mathcal{A}_2$ are defined similarly and $\mathcal{A}_2 = \mathcal{A}_{12} \cap \mathcal{A}_{23}$.

The conditional expectation $E_{12}^{123} : \mathcal{A}_{123} \to \mathcal{A}_{12}$ defined as

$$E_{12}^{123}(g)(x_1,x_2) = \int g(x_1,x_2,x_3)\frac{p(x_1,x_2,x_3)}{p(x_1,x_2)}\, dx_3. \tag{2}$$

This leaves fixed any functions $h \in \mathcal{A}_{12}$ and $E_{12}^{123}(hg) = hE_{12}^{123}(g)$ for every $g \in \mathcal{A}_{123}$. Similarly,

$$E_2^{23}(f)(x_2) = \int f(x_2,x_3)\frac{p(x_2,x_3)}{p(x_2)}\, dx_3. \tag{3}$$

If (1) holds, then

$$E_{12}^{123} \text{ restricted to } \mathcal{A}_{23} \text{ is } E_2^{23}. \tag{4}$$

It is natural to investigate the Markov chains from information theoretical point of view. The Shannon entropy $H(X_1, X_2, \dots, X_k)$ of a k-tuple of discrete random variables with values in $\mathcal{X}$ is

$$H(X_1,X_2,\dots,X_k) = -\sum_x p(x)\log p(x),$$

where summation is over $x \in \mathcal{X}^k$. The Shannon entropy has many properties that agree with the intuitive notion of what a measure of information should be, for example it helps us to express the dependence among the random variables. One can check easily that

$$H(X_1, X_2, X_3) - H(X_2, X_3) = H(X_1, X_2) - H(X_2)$$

if and only if $X_1 \to X_2 \to X_3$. Further details can be found in [6,23].

3. Gaussian Markov triplets

To show an important classical example we investigate the multivariate normal distribution forming a Markov chain. All these results with proofs can be found in [3]. Let $X := (X_1, X_2, \ldots, X_n)$ be an n-tuple of real or complex random variables. The (i, j) element of the n x n covariance matrix is given by

$$C_{i,j} := \mathbb{E}(X_i \overline{X_j}) - \mathbb{E}(X_i)\mathbb{E}(\overline{X_j}),$$

where $\mathbb{E}$ denotes the expectation. The covariance matrix is positive semi-definite. The mean $\mathbf{m} := (\mathbf{m}_1, \mathbf{m}_2, \ldots, \mathbf{m}_n)$ consists of the expectations $m_i = \mathbb{E}(X_i)$, $(i = 1, 2, \ldots, n)$. Let M be a positive definite n x n matrix and $\mathbf{m}$ be a vector. Then

$$f_{\mathbf{m},\mathbf{M}}(\mathbf{x}) := \sqrt{\frac{\det M}{(2\pi)^n}} \exp\left(-\frac{1}{2}\langle(\mathbf{x} - \mathbf{m}), M(\mathbf{x} - \mathbf{m})\rangle\right) \tag{5}$$

is a multivariate Gaussian probability distribution, denoted by $N(m, M^{-1})$, with expectation $\mathbf{m}$ and with quadratic matrix M. If $\mathbf{m} = \mathbf{0}$, then we write simply $f_M(\mathbf{x})$. If M is diagonal, then (5) is the product of functions of one-variable which means the independence of the random variables. It is a remarkable fact that the covariance matrix of the distribution (5) is M^{-1}. The following lemma is well known, see for example [3,10].

Lemma 3.1. *Let*

$$M = \begin{bmatrix} A & B \\ B^* & D \end{bmatrix} \tag{6}$$

be a positive definite $(m + k)$ by $(m + k)$ matrix written in block matrix form. Then the marginal of the Gaussian probability distribution

$$f_M(\mathbf{x}_1, \mathbf{x}_2) := \sqrt{\frac{\det M}{(2\pi)^{(m+k)}}} \exp\left(-\frac{1}{2}\langle(\mathbf{x}_1, \mathbf{x}_2), M(\mathbf{x}_1, \mathbf{x}_2)\rangle\right)$$

on $\mathbb{R}^m$ *is the distribution*

$$f_1(\mathbf{x}_1) := \sqrt{\frac{\det M}{(2\pi)^m \det D}} \exp\left(-\frac{1}{2}\langle \mathbf{x}_1, (A - BD^{-1}B^*)\mathbf{x}_1\rangle\right) . \quad (7)$$

The matrix $(M|D) := A - BD^{-1}B^*$ appears often in the matrix analysis, and called the Schur complement of D in M [26].

Now turn to the conditional distributions. Given the random variables $\mathbf{X}_1$ and $\mathbf{X}_2$, the conditional density is given by

$$f(\mathbf{x}_2|\mathbf{x}_1) := \frac{f_{(\mathbf{x}_1,\mathbf{x}_2)}(\mathbf{x}_1, \mathbf{x}_2)}{f_{\mathbf{x}_1}(\mathbf{x}_1)}, \quad (8)$$

which is a function of $\mathbf{x}_2$, since $\mathbf{x}_1$ is fixed. If $(\mathbf{X}_1, \mathbf{X}_2)$ is Gaussian with a quadratic matrix (6), then the conditional distribution (8) is the Gaussian $N(-D^{-1}B^*\mathbf{x}_1, D^{-1})$.

Let $(\mathbf{X}_1, \mathbf{X}_2, \mathbf{X}_3)$ be random variables with joint probability distribution $f(\mathbf{x}_1, \mathbf{x}_2, \mathbf{x}_3)$. The distribution of the appropriate marginals are $f(\mathbf{x}_1, \mathbf{x}_2)$, $f(\mathbf{x}_2, \mathbf{x}_3)$ and $f(\mathbf{x}_2)$. In accordance with the foregoing $(\mathbf{X}_1, \mathbf{X}_2, \mathbf{X}_3)$ is called a Markov triplet if

$$f(\mathbf{x}_3|\mathbf{x}_1, \mathbf{x}_2) = f(\mathbf{x}_3|\mathbf{x}_2). \quad (9)$$

We use the notation $\mathbf{X}_1 \to \mathbf{X}_2 \to \mathbf{X}_3$ for the Markov triplets as before. Let $(\mathbf{X}_1, \mathbf{X}_2, \mathbf{X}_3)$ be a Gaussian random variable with the quadratic matrix

$$M = \begin{bmatrix} A_1 & A_2 & B_1 \\ A_2^* & A_3 & B_2 \\ B_1^* & B_2^* & D \end{bmatrix} \quad (10)$$

and with expectation $\mathbf{m} = 0$. The next theorem gives the characterization of Gaussian Markov triplets [3].

Theorem 3.1. *For the Gaussian triplet* $(\mathbf{X}_1, \mathbf{X}_2, \mathbf{X}_3)$ *with quadratic matrix (10) and with expectation 0, the following conditions are equivalent*

(i) $\mathbf{X}_1 \to \mathbf{X}_2 \to \mathbf{X}_3$
(ii) $B_1 = 0$
(iii) The conditional distribution $f(\mathbf{x}_3|\mathbf{x}_1, \mathbf{x}_2)$ *does not depend on* $\mathbf{x}_1$.
(iv) The covariance matrix of $(\mathbf{X}_1, \mathbf{X}_2, \mathbf{X}_3)$ *is of the form*

$$\begin{bmatrix} S_{11} & S_{12} & S_{12}S_{22}^{-1}S_{23} \\ S_{12}^* & S_{22} & S_{23} \\ S_{23}^*S_{22}^{-1}S_{12}^* & S_{23}^* & S_{33} \end{bmatrix} . \quad (11)$$

(v) $h(\mathbf{X}_1, \mathbf{X}_2, \mathbf{X}_3) - h(\mathbf{X}_2, \mathbf{X}_3) = h(\mathbf{X}_1, \mathbf{X}_2) - h(\mathbf{X}_2)$, *where* $h(\mathbf{X}) := -\int f(\mathbf{x}) \log f(\mathbf{x}) d\mathbf{x}$ *is the Boltzmann-Gibbs entropy.*

4. Markov states

Recall that a pair $(\mathcal{A}, \varphi)$ consisting of an C*-algebra $\mathcal{A}$ and its state φ is called an algebraic probability space. An algebraic random variable is an embedding $j : \mathcal{B} \to \mathcal{A}$ of an algebra $\mathcal{B}$ into $\mathcal{A}$. The state $\varphi \circ j$ of $\mathcal{B}$ is called the distribution of the random variable j. At the characterization of the Markov property in a non-commutative C*-algebra setting, the basic problem is that in the quantum setting conditional probabilities and conditional expectations (preserving a given state) do not exist. Indeed, states compatible with norm one projections tend to be trivial in the extremely noncommutative case, i.e. factorial case, that is algebras with trivial center.

Assume that $\mathcal{A}$ is the tensor product of matrix algebras: $\mathcal{A}_{123} = \mathcal{A}_1 \otimes \mathcal{A}_2 \otimes \mathcal{A}_3$. A state φ of $\mathcal{A}_{123}$ is called Markov if there exists a completely positive unital mapping $F : \mathcal{A}_{123} \to \mathcal{A}_1 \otimes \mathcal{A}_2 \equiv \mathcal{A}_{12}$ such that

$$F(A_1 \otimes I_{23}) = A_1 \otimes I_2$$

and

$$\varphi(A_{123}) = \varphi(F(A_{123})).$$

The first condition tells that $\mathcal{A}_1$ is in the fixed point algebra of F and the second one says that F preserves the given state. (Originally, Accardi and Frigerio called F quasi-conditional expectation and wrote about infinite tensor product [2].) If φ is a faithful, normal state with a density operator D, $(D > 0,\ \mathrm{Tr}D = 1)$, then its von Neumann entropy is given by

$$S(\varphi) \equiv S(D) = -\mathrm{Tr}D \log D.$$

Theorem 4.1. *The following conditions are equivalent, $\varphi \equiv \varphi_{123}$.*

(i) φ is Markov.

(ii) The von Neumann entropy increase is constant, i.e.

$$S(\varphi_{123}) - S(\varphi_{23}) = S(\varphi_{12}) - S(\varphi_2). \tag{12}$$

(iii) There exist a subalgebra $\mathcal{B}$ and a conditional expectation from $\mathcal{A}_{123}$ onto $\mathcal{B}$ such that

$$\mathcal{A}_1 \subset \mathcal{B} \subset \mathcal{A}_{12}$$

and E leaves the state φ invariant.

(iv) There is a state transformation

$$\mathcal{E} : \mathcal{A}_2 \to \mathcal{A}_{23}$$

such that $(Id_1 \otimes \mathcal{E})(\varphi_{12}) = \varphi_{123}$.

For the details, see Chapter 9 of [23]. Moreover, it was shown that a state is Markov if and only if it is the convex combination of orthogonal product type states [11].

Theorem 4.1 inspire us to call a triplet $(\mathcal{A}_1, \mathcal{A}_2, \mathcal{A}_3)$ of subalgebras of a C*-algebra $\mathcal{A}$ to Markovian with respect to the state φ of $\mathcal{A}$ if $\mathcal{A}_3$ and $\mathcal{A}_1$ are conditionally independent with respect to $\mathcal{A}_2$ and (12) holds.

5. CCR algebra and Gaussian states

Let $\mathcal{H}$ be a Hilbert space. Assume that for every $f \in \mathcal{H}$ a unitary operator $W(f)$ is given such that the relations

$$W(f_1)W(f_2) = W(f_1 + f_2)\exp(\mathrm{i}\,\sigma(f_1, f_2)),$$

$$W(-f) = W(f)^*$$

hold for $f_1, f_2, f \in \mathcal{H}$ with $\sigma(f_1, f_2) :=\mathrm{Im}\langle f_1, f_2\rangle$. The C*-algebra generated by these unitaries is unique and denoted by $CCR(\mathcal{H})$ [19,27].

The C*-algebra $CCR(\mathcal{H})$ is not separable, but nuclear [8], therefore its tensor product with any other C*-algebra is uniquely defined [15] . The relations show that $W(f_1)$ and $W(f_2)$ commute if f_1 and f_2 are orthogonal. It follows that $CCR(\mathcal{H}_1) \otimes CCR(\mathcal{H}_2)$ is isomorphic to $CCR(\mathcal{H}_1 \oplus \mathcal{H}_2)$.

The C*-algebra $CCR(\mathcal{H})$ has a very natural state

$$\omega(W(f)) := \exp\left(-\|f\|^2/2\right) \tag{13}$$

which is called Fock state. The GNS-representation of $CCR(\mathcal{H})$ is called Fock representation and it leads to the the Fock space $\mathcal{F}(\mathcal{H})$ with cyclic vector Φ. Since ω is actually a product state, the GNS Hilbert space is a tensor product. We shall identify the abstract unitary $W(f)$ with the representing unitary acting on the Fock space $\mathcal{F}(\mathcal{H})$. The map

$$t \mapsto \Phi(W(tf))$$

is a strongly continuous 1-parameter group of unitaries and according to the Stone theorem we have

$$W(tf) = \exp(\mathrm{i}tB(f)) \quad \text{and} \quad \frac{\partial}{\partial t}\Big|_{t=0} W(tf) = \mathrm{i}B(f)$$

for a self-adjoint operator $B(f)$, called field operator. The distribution of a field operator is Gaussian with respect to the Fock state.

The Fock state (13) can be generalized by choosing a positive operator $A \in B(\mathcal{H})$:

$$\omega_A(W(f)) := \exp\left(-\|f\|^2/2 - \langle f, Af\rangle\right). \tag{14}$$

This is called Gaussian or quasi-free state. By derivation we get

$$\begin{aligned}\omega_A(B(f)B(g)) &= -\mathrm{i}\sigma(f,g) + \frac{1}{2}\left(\langle f,(I+2A)g\rangle + \langle g,(I+2A)f\rangle\right)\\ &= \mathrm{Re}\,\langle f,(I+2A)g\rangle - \mathrm{i}\mathrm{Im}\,\langle f,g\rangle\ ,\end{aligned}$$

and all higher order correlation functions are expressed by this two-point functions [5]. Moreover,

$$\omega_A(B^+(f)B^-(g)) = \langle g, Af\rangle\ . \tag{15}$$

For $0 \le A \in B(\mathcal{H})$, the statistical operator of the Gaussian state ω_A of $CCR(\mathcal{H})$ in the Fock representation is

$$\rho_A := \frac{\Gamma(A(I+A)^{-1})}{\mathrm{Tr}\Gamma(A(I+A)^{-1})},$$

where Γ is the second quantization of operators. If λ_i are the eigenvalues of A, then ρ_A has the eigenvalues

$$\prod_i \frac{1}{1+\lambda_i}\left(\frac{\lambda_i}{1+\lambda_i}\right)^{n_i},$$

where $n_i \in \mathbb{Z}_+$. Therefore the von Neumann entropy is

$$S(\omega_A) = \mathrm{Tr}\kappa(A),$$

where $\kappa(t) = -t\log t + (t+1)\log(t+1)$.

Assume that $\mathcal{H} = \mathcal{H}_1 \oplus \mathcal{H}_2$ and write the positive mapping $A \in B(\mathcal{H})$ in the form of block matrix:

$$A = \begin{bmatrix} A_{11} & A_{12} \\ A_{21} & A_{22} \end{bmatrix}.$$

If $f \in \mathcal{H}_1$, then

$$\omega_A(W(f \oplus 0)) = \exp\left(-\|f\|^2/2 - < f, A_{11}f\right)\ .$$

Therefore the restriction of the Gaussian state ω_A to $CCR(\mathcal{H}_1)$ is the Gaussian state $\omega_{A_{11}}$.

6. Markov property on CCR

Let $\mathcal{H}_{123} = \mathcal{H}_1 \oplus \mathcal{H}_2 \oplus \mathcal{H}_3$ be a finite dimensional Hilbert space and consider the CCR-algebras $\mathcal{A}_i := CCR(\mathcal{H}_i)$. Then $\mathcal{A}_{123} := CCR(\mathcal{H}_{123}) = \mathcal{A}_1 \otimes \mathcal{A}_2 \otimes \mathcal{A}_3$ holds. Assume that D_{123} is a statistical operator in $\mathcal{A}_{123}$ and we denote by D_{12}, D_2, D_{23} its reductions into the subalgebras $\mathcal{A}_{12}, \mathcal{A}_2, \mathcal{A}_{23}$, respectively. As before, these subalgebras form a Markov triplet with respect to the state D_{123} if

$$S(D_{123}) - S(D_{23}) = S(D_{12}) - S(D_2), \tag{16}$$

where S denotes the von Neumann entropy and we assume that both sides are finite in the equation. Actually, the strong additivity of von Neumann entropy (16) is strongly related to statistical sufficiency and we have further equivalent conditions.

(i) $D_{123}^{it} D_{23}^{-it} = D_{12}^{it} D_2^{-it}$ for every real t [13,22]
(ii) $D_{123}^{1/2} D_{23}^{-1/2} = D_{12}^{1/2} D_2^{-1/2}$,
(iii) $\log D_{123} - \log D_{23} = \log D_{12} - \log D_2$,
(iv) There are positive matrices $X, Y \in \mathcal{A}_{12}$ and $0 \leq Z \in \mathcal{A}_{123}$, such that $D_{123} = XZ$, $D_{12} = YZ$ and the commutation relation $ZX = XZ$ and $ZY = YZ$ hold.

Remark that some of the equivalences are valid also in infinite dimensional Hilbert space, for example, the equivalence of (i) and (iv) is obtained in [13]. Mostly we study the Markov property of a Gaussian state $\omega_A \equiv \omega_{123}$ with a density matrix D_{123}, where A is a positive operator acting on $\mathcal{H}_{123}$, given in a block matrix form

$$A = \begin{bmatrix} A_{11} & A_{12} & A_{13} \\ A_{21} & A_{22} & A_{23} \\ A_{31} & A_{32} & A_{33} \end{bmatrix}. \tag{17}$$

Then the restrictions D_{23}, D_{12} and D_2 are also Gaussian states with the positive operators

$$D = \begin{bmatrix} I & 0 & 0 \\ 0 & A_{22} & A_{23} \\ 0 & A_{32} & A_{33} \end{bmatrix}, \quad B = \begin{bmatrix} A_{11} & A_{12} & 0 \\ A_{21} & A_{22} & 0 \\ 0 & 0 & I \end{bmatrix}, \quad \text{and} \quad C = \begin{bmatrix} I & 0 & 0 \\ 0 & A_{22} & 0 \\ 0 & 0 & I \end{bmatrix},$$

respectively. Then the equivalent conditions above implies the following[14]:

Theorem 6.1. *For a Gaussian state ω_A the Markov property (16) is equivalent to the condition*

$$A^{it}(I+A)^{-it} D^{-it}(I+D)^{it} = B^{it}(I+B)^{-it} C^{-it}(I+C)^{it} \tag{18}$$

for every real t.

Moreover, we can give also a description in terms of the operator A.

Theorem 6.2. *A Gaussian state ω_A is Markov if and only if A has the following form*

$$A = \begin{bmatrix} A_{11} & [a\,0] & 0 \\ \begin{bmatrix} a^* \\ 0 \end{bmatrix} & \begin{bmatrix} c\,0 \\ 0\,d \end{bmatrix} & \begin{bmatrix} 0 \\ b \end{bmatrix} \\ 0 & [0\,b^*] & A_{33} \end{bmatrix} = \begin{bmatrix} \begin{bmatrix} A_{11} & a \\ a^* & c \end{bmatrix} & 0 \\ 0 & \begin{bmatrix} d & b \\ b^* & A_{33} \end{bmatrix} \end{bmatrix},$$

where the parameters a, b, c, d (and 0) are matrices.

In other words A should be a block diagonal matrix. There are nontrivial Markovian Gaussian states which are not a product in the time localization. However, the first and the third subalgebras are always independent.

Theorem 6.3. *Let $\omega \equiv \omega_A$ be a Markovian Gaussian state on the CCR-algebra $\mathcal{A}_{123}$. If ψ is the state minimizing the relative entropy $S(\psi||\omega_A)$ under the constraint that $\psi|\mathcal{A}_1 = \varphi$ is fixed, then ψ is a Markov state.*

The proof and a similar result are in the paper [24]. In the probabilistic case the similar statatement is well-known, see [4], for example.

7. Connection to classical Gaussian

Let $\mathbf{X}_1, \mathbf{X}_2, \mathbf{X}_3$ be vector-valued random variables with Gaussian joint probability distribution

$$\sqrt{\frac{\det M}{(2\pi)^n}} \exp\left(-\frac{1}{2}\langle \mathbf{x}, M\mathbf{x}\rangle\right),$$

where M is positive definite matrix. Theorem 3.1 says that the triplet $(\mathbf{X}_1, \mathbf{X}_2, \mathbf{X}_3)$ has the Markov property if and only if the covariance matrix $S = M^{-1}$ of $(\mathbf{X}_1, \mathbf{X}_2, \mathbf{X}_3)$ is of the form (11), that is

$$S_{13} = S_{12}S_{22}^{-1}S_{23}, \tag{19}$$

see [3]. To show some analogy between the classical Gaussian and the CCR Gaussian case, we can formulate a somewhat similar description to (19) in the CCR setting [24].

Theorem 7.1. *The block matrix*

$$A = \begin{bmatrix} A_{11} & A_{12} & A_{13} \\ A_{21} & A_{22} & A_{23} \\ A_{31} & A_{32} & A_{33} \end{bmatrix}$$

gives a Gaussian state with the Markov property if and only if

$$A_{13} = A_{12} f(A_{22}) A_{23}$$

for any continuous function $f : \mathbb{R} \to \mathbb{R}$.

This shows that the CCR condition is much more restrictive than the classical one. Next we assume that $\mathcal{H} = \mathcal{H}_1 \oplus \mathcal{H}_2 \oplus \mathcal{H}_3$ and assume that $\dim \mathcal{H}_i = k$ $(1 \leq i \leq 3)$. Choose an orthonormal basis $\{f_j : 1 \leq j \leq 3k\}$ such that

$$f_{ik+r} \in \mathcal{H}_i, \qquad 1 \leq i \leq 3, \quad 0 \leq r \leq k-1$$

and unit vectors e_j, $1 \leq j \leq 3k$ such that

$$e_{ik+r} \in \mathcal{H}_i, \qquad 1 \leq i \leq 3, \quad 0 \leq r \leq k-1$$

and

$$\langle e_t, e_u \rangle \text{ is real for any } 1 \leq t, u \leq 3k.$$

This implies that the Weyl unitaries $W(te_j) = \exp(itB(e_j))$ commute and the field operators $B(e_j)$ have a joint distribution. If ω_A is Gaussian, then the classical (multi-valued) Gaussian triplet

$$(B(e_1), \ldots, B(e_k)), \quad (B(e_{k+1}), \ldots, B(e_{2k})), \quad (B(e_{2k+1}), \ldots, B(e_{3k})) \tag{20}$$

has Gaussian joint distribution with covariance $S^*(I + 2A)S$, where S is defined as $Sf_j = e_j$, $1 \leq j \leq 3k$. Moreover, Theorem 7.1 and 3.1 implies the following result [14].

Theorem 7.2. *If* ω_A *is a Gaussian Markov state, then the classical (multi-valued) Gaussian triplet (20) is Markovian as well, moreover,*

$$(B(e_1), \ldots, B(e_k)) \quad \text{and} \quad (B(e_{2k+1}), \ldots, B(e_{3k}))$$

are independent.

The converse is not true as some numerical computation shows [14]. However, if for every λA $(\lambda > 0)$ the classical Markov property is true, then the Markovianity of the CCR Gaussian ω_A follows [24].

8. Conclusions

A simple, but important example of a stochastic process with dependence is one in which each random variable depends on the one preceding it and is conditionally independent of all the other preceding random variables. In this paper we attended to the case of three random variables, called Markov triplet. In the classical probability three random variables form a Markov triplet if and only if we have got equality in the strong subadditivity of Shannon entropy for their distributions. This fact yields to define Markov property for a state on a CCR algebra as a state which saturate the strong subadditivity of von Neumann entropy with equality. We investigated two important special cases: the multivariate Gaussian distribution in classical probability and its analogue on CCR algebras, the so-called quasi-free states. Both of them is completely characterized by a positive matrix (quadratic or covariance matrix in the classical case and the so-called symbol in the CCR case) and we characterized the Markov property via these matrices. We found that the CCR case is much more restrictive: a CCR quasi-free Markov state implies classical Gaussian Markov triplets if we consider commuting Weyl unitaries, while the converse is not true in general.

References

1. L. Accardi, On the noncommutative Markov property (in Russian), Funkcional. Anal. i Prilozen. **9**(1975), 1–8.
2. L. Accardi and A. Frigerio, Markovian cocycles, Proc. R. Ir. Acad., **83**(1983), 251–263.
3. T. Ando and D. Petz, Gaussian Markov triplets approached by block matrices, Acta Sci. Math. (Szeged), **75**(2009), 265–281.
4. A. Beghi, On the relative entropy of discrete-time Markov processes with given end-point densities, IEEE Transactions on Information Theory **42**(1996), 1529–1535.
5. O. Bratteli and D.W. Robinson, *Operator algebras and quantum statistical mechanics I, II*, Springer, 1979, 1981.
6. T.M. Cover and J.A. Thomas, *Elements of information theory*, John Wiley & Sons, Inc. 1991.
7. A. van Daele and A. Verbeure, Unitary equivalence of Fock representations on the Weyl algebra, Comm. Math. Phys. **20**(1971), 268–278.
8. D.E. Evans and F.T. Lewis, Dilations of irreversible evolutions in algebraic quantum theory, Dublin Institute for Advanced Studies, 1977.
9. M. Fannes, B. Nachtergaele and R. Werner, Finitely correlated states on quantum spin chains, Comm. Math. Phys. **144** (1992), 443–490.
10. W. Feller,*An introduction to probability theory and its applications*, Vol.2,

John Wiley & Sons, Inc., 1966.

11. P. HAYDEN, R. JOZSA, D. PETZ AND A. WINTER, Structure of states which satisfy strong subadditivity of quantum entropy with equality, Comm. Math. Phys. **246**(2004), 359–374.
12. A.S. HOLEVO, *Probabilistic and statistical aspects of quantum theory*; Noth-Holland, 1982.
13. A. JENČOVÁ AND D. PETZ, Sufficiency in quantum statistical inference, Comm. Math. Phys. **263**(2006), 259–276.
14. A. JENČOVÁ, D. PETZ AND J. PITRIK, Markov triplets on CCR-algebras, to be published in Acta Sci. Math. (Szeged)
15. R.V. KADISON AND J.R. RINGROSE, *Fundamentals of the theory of operator algebras,* Vol. II, Academic Press, 1986.
16. E. H. LIEB AND M. B. RUSKAI, Proof of the strong subadditivity of quantum-mechanical entropy, J. Math. Phys., **14**(1973), 1938–1941.
17. F. MANUCEAU AND A. VERBEURE, Quasi-free states of the C.C.R.-algebra and Bogoliubov transformations, Comm. Math. Phys., **9**(1968), 293-302.
18. M. OHYA AND D. PETZ, *Quantum entropy and its use*, Springer-Verlag, Heidelberg, 1993.
19. D. PETZ, *An invitation to the algebra of the canonical commutation relation*, Leuven University Press, Leuven, 1990.
20. D. PETZ, Entropy, the central limit theorem and the algebra of the canonical commutation relation, Lett. Math. Phys. **24**(1992), 211–220.
21. D. PETZ, Entropy of Markov states, Riv. di Math. Pura ed Appl. **14**(1994), 33–42.
22. D. PETZ, Monotonicity of quantum relative entropy revisited, Rev. Math. Phys. **15**(2003), 79–91.
23. D. PETZ, *Quantum Information Theory and Quantum Statistics*, Springer-Verlag, Heidelberg, 2008.
24. D. PETZ AND J. PITRIK, Gaussian Markov chains on CCR-algebras, to be published
25. J. PITRIK, Markovian quasifree states on canonical anticommutation relation algebras, J. Math. Phys. **48**(2007), 112110.
26. V. V. PRASOLOV, Problems and theorems in linear algebra, American Mathematical Society, 1994.
27. F. Slawny, On factor representations and the C*-algebra of canonical commutation relation", Commun. Math. Phys. **24**(1971), 151–170

Quantum Bio-Informatics III
eds. L. Accardi, W. Freudenberg & M. Ohya

*PRE-M*RNA *INTRONS* AS A MODEL FOR CRYPTOGRAPHIC ALGORITHM: THEORY AND EXPERIMENTS

MASSIMO REGOLI

Centro Vito Volterra,
Università di Roma "Tor Vergata",
Roma I-00133, Italy
E-mail: regoli@uniroma2.it

The RNA-Crypto System (shortly RCS) is a symmetric key algorithm to cipher data. The idea for this new algorithm starts from the observation of nature. In particular from the observation of RNA behavior and some of its properties. In particular the RNA sequences have some sections called Introns. Introns, derived from the term "intragenic regions", are non-coding sections of precursor mRNA (pre–mRNA) or other RNAs, that are removed (spliced out of the RNA) before the mature RNA is formed. Once the introns have been spliced out of a pre–mRNA, the resulting mRNA sequence is ready to be translated into a protein. The corresponding parts of a gene are known as introns as well. The nature and the role of Introns in the pre–mRNA is not clear and it is under ponderous researches by Biologists but, in our case, we will use the presence of Introns in the RNA-Crypto System output as a strong method to add chaotic non coding information and an unnecessary behaviour in the access to the secret key to code the messages. In the RNA-Crypto System algorithm the introns are sections of the ciphered message with non-coding information as well as in the precursor mRNA.

1. Introduction

The RNA-Crypto System (shortly RCS) is a symmetric key algorithm to cipher data.

The idea for this new algorithm starts from the observation of nature. In particular from the observation of RNA behavior and some of its properties.

In particular the RNA sequences have some sections called *Introns*. Introns, derived from the term "intragenic regions", are non-coding sections of precursor mRNA (*pre–m*RNA) or other RNAs, that are removed (spliced out of the RNA) before the mature RNA is formed. Once the introns have been spliced out of a *pre–m*RNA, the resulting mRNA sequence is ready to be translated into a protein. The corresponding parts of a gene are known as *introns* as well.

The nature and the role of Introns in the *pre–m*RNA is not clear and it is under ponderous researches by Biologists but, in our case, we will use the presence of Introns in the RNA-Crypto System output as a strong method to add only apparently chaotic and non coding information with an unnecessary behavior in the access to the secret key to code the messages.

In the RNA-Crypto System algorithm the *introns* are sections of the ciphered message with non-coding information as well as in the precursor mRNA.

In this text a new cryptographic algorithm is described starting from a mathematical point of view.

1.1. *Interpretation*

This model can be used to code clear messages (and the main scope of this is job regard this application), but it can be used as well as to suggest to biologists to consider redundancy in the *pre–m*RNA sequences as a mechanism used by nature as a sort of protection against a possible decoding attack from the outside or, from another point of view, they can give to introns an important role in the mechanism for coding the resulting *mRNA* for example to achieve future functionalities or to archive old ones.

2. Ingredients

To introduce the RCS algorithm let us start to give some important ingredients.

2.1. *Spaces*

We introduce for the first the spaces: $\mathcal{K} := \{\kappa_i | \kappa_i \in K\}$ as the space of all finite sequences in a given space K, of length less or equal to N_K where $\infty > N_K \in \mathbb{N}$, $\mathcal{M} := \{\sigma_i | \sigma_i \in M\}$, as the space of all finite sequences in M of length less or equal to N_M, where $\infty > N_M \in \mathbb{N}$ (could be $M = K$) and $S = S_1 \cup S_2$ with $S_1 \cap S_2 = \emptyset$ and $|S| < \infty$.

Namely $\mathcal{K}$ is the set of *Secret Keys of length* $\leq N_K$, $\mathcal{M}$ is the set of *Clear Messages of length* $\leq N_M$, K and M are finite spaces of *symbols* (just to fix the ideas should be: $K = M = \{0, 1\}$ the standard binary codes). Also S is a finite set of symbols and we can consider it as the *set of all possible actions.*

At least we introduce the following space: $\mathcal{O} = \mathcal{A} \bigcup \mathcal{B}$ where $\mathcal{A} = A^{N_A}$ and $\mathcal{B} = \bigcup_i \mathcal{B}_i$ where $\mathcal{B}_i = A^{N_A} \times B_i^{N_{B_i}}$ where also in this case A and B_i

are finite sets of symbols as above. From a biological point of view the space $\mathcal{A}$ is the *exons* space and $\mathcal{B}$ is the *introns* space.

$\mathcal{O}$ is the base space for *Coded Messages* (a coded messages will be a finite sequences of elements of $\mathcal{O}$).

After this preliminary description of necessary spaces, now we can introduce some useful definitions:

Definition 2.1. Be $\kappa \in \mathcal{K}$. We define for $m \in \mathbb{N}$ and foreach $i \in \{1, ..., N_K\}$,

$$\tilde{\kappa}_{i,m} := \left\{\kappa_i, \kappa_{i\dot{+}1,...,}\kappa_{i\dot{+}(m-1)}\right\}$$

with $0 \leq m \leq N_k$ as a *subsequence* of κ
($m = 0$ represent a subsequence of length 0).

Definition 2.2. Be $\sigma \in \mathcal{M}$. We define for $n \in \mathbb{N}$ and foreach $i \in \{1, ..., N_M\}$,

$$\tilde{\sigma}_{i,n} := \left\{\sigma_i, \sigma_{i\dot{+}1,...,}\sigma_{i\dot{+}(n-1)}\right\}$$

with $0 \leq n \leq N_M$ as a subsequence of σ
($n = 0$ represent a subsequence of length 0).

In the above definitions the operator $\dot{+}$ is the sum with appropriate module and, for simplicity, in the following we will use the form $\tilde{\kappa}_{i,m} \equiv \tilde{\kappa}_i$ and $\tilde{\sigma}_{i,n} \equiv \tilde{\sigma}_i$ moreover holds: $\tilde{\kappa}_i \in \mathcal{K}$ and $\tilde{\sigma}_i \in \mathcal{M}$

Definition 2.3. Be $\mathcal{C} = \{\Sigma_i | \Sigma_i \in \mathcal{O}\}$the space of all finite sequences in $\mathcal{O}$.

$\mathcal{C}$ is the set of *Coded Messages.*

2.2. *Functions*

To reach our goal we must introduce two kinds of functions:

- Coding functions: these functions are used to:
  - transform a portion of the clear message in a portion of the coded message (*exons*)
  - insert some *apparently* redundant information in the coded message (*introns*)
- Operational functions: these functions are used to:
  - modify the local state of the coding system
  - modify the global state of the variables of the system (see below)

For the first we start with the definition of a family of *coding functions* $f_\alpha : \mathcal{M} \times \mathcal{K} \to \mathcal{O}$ with $\alpha \in S$:

$$f_\alpha(\sigma, \kappa) = \Sigma = \begin{cases} \varepsilon & \text{if } \alpha \in S_1 \\ \iota & \text{if } \alpha \in S_2 \end{cases} \tag{1}$$

where $\varepsilon \in \mathcal{A}$ and $\iota \in \mathcal{B}$.

The functions f_α must be chosen in such way to guarantee the existence of a function $\bar{f} : \mathcal{O} \times \mathcal{K} \to \mathcal{M}$ such that

$$\bar{f}(\Sigma, \kappa) = \bar{f}(f_\alpha(\sigma, \kappa), \kappa) = \begin{cases} \sigma & \text{if } \alpha \in S_1 \\ \emptyset & \text{if } \alpha \in S_2 \end{cases} \tag{2}$$

Now we introduce the definition of some operational functions: be $g_\alpha : \mathcal{K} \times \mathcal{O} \to \mathcal{K}$ as follow:

$$g_\alpha(\kappa, \Sigma) = \bar{g}(\kappa, \Sigma) = \kappa^* \tag{3}$$

the existence of a function $\bar{g}$ that does not depend from α ensures the chaoticity of the system.

For the last, we need, as a member of the set of the operational functions, the trivial characteristic function $\chi_{S_1} : S \to \{0, 1\}$

2.3. *Global variables and further definitions*

Now let us introduce some further definitions:

Definition 2.4. Be $\kappa \in \mathcal{K}$ a Pre Shaked Key (PSK) of length N_K.

In cryptography, a pre–shared key or PSK is a shared secret which was previously shared between the two parties using some secure channel before it needs to be used.

Definition 2.5. Be $\{\alpha\} = \{\alpha_i | \alpha_i \in S\}$ a random sequence of arbitrary length.

A random sequence is a kind of stochastic process. In short, a random sequence is a sequence of random variables. In computer science it comes from a random number generator, often abbreviated as RNG, that is a computational device designed to generate a sequence of numbers or symbols that lack any pattern, i.e. **appear random**.

Definition 2.6. Be $\sigma \in \mathcal{M}$ a message of length N_M.

In communications science, a message is information which is sent from a source to a receiver.

Definition 2.7. $\Sigma \in \mathcal{C}$ will be the coded message of length that will depend on the system and the variables state.

In cryptography, a coded message is a clear message transformed into an obscured form, preventing those who do not possess special information, or *key*, required to apply the transform from understanding what is actually transmitted.

2.4. *Biological parallelism*

From the biological point of view the PSK and the random sequence $\{\alpha\}$ represent the whole mechanism of *splicing* (this is the process by which *pre–m*RNA is modified to remove certain stretches of non-coding sequences called exactly *introns*), while the coded message is the *pre–m*RNA itself and the clear message is the final *mRNA*.

3. The algorithm

3.1. *Coding*

Suppose to have $\kappa \in \mathcal{K}$ as a Pre Shared Key (PSK) of length N_K and a message $\sigma \in \mathcal{M}$ of length N_M and that we have fixed two numbers $n, m \in \mathbb{N}$. (For simplicity and without affecting the generality of the system, we can suppose that N_M is a multiple of n).

Using the above definitions for α_i, $\tilde{\sigma}_{l_i}$ and $\tilde{\kappa}_{j_i}$ we can define the following algorithm:

$$\begin{cases} f_{\alpha_i}(\tilde{\sigma}_{l_i}, \tilde{\kappa}_{j_i}) & = \Sigma_i = \begin{cases} \varepsilon_i & \text{if } \alpha_i \in S_1 \\ \iota_i & \text{if } \alpha_i \in S_2 \end{cases} \\ l_{i+1} & = l_i + \chi_{S_1} n \\ \tilde{\kappa}_{j_{i+1}} & = g_{\alpha_i}(\tilde{\kappa}_{j_i}, \Sigma_i) \end{cases} \tag{4}$$

where $\varepsilon_i \in \mathcal{A}$ and $\iota_i \in \mathcal{B}$.

Definition 3.1. The sequence $\Sigma = \{\Sigma_1, ..., \Sigma_N\} \in \mathcal{C}$ is the coded message.

3.2. *Decoding*

For the decoding phase suppose to have $\kappa \in \mathcal{K}$ as a Pre Shared Key (PSK) of length N_K and a coded message $\Sigma \in \mathcal{C}$ of arbitrary length, and suppose that we have fixed two numbers $n, m \in \mathbb{N}$.

Using the above definitions for $\tilde{\sigma}_{l_i}$ and $\tilde{\kappa}_{j_i}$ we can define the following algorithm:

$$\begin{cases} \begin{cases} \bar{f}(\Sigma_i, \tilde{\kappa}_{j_i}) & = \tilde{\sigma}_i \\ \tilde{\kappa}_{j_{i+1}} & = \bar{g}(\tilde{\kappa}_{j_i}, \Sigma_i) \end{cases} & \text{if } \Sigma_i \in \mathcal{A} \\ \\ \begin{cases} \bar{f}(\Sigma_i, \tilde{\kappa}_{j_i}) & = \emptyset \\ \tilde{\kappa}_{j_{i+1}} & = \bar{g}(\tilde{\kappa}_{j_i}, \Sigma_i) \end{cases} & \text{if } \Sigma_i \in \mathcal{B}_* \subset \mathcal{B} \end{cases} \tag{5}$$

Note that, thanks to the definition of $\bar{g}$, in the decoding phase is not necessary to know the random $\{\alpha_i\}$ sequence.

3.3. *Preliminary observations*

The coded message depends on:

- The clear message;
- The secret Key (PSK);
- The random sequence $\{\alpha\}$;

and it important to understand that the algorithm during the coding phase using the same message and the same PSK can supply different coded messages simply using different $\{\alpha\}$.

It is also important to underline the fact that it is no necessary, during the decoding phase, to know the sequence $\{\alpha\}$ used in the coding phase. This because the information about the $\{\alpha\}$ sequence are intrinsic in the couple $(\Sigma_i, \tilde{\kappa}_{j_i})$.

In this circumstance we want to underline the possible role of *introns* to carry important information for the decoding phase, for example the function $\bar{g}(\tilde{\kappa}_j, \Sigma)$ can use the information present in $\Sigma \in \mathcal{B}$ to produces a change in the PSK.

Moreover, during the coding phase there is an expansion of the original messages into the coded message, and this expansion comes from several factors, for example it comes from the dimensions of the $\mathcal{B}_i$ and from the number of elements of S_2 in the $\{\alpha\}$ sequence.

4. The first implementation

4.1. *The environment*

Now let us introduce the first realization of the above algorithm.

Step by step we will substitute each ingredient of the algorithm, with its representative object in the implementation.

Starting from the definition of $K = M = A = B_1 = \{0,1\}$, with $N_K, N_M \in \mathbb{N}$, $N_A = 3$, $N_{B_1} = 2$ and $S = S_1 \cup S_2$ with $S_1 = \{a, b\}$ and $S_2 = \{c, d\}$. $\mathcal{O} = \{0,1\}^{N_A} \times \left(\{0,1\}^{N_A} \times B_1^{N_{B_1}}\right)$. Finally $n = 1$ and $m = 3$.

4.2. *Functions and implementation*

To introduce the functions f_α the table 1 $T : \mathcal{A} \times K^n \to S$ will be useful using the following rules:

1) if $\alpha \in S_1$
   - be $c = \tilde{\kappa}_j$
   - if $\tilde{\sigma}_i = \sigma_i = 0$ then localize a binary number $000 \le r \le 111$ such that $T_{r,c} = a$ (or $T_{r,c} = b$)
   - if $\tilde{\sigma}_i = \sigma_i = 1$ then localize a binary number $000 \le r \le 111$ such that $T_{r,c} = b$ (or $T_{r,c} = a$)
   - now let $\Sigma_i = (\Sigma_{i,1}, \Sigma_{i,2}, \Sigma_{i,3}) = (r_1, r_2, r_3)$ (i.e. an exon $\Sigma_i \in \mathcal{A}$).
2) if $\alpha \in S_2$
   - be $c = \tilde{\kappa}_j$
   - localize a binary number $000 \le r \le 111$ such that $T_{r,c} = \alpha$
   - now let $\Sigma_i = ((\Sigma_{i,1}, \Sigma_{i,2}, \Sigma_{i,3}), (\Sigma_{i,4}, \Sigma_{i,5})) = ((r_1, r_2, r_3), (\Sigma_{i,4}, \Sigma_{i,5}))$, (i.e. an intron $\Sigma_i \in \mathcal{B}$)
   - the other 2 components of the vector (i.e. $\Sigma_{i,4}, \Sigma_{i,5}$) will be given using a random choice in the binary range $\{00, \ldots, 11\}$

		$\tilde{\kappa}_j$							
		000	001	000	011	100	101	110	111
$(\Sigma_{i,1}, \Sigma_{i,2}, \Sigma_{i,3})$	000	a	b	c	d	a	b	c	d
	001	b	c	d	a	b	c	d	a
	010	c	d	a	b	c	d	a	b
	011	d	a	b	c	d	a	b	c
	100	a	b	c	d	a	b	c	d
	101	b	c	d	a	b	c	d	a
	110	c	d	a	b	c	d	a	b
	111	d	a	b	c	d	a	b	c

Now for the $\bar{g}$ function we will use the following rules:

- obtain r, c as above
- if $T_{r,c} \in \{a, b\}$ then $\bar{g}$ will return $\tilde{\kappa}_{j_i+m}$
- if $T_{r,c} = c$ then $\bar{g}$ will return $\tilde{\kappa}_{j_i+(\Sigma_{i,4}\Sigma_{i,5})_{10}}$
- if $T_{r,c} = d$ then $\bar{g}$ will return $\tilde{\kappa}_{j_i-(\Sigma_{i,4}\Sigma_{i,5})_{10}}$

where the subscript 10 means the decimal representation of the binary number in parenthesis.

The last function we must define is $\bar{f} : \mathcal{O} \times \mathcal{K} \to \mathcal{M}$. Remember that conditions in equation (2) must hold.

The function $\bar{f}(\Sigma_i, \tilde{\kappa}_j)$ could be able to understand if $\Sigma_i \in \mathcal{A}$ or $\Sigma_i \in \mathcal{B}$. Starting from the fact that Σ (i.e. the entire coded message) is a sequence of 0 and 1, we can extract the single element in terms of bits (i.e. $\tilde{\Sigma}_{l_i}, \tilde{\Sigma}_{l_i+1}, \ldots$). Following this strategy we can define:

$$\bar{f}(\Sigma_i, \tilde{\kappa}_{j_i}) := \bar{f}((\tilde{\Sigma}_{l_i}, \tilde{\Sigma}_{l_i+1}, \ldots), \tilde{\kappa}_{j_i}) \tag{6}$$

the new function act in this way:

- be $r = (\tilde{\Sigma}_{l_i}, \tilde{\Sigma}_{l_i+1}, \tilde{\Sigma}_{l_i+2})$
- be $c = \tilde{\kappa}_{j_i}$
- if $T_{r,c} = a$ then $\bar{f}$ returns $\sigma_i = \{0\}$ (or $\sigma_i = \{1\}$)
- if $T_{r,c} = b$ then $\bar{f}$ returns $\sigma_i = \{1\}$ (or $\sigma_i = \{0\}$)
- else $\bar{f}$ return $\sigma_i = \emptyset$

The function $\bar{g}$ acts as above but using $(\tilde{\Sigma}_{l_i}, \tilde{\Sigma}_{l_i+1}, \tilde{\Sigma}_{l_i+2})$ if $T_{r,c} = \{a, b\}$ or $(\tilde{\Sigma}_{l_i}, \tilde{\Sigma}_{l_i+1}, \tilde{\Sigma}_{l_i+2}, \tilde{\Sigma}_{l_i+3}, \tilde{\Sigma}_{l_i+4})$ if $T_{r,c} = \{c, d\}$.

4.3. *Observation*

In this simple implementation of the RNA algorithm, redundancy is given by the following considerations:

- for each processed bit in the clear message 3 bits will be inserted in the coded one
- each *intron* will insert in the coded message 5 bits
- we use an uniform distribution for the generation of the sequence $\{\alpha_i\}$
- then we can suppose that the length of the sequence $\{\alpha_i\}$ is $\|\{\alpha_i\}\| = \|\{\alpha_i | \alpha_i \in S_1\}\| + \|\{\alpha_i | \alpha_i \in S_2\}\|$
- of course, in this implementation, must be $\|\{\alpha_i | \alpha_i \in S_1\}\| = N_M$

Then, if the original message has a length of N_M bits, the length of coded one is $l_c \approx 3N_M + 5N_M = 8N_M$.

But it is important to underline that this implementation uses a redundant table $T_{r,c}$, in fact for each c there are two r that satisfy the condition $T_{r,c} = \alpha$.

But of course if we want to reduce the expansion of the message we can replace it with the one in table 2 (in this version must be $N_A = 2$ and $m = 2$).

		$\tilde{\kappa}_j$			
		00	01	10	11
$(\Sigma_{i,1}, \Sigma_{i,2})$	00	a	b	c	d
	01	b	c	d	a
	10	c	d	a	b
	11	d	a	b	c

in this case the message explosion will be $l_c \approx 2N_M + 4N_M = 6N_M$.

[4] Another solution in order to diminish the explosion of the coded message is that to use an asymmetric probability function for the generation of the $\{\alpha_i\}$ for example:

α	p
a	3/8
b	3/8
c	1/8
d	1/8

Using these last two optimizations the approximate length of the coded message will be reduced to $l_c \approx 3N_M$.

5. A biological implementation

Now, as an example, we want to create an implementation of our model nearer the biology. For this let us introduce again the ingredients for the new model:

Starting from the definition of $K = \mathbb{N}$, $M = A = \{a, b, c, d\}$, with $N_K, N_M \in \mathbb{N}$, and $S = S_1 \cup S_2$ with $S_1 = \{c\}$ and $S_2 = \{nc\}$. $\mathcal{O} =$

$\{A^{N_A}\backslash(a,a,a)\} \cup \left(\{a,a,a\} \times \left(\bigcup_i A^{N_{B_i}}\right)\right)$, $N_A = n = 3$, $N_{B_i} \in \mathbb{N}$ and $m = 1$.

The function f will be:

$$\Sigma_{i^*} = f_\alpha(\tilde{\sigma}_i, \tilde{\kappa}_{j_i}) = \begin{cases} \tilde{\sigma}_i & \text{if } \alpha \in S_1 \\ (a,a,a,B) \text{ with } B \in A^{\tilde{\kappa}_{j_i}} & \text{if } \alpha \in S_2 \end{cases} \tag{7}$$

in this case the element $(a,a,a) \in A^3$ is the codon to localize the *beginning of the introns* in the sequence (it will never appear in the clear code) and the function $\bar{f}$ will be:

$$\sigma_{i^*} = \bar{f}(\Sigma_i, \tilde{\kappa}_{j_i}) = \begin{cases} (\Sigma_{i,1}, \Sigma_{i,2}, \Sigma_{i,3}) & \text{if } (\Sigma_{i,1}, \Sigma_{i,2}, \Sigma_{i,3}) \neq (a,a,a) \\ \emptyset & \text{otherwise} \end{cases} \tag{8}$$

In both cases holds $\tilde{\kappa}_{j_{i+1}} = \tilde{\kappa}_{j_i+1}$.

6. Conclusion and future implementation

This paper shows how it is possible to use a possible interpretation of the behavior of the mRNA and the relative mechanism of *splicing* to create a cryptographic redundant model.

This model can also be interpreted as a coding/decoding machine in which the *input* is divided in *instructions* and *data*, *instructions* can be (as showed before) divided in *coding* function and *operational* one, and data divided in *message* and *variables*.

Following this fact we can imagine different instructions (in particular different operational function) that are able to modify, insert redundancy or at least add a real chaoticity to the system inserting random behavior.

For example some operational function could acts for:

- Change the PSK
- Insert a variable number of random data in the output
- Reset the original PSK

and so on.

This algorithm can be used in all those situations where the size of the output is not important, or the size of the input is small enough to be able to afford a significant expansion.

In conclusion our next step is to collect several kind of operational function and to create a new complete coding algorithm.

References

1. Lewin, B.,"Gene VI" *Oxford University Press*, 1997.
2. Menezes, A., van Oorschot, P. and Vanstone, S., "Handbook of Applied Cryptography", CRC Press, 1996
3. Regoli, M., "Bio-Cryptography: A Possible Coding Role for RNA Redundancy", Foundations of Probability and Physics-5 (AIP Conference Proceedings), Accardi EDT, 2008

Quantum Bio-Informatics III
eds. L. Accardi, W. Freudenberg & M. Ohya

DUALITY ARISING FROM MULTIPLE MARKOV GAUSSIAN PROCESSES

SI SI

Faculty of Information Science and Technology
Aichi Prefectural University
Nagoya, Japan

WIN WIN HTAY

Department of Engineering Mathematics
Yangon Technological University
Yangon, Myanmar

2000 AMS Subject Classification 60H40

We first recall the canonical representation theory of Gaussian processes (Hida[2]) which plays an important role in our study. For a given multiple Markov Gaussian process in a restricted sense, a new multiple Gaussian process is obtained using the differential operator in a sense of duality. We introduce a new concept to uniformly multiple Markov Gaussian process.

1. Preliminaries

In this section we first recall the canonical representation theory of Gaussian processes from the view point of innovation theory. The innovation in the Gaussian case is white noise $\dot{B}(t)$. It should be noted that random complex system would be studied in line with the reductionism explained in the monograph [7] with the causality in mind.

Canonical representation

We now briefly recall the general theory of the representation of Gaussian processes.

Definition 1.1. Let X be a Gaussian process. Assume that there exists a Gaussian process $B = \{B(t);\ t \in I\}$ with independent increments such that

$$B(t) = (B_i(t), 1 \leq i \leq N; B_j^l(t), 1 \leq l \leq L_j, 1 \leq j \leq J) \tag{1}$$

with $B(0) = 0, \quad N \leq \infty, L_j \leq \infty, J \leq \infty$, satisfying the following conditions.

(1) Each $B_i(t)$ has independent increments and

$$E(|dB_i(t)|^2) = m_i(dt)$$

defines a continuous measure. In addition, m_{i+1} is absolutely continuous with respective to $m_i : m_i(dt) \gg m_{i+1}(dt)$ for every i.

(2) Each $B_j^l(t)$ is a process of the form

$$B_j^l(t) = \begin{cases} B_j^l, & t \geq t_j \\ 0, & \text{otherwise} \end{cases}$$

where each B_j^l is subject to the standard Gaussian distribution $N(0,1)$.

(3) The Gaussian processes B_i and the B_j^l are independent.

(4) For every t, $X(t)$ has the same distribution as $\widetilde{X}(t)$ given by

$$\widetilde{X}(t) = \sum_{i=1}^{N} \int_0^t F_i(t,u)dB_i(u) + \sum_{t_j \leq t} \sum_{l=1}^{L_j} b_j^l(t) B_j^l(t_j). \tag{2}$$

where the kernel functions $F_i(t,u)$ satisfy the condition

$$\int (F_i(t,u))^2\, m_i(du) < \infty, \qquad i = 1, 2, \cdots N, \ \forall t,$$

and where the function $b_j^l(t)$ vanishes for $t < t_j$ and satisfies

$$\sum_{t_j \leq t} \sum_{l=1}^{L_j} b_j^l(t)^2 < \infty, \text{ for every } t.$$

Then $\{F_i(t,u), B_i(u); b_j^l(t), B_j^l(t)\}$ is called a representation of X.

Definition 1.2. The representation (2) is called a generalized canonical representation, if

$$E[X(t)|\mathcal{B}_s(X)] = \sum_{i=1}^{N} \int^s F_i(t,u)dB(u) + \sum_{t_j \leq s} \sum_{l} b_j^l(t) B_j^l(t_j). \tag{3}$$

holds for all $s \leq t$.

To fix the idea, assume that $N = 1$ and that there is no discrete part of the spectrum. Thus the representation

$$X(t) = \int^t F(t,u)dB(u) \tag{4}$$

is called the canonical representation if

$$E[X(t)|\mathcal{B}_s(X)] = \int^s F(t,u)dB(u),\ s \le t \tag{5}$$

The kernel $F(t,u)$ is called a canonical kernel.

Definition 1.3. The canonical representation (4) is called proper canonical representation if

$$\mathcal{B}_t = \mathcal{B}_t(X), \text{ for every } t \in T, \tag{6}$$

where B_t is the σ-field generated by $\{B(s), s \le t\}$.

There are many important cases which suggested to claim that $B(t)$ in the above expression is a standard Brownian motion, so that $dB(u)$ may be written as $\dot{B}(u)du$, where $\dot{B}(u)$ is a white noise.

Theorem 1.1. $(Canonical\ criterion)$

A representation in the form (4) is a proper canonical representation if and only if, for any fixed $t_0 \in T$ *and* f *in* $L^2(v, t_0)$

$$\int^t F(t,u)\varphi(u)dm(u) = 0 \quad \text{for every } t \le t_0$$

implies

$$\varphi(u) = 0 \text{ a.e } (m) \text{ on } (-\infty, t_0]^T.$$

2. General Multiple Markov Gaussian process

We first recall the simple Markov property of Gaussian process which can be given by the equivalent condition as follows.

A Gaussian process is simple Markov if

$$E[X(t)|\mathcal{B}_s(X)] = \varphi(t,s)X(s), \quad s \le t, \tag{7}$$

where $\varphi(t,s)$ is a real ordinary function of (t,s).

Here we can prove that

$$\varphi(t,s) = \frac{f(t)}{f(s)} \text{ for } t > s.$$

With some regularity assumptions, we can prove that a martingale $U(t)$ has a canonical representation of the form

$$U(t) = \int^t g(u)\dot{B}(u)du,$$

where g is locally square integrable. For a Markov process $X(t)$ we have

$$X(t) = f(t)U(t) = f(t)\int^t g(u)\dot{B}(u)du,$$

which is a canonical representation.

We come to the study of general multiple Markov processes by using the canonical representation.

The canonical representation is obtained by Reproducing Kernel Hilbert Space technique. It is unique and is of the form

$$X(t) = \int^t F(t,u)\dot{B}(u)du. \tag{8}$$

This means that a Gaussian process can be completely characterized by the kernel F and white noise. The former is investigated by non-random functional analysis. The latter is the most basic stochastic process (generalized process, indeed) and its probabilistic properties are well-known.

Hence the combination of the studies of F and referring the probabilistic properties of $\dot{B}$. we can investigate Gaussian process.

There is an interesting case where dependence (probabilistic property) determines analytic structure of the kernel.

We first give the definition of multiple Markov Gaussian process.

Definition 2.1. For any choice of $\{t_i\}$ and $t_0 \le t_1 < ... < t_N < t_{N+1}$,

i) $E[X(t_i)|B_{t_0}(X)], i = 1, 2, ..., N$, are linearly independent and
ii) $E[X(t_i)|B_{t_0}(X)], i = 1, 2, ...N + 1$ are linearly dependent

then, $X(t)$ is said to be N-ple Markov Gaussian process.

This property can be characterized by the expression of $F(t,u)$.

Theorem 2.1. *(Hida 1960) The process $X(t)$ is N-ple Markov if and only if the kernel F, in the expression (8) is a Goursat kernel of order N. Namely,*

$$F(t,u) = \sum_1^N f_i(t)g_i(u) \tag{9}$$

where $det(f_i(t_j)) \neq 0$ for different t_j's and $\{g_i(u), i = 1, 2, \cdots\}$ are linearly independent on $L^2([0,t])$, for any t.

3. Restricted multiple Markov processes

We then come to a more restricted multiple Markov processes. Note that we are discussing multiple Markov property only for Gaussian processes.

Definition 3.1. A Gaussian process $X(t)$ is said to be N-ple Markov in the restricted sense, if $X(t)$ is differentiable in $L^2(\Omega, P)$-norm up to $(N-1)$ times and satisfy

$$L_t X(t) = \dot{B}(t)$$

for an N-th order differential operatorL_t. The N-th derivative is taken in the topology of Hida distribution.

Let $D = \frac{d}{dt}$ and L_t be an N-th order ordinary differential operator such that

$$L_t = \sum_{k=0}^{N} a_k(t) D^{N-k},$$

where a_k's are sufficiently smooth.

Assume that a Gaussian process $X(t)$ satisfies

$$L_t X(t) = \dot{B}(t), \tag{10}$$

and $X(0) = 0$.

Note that the equation has meaning in the space $H_1^{(1)}$, where $\dot{B}(t)$ lives in.

Now we have

Theorem 3.1. *There is a solution to the equation (10) expressed in the form*

$$X(t) = \int_0^t R(t,u)\dot{B}(u)du, \tag{11}$$

where $R(t,u)$ is the Riemann function for an ordinary differential operator L_t. The representation of $X(t)$ is canonical.

Obviously the solution $X(t)$ satisfies the N-ple Markov property, so that it is said to be an N-ple Markov Gaussian process in the restricted sense.

We can take real valued functions $v_i(t), i = 0, 1, 2, \cdots, N$, such that (see Hida[2])

$$v_i(t) \in \mathbf{C^N},\ i = 0, 1, 2, \cdots, N$$
$$v_i(t) \neq 0,\ t > 0,\ i = 0, 1, 2, \cdots, N$$

and have

$$f_1(t) = v_N(t)$$
$$f_i(t) = v_N(t) \int_0^t v_{N-1}(t_1)dt_1 \int_0^t v_{N-2}(t_2)dt_2 \int_0^{t_2}$$
$$\cdots \int_0^t v_{N-i+1}(t_{i-1})dt_{i-1}, i = 1, 2, \cdots, N.$$

Also we have

$$g_i(u) = (-1)^{N-i} v_0(u) \int_0^u v_1(u_1)du_1 \int_0^{u_1} v_2(u_2)du_2 \int_0^{u_2}$$
$$\cdots \int_0^{u_{N-i-1}} v_{N-i}(u_{N-i})du_{N-1}, i = 1, 2, \cdots, N-1$$
$$g_N(u) = v_0(u).$$

The operator L_t itself can be expressed in the form

$$L_t = \frac{1}{v_0(t)} D \frac{1}{v_1(t)} D \cdots \frac{1}{v_{N-1}(t)} D \frac{1}{v_N(t)}. \tag{12}$$

We continue to have observation from somewhat different viewpoint. The canonical kernel of a general N-ple Markov Gaussian process is a Goursat kernel. The f_i's and the g_i's have had different analytic properties. If, in particular, the process is N-ple Markov in the restricted sense, g_i's are not only linearly independent but have the same property as f_i's (see the expression in terms of the functions v_i's and exchange v_i with v_{N-i}). So, f_i's and g_i's have the same analytic properties.

Note 1. From now on, we will use the notation $(X(t), L_t, R)$ for a triple of a Gaussian Markov process (in a restricted sense) $X(t)$, the corresponding differential operator L_t and the Riemann function $R(t, u)$.

Note 2. Since L_t is a local operator, we may restrict t to be in a finite interval. However, in the stationary case we take the wholeR^1 through which t runs.

Theorem 3.2. *If a triple $(X(t), L_t, R), t \in [0, 1]$, is given, then N-ple Markov process $X^*(t)$ can be constructed such that $(X^*(t), L_t^*, R^*), t \in [0, 1]$ is a dual system to $(X(t), L_t, R)$ in the sense that*

i) L_t^* is the formal adjoint of L_t,
ii) R^* is the Riemann function of L_t^*,

iii) $X^*(t) = \int_t^1 R^*(t,u)\dot{B}(u)du$ *is the solution of*

$$L_t^* X^*(t) = \dot{B}(t),$$

where $\dot{B}(t)$ *is the same* $\dot{B}$ *given in (11) up to sign.*

Further, $(X^*)^*(t) = X(t)$ *holds.*

Proof. i) The formal adjoint L_t^* of L_t, which is given by (12), is of the form (ignoring $(-1)^n$)

$$L_t^* = \frac{1}{v_N(t)} D \frac{1}{v_{N-1}(t)} D \cdots \frac{1}{v_1(t)} D \frac{1}{v_0(t)}. \tag{13}$$

Namely, for $f, g \in S$, (S : Schwartz space)

$$(L_t f, g) = (f, L_t^* g), \ (\cdot,\cdot) : \text{inner product} \tag{14}$$

We see that

$$L_t f_i(t) = 0, \ L_u^* g_j(u) = 0 \tag{15}$$

ii) There exists a Riemann function of L_t^*, let it be denoted by $R^*(t,u)$. i.e. $L_t^* R^*(t,u) = 0$.

iii) Construct $X^*(t)$ as

$$X^*(t) = \int_t^1 R^*(t,u)\dot{B}(u)du.$$

Then we have

$$L_t^* X^*(t) = \dot{B}(t).$$

Here $\dot{B}(t)$ is the same as given $\dot{B}$ in (11) up to sign.

From the construction of $X^*(t)$, it is easy to see that

$$(X^*)^*(t) = X(t),$$

since $(L_t^*)^* = L_t$.

We can easily prove the following proposition with the same idea in the stationary case as above. The time interval is taken to be $(-\infty, \infty)$. (Ref. See Win Win Htay [19].)

Proposition 3.1. *If $X(t), t \in (-\infty, \infty)$ is a stationary N-ple Markov in a restricted sense such that*

$$X(t) = \int_{-\infty}^{t} R(t-u)\dot{B}(u)du. \tag{16}$$

then L_t is an ordinary operator of order N with constant coefficients and the dual process $X^*(t)$ is defined as

$$X^*(t) = \int_{t}^{\infty} R^*(t-u)\dot{B}(u)du. \tag{17}$$

where $R^*(u) = R(-u)$.

4. Uniformly Multiple Markov process

In this section we will introduce a uniformly multiple Markov process which satisfies somewhat stronger assumption than the above multiple Markov processes. The time parameter space is taken to be a sub interval of $[0, \infty)$.

Assume that $X(t), t \in [0, \infty)$ has the canonical representation of the form

$$X(t) = \int_{0}^{t} F(t,u)\dot{B}(u)du.$$

Take an interval $[a, b] \subset [0, \infty)$, and define the restriction $X(t; a, b)$ of $X(t)$ by

$$X(t; a, b) = X(t) - E\left(X(t)|\mathbf{B}_a(X)\right),\ a \leq t \leq b. \tag{18}$$

Definition 4.1. If a Gaussian process $X(t; a, b), t \in [a, b]$ is N-ple Markov for any $a, b \in R$ such that $0 < a < b < \infty$, then $X(t)$ is called a uniform N-ple Markov Gaussian process.

Assume that $X(t)$ has canonical representation

$$X(t) = \int_{0}^{t} F(t,u)\dot{B}(u)du,\ t \geq 0, \tag{19}$$

where $F(t, u)$ is a Goursat kernel. That is

$$F(t,u) = \sum_{i=1}^{N} f_i(t)g_i(u). \tag{20}$$

We see that

$$X(t; a, b) = \int_{a}^{t} \left(\sum_{i=1}^{N} f_i(t)g_i(u)\right) \dot{B}(u)du,\ t \geq 0, \tag{21}$$

is the canonical representation. It is important to note that for any choice of [a,b], we can choose f_i, g_i to be those in (20).

Proposition 4.1. *The uniform Markov property asserts that g_i's are linearly independent in any $L^2([a,b])$.*

5. Concluding remark

The functions f_i and g_i have similar analytic properties, but to come to claim the same property we have to generalize the Gaussian process. This will be discussed in the forth coming paper.

Acknowledgement

The authors would like to express their gratitude to the organizers of the International Conference QBIC09 and also to Professors L. Accardi and T. Hida for their valuable advices. This research was supported in part by Grant-in-Aid for Scientific Research (C)(2) 20540128 and Hibi Foundation. The authors are grateful for their supports.

References

1. L. Accardi et al eds, Selected papers of Takeyuki Hida. World Sci. Pub. Co. 2001.
2. T. Hida, Canonical representations of Gaussian processes and their applications. Mem. Coll. Sci. Univc. Kyoto, 34 (1960), 109-155.
3. T. Hida and N. Ikeda, Analysis on Hilbert space with reproducing kernel arising from multiple Wiener integral. Proc, 5th Berkeley Symp. on Math. Statistics and Probability. vol. 2, (1967) 117-143.
4. T. Hida, Analysis of Brownian functionals. Carleton Math. Lecture Notes no. 13, Carleton University, 1975.
5. T. Hida, Theory of Probability. Foundation and Developments. Kyouritsu Pub. Co. in Japanese 2008.
6. T. Hida and Si Si, An innovation approach to random fields. Application of white noise theory. World Scientific Pub. Co. 2004.
7. T. Hida and Si Si, Lectures on white noise functionals. World. Sci. Pub. Co. 2008.
8. T. Hida, Si Si and T. Shimizu, The $\dot{B}(t)$'s as idealized elemental random variables. Volterra Center Notes N.614. 2008.
9. E. L. Ince, Ordinary differential equations, New Dover edition. 1956. §5-2. The Wronskian. pp 116-119.
10. K. Karhunen, Ueber die Struktur statinaer zufaelliger Funktionen. Ark. Mat. 1 (1950), 141-160.
11. P. Lévy, Processus stochastiques et mouvement brownien. Gauthier-Villars. 1948. 2ème ed. with supplement 1965.
12. P. Lévy, Problèmes concrets d'analyse fonctionnelle. Gauthier-Villars. 1951.

13. J. Mikusiński, On the square of the Dirac delta-distribution. Bulletin de l'Academie Polonaise des Sciences. Ser. math, astro et Phys. 14 (1966), 511-513.
14. Si Si and T. Hida, Some aspects of quadratic generalized white noise functionals. Proc. QBIC08 held at Tokyo Univ. of Science. 2008, to appear.
15. Si Si, Effective determination of Poisson noise. IDAQP 6 (2003), 609-617.
16. Si Si, An aspect of quadratic Hida distributions in the realization of a duality between Gaussian and Poisson noises. IDAQP 11 (2008) 109-118.
17. Si Si, Introduction to Hida distributions. World Sci. Pub. Co. 2009. tp appear
18. Si Si, L. Accardi and Win Win Htay, T-transform of Hida distribution and factorization, Volterra center publication, N.625, 2008, 1-14.
19. Win Win Htay, Multiple Markov Property and Entropy of a Gaussian Process (preprint).

Quantum Bio-Informatics III
eds. L. Accardi, W. Freudenberg & M. Ohya

NOVEL COMPUTATIONAL APPROACHES TO DRUG DISCOVERY

JEFFREY SKOLNICK†

Center for the Study of Systems Biology, Georgia Institute of Technology, 250 14th Street NW, Atlanta, GA, 30076 USA

MICHAL BRYLINSKI

Center for the Study of Systems Biology, Georgia Institute of Technology, 250 14th Street NW, Atlanta, GA, 30076 USA

New approaches to protein functional inference based on protein structure and evolution are described. First, FINDSITE, a threading based approach to protein function prediction, is summarized. Then, the results of large scale benchmarking of ligand binding site prediction, ligand screening, including applications to HIV protease, and GO molecular functional inference are presented. A key advantage of FINDSITE is its ability to use low resolution, predicted structures as well as high resolution experimental structures. Then, an extension of FINDSITE to ligand screening in GPCRs using predicted GPCR structures, FINDSITE/QDOCKX, is presented. This is a particularly difficult case as there are few experimentally solved GPCR structures. Thus, we first train on a subset of known binding ligands for a set of GPCRs; this is then followed by benchmarking against a large ligand library. For the virtual ligand screening of a number of Dopamine receptors, encouraging results are seen, with significant enrichment in identified ligands over those found in the training set. Thus, FINDSITE and its extensions represent a powerful approach to the successful prediction of a variety of molecular functions.

1. Introduction

One of the key goals of Systems Biology is to understand the function of all molecules in a cell and how they interact on a system-wide level [1]. In that respect, we specifically focus on computational tools designed to elucidate biochemical function. By detecting evolutionary relationships between proteins, sequence-based methods can provide insights into the function of about 50% of the ORFs in a given proteome [2], with the remainder being too evolutionarily distant to accurately infer function based on sequence information alone [3]. Thus, the prediction of the function of these unannotated ORFs is a significant challenge. However, since protein structure is more conserved than protein sequence [4], it can play an essential role in annotating genomes [5], including lead compound identification for subsequent use in drug discovery [6]. Of course, the key question is whether one can use low-to-moderate resolution predicted structures or if high-resolution experimental structures are required [7, 8]. This issue also has implications for the requisite scope of structural genomics that aims for high-throughput protein structure determination [9]. If low-to-moderate resolution models were to prove useful for functional inference, then the value of contemporary protein structure prediction approaches would be significantly enhanced [10].

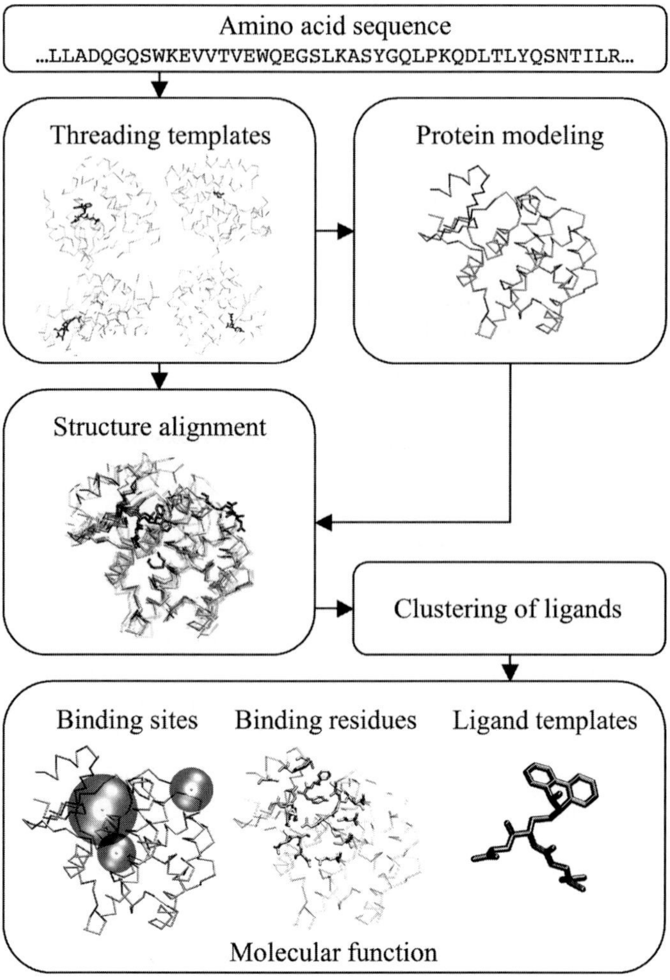

Figure 1. Overview of the FINDSITE threading-based protein functional inference algorithm.

2. FINDSITE: A threading based approach to protein functional inference

A systematic analysis of known protein structures grouped according to SCOP [11] reveals a general tendency of certain protein folds to bind substrates at a similar structural location, suggesting that evolution tends to conserve the functionally important region and a subset of ligand binding features [12]. If so, it should be possible to develop an approach for ligand binding site identification that is less sensitive than pocket-detection methods to distortions in the modeled structure. Thus, we developed FINDSITE [7], a method for the prediction of ligand-binding sites and protein functional annotation based on binding site similarity among superimposed groups of template structures identified from threading [13]. A schematic overview of the methodology is shown in Figure 1. For a given target protein, PROSPECTOR_3.5 [14] identifies ligand-bound structure templates. Then, holo-templates are superimposed onto the predicted (or experimental, if available) target protein structure using the structural alignment algorithm TM-align [15]. Upon superimposition, the clustered centers of mass of the ligands bound to the threading templates identify putative binding sites, and the predicted sites are ranked according to the number of templates that share a common binding pocket. FINDSITE also specifies the chemical properties of the ligands that likely occupy the binding site. To assess its validity, we employed a representative set of 901 proteins with < 35% sequence identity to their templates and generated models using the structure prediction algorithm TASSER [16].

2.1. *Ligand binding site prediction*

We evaluated the performance of both the LIGSITE^{CSC} [17] pocket-detection and FINDSITE threading-based approaches on a non-redundant benchmark set of 901 proteins in terms of the accuracy of ligand-binding site prediction and the ability to correctly rank identified pockets in both crystal structures and protein models. The results of ligand binding site prediction are shown in Figure 2. In Figure 2A, we employ the target protein's crystal structure. For LIGSITE^{CSC}, the native structure is scanned to identify binding pockets. For FINDSITE, predicted template models (with sequence identity <35% to the target) are superimposed onto their crystal structure and the most populated binding site is selected. FINDSITE performs better than the pocket-detection method in both overall accuracy and ranking ability of identified pockets. When the native crystal structure is used, using the best of top five identified binding pockets, the success rate is 70.9% and 51.3% for FINDSITE and LIGSITE^{CSC}, respectively.

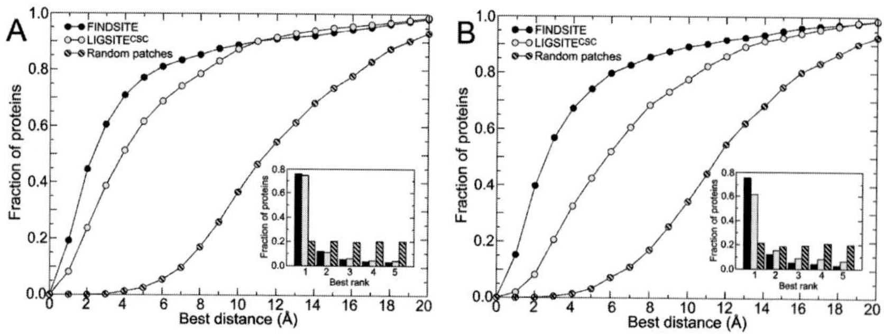

Figure 2. Performance of FINDSITE and LIGSITE^{CSC} compared to randomly selected patches on a target protein surface using target crystal structures (A) and TASSER models (B). Results are presented as the cumulative fraction of proteins with a distance between the center of mass of a ligand in the native complex and the center of the best of top five predicted binding sites ≤ the distance displayed on the x-axis and the rank of the best pocket selected from the top five predictions (inset). The results are much worse when random patches are chosen.

As shown in Figure 2B, using LIGSITE^{CSC}, the prediction accuracy falls off considerably when modeled protein structures rather than the experimental structures are used. LIGSITE^{CSC}'s success rate decreases from 51.3%, when crystal structures are used to 27.2% and 32.5% for PROSPECTOR_3 template structures and protein models generated by TASSER, respectively. This decrease is accompanied by deterioration in the ability to correctly rank the binding site. For the top-ranked threading templates and TASSER models, only 59.1% and 61.4% of the best pockets are assigned rank 1 by LIGSITE^{CSC}. In contrast, with FINDSITE both the high accuracy of ligand binding site prediction and the ability to correctly rank the identified binding sites are sustained if models instead of native structures are used as reference structures for holo-template superposition. The success rate is 67.0% and 67.3% for the top-ranked PROSPECTOR_3 templates and TASSER models, respectively, with a corresponding top ranking accuracy of 75.9%, 75.5% and 75.7%. In practice, FINDSITE tolerates structures with a global RMSD from native of 8-10 Å, provided that the local binding site has a backbone RMSD from native in the range of 2-3 Å.

2.2. *Ligand screening*

FINDSITE also provides information on the chemical properties of the binding ligands, termed here "template ligands". Subsequently, these molecules are used as ligand templates by fingerprint-based similarity searching [18] in a simple ligand-based virtual screening experiment against the KEGG compound library that contains 12,478 compounds [19]. We note that for a given target protein, template ligands can be selected even when the crystal structure is unavailable and its molecular function is unknown. Figure 3 presents the cumulative distribution of enrichment factors calculated for the 901 representative target proteins that have <35% sequence identity to the closest template protein. For accurately predicted binding sites (whose center of mass is ≤ 4 Å from the experimental one; this holds for 70.9% of the target proteins), in 78% of the cases, FINDSITE performs better than random. The ideal enrichment factor (all native-like compounds in the top 1% of the ranked library [7]) was observed for 50% of target proteins. For less accurately predicted binding pockets, ligand selection is notably worse (the ideal enrichment factor was obtained for 12% of the cases and is better than random for 34%). Finally, a case study examined the performance of FINDSITE in virtual screening for 895 active *HIV-1 protease inhibitors* in a 123,331 compound library. Again, if only templates with <35% sequence identity to the target are used, the enrichment factor of the top 1% of compounds is 40.

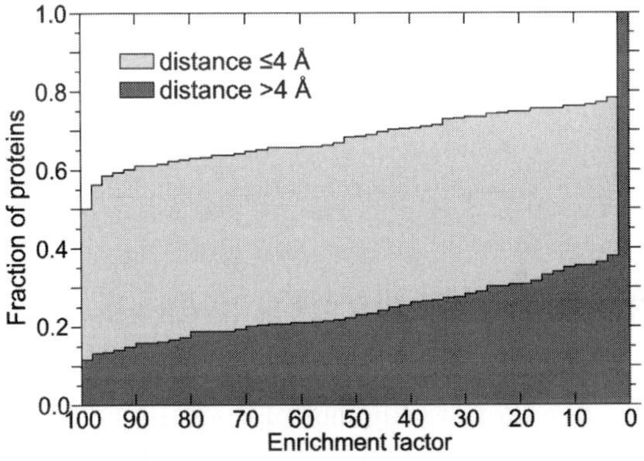

Figure 3. Cumulative distribution of enrichment factors resulting from the ligand-based virtual screening against the KEGG compound library using ligand templates selected by FINDSITE. Target proteins are divided into the two subsets with respect to binding pocket prediction accuracy (the distance between the top-ranked pocket and the center of mass of the native ligand ≤4 Å and >4 Å).

2.3. *Molecular Function Inference*

The relatively high accuracy of the ligand selection procedure encouraged us to investigate the transferability of specific functions of the threading templates to the target. We use Gene Ontology [20], GO, to describe molecular protein function. From the benchmark set, we selected 753 proteins for which a GO annotation is provided by Gene Ontology [20] or UniProt [21]. The procedure for molecular function prediction employs the superimposed group of holo-templates selected by threading as previously used for binding site detection and ligand selection. For benchmarking purposes, only predicted threading templates with <35% sequence identity to a target protein are used. For each target protein, all GO annotations are identified for the threading templates that share the top-ranked predicted binding site using the GO and UniProt databases. Then, the target protein is assigned a function with a probability that corresponds to the fraction of threading templates annotated with that molecular function. For a probability threshold of 0.5 (i.e. at least one half of the threading holo-templates must be annotated with the same GO term to transfer it to the target protein), the maximal Matthew's correlation coefficient of 0.64 is found. This corresponds to a precision (True Positives / (True Positives + False Positives)) of 0.76 with a recall (True Positives / (True Positives + False Negatives)) of 0.54. In addition, we calculated predictive metrics with respect to individual GO identifiers. When the closest template has <35% sequence identity, FINDSITE distinguishes between enzymatic and non-enzymatic function, with a precision and sensitivity of 0.93 and 0.89, respectively. Moreover, many molecular functions that cover a broad spectrum of molecular events including both enzymatic and binding activities are accurately transferable from the templates selected by threading to the target proteins. See SI, Table 1 of ref [7] for an assessment of the best predictions, with the full set of predicted functions found in [22]. Thus, FINDSITE is an encouraging approach for the prediction of protein function, ligand binding sites and ligand screening.

3. Application of FINDSITE/QDOCKX to GPCR ligand screening

For a given target protein, a binding pocket-specific potential can be derived by FINDSITE from a set of weakly homologous ligand-bound templates to improve the ligand docking accuracy [7, 23]. For most globular proteins, the requisite evolutionary related template structures can be identified by a threading approach that employs a strong sequence profile component [24]. For GPCRs, the limited number of experimentally solved structures requires construction of a synthetic protein structure/ligand library compiled from GPCR models complexed with small molecules as predicted by ligand docking. For selected GPCRs, known binders extracted from ligand databases (Drug Bank [25] and the MDL Drug Data Report [26]) are used to model the receptor-ligand complexes by low-resolution ligand docking using the Q-Dock ligand docking/screening algorithm [27]. The first pass of the docking simulations employs a generic force-field for protein-ligand interactions derived from the non-redundant dataset of ligand-bound proteins from the Protein Data Bank, PDB [28]. To improve binding mode prediction, the generic contact potential is amplified at highly conserved amino acid positions (taken from GPCR class-specific multiple sequence alignments provided by GPCRDB (information system for G-protein-coupled receptors) [29]) to enforce the experimentally known interactions with ligands. Subsequently, the known bioactive molecules are clustered using a SIMCOMP (a chemical compound-matching algorithm that provides atom equivalences) [30] similarity cutoff of 0.7. For each cluster of similar compounds, a common molecular structure, "the anchor", is then identified. The anchor substructure is defined as a maximum set of functional groups present in at least 90% of the ligands from a single cluster. Having a well-defined anchor substructure, seed molecules are extracted from ligands docked into the receptor binding pocket. Seed molecules comprise the largest set of compounds that have their common substructures docked within a 4 Å RMSD from each other. The consensus-binding mode is derived by averaging the anchor substructure pose in the seed molecules. Finally, in the second pass of docking simulations, the known active molecules are re-docked into the receptor binding pocket, but now with harmonic RMSD restraints imposed on the consensus binding mode of the anchor substructure. The synthetic GPCR-ligand library consists of TASSER-generated protein models with the bioactive molecules placed into the binding pockets during the second pass of constrained docking. Instead of the library of ligand-bound crystal structures, the synthetic library is then used by FINDSITE to derive pocket-specific potentials for high-throughput ligand docking/screening.

3.1. *Virtual Screening Experiments*

For a preliminary assessment of the ranking capability of our approach, using the identical synthetic protein/ligand library protocol as above, we performed a simplified virtual screening experiment against relatively small ligand libraries for the following three dopamine receptors: D_2, D_3, and D_4. For each target, known bioactive molecules were extracted from the MDL Drug Data Report [26] and divided into a training and benchmark subsets. The training subset was used to construct 1024-bit Daylight fingerprints [18] for ligand-based virtual screening as well as to compile the synthetic GPCR library in order to derive pocket-specific potentials and the consensus binding modes of the common ligand substructures. In ligand-based virtual screening, we followed the protocol previously applied in virtual screening of HIV-1 protease [7]. For each active compound from the benchmark set, 10 background compounds were randomly selected from the Asinex compound libraries [31]. The goal of this study was to examine whether virtual screening is capable of ranking the known bioactive molecules over the background ligands. We carried out ligand- and structure-based screening separately for each target and then we combined the results using data fusion techniques [32]. The results were assessed in terms of the enrichment behavior, i.e. the fraction of known active compounds recovered in the top-ranked fraction of the library.

We present the results for dopamine D_2, D_3, and D_4 receptor screening in Figure 4. For these GPCR targets, virtual screening performed significantly better than random ligand selection. Moreover, we observed that the enrichment calculated for the combined ranks from ligand- and receptor-based screenings is typically higher than that calculated for each method alone. These results suggest that the proposed protocol for virtual screening GPCRs may be useful in the discovery of new biopharmaceuticals.

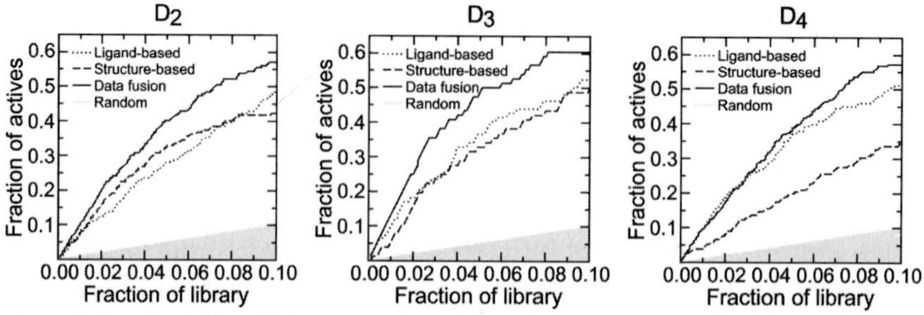

Figure 4. Results of ligand library screening of three dopamine receptors, D_2, D_3, and D_4, using known binders to construct the ligand fingerprint (dotted), structure based (dash), Data fusion (solid) and random (gray) approaches.

4. Discussion

Many methods for protein function prediction employ functional inference by homology [33, 34]; that is, if two proteins are evolutionary related, then the function of the protein of known function is assigned to that of unknown function. However, for enzymes [2], because the extent of function conservation depends on the protein family, care must be taken if the goal is both high accuracy and coverage of sequence space. To address this issue, a number of structure-based approaches based on three-dimensional geometric descriptors of enzymatic function have been developed [5]; In practice, such approaches have been restricted to enzymes and require extensive manual intervention. To eliminate these limitations, we developed the FINDSITE algorithm [7] which exploits the fact that binding sites are strongly conserved among evolutionary distant proteins. The conservation of binding sites among threading identified templates can be used to predict the target binding site, the ligands that bind to this site and consensus GO molecular functions [20]. Since these observations hold for evolutionary quite distant proteins (well below 35% sequence identity), this has profound implications as to how protein molecular function evolved.

Not only is the protein's structure conserved, but the chemical features of the ligands that bind to the protein are also conserved. This provides a type of signal averaging that can be exploited in ligand screening.

A clear advantage of FINDSITE is that predicted as well as experimental structures can be used with comparable results. This is important because state-of-the-art approaches provide useful predicted structures for > 2/3 of protein domains in a given proteome [16, 35]. It is quite likely that there are other functional properties that can be detected by extensions of the FINDSITE approach, one of which, the extension to GPCR virtual ligand screening, was described here with encouraging preliminary results. The idea is to identify distantly related structures with common functional features and then transfer these features to the protein of interest. Thus, this promising avenue of investigation should help extend the range of applicability of structure-based approaches to protein function prediction.

Acknowledgments

This work is supported by NIH grant Nos. GM-48835, GM-37408, GM-084222 and RR-12255.

References

1. L. Hood *et al.*, Science **306**, 640 (2004).
2. W. Tian, and J. Skolnick, J Mol Biol **333**, 863 (2003).
3. R. Kolodny, P. Koehl, and M. Levitt, J Mol Biol **346**, 1173 (2005).
4. A. Andreeva *et al.*, Nucleic Acids Res **36**, D419 (2008).
5. J. S. Fetrow, and J. Skolnick, J Mol Biol **281**, 949 (1998).
6. D. M. Schnur, Curr Opin Drug Discov Devel **11**, 375 (2008).
7. M. Brylinski, and J. Skolnick, Proc Natl Acad Sci U S A **105**, 129 (2008).
8. M. Brylinski, and J. Skolnick, Proteins **70**, 363 (2008).
9. M. Gerstein *et al.*, Science **299**, 1663 (2003).
10. A. Kryshtafovych, K. Fidelis, and J. Moult, Proteins 69 Suppl **8**, 194 (2007).
11. A. G. Murzin *et al.*, J Mol Biol **247**, 536 (1995).
12. R. B. Russell, J Mol Biol **279**, 1211 (1998).
13. D. T. Jones, and C. Hadley, in Bioinformatics: Sequence, structure and databanks, edited by D. Higgins, and W. R. Taylor (Springer-Verlag, Heidelberg, 2000), pp. 1.
14. S. Y. Lee, and J. Skolnick, Biophys J (2008).
15. Y. Zhang, and J. Skolnick, Nucleic Acids Res **33**, 2302 (2005).
16. Y. Zhang, and J. Skolnick, Proc Natl Acad Sci U S A **101**, 7594 (2004).
17. B. Huang, and M. Schroeder, BMC Struct Biol **6**, 19 (2006).
18. in Daylight Theory Manual, (Daylight Chemical Information Systems, Inc., Aliso Viejo, CA, 2007).

19. M. Kanehisa, and S. Goto, Nucleic Acids Res **28**, 27 (2000).
20. M. Ashburner *et al.*, Nat Genet **25**, 25 (2000).
21. C. H. Wu *et al.*, Nucleic Acids Res **34**, D187 (2006).
22. M. Brylinski, and J. Skolnick, 2008).
23. M. Brylinski, and J. Skolnick, J Comput Chem **29**, 1574 (2008).
24. J. Skolnick, and D. Kihara, Proteins **42**, 319 (2001).
25. D. S. Wishart *et al.*, Nucleic Acids Res **34**, D668 (2006).
26. A. E. Fournier, T. GrandPre, and S. M. Strittmatter, Nature **409**, 341 (2001).
27. M. Brylinski, and J. Skolnick, J Comput Chem (2008).
28. H. M. Berman *et al.*, Nucleic Acids Res **28**, 235 (2000).
29. F. Horn *et al.*, Nucleic Acids Res **31**, 294 (2003).
30. M. Hattori *et al.*, Genome Inform **14**, 144 (2003).
31. A. Gaulton, and T. K. Attwood, Nucleic Acids Res **31**, 3333 (2003).
32. M. Whittle *et al.*, J Chem Inf Model **46**, 2193 (2006).
33. D. Groth, H. Lehrach, and S. Hennig, Nucleic Acids Res **32**, W313 (2004).
34. G. Zehetner, Nucleic Acids Res **31**, 3799 (2003).
35. M. A. Marti-Renom *et al.*, Annu Rev Biophys Biomol Struct **29**, 291 (2000).

Quantum Bio-Informatics III
eds. L. Accardi, W. Freudenberg & M. Ohya
 (pp. 337–347)

FEYNMAN TYPE FORMULAE FOR QUANTUM EVOLUTION AND DIFFUSION ON MANIFOLDS AND GRAPHS

O.G.SMOLYANOV
Moscow state university
E-mail:smolyanov@yandex.ru

The Feynman formula is a representation of either a Schrödinger semigroup $e^{t\hat{H}}$ or a Schrödinger group $e^{it\hat{H}}$ by limits of superpositions of some multiple integrals, over Cartesian powers of some space E, and elementary functions; here H is a classical Hamilton function and $\hat{H}$ is a corresponding quantum mechanical Hamiltonian, i.e. a pseudo-differential operator whose Weyl symbol is H. In the former case the multiple integrals in the Feynman formulae approximate some integrals with respect to some diffusion type measure on a set of functions which take values in E and are defined on a real interval (such functions are also called paths, or trajectories, in E); in the latter case the multiple integrals approximate integrals with respect to a pseudo-measure on the same set. The pseudo-measure is called the Feynman pseudo-measure and the corresponding integrals are called the Feynman path integrals[a].

Hence Feynman formulae imply some representations of Schrödinger groups and semigroups by integrals over paths in E; such representations are called Feynman-Kac formulae[b]. Also the Feynman formulas give some explicit approximations for these groups and semigroups. Such finite-dimensional approximations for Feynman type pseudo-measures can also be considered as giving definitions of such pseudo-measures.

[a] A pseudo-measure is, by definition, a linear functional Φ on a suitable space of functions defined on the space of paths in E; the integral of a function f with respect to this pseudo-measure is defined as $\Phi(f)$.

[b] If E is the phase, resp. configuration, space of the corresponding classical Hamiltonian system then one can speak of the Feynman and the Feynman-Kac formulae in the phase, resp. configuration, space.

The famous Feynman's papers [1,2], on the Feynman path integrals over trajectories, contain three main observations. Firstly, it is shown that the solution of the Cauchy problem for a Schrödinger equation can be represented by a limit of a sequence of integrals over Cartesian products of either the classical configuration space or the classical phase space when the multiplicity of integrals tends to infinity. Our Feynman formulae give a formalization of this observation. Secondly the limits are interpreted as some integrals over trajectories in the corresponding space. The formalization of this observation is given by the Feynman path integrals. And finally it is noticed that the integrand contains the exponent of the classical action expressed in the first case in the Lagrangian form and in the second case in the Hamiltonian form (because of that it is natural to denote the Feynman formulae and the Feynman-Kac formulae in the configuration space as the Lagrangian Feynman and Feynman-Kac formulae and the corresponding formulae in the phase space as the Hamiltonian formulae). These integrands can be interpreted as generalized densities [16] of the Feynman pseudo-measures.

It is worth stressing that both the Feynman path integrals and the Feynman formulas are important. The Feynman path integrals connect the theory of evolutionary equations with stochastic analysis and also can be used to investigate some transformations and approximations of the solutions of the equations. Typical examples: analysis of the origin of the so called quantum anomalies, representations of the invariants of knots and the quasi-classical asymptotic. On the other hand the Feynman formulae give a method to calculate the Feynman path integrals.

In the present paper some Lagrangian Feynman formulae representing solutions for initial-boundary problems for Schrödinger type and heat equations are discussed. To get these formulae one uses the Chernoff formula [17] which is as related to the Feynman formulae under consideration as the famous Trotter formula (independently proved by Yu. L. Daleckii) is related to Feynman formulas discussed in the paper [18] of E.Nelson.

Some information about Feynman pseudomeasures and Feynman-Kac formulae can be found in [3,4,15]. In particular, some Hamiltonian Feynman formulae are discussed in [5,15] and [20] (in the latter paper some Hamiltonian Formulae for an equation with respect to functions of p-adic variable are obtained).

Some results of the paper are obtained in collaboration with Professor H.von Weizsäcker and Dr. O.Wittich.

In some places of the paper the presentation is formal: some analytical assumptions are omitted.

1. Chernoff's formula.

Let X be a Banach space, $L(X)$ be the space of all continuous linear operators in X, $\|\cdot\|$ — the operator norm on $L(X)$, I — the identity operator in X, for any linear operator A in X let $D(A)$ be the domain of A and for any function $F : [0,\infty) \to L(X)$ let $F'(0)$ be the right strong derivative of F at 0.

Theorem 1.1. *(Chernoff, [17]). Let $F : [0,\infty) \to L(X)$ be a strongly continuous function such that $F(0) = I$, $\|F(t)\| \leq \exp(at)$ for some $a \in R$, D be a linear subspace in $D(F'(0))$ and the restriction of $F'(0)$ to D be a closable operator whose closure we denote by C. If C is the generator of a strongly continuous semigroup $\exp(tC)$, then $F(t/n)^n$ converges to $\exp(tC)$ as $n \to \infty$ in the strong operator topology uniformly with respect to $t \in [0,T]$ for each $T > 0$.*

Example 1.1. (Trotter formula). Let $F(t) = e^{tA}e^{tB}$; then $F'(0) = A + B,\ F(0) = I$, and

$$e^{t(A+B)} = \lim(e^{\frac{t}{n}A}e^{\frac{t}{n}B})^n.$$

The Chernoff formula is surprisingly useful. Actually there exist a variety of applications of this formula. If one wants to apply this formula to get a Feynman type formula one needs to construct an operator-valued function F whose values are integral operators in a proper space of functions and to check that the assumptions of the Chernoff theorem are satisfied. Then the identity

$$\lim F(t/n)^n = \exp(tC)$$

is just a Feynman formula. Now we list briefly some topics in mathematical physics where the Chernoff theorem can be applied.

(1) Equations on Riemannian manifolds.
(2) Random fields in Riemannian manifolds.
(3) Feynman formulae in phase spaces.
(4) Equations in bounded domains.
(5) Quasiparticles with position-dependent mass.
(6) Selfadjoint extensions of operators and the Feynman formulae.
(7) Diffusion and quantum dynamics on graphs.

(8) Equations on infinite dimensional manifolds of paths in Riemannian manifolds.

(9) Evolutionary equations in p-adic analysis. There exist also stochastic analogs of these problems (cf. [6]).

Of course we consider in the paper only a part of the collections of the problems. Some additional information can be found in the papers sited in the list of references.

2. Quantum dynamics and diffusion on $[0, \infty)$.

We consider here the quantum dynamics and diffusion on $[0, \infty)$ generated by a selfadjoint extension of $\Delta + V$ in $L_2(0, \infty)$; here V is a bounded continuous function on $(0, \infty)$ and $\Delta = \frac{d^2}{dx^2}$ on $D((0, \infty])$. The collection H of selfadjoint extensions of Δ is labelled by elements of $(-\infty, \infty]$. If $a \in (-\infty, \infty]$, then the selfadjoint extension $\hat{H}_a \in H$ is defined as follows. Let $W = W(0, \infty)$ be the set of those differentiable functions on $[0, \infty)$ whose derivatives are abs. continuous and in turn have derivatives belonging to $L_2(0, \infty)$. Then $dom\hat{H}_a = \{f \in W : f'(0) = af(0)\}$ if $a \neq \infty$ and $dom\hat{H}_\infty = \{f \in W : f'(0) = 0\}$ if $f \in dom\hat{H}_a$, then $\hat{H}_a f(x) = f''(x)$. The operator $f \mapsto \hat{H}_a f + V(\cdot)f$ with the same domain is also selfadjoint; we denote this operator by $\hat{H}_a + V$.

Let $X = L_2(0, \infty)$, $a \in (-\infty, \infty]$ and, for $t > 0$, the mappings $F_1(t) : X \to L_2(-\infty, \infty)$, $F_2(t) : L_2(-\infty, \infty) \to X$, $F_3(t) : X \to X$, $F^a(t) : X \to X$ are defined by:

$$F_{01}(t)(f)(x) = \frac{\frac{1}{2\sqrt{t}} \int_0^{2\sqrt{t}} f(z)dz(1 + ax)}{1 + a\sqrt{t}} \psi_t(x) + f(x)\varphi_t(x), \; F_1(t) = F_{01}(\sqrt{t}),$$

where $\varphi_t = \eta_t - \psi_t$, $\eta_t : R^1 \to [0, 1]$ is a smooth function such that $\eta_t(x) = 0$ for $x < -3t$, $\eta_t(x) = 1$ for $x > -2t$ and $\psi_t(x) = \eta_t(x)\eta_t(-x)$ for x ;

$$F_2(t)(f)(x) = \frac{1}{\sqrt{2\pi t}} \int_{-\infty}^{\infty} \exp(-\frac{(x - z)^2}{2t}) f(z)dz;$$

$$F_3(t)(f)(x) = e^{tV(x)} f(x), F^a(t) = F_3(t)F_2(t)F_1(t).$$

Theorem 2.1. *For any $\varphi \in X$, $t > 0$ the following Feynman formula holds:*

$$e^{t(\hat{H}_a + V)} \varphi = \lim_{n \mapsto \infty} \left(F^a \left(\frac{t}{n} \right) \right)^n \varphi$$

in X.

Let $F_2^1(t)$, $t \in (0, \infty)$ be a continuous map of $L_2(-\infty, \infty)$ into X whose restriction to $D(-\infty, \infty)$ is defined by

$$F_2^1(t)(f)(x) = \frac{e^{-i\frac{\pi}{4}}}{\sqrt{2\pi t}} \int_{-\infty}^{\infty} \exp(i\frac{(x-z)^2}{2t})f(z)dz,$$

$$F_3^1(t)(f)(x) = e^{iV(x)}f(x), F^{1a}(t) = F_3^1(t)F_2^1(t)F_1(t).$$

Theorem 2.2. *For any $\varphi \in L^2(0, \infty)$, $t > 0$, the following Feynman formula holds:*

$$e^{it(\hat{H}_a+V)}\varphi = \lim_{n\mapsto\infty} \left(F^{1a}\left(\frac{t}{n}\right)\right)^n \varphi$$

in $L^2(0, \infty)$.

Remark 2.1. If $a = \infty$, which correspond to the Cauchy-Dirichlet problem, then one can also define F^a and F^{1a} by

$$F^a(t)(f) = \frac{1}{\sqrt{2\pi t}} \int_0^{\infty} \exp(-\frac{(x-z)^2}{2t})f(z)dz$$

and

$$\frac{e^{-i\frac{\pi}{4}}}{\sqrt{2\pi t}} \int_0^{\infty} \exp(i\frac{(x-z)^2}{2t})f(z)dz.$$

3. Feynman formulae for quantum dynamics and diffusion of particle with position-dependent mass.

In this section we consider the quantum dynamics and diffusion governed by selfadjoint extensions of the operator $g(\cdot)\Delta + V$ in $L_2(-\infty, \infty)$. Here V is a bounded continuous function on $(-\infty, \infty)$, $\Delta = \frac{d^2}{dx^2}$ on $D(-\infty, \infty)$ and g is defined by: $g(x) = c_1 > 0$ for $x < 0$ and $g(x) = c_2 > 0$ for $x > 0$.

We use the description of selfadjoint extensions of $g(\cdot)\Delta$ obtained by Gadella and his collaborators [14]. Any such extensions is defined by an invertible operator $T = (t_{ij})$ in C^2 as follows: the domain is the collection of functions $h \in W(-\infty, \infty)$ such that if $h^+ = (h^+(0), h'^+(0))(\in C^2)$, $h^- = (h^-(0), h'^-(0))$, then $h^+ = Th^-$. This extension is denoted by $\hat{H}_T$. The operator $f \mapsto \hat{H}_T f + V(\cdot)f$ with the same domain is also selfadjoint; we denote this operator by $\hat{H}_T + V$.

Let $X = L_2(-\infty, \infty)$, T be the operator defined a selfadjoint extension and let for each $t > 0$, the mapings $F_4(t) : X \to X \oplus X, F_5(t) : X \oplus$

$X \to X$, $F_6(t)\ :\ X \to X$, $F^T(t)\ :\ X \to X$ are defined as follows. $F_4(t)(f) = (h,k) \in X \oplus X$, where $h(x) = (a_+x + z_+)\psi_t(x) + f(x)\varphi_t$, $k(x) = (a_-x + z_-)\psi_t(x) + \varphi_t(-x)f(x)$, the functions φ_t and ψ_t are the same as above and a_+, a_-, z_+, z_- are derined by
$a_+\sqrt{t} + z_+ = \frac{1}{2\sqrt{t}}\int_0^{2\sqrt{t}} f(z)dz$, $-a_-\sqrt{t} + z_- = \frac{1}{2\sqrt{t}}\int_{-2\sqrt{t}}^0 f(z)dz$
$z_+ = t_{11}z_- + t_{12}a_-$, $a_+ = t_{21}z_- + t_{22}a_-$. Moreover,

$$F_5(t)((h,k))(x) = \frac{1}{\sqrt{2\pi tc_1}}\int_{-\infty}^{\infty} \exp(-\frac{(x-z)^2}{2tc_1})h(z)dz$$

for $x > 0$,

$$F_5(t)((h,k))(x) = \frac{1}{\sqrt{2\pi tc_2}}\int_{-\infty}^{\infty} \exp(-\frac{(x-z)^2}{2tc_2})k(z)dz$$

for $x < 0$;

$$F_6(t)(f)(x) = e^{tV(x)}f(x); F^T(t) = F_6(t)F_5(t)F_4(\sqrt{t}).$$

Theorem 3.1. *For any $\varphi \in X$ and $t > 0$ the following Feynman formula holds:*

$$e^{t(\hat{H}_T+V)}\varphi = \lim_{n\mapsto\infty}\left(F^T\left(\frac{t}{n}\right)\right)^n\varphi$$

in X.

Let $F_5^1(t)$, $t \in (0,\infty)$ be a continuous mapping of $X \otimes X$ into X, whose restriction to $D(-\infty,\infty)$ is defined by:

$$F_5^1(t)((h,k))(x) = \frac{e^{-i\frac{\pi}{4}}}{\sqrt{2\pi tc_1}}\int_{-\infty}^{\infty} \exp(i\frac{(x-z)^2}{2tc_1})h(z)dz$$

if $x > 0$,

$$F_5^1(t)((h,k))(x) = \frac{e^{-i\frac{\pi}{4}}}{\sqrt{2\pi tc_2}}\int_{-\infty}^{\infty} \exp(i\frac{(x-z)^2}{2tc_2})k(z)dz$$

if $x < 0$ and let

$$F_6^1(t)(f)(x) = e^{itV(x)}f(x);\ \ F^{1T}(t) = F_6^1(t)F_5^1(t)F_4(\sqrt{t}).$$

Theorem 3.2. *For any $\varphi \in X$, $t > 0$ the following Feynman formula holds:*

$$e^{it(\hat{H}_T+V)}\varphi = \lim_{n\mapsto\infty}\left(F^{1T}\left(\frac{t}{n}\right)\right)^n\varphi$$

in X.

Remark 3.1. There exists a generalization of results of this section to the case of diffusion and quantum dynamics of a particle, with position dependent mass, on a Riemannian manifold. To get the generalization one need to use some results of Section 5 of the present paper.

4. Diffusion on graphs.

In this section we consider Feynman formulae for a heat equation with a potential, on a simple graph. Let $A_1 = \{(a,0,0) : a \in [0,\infty\}$, $A_2 = \{(0,a,0) : a \in [0,\infty)\}$, $A_3 = \{(0,0,a) : a \in [0,\infty)\}$, $G = A_1 \cup A_2 \cup A_3$. We assume that G is equipped with the measure generated by the Lebesgue measures on the real line and with the natural metric. Let $X = C_b(G)$ be the space of bounded functions on G which are continuous outside of $(0,0,0)$. We identify, in the natural way, G with the following subset of $R^2 = \{(x_1,x_2)\}$: $\{(s\cos\frac{\pi}{3}, s\sin\frac{\pi}{3}) : s \in [0,\infty)\} \cup \{(s\cos\frac{\pi}{3}, -s\sin\frac{\pi}{3}) : s \in [0,\infty)\} \cup \{(-s,0) : s \in [0,\infty)\}$. A (vector-valued) function f on G is called differentiable at $x \in G, x \neq 0$ if it is differentiable, as a function on the subset G of R^2, along G. In this case the derivative of f is defined as the derivative along the corresponding unit vector from G. The derivatives of highest orders are defined in a similar way. The function f is called continuously differentiable if there exists a continuously differentiable everywhere function g_f on R^2 coinciding with f on G.

Let V be a continuous function on G. Our aim in this section is to develop a Feynman formula for the operator $\hat{H}$ defined by $\hat{H}f = \Delta f + Vf$ whose domain is the set of all twice continuously differentiable real-valued functions on X.

Let the functions $F_7(\cdot), F_8(\cdot)$ whose values are operators in $C_b(G)$ be defined by

$$F_7(t)f(x,0,0) = \sqrt{\frac{2}{\pi t}} \int_0^\infty \exp(-\frac{(x-z)^2}{2t}) f(z,0,0)dz$$

and so on; $F_8(t)f = e^{tV}f$. The function $F_9(\cdot)$ is defined as follows. Let $t > 0$, $f \in C_b(G)$ and Ψ be the function on R^2 whose graph is the plane which pass through the points $((t\cos\frac{\pi}{3}, t\sin\frac{\pi}{3}), f((t\cos\frac{\pi}{3}, t\sin\frac{\pi}{3}))$, $((t\cos\frac{\pi}{3}, -t\sin\frac{\pi}{3}), f((t\cos\frac{\pi}{3}, -t\sin\frac{\pi}{3})), ((-t,0), f(-t,0))$. Then

$$F_9(t)f(x) = \Psi(x)\psi_t(\|x\|^2) + f(x)\varphi_t(\|x\|^2)$$

where ψ_t and φ_t are the same as above. Let $F(t) = F_8(t)F_7(t)F_9(\sqrt{t})$.

Theorem 4.1. *For any $\varphi \in X$, $t > 0$ the following Feynman formula holds:*
$e^{t\hat{H}}\varphi = \lim_{n\mapsto\infty} \left(F\left(\frac{t}{n}\right)\right)^n \varphi$ *in* X.

Remark 4.1. A similar statement can be formulated for the corresponding Schrödinger group.

5. Feynman formulae for equations on Riemannian manifolds.

In this section we discuss briefly some Feynman formulae for the heat equation on Riemannian manifolds. Some additional information can be found in [11]. Let S be a compact Riemannian manifold realized as a smooth submanifold in R^n. We consider the following Cauchy problem for the heat equation on S.

$$\frac{\partial f}{\partial t} = \frac{1}{2}\Delta f; \quad f : [0,\infty) \times S \to C; \quad f(0,\cdot) = g$$

Let $F_R(t), t > 0$ and $F_E(t), t > 0$ be linear operators in the space $C(S)$ of continuous functions on S defined as follows. Let $p_E(t,x,z) = e^{-\frac{\|x-z\|^2}{2t}}$, where $\|\cdot\|$ is the norm in R^n and let $p_R(t,x,z) = e^{-\frac{(\rho(x,z))^2}{2t}}$.

$$F_R(t)f(x) = \frac{\int_S p_R(t,x,z)f(z)dz}{\int_S p_R(t,x,z)dz}$$

and

$$F_E(t)f(x) = \frac{\int_S p_E(t,x,z)f(z)dz}{\int_S p_E(t,x,z)dz}.$$

Theorem 5.1. *The solution $f(\cdot,\cdot)$ of the above Cauchy problem is given by the following Feynman formulae:*

$$f(t,\cdot) = \lim_{n\mapsto\infty} \left(F_R\left(\frac{t}{n}\right)\right)^n g = \lim_{n\mapsto\infty} \left(F_E\left(\frac{t}{n}\right)\right)^n g.$$

Remark 5.1. Here the Feynman formulae with different functions F's represent the sane semigroups. In this case the corresponding functions F's are called Chernoff equivalent [11].

Let now $F_R^1(t)f = \int_S p_R(t,x,z)f(z)dz$ and $F_E^1(t)f = \int_S p_E(t,x,z)f(z)dz$.

Theorem 7. Let the functions f_1 and f_2 on $[0,\infty)\times S$ be defined by

$$f_1(t,x) = \lim_{n\mapsto\infty} \frac{(F_E^1(\frac{t}{n}))^n g(x)}{(F_E^1(\frac{t}{n}))^n 1(x)}$$

and

$$f_2(t,x) = \lim_{n\mapsto\infty} \frac{(F_R^1(\frac{t}{n}))^n g(x)}{(F_R^1(\frac{t}{n}))^n 1(x)}$$

Then $f_k(t,x) = c_k(t,x)g_k(t,x)$ where g_k, $k=1,2$ is the solution of the Cauchy problem for the equation

$$\frac{\partial f}{\partial t} = \frac{1}{2}\Delta f + V_k;\quad f:[0,\infty)\times S\to\mathbb{C};\quad f(0,\cdot)=g,$$

with $V_1(x) = \frac{1}{8}\|a\|^2(x) - \frac{1}{4}s(x)$ and $V_2(x) = -\frac{1}{6}s(x)$, $s(x)$ is the scalar curvature at x (we use the normalization in which the scalar curvature of the sphere of radius r is equal to $n(n-1)r^{-n}$) and $a(x)$ is the vector-valued mean curvature at x times the dimension of the manifold; $c_1(t,x) = \int_G \exp(\int_0^t(\frac{1}{8}\|a\|^2(h(\tau)) - \frac{1}{4}s(h(\tau)d\tau)w_x(dh)$, $c_2(t,x) = \int_G \exp(\int_0^t(-\frac{1}{6}s(h(\tau)d\tau)w_x(dh)$ and w_x is the Wiener measure concentrated on functions (on $[0,t]$ taking values in S) which takes value x at 0. This statement can be deduced from results of [11].

6. Concluding remarks.

In the paper we have considered only Feynman formulae. As it have already been mentioned each of them can be used to define a Feynman pseudomeasure. But there exist also a collection of other definitions of the Feynman pseudomeasures. Here is a list of most important of them.

(1) Definitions via limits of integrals over Cartesian products of a space(Nelson, 1964, Lagrangian pseumomeasure; Smolyanov-Tokarev-Truman, 2002, Hamiltonian pseudomeasure. Just these definitions are most related to the Feynman formulae.
(2) Definitions in the frame of the Hida White noise analysis (Hida-Streit, 1975-80).
(3) By analytical continuation (Gelfand-Yaglom, 1956, Lagrangian case; Smolyanov-Shavgulidze, 1990, Hamiltonian case).
(4) By the Fourier transform (Cecile deWitt-Morette , 1974).

(5) By the Parceval identity (Maslov-Chebotarev, 1976, Albeverio-Hoeg-Krohn, 1976).
(6) By the central limit theorem (Smolyanov-Khrennikov, 1985).

It would be interesting to define and investigate the measures and pseudomeasures in the Feynman-Kac formulae corresponding to the Feynman formulae in Theorems 1-5. In particular it is interesting to discuss the stochastic differential equations generating the measures.

References

1. R.P. Feynman, "Space-time approach to nonrelativistic quantum mechanics," Rev.Mod.Phys. 20 367–387 (1948).
2. R.P. Feynman, "An operation calculus having applications in quantum electrodynamics," Phys. Rev. 84 108–128. (1951)
3. O.G. Smolyanov, E.T. Shavgulidze, Functional integrals (Moscow State University Press, Moscow, 1990, in Russian).
4. O. G. Smolyanov, M. O. Smolyanova, "Transformations of the Feynman integral under nonlinear transformations of the phase space," Theoret. and Math. Phys. 100 (1994), no. 1, 803–810 (1995).
5. O.G.Smolyanov, A.G.Tokarev, A.Truman. *Hamiltonian Feynman path integrals via the Chernoff formula J.Math.Phys.* 43, 10, (2002) 5161-5171.
6. J.Gough, O.O.Obrezkov, O.G.Smolyanov. *Randomized Hamiltonian Feynman integrals and stochastic Schrödinger-Ito equations Izvestia Mathematics* 69, 6 (2005), 3-20.
7. O.G.Smolyanov, H. von Weizsaecker, O.Wittich. The Feynman Formula for the Cauchy Problems in Domains with Boundary. Doklady Mathematics, 69, 2 (2004), 257-261.
8. O.G.Smolyanov, H. von Weizsaecker, O.Wittich. Constructon of diffusions on sets of Mappings from an Interval to Compact Riemannian Manifolds. Doklady Mathematics, 71, 3 (2005), 390 - 395.
9. L.Accardi, O.G.Smolyanov. Feynman Formulae for Evolution Equations with Levy Laplacians on Infinite-Dimensional Manifolds. Doklady Mathematics, 73, 2 (2006), 252 - 257.
10. O.G.Smolyanov, H. von Weizsaecker, O.Wittich. Surface integrals in Riemannian Spaces and Feynman Formulae. Doklady Mathematics, 73, 3 (2006), 432 - 437.
11. O.G.Smolyanov, H. von Weizsaecker, O.Wittich Chernoff's Theorem and Discrete Time Approximations of Brownian Motion on Manifolds. Potential Anal. 26 (2007), 1-29.
12. O.G.Smolyanov, H. von Weizsaecker, O.Wittich. Surface Measures and Initial Boundary Value Problem Generated by Diffusion with Drift. Doklady Mathematics, 76, 1 (2007), 606-610.
13. S. Albeverio, R. Hoegh-Krohn, Mathematical theory of Feynman path inte-

grals Lecture notes in math 523. (Berlin: Springer, 1976; the second edition, 2008).
14. M. Gadella, S. Kuru, and J. Negro, Phys. Lett. A 362, 265-68 (2007).
15. O.G.Smolyanov. Feynman formulae for evolutionary equations. in:Trends in Stochastic Analysis. Edited by J.Blath, P.Mörters, M.Scheutzow, 284-303, Cambridge University Press, 2009.
16. O.G.Smolyanov, H. von Weizsaecker. Smooth probability measures and associated differential operators. Infinite dimensilonal Analysis, Quantum probability and related topics 2:1, 51-78, 1999.
17. P.R. Chernoff, "Note on Product Formulas for Operator Semigroups," J. Funct. Ana. 84, 238–242 (1968).
18. E. Nelson, "Feynman integrals and the Schrödinger equation," J. Math. Phys. 5 N 3 332–343 (1964).
19. V. P. Maslov, (Russian) Complex Markov chains and the Feynman path integral for nonlinear equations (Moscow: Nauka, 1976.).
20. O,G,Smolyanov, N.N.Shamarov. Hamiltonian Feynman integrals for equation with the Vladimirov operator. Doklady Mathematics, 2010 (to appear).

Quantum Bio-Informatics III
eds. L. Accardi, W. Freudenberg & M. Ohya

POISSON NOISE AND THE DYNAMICS OF INFINITE PARTICLE SYSTEMS

LUDWIG STREIT

BiBoS, Univ. Bielefeld and CCM, Univ. da Madeira

We introduce Poisson analysis as a means to study processes of particle configurations in the continuum and present the free Kawasaki dynamics as illustration..

1. Poisson versus Gaussian White Noise - Setting the stage

Gaussian white noise is most concisely characterized by the Fourier transform of its (probability) measure. Given

$$C(f) = \exp\left(-\frac{1}{2}\int f^2(x)\,dx\right),$$

the Bochner-Minlos theorem, - see e.g. [6] - ensures the existence of a probability measure μ on distribution space such that

$$C(f) = \int_{S^*(R)} d\mu(\omega)\exp\left(i\int \omega(x)\,f(x)\,dx\right)$$

for test functions f. The L^2space w.r. to this measure

$$(L^2) \equiv L^2(S^*(R), d\mu)$$

is then the basis of Gaussian white noise analysis.

In the Poisson case we could proceed similarly, starting from the characteristic function

$$C_\pi(f) = \exp\left(\int_{R^d}\left(e^{if(x)} - 1\right)dx\right) = \int_{D^*(R)} d\pi(\omega)\exp\left(i\int \omega(x)\,f(x)\,dx\right)$$

to obtain the Poisson measure π, again by the Bochner-Minlos theorem. However, it is interesting to proceed differently.

1.1. *Configurations*

We want to describe infinite systems of particles: "configurations" of indistinguishable point particles in R^d or in some subset $X \subseteq R^d$.

The configuration space $\Gamma := \Gamma_X$ is the set of all **locally finite subsets** of X, *i.e.*,

$$\Gamma := \{\gamma \subset X : \#(\gamma \cap K) < \infty \text{ for bounded } K \subset X\}.$$

For a given configuration $\gamma = \{x_1, x_2, \ldots\}$ we denote

$$\langle \gamma, f \rangle = \sum_{x \in \gamma} f(x) = \sum_{x \in \gamma} \int \delta(x - x') f(x') dx'.$$

This is well defined if f is continuous and zero outside a finite volume: the sum is then finite - no problem of convergence arises.

1.2. *Poisson Measures*

We begin by considering configurations in a finite volume:

$$|X| = V < \infty$$

For configurations of only one point $x \in R^d$ the obvious choice will be a probability proportional to the volume element dv. For n-point configurations, elements of $\Gamma_X^{(n)}$ we shall use

$$dm_n = \frac{1}{n!} (dv)^n$$

the combinatorial 1/n! factor for the indistinguishability of the n particles. - But we are interested in configurations of arbitrary many particles, i.e. we want a probability measure on

$$\Gamma_X = \bigsqcup_{n=0}^{\infty} \Gamma_X^{(n)}.$$

We first extend the measures m_n to a measure m on Γ_X, simply by setting

$$m|_{\Gamma_X^{(n)}} = m_n.$$

This is not a probability measure:

$$m(\Gamma_X) = m\left(\bigsqcup_{n=0}^{\infty} \Gamma_X^{(n)}\right) = \sum_n m\left(\Gamma_X^{(n)}\right) = \sum_n \frac{1}{n!} \left(\int_X dv\right)^n = \exp(V).$$

We must normalize it to get a probability measure on Γ

$$\pi \equiv \exp(-V) \cdot m$$

with characteristic function

$$E(\exp(i\langle\gamma, f\rangle)) = \int_\Gamma \exp(i\langle\gamma, f\rangle)d\pi(\gamma) = \sum_n \int_{\Gamma^{(n)}} \exp(i\langle\gamma, f\rangle)d\pi(\gamma)$$

$$= \exp(-V)\sum_n \frac{1}{n!}\left(\int_{X^n} \exp(i\sum_{k=1}^n f(x_k))\prod {}_k (dx_k)\right)$$

$$= e^{-V}\sum_n \frac{1}{n!}\left(\int_X \exp(if(x))dx\right)^n$$

$$= e^{-V}\left(\int_X \exp(if(x))dx\right) = \exp\left(\int_X \left(e^{if(x)} - 1\right)dx\right).$$

We have (re)discovered the "characteristic function", i.e. the Fourier transform, of the Poisson White Noise probability measure:

$$E(\exp(i\langle\gamma, f\rangle)) = \exp\left(\int \left(e^{if(x)} - 1\right)dx\right) = C_\pi(f) = \int e^{i\langle\omega, f\rangle}d\pi(\omega)$$

Note: there is no need to restrict ourselves to a space of finite volume -

$$C_\pi(f) = \exp\left(\int_X \left(e^{if(x)} - 1\right)dx\right)$$

is well defined even in the limit where $X = R^d$, and we have a limiting measure

$$\pi = \lim_{X\to R^d} \pi|_{\Gamma_X}$$

Likewise for more general densities, with

$$dv = z(x)dx$$

where z is a non-negative "intensity":

$$C_{\pi_z}(f) = \exp\left(\int_{R^d} \left(e^{if(x)} - 1\right)z(x)dx\right)$$

Bochner-Minlos

Recall that the Bochner-Minlos theorem guarantees the existence of a probability measure *on the space of distributions* such that

$$C_{\pi_z}(f) = \int_{D^*} e^{i\langle\omega, f\rangle}d\pi_z(\omega)$$

In our explicit construction we have used the formula

$$\langle \gamma, f \rangle = \sum_{x \in \gamma} f(x) = \sum_{x \in \gamma} \int \delta(x - x') f(x') dx'.$$

We see from this that the measure is concentrated on only those distributions which are sums of Dirac δ-functions

$$\omega = \omega_\gamma = \sum_{x \in \gamma} \delta_x.$$

1.3. *Recall Fock Space*

Consider Fock space

$$\mathcal{F} = \bigoplus_n Sym\, L^2(R^n, n! d^n t),$$

i.e.

$$\mathcal{F} = \{\Psi : \Psi = (\Psi_0, \Psi_1, \dots, \Psi_n, \dots)\}$$

with norm

$$\|\Psi\|_{\mathcal{F}}^2 = \sum_{n=0}^{\infty} n! \, (\Psi_n, \Psi_n)_{L^2(R^n)} \, .$$

For $n = 0$ the zero-particle vectors Ψ_0 are just constants: $\Psi_0 = c$ with

$$\|\Psi_0\|^2 = |c|^2 \, .$$

Annihilation operators $a\,(f)$ are given by

$$(a\,(f)\,\Psi)_n\,(x_1, \dots, x_n) = (n+1) \int dx\, f\,(x)\, \Psi_{n+1}\,(x, x_1, \dots, x_n)$$

We consider in particular Fock space vectors $\Psi\,(g)$ with $\Psi_n\,(g) = \frac{1}{n!} g^{\otimes n}$. They are eigenvectors of the annihilation operators

$$a(f)\Psi\,(g) = (f, g)\Psi\,(g)$$

and have the scalar product

$$(\Psi\,(f)\,, \Psi\,(g))_F = e^{(f,g)}.$$

(Suitably normalized, they become the "coherent states" of quantum optics.)

1.4. *Three Isomorphisms*

It is well known, and easy to verify, that in the White Noise Hilbert space (L^2) the vectors $e(f)$, with

$$e\left(f,\omega\right)=\frac{e^{\int\omega(x)f(x)dx}}{E\left(e^{\int\omega(x)f(x)dx}\right)}$$

have the same scalar product:

$$\left(e\left(f\right),e\left(g\right)\right)_{(L^2)}=\left(\Psi\left(f\right),\Psi\left(g\right)\right)_F=e^{(f,g)}.$$

Now, in the Poisson Hilbert space $L^2\left(d\pi_z\right)$, consider the vectors

$$e_\pi(f,\omega)=\exp\left(\langle\omega,\ln(1+f)\rangle-\langle f\rangle\right),\quad \omega=\omega_\gamma, \tag{1}$$

with

$$\langle f\rangle=\int f(x)z(x)dx.$$

For $\omega=\omega_\gamma{=}\sum_{x\in\gamma}\delta_x$, find

$$e_\pi(f,\omega_\gamma)=\exp\left(-\langle f\rangle\right)\prod_{x\in\gamma}(1+f(x)).$$

Their scalar product is again computed directly from the characteristic function:

$$\left(e_\pi\left(f\right),e_\pi\left(g\right)\right)_{L^2(d\pi_z)}=e^{(f,g)_{L^2(d\nu)}}.$$

As a consequence, we have three isomorphisms

$$\begin{array}{ccc} & \mathcal{F} & \\ \nearrow & & \searrow \\ L^2\left(d\mu\right) & \leftrightarrow & L^2\left(d\pi\right) \end{array}$$

one of them the famous *Gelfand-Ito-Segal isomorphism* between the L^2-space with Gaussian measure μ and symmetric or "boson" Fock space $\mathcal{F}$.

For yet another naturally isomorphic L^2-space, on *finite* configurations ("Lebesgue-Poisson space"), see [4, 5, 9, 10]. For an overview and much more see e.g. the thesis of M. J. Oliveira [22].

1.5. *Operators*

In Fock space there is a representation of the CCR:

$$[a(f), a(g)] = 0 = [a^*(f), a^*(g)]$$
$$[a(f), a^*(g)] = \int dx\, f(x)g(x).$$

The image D_f of $a(f)$ in Gaussian White Noise space is a directional derivative ("Hida derivative")

$$D_f\varphi(\omega) = \lim_{\varepsilon\to 0}\frac{\varphi(\omega+\varepsilon f)-\varphi(\omega)}{\varepsilon}$$

and that of its adjoint $a^*(f)$ is

$$D_f^* = -D_f + \int \omega(x)\, f(x)\, dx.$$

In Poisson space the image of $a(f)$ acts as a difference operator, adding one more particle to the configuration γ:

$$(a_\pi(f)\, F)(\gamma) = \int_X (F(\gamma\cup\{x\}) - F(\gamma))\, f(x)\, dx$$

(straightforward to check for eigenvectors $e_\pi(g)$, which span $L^2(d\pi)$). For the adjoint one finds

$$(a_\pi^*(g)\, F)(\gamma) = \sum_{x\in\gamma} F(\gamma\setminus\{x\})\, g(x) - \int f(y)z(y)dy \cdot F(\gamma).$$

For later reference we finally introduce

$$e_B(f,\omega_\gamma) = \exp\langle f\rangle\, e_B(f,\omega_\gamma) = \prod_{x\in\gamma}(1+f(x)).$$

Their expectation w.r. suitable measures μ on configuration space

$$E(e_B(f)) = \int_\Gamma e_B(f,\omega_\gamma)d\mu(\gamma) = \sum_n \frac{1}{n!}\langle k_n^\mu, f^{\otimes n}\rangle$$

are called Bogoliubov functionals and are the generators of the n^{th}order correlation functions k_n^μ for the distribution μ.

2. Dynamics of Infinite Particle Systems

For discrete configuration spaces there exists a vast literature on various possible dynamical processes, see e.g . [20], used for modelling a large variety of models of population dynamics in the wider sense - epidemiology, ecology, opinion formation, spreading of information in large distributed systems, ...

Examples are independent random births and deaths of particles : "*Glauber dynamics*", or simultaneous death and birth at two lattice sites: "Kawasaki dynamics", where particles hop from one site to another and particle number is conserved.

2.1. *Continuous Configuration Space*

For this type of dynamics in the continuum much less is known. Recent results can be found for Glauber dynamics e.g. in [1 15 16], for Kawasaki in [18 14].

2.2. *Methods*

In terms of Markov processes the complication is - for infinite configurations (finite density, thermodynamic limit) - that infinitely many jumps occur in any finite time interval, and even without interactions, free jumps can produce infinite local densities in finite times ("explosion").

Hilbert Space methods - Dirichlet forms , evolution operators - are suitable for (approach to) equilibrium [2 7].

2.3. *Free Kawasaki Dynamics*[13]

As a simple model, we shall focus on "free" Kawasaki dynamics:

$$\partial_t F(\gamma) = \sum_{x \in \gamma} \int_{\mathbb{R}^d} dy\, g(x-y)\, (F(\gamma \backslash x \cup y) - F(\gamma))$$

Particles are hopping from x to y, with rate $g(x-y)$, otherwise independent of the configuration γ.

In terms of creation and annihilation operators, one finds

$$H = \int dx\, z(x) \int dy\, (g(x-y) - g_0 \delta(x-y))\, (a_\pi^*(x) a_\pi(y) - a_\pi(y)),$$

Clearly, in Fock space language this corresponds to a quadratic Hamiltonian, and time development can be calculated in closed form.

Time evolution of Bogoliubov exponentials takes on a particularly simple form:

$$e^{Ht}e_B(f) = e_B(e^{tA}f)$$

$$Af(x) := \int_{\mathbb{R}^d} dy\, g(x-y)\,(f(y)-f(x))\,.$$

Evolution of the initial (Poisson) distribution

$$\pi_z \to P_{\pi_z,t}$$

under the dual of e^{Ht} is characterized by

$$\int e_B(f,\gamma)P_{\pi_z,t}(d\gamma) = \int e_B(e^{tA}f,\gamma)\pi_z(d\gamma) = \exp\left(\int_{\mathbb{R}^d} e^{tA}f(x)z(x)dx\right).$$

Starting with a Poisson distribution the distribution at time t, $P_{\pi_z,t}(d\gamma)$ is again Poissonian, with intensity $z_t \in L^\infty(\mathbb{R}^d, dx)$, given by

$$\int_{\mathbb{R}^d} dx\, e^{tA}f(x)z(x) = \int_{\mathbb{R}^d} dx\, f(x)z_t(x), \tag{2}$$

for all $f \in L^1(\mathbb{R}^d, dx)$. Since e^{tA} is positivity preserving in $L^1(\mathbb{R}^d, dx)$, it follows that $z_t \geq 0$.

Poisson distributions are invariant under free Kawasaki dynamics iff their intensity is constant:

$H^*1 = 0$ iff the linear annihilation term in H vanishes:

$$\int dy\, a(y)\left(\int dx z\,(x)\,(g\,(x-y) - g\,(x))\right) \stackrel{!}{=} 0$$

Using Fourier transforms one sees that this requires $z = const.$

For g even and constant $z > 0$, H gives rise to a symmetric Dirichlet form on $L^2(\Gamma, \pi_z)$,

$$(F, HF) = -\frac{1}{2}\int_\Gamma \pi_z(d\gamma)\sum_{x\in\gamma}\int_{\mathbb{R}^d} dy\, g(x-y)\,|F(\gamma)|^2\,. \tag{3}$$

This allows to derive a Markov process on Γ with *cadlag* paths and having π_z as an invariant measure[18]. In this setting H is a negative essentially self-adjoint operator on $L^2(\Gamma, \pi_z)$, and the generator of a contraction semigroup on $L^2(\Gamma, \pi_z)$.

3. Asymptotics

3.1. *Large Time Asymptotics $t \to \infty$*

- Any Poisson state of constant intensity is invariant under the evolution (equilibrium).
- "Local Equilibrium": Poisson states with non-constant intensity $z = z(x)$. Recall that the state at $t \geq 0$ is again Poissonian, with intensity $z_t \in L^\infty(\mathbb{R}^d, dx)$, given by

$$\int_{\mathbb{R}^d} dx\, e^{tA} f(x) z(x) = \int_{\mathbb{R}^d} dx\, f(x) z_t(x).$$

One says that a function $z \in L^1_{\text{loc}}(\mathbb{R}^d, dx)$ has arithmetic mean whenever

$$\lim_{R \to +\infty} \frac{1}{\text{vol}(B(R))} \int_{|x|<R} dx\, z(x) \equiv mean(z) \tag{4}$$

exists.

If $z \geq 0$ is a bounded measurable function whose Fourier transform $\widetilde{z}$ is a signed measure, then $mean(z) = \widetilde{z}(\{0\})$,

$$\int_{\mathbb{R}^d} dx\, f(x) z_t(x) = \int_{\mathbb{R}^d} dx\, e^{tA} f(x) z(x) \to mean(z) \int_{\mathbb{R}^d} dx\, f(x),$$

and the distribution π_{z_t} converges weakly to $\pi_{\text{mean}(z)}$ as t goes to infinity, because of convergence of the characteristic functions[a].

Not all measurable bounded non-negative functions z have an arithmetic mean. Counterexamples are slowly oscillating functions such as

$$z(x) = c + \cos(\ln(1 + |x|)), \quad x \in \mathbb{R}^d,$$

where $c > 1$. Then for large R

$$\frac{1}{\text{vol}(B(R))} \int_{B(R)} z(x)\, dx \sim c + \frac{d}{\sqrt{1+d^2}} \sin(\ln(R) + \arctan(d)).$$

3.2. *The hydrodynamic limit*

Shall use the so-called "empirical field" corresponding to a $\varphi \in \mathcal{D}(\mathbb{R}^d)$,

$$\langle \varphi, \mathbf{X}_t \rangle = \sum_{x \in \mathbf{X}_t} \varphi(x).$$

[a]To extend the result to all bounded$z$ with arithmetic mean finer considerations are necessary. If the jump rate a has finite second moment we can check the requirements of [3]. General jump rates will be considered in.[19]

The first correlation function $\rho_t(x)$ is given by

$$E(\langle\varphi, \mathbf{X}_t\rangle) = E_{\pi_{z_t}}(\langle\varphi, \gamma\rangle) = \int \varphi(x)\rho_t(x)dx$$

Consider space-time scale transformation given by $\langle\varphi, \gamma\rangle \to \varepsilon^d \langle\varphi(\varepsilon\cdot), \gamma\rangle$, $t \to \varepsilon^{-\kappa} t$ for suitable $\kappa > 0$, $z \to z(\varepsilon\cdot)$.

1. If

$$g_i^{(1)} := \int_{\mathbb{R}^d} dx\, x_i g(x) \neq 0,$$

then for $\kappa = 1$

$$\int_{\mathbb{R}^d} dx\, \rho_t(x)\varphi(x) = \int_{\mathbb{R}^d} dx\, z(x + tg^{(1)})\varphi(x),$$

so that, if the intensity z is smooth enough

$$\frac{\partial}{\partial t}\rho_t(x) = g^{(1)} \cdot \nabla\rho_t(x) = \operatorname{div}(g^{(1)} \cdot \rho_t(x))$$

with the initial condition $\rho_0 = z$.

2. If $g^{(1)} = 0$, and

$$g_{ij}^{(2)} := \int_{\mathbb{R}^d} dx\, x_i x_j g(x)$$

then for time rescaling with $\kappa = 2$

$$\int_{\mathbb{R}^d} dx\, \rho_t(x)\varphi(x) = \frac{1}{(2\pi)^{d/2}} \int_{\mathbb{R}^d} dx\, z(x) \int_{\mathbb{R}^d} dk\, e^{ik\cdot x} e^{-\frac{t}{2}\langle g^{(2)}k, k\rangle} \hat{\varphi}(k),$$

solution of the partial differential equation

$$\frac{\partial}{\partial t}\rho_t(x) = \frac{1}{2}\sum_{i,j=1}^{d} g_{ij}^{(2)} \frac{\partial^2}{\partial x_i \partial x_j}\rho_t(x).$$

3. Consider weak asymmetries, decomposing g into a sum of an even function p and an odd function q, and use the scaling

$$g_\varepsilon := p + \varepsilon q$$

and $\kappa = 2$.

The limiting density ρ_t is solution of the partial differential equation

$$\frac{\partial}{\partial t}\rho_t(x) = \operatorname{div}(g^{(1)}\rho_t(x)) + \frac{1}{2}\sum_{i,j=1}^{d} g_{ij}^{(2)} \frac{\partial^2}{\partial x_i \partial x_j}\rho_t(x).$$

3.3. *Far from Equilibrium*

The construction of the free Kawasaki process and its scaling limits are not restricted to Poissonian initial distributions. Sufficient conditions for admissible measures can be stated in terms of their correlation functions and are in particular fulfilled for Gibbs measures at high temperatures. For more details see [13].

3.4. *Yet another Scaling Limit*

For interacting Kawasaki processes, as the jump range becomes infinite, Glauber dynamics appears as scaling limit [8]. for the free Kawasaki dynamics, in the limit $g \to 1$, we obtain as generator the image of the Fock space number operator $N = \int dx\, a^*(x)\, a(x)$.It is straghtforward to compute that this is the generator of a birth-and-death process with unit death rate and birth rate $b = z$:

$$\partial_t F(\gamma) = \int_{\mathbb{R}^d} dy\, z\ \left(F(\gamma \cup y) - F(\gamma)\right) + \sum_{x \in \gamma} \left(F(\gamma \backslash x) - F(\gamma)\right),$$

and from its isomorphism with N in Fock space we see immediately that this generator, contrary to the original one in (3), now has a spectral gap.

References

1. L. Bertini, N. Cancrini, F. Cesi, The spectral gap for a Glauber-type dynamics in a continuous gas, Ann. Inst. H. Poincaré Probab. Statist. 38, 91-108 (2002).
2. V. Choi, Y. M. Park, H.J. Yoo, Dirichlet forms and Dirichlet operators for infinite particle systems: essential self-adjointness, J. Math. Phys. 39, 6509–6536 (1998).
3. R. L. Dobrushin, R. Siegmund-Schultze: The hydrodynamic limit for systems of particles with independent evolution. Math. Nachr. 105, 199-224 (1982)
4. K.-H. Fichtner, W. Freudenberg. Characterization of states of infinite Boson systems I. On the construction of states of Boson systems. Comm. Math. Phys., 137, 315–357, 1991.
5. K.-H. Fichtner, G. Winkler. Generalized Brownian motion, point processes and stochastic calculus for random Fields. Math. Nachr., 161, 291-307, 1993.
6. I.M. Gel'fand,.N. Ya.Vilenkin. Generalized Functions. Academic Press (1964)
7. M. Grothaus, Y. Kondratiev, E. Lytvynov, M. Roeckner: Scaling limit of Stochastic Dynamics in Classical continuum systems. arXiv:math/0211323v1 [math.PR] 20 Nov 2002
8. D. Finkelshtein, Y. Kondratiev, E. Lytvynov. On convergence of dynamics of hopping particles to a birth-and-death process in continuum. Bielefeld preprint http://arxiv.org/abs/0803.3551v1

9. Yu.G. Kondratiev, T. Kuna, M.J. Oliveira, Analytic aspects of Poissonian white noise analysis. Methods Funct. Anal. Topology 8, no. 4, 15–48 (2002)
10. Yu. G. Kondratiev, T. Kuna, M. J. Oliveira. On the relations between Poissonian white noise analysis and harmonic analysis on configuration spaces. J. Funct. Anal., 213(1), 1–30, 2004.
11. Yu. G. Kondratiev, T. Kuna, M. J. Oliveira. Non-Equilibrium stochastic dynamics of continuous systems and Bogoliubov generating functionals. Bielefeld preprint 2005 http://arxiv.org/abs/math-ph/0505049v1
12. Yu. G. Kondratiev, T. Kuna, M. J. Oliveira. Holomorphic Bogoliubov functionals for interacting particle systems in continuum. J. Funct. Anal., 238(2), 375–404, 2006.
13. Yu. Kondratiev, T. Kuna, M.J. Oliveira, J.L. Silva, L. Streit, Hydrodynamic limits for free Kawasaki dynamics of continuous particle systems. Bielefeld preprint, 2008. http://www.math.uni-bielefeld.de/sfb701/preprints/sfb08082.pdf
14. Yu. G. Kondratiev, O. V. Kutoviy, E. Lytvynov. Diffusion approximation for equilibrium Kawasaki dynamics in continuum. Preprint, 2007
15. Yu. G. Kondratiev, O. V. Kutoviy, R. Minlos. Ergodicity of Non-Equilibrium Glauber Dynamics in Continuum. Bielefeld preprint, 2007. http://www.math.uni-bielefeld.de/sfb701/preprints/sfb08124.pdf
16. Yu. G. Kondratiev, O. V. Kutoviy, E. Zhizhina, Nonequilibrium Glauber-type dynamics in continuum. J. Math. Phys. 47, No.11 (2006)
17. Yu. G. Kondratiev, E. Lytvynov, Glauber dynamics of continuous particle systems. Ann. Inst. H.Poincaré Probab. Statist. 41, No.4, 685-702, 2005.
18. Yu. G. Kondratiev, E. Lytvynov, M. Röckner. Equilibrium Kawasaki dynamics of continuous particle systems. IDAQP, 7(4):185-209, 2007.
19. T. Kuna, J. L. Silva, Large time asymptotic and dynamical law of large numbers for jump type infinite particle processes with long tails, in preparation.
20. T. M. Liggett, "Interacting Particle Systems" (Springer, 1985).
21. J. Mecke. Stationäre zufällige Maße auf lokalkompakten Abelschen Gruppen. Z. Wahrsch. verw. Gebiete, 9:36-58, 1967.
22. M. J. Oliveira: Configuration Space Analysis and Poissonian White Noise Analysis PhD thesis, Lisbon, 2002, see "http://www.math.uni-bielefeld.de/igk/study-materials/corpo_2.pdf"

Quantum Bio-Informatics III
eds. L. Accardi, W. Freudenberg & M. Ohya

REPLICA-EXCHANGE MOLECULAR DYNAMICS SIMULATIONS OF AMYLOID PRECURSOR PROTEIN DIMER IN MEMBRANE*

NAOYUKI MIYASHITA

RIKEN Computational Science Research Program, Molecular Scale Team, 2-1 Hirosawa, Wako, Saitama 351-0198, Japan

YUJI SUGITA

RIKEN Advanced Science Institute, 2-1 Hirosawa, Wako, Saitama 351-0198, Japan

Aggregation of amyloid β peptide (Aβ) in the brain is the primary element in the pathogenesis of Alzheimer's disease. Aβ is derived from amyloid precursor protein (APP) in the membrane due to the cleavages by β- and γ-secretases. Here, we predict the transmembrane structures of the wild-type and mutant APP in the biological membrane by replica-exchange molecular dynamics simulations. The simulations illustrate large conformational differences between the wild type and mutant APP fragments in the membrane. Dimerization of the wild type occurs due to the Cα-H···O hydrogen bonds at the Gly-XXX-Gly motifs between two APP fragments, whereas the mutant dimer is stabilized by the interactions between hydrophobic side chains. We also observe the downward shift of γ-cleavage site in the mutant APP, which may cause the prohibition of Aβ production.

1. INTRODUCTION

Alzheimer's disease (AD) is one of the most well-known neurodegenerative disorders. Its main symptom is a disturbance of cognitive functions. The death of nerve cells or neurons in the brains of the patients occurs in the progressive process of this disease. The most widely accepted hypothesis for the cause is the

* This work is supported, in part, by the Development and Use of the Next-Generation Supercomputer Project of the Ministry of Education, Culture, Sports, Science and Technology (MEXT).
This work is also supported by BIRD, Japan Science and Technology Agency (JST).

amyloid hypothesis, in which AD results from the aggregation and accumulation of amyloid β peptide (Aβ) in the brain [1]. Therefore, it is vitally important to acquire better knowledge of the structures of Aβ in understanding AD [2,3].

Aβ is produced from amyloid precursor protein (APP), which is composed of approximately 700 amino-acid residues in the biological membrane of neurons. Aβ denotes a series of approximately 40 residues in APP and is derived due to the sequential cleavages at both ends of these consecutive residues by β- and γ-secretases, respectively. $A\beta_{1\text{-}40}$ consisting of 40 amino-acid residues is a primary isoform, whereas $A\beta_{1\text{-}42}$, which is a major component of senile plaques, is the second major isoform. To understand the structural differences between $A\beta_{1\text{-}40}$ and $A\beta_{1\text{-}42}$ in solution is one of the central issues in understanding AD [4-6]. However, the transmembrane structures of APP and how Aβ is produced from APP in the transmembrane have not been fully understood so far.

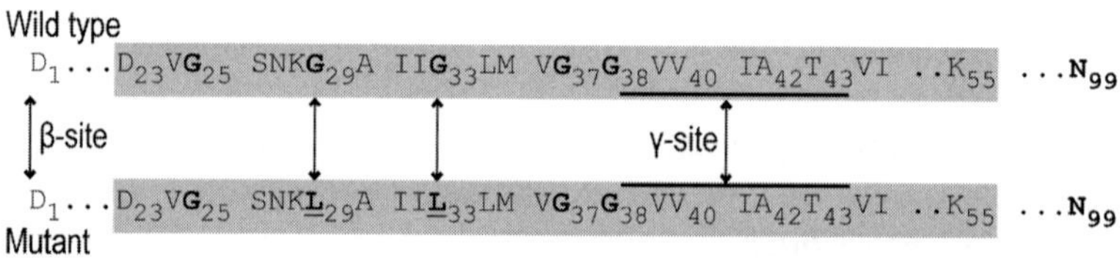

Figure 1. The amino-acid sequences of the wild type APP-C99 and a mutant, in which G29 and G33 are replaced with L29 and Leu33, respectively. Gly (G) in the transmembrane and juxtamembrane regions are shown in bold characters, whereas the substituted Leu are shown in bold and underlined characters. The peptide sequences used in this study (from D23 to K55) are in the gray boxes. β- and γ-cleavage sites are also indicated.

We have recently studied the structures of $A\beta_{23\text{-}55}$ located in the transmembrane (TM) and juxtamembrane regions (see the amino-acid sequence in Figure 1) [7,8]. $A\beta_{23\text{-}55}$ contains key residues in APP. There exist five Gly at the juxtamembrane regions, which appear to contribute the dimer formation of APP in the membrane. The Gly-XXX-Gly motif is known to promote dimerization of polypeptides via Cα-H···O hydrogen bonds between two segments in a membrane environment [9,10]. Recently, biochemical experiments have demonstrated the dimerization of the wild-type APP in the membrane [11,12]. Interestingly, it has also been reported that when Gly29 and Gly33 are replaced with Leu29 and Leu33, the mutant APP also forms a dimer as in the wild-type APP but cannot be cleaved by γ-secretase (which results in no production of Aβ). To resolve the apparent discrepancy, we predict dimer structures of both the wild-type and mutant APP by replica-exchange molecular dynamics (REMD) simulations [7,8,13]. Using the simulation data, we discuss molecular mechanisms underlying this mutation effect.

2. METHODS

2.1. *Replica-exchange molecular dynamics (REMD) method*

Here, we briefly describe the algorithm of REMD method [13] and discuss how the method is able to enhance conformational sampling of proteins or other systems with rugged energy landscapes. In REMD simulations, a number of non-interacting copies of the original system (or replicas) at different temperatures are simulated independently and simultaneously by conventional MD method. Every few steps, temperatures for pairs of replicas are exchanged with a transition probability according to the Metropolis criteria. This exchange process generates a random walk in the temperature space, which, in turn, induces a random walk in the potential energy space. Because the random walk in the potential energy space avoids trapping at one of the local energy minima of the simulated systems, REMD simulations enable sampling of a wider conformational space than conventional MD or Monte Carlo simulations. Thereby, REMD ensures statistically more reliable thermal averages at each temperature by using single or multiple histogram reweighting techniques.

2.2. *Simulation Procedures*

In this study, we used and modified MMTSB toolset [14] with CHARMM19 EEF1.1 force field parameters [15-18]. The effect of solvent and membrane was introduced using effective energy function 1 (EEF1; its extension to membrane is denoted as Implicit Membrane Model 1 (IMM1)) [16, 17], which is based on Gaussian solvent-exclusion approximation. In IMM1, (a) solvation parameters for a nonpolar phase have been introduced by using experimental data for the transfer amino acid side-chains from water to cyclohexane and (b) the distance-dependent dielectric model has been modified to account for reduced screening of electrostatic interactions in the membrane. In the preceding study [18], this model was used for predicting the dimeric structure of Glycophorin A and could reproduce the Gly-XXX-Gly interaction at the dimer interface. However, because of the approximate nature of implicit solvent model, IMM1 does not reproduce the direct interactions between lipid and polypeptide, which would be important for obtaining the potential of mean force as a function of helix-helix interaction. In the current study, we focus on the dimer structures of the wild type and mutant APP fragments obtained with the implicit membrane model [7].

The membrane thickness was set to 30 Å. In REMD simulations, we provided 32 replicas spanning temperatures from 300 K to 500 K. Each replica was run for 22 ns (in total 704 ns). The time step was set to 2 fs and replica exchanges were performed every 2 ps. The SHAKE method was applied to bonds involved hydrogen atoms. The Nosé-Hoover algorithm was used to impose a thermostat. A cylindrical wall boundary with a radius of using 30 Å was added around two APP fragments using the MMFP module of CHARMM to avoid the fragments from going far and away from each other.

The most stable conformation in the simulation of the APP monomer [8] was taken as the initial structure of an APP fragment in the current simulations. The two APP fragments were initially placed in the following manner: the initial inter peptide distance between the center of mass of two APP fragments was set to 20 Å.

The first fragment was placed at the center of a circle with a 20 Å radius and the second fragment was placed on the circle every 360/32 degrees, preserving the same orientation. The illustrations of peptides in all figures were prepared using VMD [19] and POV-Ray.

3. RESULTS AND DISCUSSION

We first check whether the REMD algorithm works properly or not in the current simulations, examining a random walk in the potential energy space. Figure 2 demonstrates how good the random walk is in the 10-th replica of mutant APP simulations. During the production dynamics (for 20 ns), the potential energy goes up and down between the average energies at the highest (500 K) and lowest (300 K) temperatures. Due to the random walk, the mutant APP fragments take a large number of different configurations in the simulation as shown in Figure 2. Thus, we successfully sample all the possible configurations of the APP fragments in the biological membrane.

Next, we investigate the differences in conformations of the wild-type and the mutant APP fragment at 300 K by principal component analysis (PCA) [20, 21] of the backbone atoms in the region from Gly29 to Thr43. Figure 3 shows the projection of the simulation trajectories at 300 K to the space spanned with the first (PC1) and third (PC3) principal component axes. By using the space, we obtain two major peaks for the mutant (PM1: 64.4 %; PM2: 28.5 %) and several other peaks for the wild type. Because of the backbone flexibility in the wild type, PCA cannot classify its major dimer structures. Instead, we use the Cα-H···O hydrogen bonds between the two fragments as a measure to classify the structures of the wild type APP.

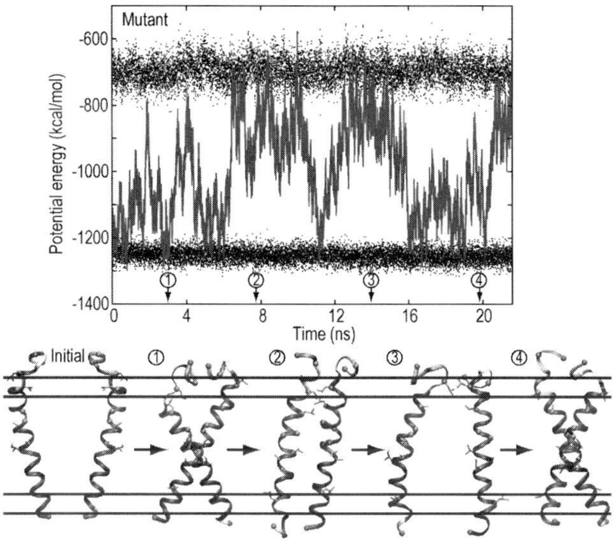

Figure 2. Time-series of potential energies (a) and conformations (b) of mutant APP fragments in the REMD simulations. Black dots in the time-series represent the potential energies at the highest (500 K) and the lowest (300 K) temperatures. The straight line shows the potential energies in the 10th replica. The five conformations in Figure 2(b) represent the snapshots obtained at the 0 ns (initial), 2.86 ns (1), 7.60 ns (2), 13.90 ns (3), and 19.88 ns (4), in the 10th replica's trajectory.

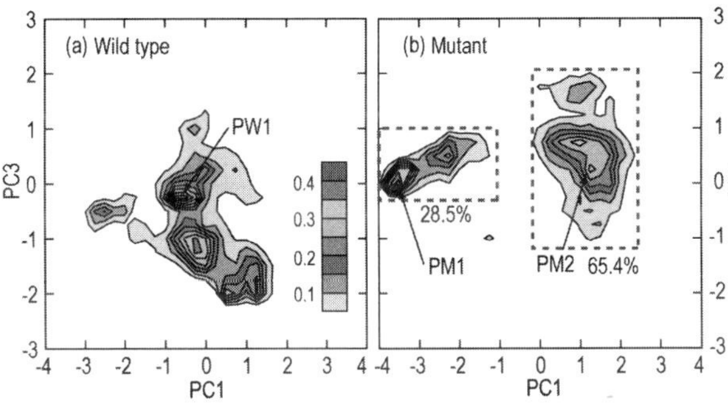

Figure 3. Projection of the simulation trajectories at 300 K to the principal component axes for the wild type (a) and the mutant (b) proteins.

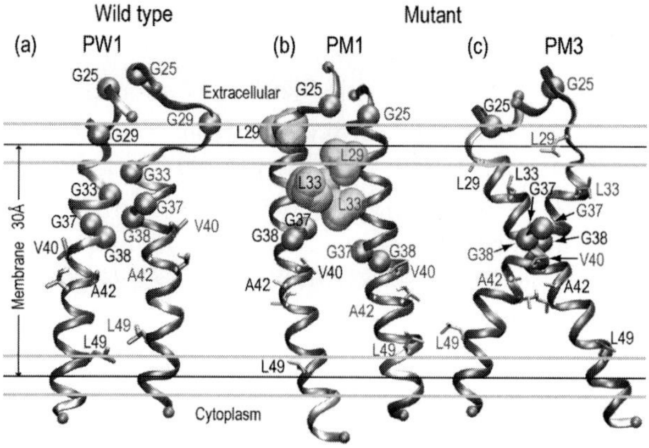

Figure 4. Structures of major dimer conformations of the wild-type (a), and the mutant (b and c). Gray boxes indicate specific dimerization sites, and are enlarged at the bottom. Red and orange dashes indicate strong and medium Cα-H···O hydrogen bonds, respectively.

As shown in Figure 4, we observe three different types of homodimer conformations of APP fragments in the REMD simulations. The first type observed only in the wild-type simulation is stabilized by Cα-H···O hydrogen bonds at the Gly-XXX-Gly motifs between two fragments. Of three Gly-XXX-Gly motifs in the wild-type APP, the hydrogen bonds involving Gly33 and Gly38 are the most essential interactions for the dimerization (Figure 4(a)). This causes a partial unwinding of the α-helices at the juxtamembrane region of the wild-type APP fragments (data not shown). Solid-state NMR spectroscopy has also shown that the Gly in the Gly-XXX-Gly motif lie at the dimer interface of the wild-type APP fragments [12].

In the second type of dimer conformation (Figure 4(b)), hydrophobic residues, namely, the mutated Leu29 and Leu33, face with each other to contribute the dimer formation. In addition to the mutated residues, Ile31, Leu34, or Val36 intervene between two APP fragments. In contract, two α-helices in the third type (Figure 4(c)) cross with each other at Gly38. This shape is similar to the homodimer structure of Glycophorin A observed with NMR experiments. Thus, the REMD simulations suggest that both the wild-type and mutant APP fragments form homodimer conformations in the biological membrane, but the structure of the mutant is significantly different from that of the wild type.

To produce Aβ peptides from APP, β- and γ-secretases bind APP and cleave it at the distinct cleavage sites. How do the conformational differences between the wild-type and the mutant APP fragments affect the secretion of Aβ or the cleavage by γ-secretase? Figure 5 demonstrates that the mutated two Leu (Leu29 and Leu33) are located more inside of the membrane than Gly29 and

Gly33 in the wild type, presumably due to the hydrophobic nature of the mutated amino-acid residues (Leu). Since Leu49 near the C-terminal for the wild-type and mutant APP are located almost at the same positions (Figure 5), α-helices for the mutant are tilted more than those for the wild type. As shown in Figure 5 (arrow near Gly38), the γ-site (Gly38-Thr43) in the mutant is shifted toward the center of the biological membrane by approximately 3 Å along the bilayer normal. In addition, the conformational flexibility of the γ-site might be increased in the mutant, because interfragment Cα-H···O hydrogen bonds at Gly38 become weaker than those in the wild type. These changes in the mutant APP fragments likely induce mismatched interactions between the γ-site of APP and the active site of γ-secretase, which would prohibit the secretion of $A\beta_{1\text{-}40}$ or $A\beta_{1\text{-}42}$.

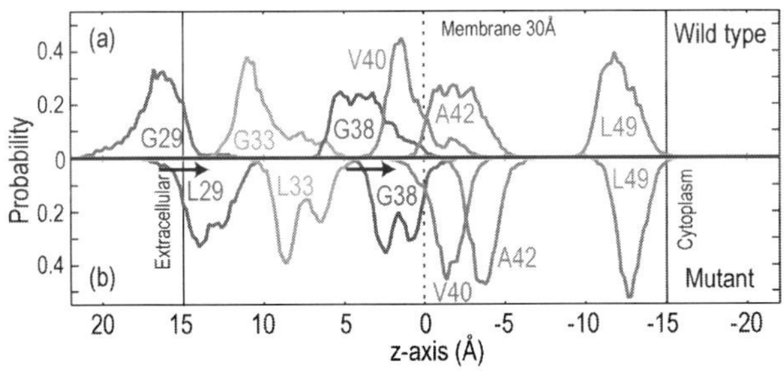

Figure 5. The distributions of the Cα positions of G29 (L29 in the mutant), G33 (L33 in the mutant), G38, V40, A42, and L49 along the z-axis (the bilayer normal) for the wild-type (a) and the mutant (b) proteins.

In summary, we have predicted the APP fragment ($A\beta_{23\text{-}55}$) dimer structures of the wild type and a mutant protein using REMD simulations and observed significant changes in the structures due to the mutation. The simulation results agree with the existing experimental data and provide fundamental insight into the initial steps in the amyloid formation. In addition to AD, it has been known that many human diseases are associated with the behavior of membrane proteins. We hope that the simulations in this direction will lead to a better understanding of the biological activities associated with the diseases and can be utilized for their treatments in the future.

Acknowledgments

This research was supported in part by a Grant for Scientific Research on a Priority Areas 'Membrane Interface' (to YS) and the Development and Use of the Next-Generation Supercomputer Project of the Ministry of Education

Culture, Sports, Science and Technology (MEXT), and by CREST & BIRD, Japan Science and Technology Agency (JST) (to YS). We thank the RIKEN Super Combined Cluster (RSCC) for providing computational resources.

References

1. C. Haass and D. J. Selkoe, *Nature Reviews Molecular Cell Biology* **8**, 101 (2007).
2. A. T. Petkova, W-M. Yau and R. Tycko, *Biochemistry* **45**, 498 (2006).
3. R. Nelson and D. Eisenberg, *Curr. Opin. Struc. Biol.* **16**, 260 (2006).
4. B. Tarus, J. E. Straub and D. Thirumalai, *J. Am. Chem. Soc.* **128**, 16159 (2006).
5. B. Tarus, J. E. Straub and D. Thirumalai, *J. Mol. Biol.* **345**, 1141 (2005).
6. F. Massi, J. W. Peng, J. P. Lee and J. E. Straub, *Biophys. J.* **80**, 31 (2001).
7. N. Miyashita, J. E. Straub, D. Thirumalai and Y. Sugita, *J. Am. Chem. Soc.* **131**, 3438 (2009).
8. N. Miyashita, J. E. Straub and D. Thirumalai, *J. Am. Chem. Soc.* in press.
9. R. A. MacKenzie, J. H. Prestegard and D. M. Engelman, *Science* **276**, 131 (1997).
10. A. Senes, I. Ubarretxena-Belandia and D. M. Engelman, *Proc. Natl. Acad. Sci. USA* **98**, 9056 (2001).
11. P. Kienlen-Campard, B. Tasiaux, J. V. Hees, M. Li, S. Huysseune, T. Sato, J. Z. Fei, S. Aimoto, P. J. Courtoy, S. O. Smith, S.N. Constantinescu and J. N. Octave, *J. Biol. Chem.* **283**, 7733 (2008).
12. T. Sato, T. Tang, G. Reubins, J.Z. Fei, T. Fujimoto, P. Kienlen-Campard, S. N. Constantinescu, J. N. Octave, S. Aimoto and S. O. Smith, *Proc. Natl. Acad. Sci. USA* **106**, 1421 (2009).
13. Y. Sugita and Y. Okamoto, *Chem. Phys. Lett.* **314**, 141 (1999).
14. M. Feig, J. Karanicolas and C. L. Brooks III, *J. Mol. Graphics and Modeling* **22**, 377 (2004).
15. A. D. MacKerell Jr., D. Bashford, M. Bellott., R. L. Dunbrack, J. D. Evanseck, M. J. Field, S. Fischer, J. Gao, H. Guo, S. Ha, D. Joseph-McCarthy, L. Kuchnir, K. Kuczera, F. T. K. Lau, C. Mattos, S. Michnick, T. Ngo, D. T. Ngyuen, B. Prodhom, W. E. Reiher III, B. Roux, M. Schelenkrich, J. C. Smith, R. Stote, J. E. Straub, M. Watanabe, J. Wiókiewicz-Kuczera, D. Yin and M. Karplus, *J. Phys. Chem. B* **102**, 3586 (1998).
16. T. Lazaridis and M. Karplus, *Proteins* **35**, 133 (1999).
17. T. Lazaridis, *Proteins* **52**, 176 (2003).
18. M. Mottamal, J. Zhang and T. Lazaridis, *Proteins* **62**, 996 (2006).
19. W. Humphrey, A. Dalke and K. Schulten, *J. Mol. Graphics and Modeling* **14**, 33 (1996).
20. A. Kitao and N. Go, *Curr. Opin. Struc. Biol.* **9**, 164 (1999).
21. A. E. García, *Phys. Rev. Lett.* **68**, 2696 (1992).

Quantum Bio-Informatics III
eds. L. Accardi, W. Freudenberg & M. Ohya

COMPARISON OF SQUARE CONTINGENCY TABLES USING MEASURE OF DEPARTURE FROM MARGINAL HOMOGENEITY

KOUJI TAHATA, KOUJI YAMAMOTO, NOBUKO MIYAMOTO, AND SADAO TOMIZAWA*

Department of Information Sciences, Tokyo University of Science, Noda City, Chiba 278-8510, Japan
**E-mail: tomizawa@is.noda.tus.ac.jp*

For the analysis of square contingency tables with same row and column classifications, Stuart (1955) considered the marginal homogeneity model which indicates that the row marginal distribution is identical with the column marginal distribution. Tahata, Iwashita and Tomizawa (2006, 2008) proposed the measure of departure from the marginal homogeneity model for square contingency tables with ordered categories. The present paper (1) reviews these measures and (2) compares between several Japanese social mobility data using these measures.

Keywords: Father-son data; Marginal homogeneity; Measure; Social-mobility; Square contingency table.

1. Introduction

The data in Table 1 are taken from Hashimoto (1999, p. 151). These data describe the cross-classification of father's and his son's social class in Japan which were examined in 1955, 1965, 1975, 1985 and 1995.

For father's and his son's social class data as the data in Table 1, many father-son pairs concentrate on (or near) the main diagonal cells. Therefore the independence between a father's social class and his son's social class does not hold. Namely, a son's social class is strongly associated with his father's social class. Thus, we want to see symmetry of father's and his son's social class instead of independence of them. For instance, we are interested in whether or not the distribution of son's social class is identical with the distribution of his father's social class.

For the analysis of square contingency tables with same row and column classifications as father-son social class data, the marginal homogeneity

Table 1. Cross-classification of father's and his son's social class; taken from Hashimoto (1999, p. 151).

(a) Examined in 1955

Father's class	Son's class Capitalist	New-Middle	Labor	Self-Support	Peasantry	Total
Capitalist	39	39	39	57	23	197
New-Middle	12	78	23	23	37	173
Labor	6	16	78	23	20	143
Self-Support	18	80	79	126	31	334
Peasantry	28	106	136	122	628	1020
Total	103	319	355	351	739	1867

(b) Examined in 1965

Father's class	Son's class Capitalist	New-Middle	Labor	Self-Support	Peasantry	Total
Capitalist	64	51	26	30	6	177
New-Middle	18	102	51	18	6	195
Labor	6	48	138	24	11	227
Self-Support	35	82	107	112	13	349
Peasantry	30	140	279	106	306	861
Total	153	423	601	290	342	1809

(c) Examined in 1975

Father's class	Son's class Capitalist	New-Middle	Labor	Self-Support	Peasantry	Total
Capitalist	29	43	25	31	4	132
New-Middle	23	159	89	38	14	323
Labor	11	69	184	34	10	308
Self-Support	42	147	148	184	17	538
Peasantry	42	176	377	114	298	1007
Total	147	594	823	401	343	2308

model which indicates that the row marginal distribution is identical with the column marginal distribution is available (Stuart, 1955; Bhapkar, 1966; Agresti, 2002, p. 422). When the marginal homogeneity model does not hold for the given data, we are also interested in seeing what degree the departure from the marginal homogeneity is. For square contingency tables

Table 1. (Continued)

(d) Examined in 1985

Father's class	Son's class					Total
	Capitalist	New-Middle	Labor	Self-Support	Peasantry	
Capitalist	46	59	34	39	3	181
New-Middle	20	193	79	29	2	323
Labor	9	122	202	44	4	381
Self-Support	23	122	127	144	6	422
Peasantry	24	148	285	104	126	687
Total	122	644	727	360	141	1994

(e) Examined in 1995

Father's class	Son's class					Total
	Capitalist	New-Middle	Labor	Self-Support	Peasantry	
Capitalist	68	48	36	23	1	176
New-Middle	33	191	102	33	3	362
Labor	25	147	229	34	2	437
Self-Support	48	119	146	129	5	447
Peasantry	40	126	192	82	88	528
Total	214	631	705	301	99	1950

with ordinal categories, Tahata, Iwashita and Tomizawa (2006, 2008) considered the measures which represent the degree of departure from marginal homogeneity.

The purpose of the present paper is (1) to review the marginal homogeneity model and two measures which represent the degree of departure from the marginal homogeneity and (2) to compare between five Japanese social mobility data in Table 1 using these measures.

2. Review of model and measures

We shall review briefly the marginal homogeneity model and the measure which represents the degree of departure from marginal homogeneity.

For an $r \times r$ square contingency table with ordinal categories, let p_{ij} denote the probability that an observation will fall in the row i and the column j of the table ($i = 1, \ldots, r; j = 1, \ldots, r$), and let X and Y denote the row and column variables, respectively. For the father's and his son's social class data, X is the father's social class and Y is the son's social class.

2.1. *Marginal homogeneity model*

The marginal homogeneity model is defined by

$$p_{i\cdot} = p_{\cdot i} \quad (i = 1, \ldots, r),$$

where $p_{i\cdot} = \sum_{t=1}^{r} p_{it}$ and $p_{\cdot i} = \sum_{s=1}^{r} p_{si}$ (Stuart, 1955). This model indicates that the row marginal distribution is identical with the column marginal distribution. For the father's and his son's social class data, this model indicates that the probability that a father's social class is i $(i = 1, \ldots, 5)$ is equal to the probability that the son's social class is i.

Let F_i^X and F_i^Y denote the cumulative marginal probabilities of X and Y, respectively; those are $F_i^X = \sum_{k=1}^{i} p_{k\cdot}$ and $F_i^Y = \sum_{k=1}^{i} p_{\cdot k}$ for $i = 1, \ldots, r-1$. Then the marginal homogeneity model may also be expressed as

$$F_i^X = F_i^Y \quad (i = 1, \ldots, r-1).$$

This states that the row cumulative marginal distribution is identical with the column cumulative marginal distribution.

Let S_i^X and S_i^Y denote the conditional cumulative marginal probabilities of X and Y, respectively; those are $S_i^X = \sum_{k=1}^{i} p_{k\cdot}^c$ and $S_i^Y = \sum_{k=1}^{i} p_{\cdot k}^c$ for $i = 1, \ldots, r-1$, where $p_{i\cdot}^c = (p_{i\cdot} - p_{ii})/\delta$, $p_{\cdot i}^c = (p_{\cdot i} - p_{ii})/\delta$, and $\delta = \sum\sum_{i \neq j} p_{ij}$. Then the marginal homogeneity model may be expressed as

$$S_i^X = S_i^Y \quad (i = 1, \ldots, r-1).$$

This states that the conditional row cumulative marginal probabilities are identical with the conditional column cumulative marginal probabilities, given that an observation will fall in one of the off-diagonal cells of the table.

2.2. *Measure of marginal homogeneity*

First Tahata et al. (2006) considered the measure which represents the degree of departure from marginal homogeneity below.

Assume that $F_1^X + F_1^Y \neq 0$ and $G_{r-1}^X + G_{r-1}^Y \neq 0$ where $G_i^X = \sum_{k=i+1}^{r} p_{k\cdot}$ and $G_i^Y = \sum_{k=i+1}^{r} p_{\cdot k}$ for $i = 1, \ldots, r-1$. The measure considered by Tahata et al. (2006), is defined by

$$\Phi^{(\lambda)} = \frac{1}{2}\left(\Phi_1^{(\lambda)} + \Phi_2^{(\lambda)}\right) \quad (\lambda > -1),$$

where

$$\Phi_1^{(\lambda)} = \frac{\lambda(\lambda+1)}{2^\lambda - 1} I^{(\lambda)}, \quad I^{(\lambda)} = \frac{1}{\lambda(\lambda+1)} \sum_{k=1}^{2} \sum_{i=1}^{r-1} \left[F_{k(i)}^* \left\{ \left(\frac{F_{k(i)}^*}{Q_{1(i)}^*} \right)^\lambda - 1 \right\} \right],$$

$$\Delta_1 = \sum_{i=1}^{r-1} (F_i^X + F_i^Y), \; F_{1(i)}^* = \frac{F_i^X}{\Delta_1}, \; F_{2(i)}^* = \frac{F_i^Y}{\Delta_1}, \; Q_{1(i)}^* = \frac{1}{2} \sum_{k=1}^{2} F_{k(i)}^*,$$

and $\Phi_2^{(\lambda)}$ is defined by $\Phi_1^{(\lambda)}$ with $\{F_i^X\}$ and $\{F_i^Y\}$ replaced by $\{G_i^X\}$ and $\{G_i^Y\}$, respectively, and the value at $\lambda = 0$ is taken to be the limit as $\lambda \to 0$. Thus

$$\Phi^{(0)} = \frac{1}{2} \left(\Phi_1^{(0)} + \Phi_2^{(0)} \right),$$

where

$$\Phi_1^{(0)} = \frac{1}{\log 2} I^{(0)}, \quad I^{(0)} = \sum_{k=1}^{2} \sum_{i=1}^{r-1} \left[F_{k(i)}^* \log \left(\frac{F_{k(i)}^*}{Q_{1(i)}^*} \right) \right],$$

and $\Phi_2^{(0)}$ is defined by $\Phi_1^{(0)}$ with $\{F_i^X\}$ and $\{F_i^Y\}$ replaced by $\{G_i^X\}$ and $\{G_i^Y\}$, respectively.

We see that the measure $\Phi^{(\lambda)}$ must lie between 0 and 1, and for each λ, (i) $\Phi^{(\lambda)} = 0$ if and only if there is a structure of marginal homogeneity in the $r \times r$ table, and (ii) $\Phi^{(\lambda)} = 1$ if and only if the degree of departure from marginal homogeneity is the largest, in the sense that $F_i^X = 0$ (then $G_i^X = 1$) and $F_i^Y = 1$ (then $G_i^Y = 0$), or $F_i^X = 1$ (then $G_i^X = 0$) and $F_i^Y = 0$ (then $G_i^Y = 1$), for arbitrary cut point i $(i = 1, \ldots, r-1)$.

Next Tahata et al. (2008) considered the measure to represent the degree of departure from marginal homogeneity below.

Assuming that $S_1^X + S_1^Y \neq 0$ and $T_{r-1}^X + T_{r-1}^Y \neq 0$ where $T_i^X = \sum_{k=i+1}^{r} p_{k\cdot}^c$ and $T_i^Y = \sum_{k=i+1}^{r} p_{\cdot k}^c$ for $i = 1, \ldots, r-1$, the measure, proposed by Tahata et al. (2008), is defined by

$$\Psi^{(\lambda)} = \frac{1}{2} \left(\Psi_1^{(\lambda)} + \Psi_2^{(\lambda)} \right) \quad (\lambda > -1),$$

where

$$\Psi_1^{(\lambda)} = \frac{\lambda(\lambda+1)}{2^\lambda - 1} I^{(\lambda)}, \quad I^{(\lambda)} = \frac{1}{\lambda(\lambda+1)} \sum_{k=1}^{2} \sum_{i=1}^{r-1} \left[S_{k(i)}^* \left\{ \left(\frac{S_{k(i)}^*}{Q_{2(i)}^*} \right)^\lambda - 1 \right\} \right],$$

$$\Delta_2 = \sum_{i=1}^{r-1} (S_i^X + S_i^Y), \; S_{1(i)}^* = \frac{S_i^X}{\Delta_2}, \; S_{2(i)}^* = \frac{S_i^Y}{\Delta_2}, \; Q_{2(i)}^* = \frac{1}{2} \sum_{k=1}^{2} S_{k(i)}^*,$$

and $\Psi_2^{(\lambda)}$ is defined by $\Psi_1^{(\lambda)}$ with $\{S_i^X\}$ and $\{S_i^Y\}$ replaced by $\{T_i^X\}$ and $\{T_i^Y\}$, respectively, and the value at $\lambda = 0$ is taken to be the limit as $\lambda \to 0$. Thus,

$$\Psi^{(0)} = \frac{1}{2}\left(\Psi_1^{(0)} + \Psi_2^{(0)}\right),$$

where

$$\Psi_1^{(0)} = \frac{1}{\log 2} I^{(0)}, \quad I^{(0)} = \sum_{k=1}^{2}\sum_{i=1}^{r-1}\left[S_{k(i)}^* \log\left(\frac{S_{k(i)}^*}{Q_{2(i)}^*}\right)\right],$$

and $\Psi_2^{(0)}$ is defined by $\Psi_1^{(0)}$ with $\{S_i^X\}$ and $\{S_i^Y\}$ replaced by $\{T_i^X\}$ and $\{T_i^Y\}$, respectively.

We see that $\Psi^{(\lambda)}$ must lie between 0 and 1, and for each λ, (i) $\Psi^{(\lambda)} = 0$ if and only if there is a structure of marginal homogeneity in the $r \times r$ table, and (ii) $\Psi^{(\lambda)} = 1$ if and only if the degree of departure from marginal homogeneity is the largest, in the sense that $S_i^X = 0$ (then $T_i^X = 1$) and $S_i^Y = 1$ (then $T_i^Y = 0$), or $S_i^X = 1$ (then $T_i^X = 0$) and $S_i^Y = 0$ (then $T_i^Y = 1$), for $i = 1, \ldots, r-1$.

3. Approximate confidence interval for measures

Let n_{ij} denote the observed frequency in the ith row and jth column of the table ($i = 1, \ldots, r; j = 1, \ldots, r$). Assuming that a multinomial distribution applies to the $r \times r$ table, consider an approximate standard error and large-sample confidence interval for the measure $\Phi^{(\lambda)}$ using the delta method as described by Bishop et al. (1975, Sec. 14.6) and Agresti (1990, Sec. 12.1). The sample version of $\Phi^{(\lambda)}$, i.e., $\widehat{\Phi}^{(\lambda)}$, is given by $\Phi^{(\lambda)}$ with $\{p_{ij}\}$ replaced by $\{\widehat{p}_{ij}\}$, where $\widehat{p}_{ij} = n_{ij}/n$ and $n = \sum\sum n_{ij}$. Using the delta method, $\sqrt{n}(\widehat{\Phi}^{(\lambda)} - \Phi^{(\lambda)})$ has asymptotically (as $n \to \infty$) a normal distribution with mean zero and variance $\sigma^2[\Phi^{(\lambda)}]$. For the detail of $\sigma^2[\Phi^{(\lambda)}]$, see Tahata et al. (2006).

Let $\widehat{\sigma}^2[\Phi^{(\lambda)}]$ denote $\sigma^2[\Phi^{(\lambda)}]$ with $\{p_{ij}\}$ replaced by $\{\widehat{p}_{ij}\}$. Then $\widehat{\sigma}[\Phi^{(\lambda)}]/\sqrt{n}$ is an estimated approximate standard error for $\widehat{\Phi}^{(\lambda)}$ and $\widehat{\Phi}^{(\lambda)} \pm z_{p/2}\widehat{\sigma}[\Phi^{(\lambda)}]/\sqrt{n}$ is an approximate $100(1-p)$ percent confidence interval for $\Phi^{(\lambda)}$, where $z_{p/2}$ is the percentage point from the standard normal distribution corresponding to a two-tail probability equal to p.

Also, we can get an approximate standard error and large-sample confidence interval for the measure $\Psi^{(\lambda)}$ in a similar way to the case of $\Phi^{(\lambda)}$ (for the details, see Tahata et al., 2008).

4. Comparison between Japanese social class data

Consider the data in Table 1. We denote the power-divergence statistic for testing goodness-of-fit of the marginal homogeneity model with $r-1=4$ degrees of freedom by $W_M^{(\lambda)}$, where

$$W_M^{(\lambda)} = \frac{2}{\lambda(\lambda+1)} \sum_{i=1}^{r} \sum_{j=1}^{r} n_{ij} \left[\left(\frac{n_{ij}}{\widehat{m}_{ij}} \right)^{\lambda} - 1 \right] \quad (-\infty < \lambda < \infty),$$

and $\widehat{m}_{ij}$ is a maximum likelihood estimator of expected frequency m_{ij}. Note that the values at $\lambda = -1$ and $\lambda = 0$ are taken to be the limit as $\lambda \to -1$ and $\lambda \to 0$, respectively. See Cressie and Read (1984) and Read and Cressie (1988, p.15) for details of the power-divergence test statistic. In particular, $W_M^{(0)}$ and $W_M^{(1)}$ are the likelihood ratio and the Pearson's chi-squared statistics, respectively. Although the details are omitted, each data in Table 1 fit the marginal homogeneity model very poorly. So, we now consider the degree of departure from marginal homogeneity using two measures.

First, we analyze the data in Table 1 using the measure $\Phi^{(\lambda)}$ proposed by Tahata et al. (2006). Since the confidence intervals for $\Phi^{(\lambda)}$ applied to the data in each of Tables 1a, 1b, 1c, 1d and 1e do not include zero for all λ (see Table 2), these would indicate that there is not a structure of marginal homogeneity in each table.

We shall investigate the degree of departure from marginal homogeneity in more details. For instance, when $\lambda = 0$, the estimated measure $\Phi^{(0)}$ equals 0.013 for Table 1a, 0.040 for Table 1b, 0.045 for Table 1c, 0.043 for Table 1d and 0.040 for Table 1e (see Table 2). Thus, (i) for Table 1a, the degree of departure from marginal homogeneity is estimated to be 1.3 percent of the maximum degree of departure from marginal homogeneity, (ii) for Table 1b, it is estimated to be 4.0 percent of the maximum degree of departure from marginal homogeneity, (iii) for Table 1c, it is estimated to be 4.5 percent of the maximum degree of departure from marginal homogeneity, (iv) for Table 1d, it is estimated to be 4.3 percent of the maximum degree of departure from marginal homogeneity, and (v) for Table 1e, it is estimated to be 4.0 percent of the maximum degree of departure from marginal homogeneity.

When the degrees of departure from marginal homogeneity in Tables 1a, 1b, 1c, 1d and 1e are compared using the confidence interval for $\Phi^{(\lambda)}$, it would be greater for Tables 1b, 1c, 1d and 1e than for Table 1a. Namely, for the father-son pairs, the distribution for a father's social class is different

Table 2. Estimate of measure $\Phi^{(\lambda)}$, estimated approximate standard error for $\widehat{\Phi}^{(\lambda)}$, and approximate 95% confidence interval for $\Phi^{(\lambda)}$, applied to Tables 1(a), 1(b), 1(c), 1(d) and 1(e).

(a) For Table 1(a)

λ	Estimated measure	Standard error	Confidence interval
−0.4	0.009	0.001	(0.007, 0.011)
0.0	0.013	0.002	(0.010, 0.016)
0.4	0.016	0.002	(0.012, 0.020)
1.0	0.018	0.002	(0.014, 0.023)

(b) For Table 1(b)

λ	Estimated measure	Standard error	Confidence interval
−0.4	0.028	0.002	(0.024, 0.032)
0.0	0.040	0.003	(0.034, 0.047)
0.4	0.049	0.004	(0.041, 0.056)
1.0	0.055	0.004	(0.047, 0.063)

(c) For Table 1(c)

λ	Estimated measure	Standard error	Confidence interval
−0.4	0.032	0.002	(0.027, 0.036)
0.0	0.045	0.003	(0.039, 0.052)
0.4	0.055	0.004	(0.047, 0.062)
1.0	0.062	0.004	(0.053, 0.070)

(d) For Table 1(d)

λ	Estimated measure	Standard error	Confidence interval
−0.4	0.030	0.002	(0.026, 0.034)
0.0	0.043	0.003	(0.037, 0.048)
0.4	0.051	0.003	(0.044, 0.058)
1.0	0.057	0.004	(0.050, 0.064)

(e) For Table 1(e)

λ	Estimated measure	Standard error	Confidence interval
−0.4	0.028	0.002	(0.024, 0.032)
0.0	0.040	0.003	(0.034, 0.045)
0.4	0.047	0.003	(0.040, 0.054)
1.0	0.053	0.004	(0.045, 0.060)

from that for his son's social class, and the degrees of difference are greater for Tables 1b, 1c, 1d and 1e than for Table 1a.

Secondly, we analyze the data in Table 1 using the measure $\Psi^{(\lambda)}$ proposed by Tahata et al. (2008). Since the confidence intervals for $\Psi^{(\lambda)}$ applied to each of Tables 1a, 1b, 1c, 1d and 1e do not include zero for all λ (see Table 3), these would indicate that there is not a structure of marginal homogeneity in each table.

We shall investigate the degree of departure from marginal homogeneity in more details. For instance, when $\lambda = 0$, the estimated measure $\Psi^{(0)}$ equals 0.043 for Table 1a, 0.120 for Table 1b, 0.122 for Table 1c, 0.108 for Table 1d and 0.098 for Table 1e (see Table 3). Thus, (i) for Table 1a, the degree of departure from marginal homogeneity is estimated to be 4.3 percent of the maximum degree of departure from marginal homogeneity, (ii) for Table 1b, it is estimated to be 12.0 percent of the maximum degree of departure from marginal homogeneity, (iii) for Table 1c, it is estimated to be 12.2 percent of the maximum degree of departure from marginal homogeneity, (iv) for Table 1d, it is estimated to be 10.8 percent of the maximum degree of departure from marginal homogeneity, and (v) for Table 1e, it is estimated to be 9.8 percent of the maximum degree of departure from marginal homogeneity.

When the degrees of departure from marginal homogeneity are compared using the confidence interval for $\Psi^{(\lambda)}$, it would be greater for Tables 1b, 1c, 1d and 1e than for Table 1a. Namely, for the father-son pairs such that a son's social class moved from his father's social class, the (conditional) distribution for a father's social class is different from that for his son's social class, and the degrees of difference between father's and his son's social class are greater for Tables 1b, 1c, 1d and 1e than for Table 1a.

5. Concluding remarks

The measures $\Phi^{(\lambda)}$ and $\Psi^{(\lambda)}$ always ranges between 0 and 1 independent of the dimension r and sample size n. Therefore, these measures may be useful for comparing the degrees of departure from marginal homogeneity in several tables (see Tahata et al., 2006, 2008).

For the father-son social class data which were studies in different five years, we have compared the degree of departure from marginal homogeneity by using the estimated measures $\widehat{\Phi}^{(\lambda)}$ and $\widehat{\Psi}^{(\lambda)}$, and the confidence intervals of the measures $\Phi^{(\lambda)}$ and $\Psi^{(\lambda)}$.

Lastly we point out that the measures $\Phi^{(\lambda)}$ and $\Psi^{(\lambda)}$ are not invariant

Table 3. Estimate of measure $\Psi^{(\lambda)}$, estimated approximate standard error for $\widehat{\Psi}^{(\lambda)}$, and approximate 95% confidence interval for $\Psi^{(\lambda)}$, applied to Tables 1(a), 1(b), 1(c), 1(d) and 1(e).

(a) For Table 1(a)

λ	Estimated measure	Standard error	Confidence interval
−0.4	0.030	0.004	(0.023, 0.037)
0.0	0.043	0.005	(0.033, 0.053)
0.4	0.051	0.006	(0.039, 0.063)
1.0	0.058	0.007	(0.044, 0.071)

(b) For Table 1(b)

λ	Estimated measure	Standard error	Confidence interval
−0.4	0.088	0.006	(0.076, 0.101)
0.0	0.120	0.008	(0.105, 0.136)
0.4	0.140	0.009	(0.122, 0.157)
1.0	0.153	0.010	(0.134, 0.172)

(c) For Table 1(c)

λ	Estimated measure	Standard error	Confidence interval
−0.4	0.090	0.006	(0.079, 0.101)
0.0	0.122	0.007	(0.108, 0.136)
0.4	0.141	0.008	(0.125, 0.157)
1.0	0.155	0.009	(0.138, 0.172)

(d) For Table 1(d)

λ	Estimated measure	Standard error	Confidence interval
−0.4	0.082	0.005	(0.072, 0.091)
0.0	0.108	0.006	(0.096, 0.120)
0.4	0.123	0.007	(0.109, 0.137)
1.0	0.134	0.007	(0.119, 0.149)

(e) For Table 1(e)

λ	Estimated measure	Standard error	Confidence interval
−0.4	0.075	0.005	(0.065, 0.085)
0.0	0.098	0.006	(0.086, 0.111)
0.4	0.112	0.007	(0.099, 0.126)
1.0	0.122	0.008	(0.107, 0.137)

under the same arbitrary permutations of row and column categories. So, these measures can be applied to square tables with ordinal categories.

References

1. Agresti, A. (1990). *Categorical Data Analysis.* Wiley, New York.
2. Agresti, A. (2002). *Categorical Data Analysis*, second edition. Wiley, New York.
3. Bhapkar, V. P. (1966). A note on the equivalence of two test criteria for hypotheses in categorical data. *Journal of the American Statistical Association*, **61**, 228-235.
4. Bishop, Y. M. M., Fienberg, S. E. and Holland, P. W. (1975). *Discrete Multivariate Analysis: Theory and Practice.* The MIT Press, Cambridge, Massachusetts.
5. Cressie, N. and Read, T. R. C. (1984). Multinomial goodness-of-fit tests. *Journal of the Royal Statistical Society, Ser. B*, **46**, 440-464.
6. Hashimoto, K. (1999). *Gendai Nihon no Kaikyuu Kouzou* (*Class Structure in Modern Japan: Theory, Method and Quantitative Analysis*). Toshindo Press, Tokyo (in Japanese).
7. Read, T. R. C. and Cressie, N. (1988). *Goodness-of-Fit Statistics for Discrete Multivariate Data.* Springer, New York.
8. Stuart, A. (1955). A test for homogeneity of the marginal distributions in a two-way classification. *Biometrika*, **42**, 412-416.
9. Tahata, K., Iwashita, T. and Tomizawa, S. (2006). Measure of departure from symmetry of cumulative marginal probabilities for square contingency tables with ordered categories. *SUT Journal of Mathematics*, **42**, 7-29.
10. Tahata, K., Iwashita, T. and Tomizawa, S. (2008). Measure of departure from conditional marginal homogeneity for square contingency tables with ordered categories. *Statistics*, **42**, 453-466.

Quantum Bio-Informatics III
eds. L. Accardi, W. Freudenberg & M. Ohya

ON THE STATICS FOR MICRO-ARRAY DATA ANALYSIS

TOMOKO URUSHIBARA, SHIZU AKASAKA, MAKIKO ITO, TOMONORI SUZUKI AND SATORU MIYAZAKI

Department of Medical and Life Science, Faculty of Pharmaceutical Sciences, Tokyo University of Science, 2641 Yamazaki, Noda, Chiba, 278-8510, Japan

Recently after human genome sequence has been determined almost perfectly, more and more researchers have been studying genes in detail. Therefore, we are sure that accumulated gene information for human will be getting more important in the near future to develop customized medicine and to make gene interactions clear. Among plenty of information, micro array might be one of the most important analysis method for genes because it is the technique that can get big amount of the gene expressions data from one time experiment and also can be used for DNA isolation. To get the novel knowledge from micro array data, we need to enrich statistical tools for its data analysis. So far, many mathematical theories and definition have been proposing. However, many of those proposals are tested with strict conditions or customized to data for specific species. In this paper, we reviewed existing typical statistical methods for micro array analysis and discussed the repeatability of the analysis, construction the guideline with more general procedure. First we analyzed the micro array data for TG rats, with statistical methods of family-wise error rate (FWER) control approach and False Discovery Rate (FDR) control approach. As existing report, no significantly different gene could be detected with FWER control approach. On the other hand, we could find several genes significantly with FDR control approach even q=0.5. To find out the reliability of FDR control approach with micro array conditions, we have analyzed 2 more pieces of data from Gene Expression Omnibus (GEO) public database on the web site with SAM in addition to FWER and FDR control approaches. We could find a certain number of significantly different genes with BH method and SAM in the case of q=0.05. However, we have to note that the number and kinds of detected genes are different when we compare our result with the one from the published paper. Even if the same approach is used to analyze the same micro array data, we might get a different result because the distinct definition for micro array data has not been set yet. It means that from the same data we will get different results depending on researchers. We are afraid that this problem will have a big effect on developing new medicines and to progress the next step, like a 2nd screening. So, we suggest that we should have certain guidelines to analyze Micro-Array data validly with statistic method and it will surely be helpful for Micro-Array analysis for medical studies in the future.

1. Introduction

In 2003, we got the announcement that Human Genome Project was finished completely [1]. It took 50 years since double-helix structure of DNA was found in 1953. These days many kinds of genes in this field have been studied by many researchers. And for customized medicines and breakthrough for gene interactions, gene analysis is getting more important. To analyze the genes that have been read, individual treatments of a huge number of them are inefficient and actually might be impossible. Therefore micro array experiment is commonly used for the 1st. screening. And then, to evaluate the data objectively to be obtained from micro array experiments, it is necessary to employ statistic tools. In the case of a static test with more than three hypotheses, FWER control approaches are widely used. However, we have more hypothesis, we have less detection powers in general. So, we have a care about selecting static test method with taking multiple hypotheses into account. To solve the problem, FDR control approach has been developed. FDR controls the ratio between wrongly rejected hypothesis and rejected hypothesis. As a result, we can expect that the detection power is increased. At this point in time, there is no defined rule and few guidelines to choose analysis tools for micro array data, that is, it depends on researchers. So, we have high possibility that the different significant tools predict different number of the significant genes even if the data set is the same. This must be a big matter and something to solve it should be shown. We are afraid that we make no significance of truly significant gene instead of opposite occasions, because we might lose something important. We need to think how to make this type errors reduced. In this study, at first, we have shown to compare and validate the evidence of features on FWER control approach and FDR control approaches with in-house rat micro array data. And then, we have analyzed the data sets obtained from public database on-line which are mice and rat micro array data and discussed to compare the results in published paper and ours. We have had some questions and discussed better choice of statistic methods for analysis especially for micro array data including the dependency on species and experimental conditions.

2. Materials

3 Micro array data sets have been studied as below.

The first one is rat micro array data (data1), which has 31099 genes and was kindly provided by Dr. Higami. In the present study, 3 groups were prepared: nontransgenic male Wistar rats ad libitum intake (AL), calorie restricted (CR) rats and heterotransgenic rats (tg/-: TG) were used, whereas homozygous (tg/tg) rats were not included due to their premature death from neoplastic diseases [2]. Each group had 4 animals. AL group rats continued to receive food at libtium, and CR group rats were provided with 140% of the mean daily intake of AL group in each rat 30 min before the light were turned off every other day. All rats were killed at 6 months of age to collect tissues. For this micro array experiments, epididymal white adipocyte was provided, rapidly frozen on dry ice and stored in liquid nitrogen until micro array experiments (Affymetrix GeneChip) were carried out [3].

And next, we got 2 micro array data sets form public database Gene Expression Omnibus (GEO) on the web site [4].

1. GDS-1261 (data2)

This one is mice micro array data and also CR experiments, which has 12488 genes tested. 2 groups were prepared: nontransgenic female Ames NL mice (control), and CR mice. Each group had 8 animals. Starting at the age of 2 months, 16 mice were divided into 2 groups randomly. Mice were killed at 6 months of age, and their livers were removed, rapidly frozen on dry ice and stored in liquid nitrogen [5].

2. GDS-851 (data3)

This is rat micro array data and 15923 probe sets were put to the test. It is again rat experiment's data but not CR experiment's. 2 adult Sparaue-Dawley rats were prepared for this experiment. Orbital and Global layers (OL and GL) of extraocular muscles (EOM) for both sides of rats were removed for sampling under sterile conditions [6]. That is to say, EOM were sectioned then, the OL and GL layers of EOM were selectively dissected out and were stored at -80°C until for micro array experiment.

3. Methods

3.1. *Multiple Comparisons*

3.1.1. *FWER control approach*

Type I errors with multiple comparison is controlled in FWER control approach. Bonferroni method is widely and commonly used as a multiple comparison, FWER control approach. Calculating formulas of Bonferroni method were indicated as we see below [7],

$$P=n*p\ (\alpha)$$

P: α error (level of significance)
N: total number of the hypothesis
p: calculated level of significance

Statistics were carried out with calculated p (α) for each gene data.

3.1.2. *FDR control approaches*

In this study, as FDR control approaches, BH, ABH BY method and SAM were selected as below.

Benjamini-Hochberg (BH) method:

BH method was developed as FDR control approach by Benjamin et.al. in 1995. It has been used in the case of the independent hypothesis. Its calculating formulas are described in [8].

Adaptive Benjamini-Hochberg (ABH) method:

ABH method is also an FDR control approach which is modified from BH method, and developed in 2000 by Benjamin et.al.. ABH has sensitive detection compare to BH method. Its calculating formulas are described in [9].

Benjamini-Yekurieli (BY) method:

BY method is also an FDR control approach based on BH method. It was established in 2001 by Benjamini et.al.. This is for treating the hypothesis with dependent hypothesis. Its calculating formulas are described in [10].

Significance Analysis of Microarray (SAM):

SAM was developed by Gilbert C. at Stanford Univ. in 2000 [11]. The SAM is a testing procedure for micro array analysis, which estimates the FDR with permutations under the assumption that all null hypotheses are true. The procedure consists of three components: (1) the adjusted test statistics, (2) approximation of the distribution of the test statistic based on permutations, (3) the control of FDR. The t-test statistics is modified to the SAM procedure. Calculation formulas were described in [12] at length and [14] was also referred.

3.2. *How to get p scores*

It was supposed that micro array experiments had been performed for each gene group on separate days. Thus, there might be some possibilities that data has any biases, for example like different operators, weather conditions, time and so on, among each gene group. And so, to get p scores, we used 3 ways as follows in this study.

1. To hypothesize that all genes distributions were equal, that is to say, no bias is existed. In this case, Student t-test was used to get p scores for all genes.
2. This was totally opposite from the first. It was presumed that all genes distributions are not equal, that means there were some bias there. In this situation, Welch t-test was used for all genes.

3. To screen whether the distribution of the gene was equal or not equal for each gene with F-test (α=0.05). If the result of F-test says its distribution is equal, Student t-test will be selected to get p score for the gene. If it is not equal by F-test, Welch t-test will be selected. Multiple comparisons were carried out with p scores calculated as described above.

3.3. *Data analysis environment*

Software "R" was used for data analysis in this study. R has been developed since 1993 by Ross Ihaka and Robert Gentleman and this software is free and it is for statistic calculation and graphics. R is easy to get, its calculation speed is fast and is able to not only calculate but also create graphics and practice statistic tests. The URL for R ver. 2.3.0 software is http://www.r-project.org/.

Our computer simulations are conducted by personal computers with following specs;

OS: Mac OS X
CPU: Intel Core 2 Duo
Main memories: 2Gbyte
Hard disk size: 150Gbyte

3.4. *Simulations*

To analysis micro array data on the programmed R, we programmed some scripts for carrying out FDR control approaches like BH, ABH and BY method. Practically we calculated q value in variously setting conditions for optional 2 groups data as below whether it could retain criterion value of FDR, q<=0.05 to evaluate the capability of this programmed R. Practically. Simulation conditions were set by reference to the paper [14]. Table 1 showed that 8 patterns of simulations for each BH, ABH and BY method.

It was hypothesized that all hypothesis were independent and the number of simulation was 10000 times in all patterns.

Table 1 Simulating conditions for optional 2 groups to evaluate the capability of programmed R

m0	**5**	**15**	**25**	**10**	**20**	**5**	**15**	**20**
m1	**5**	**5**	**5**	**10**	**10**	**15**	**15**	**20**

m0 the number of true positive
m1 the number of false positive

4. Result, Conclusion and future work

Before data analysis, we carried out simulations to check R programmed by us. The result was shown in Table 2. All p values in the different approaches could be controlled around 0.05, and they were almost the same as described in [14]. It has some features that all FDR values figured by ABH were near 0.05. They were controlled better than FDR values by BH and BY. All FDR values in BY are quite low, that is, it was shown that BY method is very conservative one. On these 8 patterns of simulation we could get almost the same values in every 3 FDR control approaches from the original result, now we are sure that our program is accurately programmed and software quality of this R script can be used for further micro array data analysis. Second, we analyzed CR rat micro array data (in-house data) with this program. With FWER control approach, Bonferroni, no significant gene was detected (data not shown). This outcome of the experiment was predicted because the number of the hypotheses was over 30000. For example, in the case of 10000 hypothesis we have and α=0.05 we set, for each hypothesis α=0.000005 will be calculated with Bonfferoni approach. Actually, this is too strict to detect any genes significantly. In this case, we can avoid Type I error (False positive), however there are some probabilities to have Type II error (False negative). That would be wrong.

Table 2 Calculated FDR values under several settings of simulation with artificial data

(q-value:0.05)
(10000 Times)

Null hypothesis (n)	True False	5 5	15 5	25 5	10 10	20 10	5 15	15 15	20 20
BH		**0.025**	**0.036**	**0.041**	**0.026**	**0.013**	**0.025**	**0.025**	**0.017**
ABH		**0.043**	**0.047**	**0.049**	**0.046**	**0.042**	**0.047**	**0.047**	**0.044**
BY		**0.008**	**0.011**	**0.011**	**0.007**	**0.004**	**0.006**	**0.006**	**0.004**

Table 3 The number of genes detected by different acquisition of q-value, AL group vs. CR group

q value	Student	Welch	Mixture
0.5	**139**	**0**	**1**
0.6	**1205**	**4**	**708**
0.7	**2721**	**4**	**1979**

Table 4 The number and percentage of genes detected in the different conditions for statistic test with BH and ABH method, AL group vs. CR group

Genes expressed in total genes (n)	All genes 31099		More than 1rat 24871		More than 4 rats 21868	
q value	detection(n)	ratio(%)	detection(n)	ratio(%)	detection(n)	ratio(%)
0.50	0	0	0	0	22	0.1
0.55	29	0.09	23	0.09	23	0.11
0.60	32	0.1	6030	25.45	9019	41.24

Table 5 The number of genes detected by different acquisition of q-value, AL group vs. TG group

q value	Student	Welch	Mixture
0.5	0	0	0
0.6	32	0	30
0.7	14234	0	13728

Table 6 The number and percentage of genes detected in the different conditions for statistic test BH and ABH method, AL group vs. TG group

Genes expressed in total genes (n)	All genes 31099		More than 1rat 24871		More than 4 rats 21868	
q value	detection(n)	ratio(%)	detection(n)	ratio(%)	detection(n)	ratio(%)
0.45	1	0.003	120	0.48	525	2.4
0.50	139	0.45	693	2.79	944	4.32
0.55	736	2.37	1224	4.92	1510	6.91
0.60	1205	3.87	1921	7.72	2070	9.47

Next, we analyzed the same data with FDR control approach with our expectations for statistical power improved. The results were shown in Table 3 - 6. Table 3 and 4 show the comparison between AL group and CR group. Under $p<0.5$ conditions, any genes could be detected (data not shown). Table 3 showed the results of A: in the case of assuming all as equal variance, B: in the case of assuming all as unequal variance, C: after F-test, we set the assumption (evaluate equal or unequal variance for each). The numbers of detected genes compared with A and C, were very close, and the number of significantly different genes detected with A condition were a little bit bigger than that with C condition, that is, in this study we could assume that all can be analyzed as they are equal variances. Only 4 genes were significantly detected with condition B by BH and ABH method. We have some causes for concern that BH method was too strict to detect any genes as Bonferroni method and Type II errors might happen. You know p=0.5 or more is too big to evaluate significant genes. So we made an attempt to set cutoff value for this data analysis to make detection power higher. We set 2 types of cutoff value as described below: 1. to choose the genes that were expressed in 1 rat or more of 12 rats in the tissues (24871 genes). 2. to choose the genes that were expressed in 4 rats or more of 12 rats (21868 genes). The result was shown in Table 4. In fact the detection power was increased a little bit but not satisfied. Still setting p<0.45, nothing could be detected.

The results to compare AL group and TG group were shown in table 5 and 6. As you can see, the features of that result were very similar to the result of AL vs. CR. When all genes were analyzed, even under p=0.55 and p=0.6, only 29 and 32 significantly different genes were detected respectively, those ratios were just about 0.1 % of all. And we have a little confidence with our results under p>0.5. p=0.5 means significantly detected gene (s) would be 50% correct and 50% wrong. It is too dangerous to judge "significant" with p=0.5. It is hard to say this is properly significantly. In general, p value is set at 0.05 or 0.01, 0.5 is so far from 0.05 or 0.01 in fact. Some of 30000 genes have no change in all 12 rats actually and because this fact might make statistical power decreased, we would like to set cutoff value. So, we checked the distribution of this data we studied. Contrary to my expectation, it showed a uniform distribution (data not shown) therefore, it seemed to be hard to set the cutoff value to make statistical power increased. Thus, from our result, it was shown that FDR control approach is more useful than FWER control approach like Bonferroni method to analyze micro array data with huge number of the hypothesis and ABH method, which is modified BH method, is the most sensitive of all 3 kinds of FDR control approach.

As for variance, we analyzed the data with 3 patterns as we saw below, 1, all hypothesis were homoscedastic, 2, all hypothesis were unequal homoscedastic, and 3, for each hypotheses after we executed t-test, Student or Welch was chosen. There was no big difference between 1 and 3 we supposed that in this case, we can presume its homoscedasticity. And now the biggest problem we found is that in the case of p<0.5, nothing could be detected significantly. It is unclear that specifically it depends on this case or species tendency or the problem of the programming or something else. And so, to clear those problems, we got and analyzed CR mice data and non CR rats data that are different species CR data and rat data of a different clinical condition from public database GEO which is run by National Center for Biotechnology Information

(NCBI) in America. And in the next studies, BH method, neither ABH nor BY was performed to compare published data in addition to SAM. First we analyzed GDS1261 (total gene number 12488, mice CR) under these conditions: $\alpha<0.05$, $q<0.05$ (Table 7). Only 5 statistically different genes were detected with FWER control approach, that is, Bonfferoni method and 110 statistically different genes were detected with BH method. And 390 statistically different genes were detected with SAM, this overlapped perfectly 110 statistically different genes with BH method. In the published study [5], BH method was used and 177 statistically different genes were detected, however, just 38 statistically different genes were overlapped between the result in one of published studies and ours. It is wondering. Second, as the first one GDS851 (total gene number 15923, rat) was analyzed (Table 8). Only 1 statistically different gene was detected by FWER control approach. 50 statistically different genes were detected with BH method, they were perfectly overlapped 217 statistically different genes by SAM method and it was a comparable result from the first data. 216 genes were statistically different in the published study [6] and this number of the detected genes is very similar to 217 detected genes by SAM, but actually to compare our result by BH method, just 29 statistically different genes were identified. Compared with SAM method, BH method was likely to be strict one. Considering that micro array experiment has been often used for the 1st screening, SAM may be better for statistic analysis for micro array data in some cases. Nonetheless they said that they detected significantly different genes by FDR control approach, that is to say BH method in their original papers, the result didn't correspond with ours really, it seemed not to get analysis reproducibility. As for this reason, there are some possibilities that they set cutoff value or subjectively controlled genes for detections before objectively statistic analysis. In this study, we have validated the evidence for 2 micro array data sets, and more validations may be needed to make this matter clearer.

As the former rat CR data, by neither FWER control approach nor FDR control approaches, we couldn't detect significantly different genes sensitively on any settings, but some points at issue and lines of studies were straightened out. As latter micro array data sets, some genes could be detected significantly differently by FDR control approaches like BH method and SAM under $q<0.05$. Among them, the different point is existed. It is that the number of genes analyzed is over 30000 genes with rat CR data but others have just about half.
It is suggested that analysis for a huge kinds of genes at once had much less detection power than for around 15000 kinds of genes. Selections for statistic approach and cutoff value might be requested and might depend on the number of genes.

Statistical analysis are necessary for us to evaluate experimental data objectively, and it is suggested that the result from statistic analysis must be stable and in a reproducible fashion. Especially for the 1st screening like micro array data, the future plans, decisions and conclusions will be dependent on the result of statistic analysis, so to analyze data statistically is much more important. But the present situation is far removed from the ideal. So, we suggest that we should have a certain guideline to analyze Micro-Array data validly with statistic method and it will be very helpful for Micro-Array analysis absolutely.

Table 7 The number of genes detected with the different statistic tests for GDS-1261 ($\alpha<0.05$, $q<0.05$)

Method	Bonferroni	BH	SAM	Original Paper
Deteced Genes (n)	5	110	390	177

Table 8 The number of genes detected with the different statistic tests for GDS-851 ($\alpha<0.05$, $q<0.05$)

Method	Bonferroni	BH	SAM	Original Paper
Deteced Genes (n)	1	50	217	216

Acknowledgments

We thank Dr. Higami for providing Micro array data and his continuous supports.

References

1. Finishing the euchromatic sequence of the human genome. *Nature* 2004: **431**, 931-945
2. Zha Y, Taguchi T, Nazneen A, Shimokawa I, Higami Y, Razzaque MS. Genetic suppression of GH-IGF-1 activity, combined with lifelong caloric restriction, prevents age-related renal damage and prolongs the life span in rats. *Am J Nephrol.* 2008; **28(5)**:755-64. Epub 2008 Apr 24
3. Shimokawa I, Higami Y, Tsuchiya T, Otani H, Komatsu T, Chiba T, Yamaza H. Life span extension by reduction of the growth hormone-insulin-like growth factor-1 axis: relation to caloric restriction. *FASEB J.* 2003 Jun; **17(9)**:1108-9.

4. Edgar R, Domrachev M, Lash AE.Gene Expression Omnibus: NCBI gene expression and hybridization array data repository. *Nucleic Acids Res.* 2002 Jan 1;**30(1)**:207-10
5. Tomoshi.T, Joseph M. Dhahbi, Xinping.C, et al. Addictive regulation of hepatic gene expression by dwarfism and caloric restriction. *Physiol Genomics* 2004: **10**:307-315
6. Murat T.Budak,Sasha Bogdanovich,Martin H.J.Wiesen,et al. Layer-specific differences of gene expression in extraocular muscles identified by laser-capture microscopy. *Physiol Genomics* 2004 : **20**:50-65
7. Dunn-Sidak Ury, H. K.: A comparison of four procedures for multiple comparisons among means (pairwise contrasts) for arbitrary sample sizes. *Technometrics* 1976: **18**:89-97.
8. Benjamini.Y, Hochberg .Y. Controlling the False Discovery Rate. *J.Statist* 1995: **57**:289-300
9. Benjamini, Y., Hochberg, Y. : On the Adaptive Control of the False Discovery Rate in Multiple Testing with Independent Statistics. *J. Edu. Behavioral Statistics* 2000: **25(1)**, 60-83.
10. Benjamini.Y, Yekutieli .D. The Control of the False Discovery Rate in Multiple Testing under Dependency. *The annals of Statistic.* 2001: **4**:1165-1188
11. Tusher VG, Tibshirani R, Chu G. Significance analysis of microarrays applied to the ionizing radiation response. *PNAS* 2001: **98(9)**: 5116-21
12. Hirakawa A, Sato Y, Chikuma H, Yoshimura I. A new test statistic based on Shrunken sample variance for identifying differentially expressed genes in small microarray experiments. *Bioinfomatics and Biology Insights* 2008: **2** 145-156
13. Horiuchi K, Matusda S. *Academia* 2005: **6**:17-30.
14. Gil .C, Balasubramanian .N, Robert .T. SAM "Significance analysis of Microarrays" Users guide and technical document.

Quantum Bio-Informatics III
eds. L. Accardi, W. Freudenberg & M. Ohya

FUNCTIONAL MECHANICS AND TIME IRREVERSIBILITY PROBLEM

IGOR V. VOLOVICH
Steklov Mathematical Institute,
Gubkin St.8, 119991 Moscow, Russia
E-mail:volovich@mi.ras.ru

The time irreversibility problem is the problem of how to explain that there is the reversible microscopic dynamics and the irreversible macroscopic physics. In this paper an attempt is performed of the following solution of the irreversibility problem: a formulation of microscopic dynamics is suggested which is irreversible in time. In this way the contradiction between the reversibility of microscopic dynamics and irreversibility of macroscopic dynamics is avoided since both dynamics in the proposed approach are irreversible.

A widely used notion of microscopic state of the system at a given moment of time as a point in the phase space does not have an immediate physical meaning since arbitrary real numbers are non observable. In the approach presented in this paper the physical meaning is attributed not to an individual trajectory but only to a bunch of trajectories or to the distribution function on the phase space.

The fundamental equation of the microscopic dynamics in the proposed "functional" approach is not the Newton equation but the Liouville equation for the distribution function of the single particle. Solutions of the Liouville equation have the property of delocalization which accounts for irreversibility. It is shown that the Newton equation in this approach appears as an approximate equation describing the dynamics of the average values of the position and momenta for not too long time intervals. Corrections to the Newton trajectories are computed. Possible applications to the information and molecular dynamics are mentioned.

1. Introduction

The time irreversibility problem is the problem of how to explain the irreversible behaviour of macroscopic systems from the time-symmetric microscopic laws. The problem has been discussed by Boltzmann, Poincaré, Bogolyubov, Kolmogorov, von Neumann, Landau, Ohya, Prigogine, Feynman and many other authors [1] - [10] and it deserves a further study.

In particular, in works by Poincaré [1], Landau and Lifshiz [3], Prigogine [6], Feynman [10] it is stressed that the irreversibility problem is still an open problem. Poincaré [1] said that perhaps we will never solve the irreversibility problem. Landau and Lifshiz write about the principle of increasing entropy [3]: "Currently it is not clear whether the law of increasing entropy can be in principle derived from classical mechanics." Landau speculated that to explain the second law of thermodynamics one has to use quantum mechanical measurement arguments.

The microscopic mechanical description of a system assumes that the state of the system at a given moment of time is represented by a point in the phase space with an invariant measure and the dynamics of the system is described by a trajectory in the phase space. It is assumed that the microscopic laws of motion are known (Newton or Schrodinger equations) and there is a problem of derivation from them the macroscopic (Boltzmann, Navier-Stokes,...) equations, see for example, [3,4].

Note that the stochastic limit [9] gives a systematic method for investigation of irreversible processes.

In this paper we attempt to suggest the following approach to the irreversibility problem and to paradoxes of Loschmidt and Poincaré – Zermelo: we propose a *formulation of microscopic dynamics which is irreversible in time.* Thus the contradiction between microscopic reversibility and macroscopic irreversibility of the dynamics disappears, since both microscopic and macroscopic dynamics in the proposed approach are irreversible.

Note that the conventional widely used concept of the microscopic state of the system at some moment in time as the point in phase space, as well as the notion of trajectory and the microscopic equations of motion have no direct physical meaning, since arbitrary real numbers not observable (observable physical quantities are only presented by rational numbers, cf. the discussion of concepts of space and time in [11,12,13,14,15]).

The fundamental equation of the microscopic dynamics of the proposed functional probabilistic approach is not Newton's equation, but a Liouville equation for distribution function. It is well known that the Liouville equation is used in statistical mechanics for the description of the motions of gas. Let us stress that we shall use the Liouville equation for the description of a single particle in the empty space.

Although the Liouville equation is symmetric in relation to the reversion of time, but his solutions have the property of *delocalization*, that, generally speaking, can be interpreted as a manifestation of irreversibility. It is understood that if at some moment in time the distribution function describes a particle, localized to a certain extent, then over time the degree of localization decreases, there is the spreading of distribution function. Delocalization takes place even for a free particle in infinite space, where there is no ergodicity and mixing.

In the functional approach to classical mechanics we do not derive the statistical or chaotic properties of deterministic dynamics, but we suggest that the Laplace's determinism at the fundamental level is absent not only in quantum, but also in classical mechanics.

It is shown that Newton's equation in the proposed approach appears as an approximate equation describing the dynamics of the average values of coordinates and momenta for not too long time. Corrections to Newton's trajectories are computed. Possible applications to the information and molecular dynamics are mentioned.

2. States and Observables in Functional Classical Mechanics

Usually in classical mechanics the motion of a point body is described by the trajectory in the phase space, i.e. the values of the coordinates and momenta as functions of time, which are solutions of the equations of Newton or Hamilton.

Note, however, that this mathematical model is an idealization of the physical process, rather far separated from reality. The physical body always has the spatial dimensions, so a mathematical point gives only an approximate description of the physical body. The mathematical notion of a trajectory does not have direct physical meaning, since it uses arbitrary real numbers, i.e. infinite decimal expansions, while the observation is only possible, in the best case, of rational numbers, and even them only with some error. Therefore, in the proposed "functional" approach to classical mechanics, we are not starting from Newton's equation, but with the Liouville equation.

Consider the motion of a classical particle along a straight line in the potential field. The general case of many particles in the 3-dimensional space is discussed below. Let (q, p) be co-ordinates on the plane $\mathbb{R}^2$ (phase space), $t \in \mathbb{R}$ is time. The state of a classical particle at time t will be described by the function $\rho = \rho(q, p, t)$, it is the density of the probability that the particle at time t has the coordinate q and momentum p.

Note that the description of a mechanical system with the help of probability distribution function $\rho = \rho(q, p, t)$ does not necessarily mean that we are dealing with a set of identically prepared ensemble of particles. Usually in probability theory one considers an ensemble of events and a sample space. But we can use the description with the function $\rho = \rho(q, p, t)$ also for individual bodies, such as planets in astronomy (the phase space in this case the 6-dimensional). In this case one can think on the "ensemble" of different astronomers which observe the planet, or on the "ensemble" of different models of behaviour of a given object for one "intelligent" observer. Actually, it is implicitly always dealt with the function $\rho = \rho(q,p,t)$ which takes into account the inherent uncertainty in the coordinates and momentum of the body. An application of these remarks to quantum mechanics will be discussed in a separate work.

The specific type of function ρ depends on the method of preparation of the state of a classical particle at the initial time and the type of potential field. When $\rho = \rho(q,p,t)$ has sharp peaks at $q = q_0$ and $p = p_0$, we say that the particle has the approximate values of coordinate and momentum q_0 and p_0.

Emphasize that the exact derivation of the coordinate and momentum can not be done, not only in quantum mechanics, where there is the Heisenberg uncertainty relation, but also in classical mechanics. Always there are some errors in setting the coordinates and momenta. There are classical uncertainty relations:

$$\Delta q > 0\,, \quad \Delta p > 0\,,$$

i.e. the uncertainty (errors of observation) in the determination of coordinate and momentum is always positive (non zero). The concept of arbitrary real numbers, given by the infinite decimal series, is a mathematical idealization, such numbers can not be measured in the experiment.

Therefore, in the functional approach to classical mechanics the concept of precise trajectory of a particle is absent, the fundamental concept is a distribution function $\rho = \rho(q,p,t)$ and δ-function as a distribution function is not allowed.

We assume that the continuously differentiable and integrable function $\rho = \rho(q,p,t)$ satisfies the conditions:

$$\rho \geq 0, \quad \int_{\mathbb{R}^2} \rho(q,p,t)dqdp = 1, \; t \in \mathbb{R}\,. \tag{1}$$

If $f = f(q,p)$ is a function on phase space, the average value of f at time t is given by the integral

$$\overline{f}(t) = \int f(q,p)\rho(q,p,t)dqdp\,. \tag{2}$$

In a sense we are dealing with a random process $\xi(t)$ with values in the phase space. Motion of a point body along a straight line in the potential field will be described by the equation

$$\frac{\partial \rho}{\partial t} = -\frac{p}{m}\frac{\partial \rho}{\partial q} + \frac{\partial V(q)}{\partial q}\frac{\partial \rho}{\partial p}. \tag{3}$$

Here $V(q)$ is the potential field and mass $m > 0$.

Equation (3) looks like the Liouville equation which is used in statistical physics to describe a gas of particles but here we use it to describe a single particle.

If the distribution $\rho_0(q,p)$ for $t = 0$ is known, we can consider the Cauchy problem for the equation (3):

$$\rho|_{t=0} = \rho_0(q,p). \tag{4}$$

Let us discuss the case when the initial distribution has the Gaussian form:

$$\rho_0(q,p) = \frac{1}{\pi ab} e^{-\frac{(q-q_0)^2}{a^2}} e^{-\frac{(p-p_0)^2}{b^2}}. \tag{5}$$

At sufficiently small values of the parameters $a > 0$ and $b > 0$ the particle has coordinate and momentum close to the q_0 and p_0. For this distribution the average value of the coordinates and momentum are:

$$\overline{q} = \int q\rho_0(q,p)dqdp = q_0, \quad \overline{p} = \int p\rho_0(q,p)dqdp = p_0, \tag{6}$$

and dispersion

$$\Delta q^2 = \overline{(q-\overline{q})^2} = \frac{1}{2}a^2, \quad \Delta p^2 = \overline{(p-\overline{p})^2} = \frac{1}{2}b^2. \tag{7}$$

3. Free Motion

Consider first the case of the free motion of the particle when $V = 0$. In this case the equation (3) has the form

$$\frac{\partial \rho}{\partial t} = -\frac{p}{m}\frac{\partial \rho}{\partial q} \tag{8}$$

and the solution of the Cauchy problem is

$$\rho(q,p,t) = \rho_0(q - \frac{p}{m}t, p). \tag{9}$$

Using expressions (5), (9),

$$\rho(q,p,t) = \frac{1}{\pi ab}\exp\{-\frac{(q-q_0-\frac{p}{m}t)^2}{a^2} - \frac{(p-p_0)^2}{b^2}\}, \tag{10}$$

we get the time dependent distribution of coordinates:

$$\rho_c(q,t) = \int \rho(q,p,t)dp = \frac{1}{\sqrt{\pi}\sqrt{a^2+\frac{b^2t^2}{m^2}}}\exp\{-\frac{(q-q_0-\frac{p_0}{m}t)^2}{(a^2+\frac{b^2t^2}{m^2})}\}, \tag{11}$$

while the distribution of momenta is

$$\rho_m(p,t) = \int \rho(q,p,t)dq = \frac{1}{\sqrt{\pi}b} e^{-\frac{(p-p_0)^2}{b^2}}. \tag{12}$$

Thus, for the free particle the distribution of the particle momentum with the passage of time does not change, and the distribution of the coordinates change. There is, as one says in quantum mechanics, the spreading of the wave packet. From (11) it follows that the dispersion Δq^2 increases with time:

$$\Delta q^2(t) = \frac{1}{2}(a^2 + \frac{b^2 t^2}{m^2}). \tag{13}$$

Even if the particle was arbitrarily well localized (a^2 is arbitrarily small) at $t = 0$, then at sufficiently large times t the localization of the particle becomes meaningless, there is a *delocalization* of the particle.

4. Liouville Equation and the Newton Equation

In the functional classical mechanics the motion of a particle along the stright line is described by the Liouville equation (3). A more general Liouville equation on the manifold Γ with coordinates $x = (x^1, ..., x^k)$ has the form

$$\frac{\partial \rho}{\partial t} + \sum_{i=1}^{k} \frac{\partial}{\partial x^i}(\rho v^i) = 0. \tag{14}$$

Here $\rho = \rho(x,t)$ is the density function and $v = v(x) = (v^1, ..., v^k)$ - vector field on Γ. The solution of the Cauchy problem for the equation (14) with initial data

$$\rho|_{t=0} = \rho_0(x) \tag{15}$$

might be written in the form

$$\rho(x,t) = \rho_0(\varphi_{-t}(x)). \tag{16}$$

Here $\varphi_t(x)$ is a phase flow along the solutions of the characteristic equation

$$\dot{x} = v(x). \tag{17}$$

5. Corrections to Newton's Equations

For the free particle in the functional approach to classical mechanics the averages coordinates and momenta satisfy the Newton equations. However, when there is a nonlinear interaction, then in functional approach corrections to the Newton's equations appear.

Consider the motion of a particle along the line in the functional mechanics. Average value $\overline{f}$ of the function on the phase space $f = f(q,p)$ at time t is given by the integral (2)

$$\overline{f}(t) =< f(t) >= \int f(q,p)\rho(q,p,t)dqdp\,. \tag{18}$$

Here $\rho(q,p,t)$ has the form (16)

$$\rho(q,p,t) = \rho_0(\varphi_{-t}(q,p))\,. \tag{19}$$

By making the replacement of variables, subject to the invariance of the Liouville measure, we get

$$< f(t) >= \int f(q,p)\rho(q,p,t)dqdp = \int f(\varphi_t(q,p))\rho_0(q,p)dqdp\,. \tag{20}$$

Let us take

$$\rho_0(q,p) = \delta_\epsilon(q-q_0)\delta_\epsilon(p-p_0)\,, \tag{21}$$

where

$$\delta_\epsilon(q) = \frac{1}{\sqrt{\pi}\epsilon}e^{-q^2/\epsilon^2}\,, \tag{22}$$

$q \in \mathbb{R}$, $\epsilon > 0$.

Let us show that in the limit $\epsilon \to 0$ we obtain the Newton (Hamilton) equations:

$$\lim_{\epsilon\to 0} < f(t) >= f(\varphi_t(q_0,p_0))\,. \tag{23}$$

Proposition 5.1. *Let the function $f(q,p)$ in the expression (18) be continous and integrable, and ρ_0 has the form (21). Then*

$$\lim_{\epsilon\to 0}\int f(q,p)\rho(q,p,t)dqdp = f(\varphi_t(q_0,p_0))\,. \tag{24}$$

Proof. Functions $\delta_\epsilon(q)$ form a δ-sequence in $D^{'}(\mathbb{R})$ [16]. Hence we obtain

$$\begin{aligned}\lim_{\epsilon\to 0}\int f((q,p))\rho(q,p,t)dqdp &= \lim_{\epsilon\to 0}\int f(\varphi_t(q,p))\delta_\epsilon(q-q_0)\delta_\epsilon(p-p_0)\\ &= f(\varphi_t(q_0,p_0))\,,\end{aligned} \tag{25}$$

that was required to prove. □

Now calculate the corrections to the solution of the equation of Newton. In functional mechanics consider the equation, see (3),

$$\frac{\partial \rho}{\partial t} = -p\frac{\partial \rho}{\partial q} + \lambda q^2 \frac{\partial \rho}{\partial p}\,. \tag{26}$$

Here λ is a small parameter and we set the mass $m = 1$. The characteristic equations have the form of the following Hamilton (Newton) equations:

$$\dot{p}(t) + \lambda q(t)^2 = 0\,,\ \dot{q}(t) = p(t)\,. \tag{27}$$

Solution of these equations with the initial data

$$q(0) = q,\ \dot{q}(0) = p \tag{28}$$

for small t has the form

$$(q(t), p(t)) = \varphi_t(q,p) = (q + pt - \frac{\lambda}{2}q^2t^2 + ...,\ p - \lambda q^2 t + ...) \tag{29}$$

Use the asymptotic expansion $\delta_\epsilon(q)$ in $D^{'}(\mathbb{R})$ for $\epsilon \to 0$, compare [9,17]:

$$\delta_\epsilon(q) = \delta(q) + \frac{\epsilon^2}{4}\delta^{''}(q) + ...\,, \tag{30}$$

then for $\epsilon \to 0$ we obtain corrections to the Newton dynamics:

$$< q(t) >= \int (q + pt - \frac{\lambda}{2}q^2t^2 + ...)[\delta(q - q_0) + \frac{\epsilon^2}{4}\delta^{''}(q - q_0) + ...] \tag{31}$$

$$\cdot[\delta(p - p_0) + \frac{\epsilon^2}{4}\delta^{''}(p - p_0) + ...]dqdp = q_0 + p_0 t - \frac{\lambda}{2}q_0^2t^2 - \frac{\lambda}{4}\epsilon^2 t^2\,.$$

Denoting the Newton solution

$$q_{\text{Newton}}(t) = q_0 + p_0 t - \frac{\lambda}{2}q_0^2 t^2\,,$$

we obtain for small ϵ, t and λ:

$$< q(t) >= q_{\text{Newton}}(t) - \frac{\lambda}{4}\epsilon^2 t^2\,. \tag{32}$$

Here $-\frac{\lambda}{4}\epsilon^2t^2$ is the correction to the Newton solution received within the functional approach to classical mechanics with the initial Gaussian distribution function. If we choose a different initial distribution we get correction of another form.

We have proved

Proposition 5.2. *In the functional approach to mechanics the first correction at ϵ to the Newton dynamics for small t and λ for equation (27) has the form (32).*

6. Irreversibility in the Functional Mechanics

The reversibility of motion in classical mechanics deals with the individual trajectory. In the functional mechanics the concept of the individual trajectory of the particle has no direct physical sense. Instead, the state of the particle is described by the distribution function $\rho = \rho(q, p, t)$ which satisfies the Liouville equation (3)

$$\frac{\partial \rho}{\partial t} = -\frac{p}{m}\frac{\partial \rho}{\partial q} + \frac{\partial V(q)}{\partial q}\frac{\partial \rho}{\partial p}. \tag{33}$$

The Liouville equation is invariant under the replacement t to $-t$: if $\rho = \rho(q, p, t)$ is the solution of the equation (33), then $\sigma(q, p, t) = \rho(q, -p, -t)$ - also its solution. However this symmetry does not mean reversibility of the motion of a particle in the functional approach to mechanics, since the state of the particle is described there by the distribution function and the phenomenon of delocalization takes place.

In this way we obtain an answer to the arguments of Loschmidt and Poincaré - Zermelo. Indeed, to reverse the particle motion at the time moment $t = T$ as it is proposed in the Loschmidt argument, it is necessary make the co-ordinate and momentum measurement. But it will change the distribution $\rho(q, p, T)$. Further, it is necessary to prepare such condition of the particle that its evolution back in time would lead to the initial distribution ρ_0 that is difficult since the delocalization takes place. We will need something even better than Maxwell's demon.

Concerning the Zermelo argument related with the Poincaré recurrence theorem we note that this argument can not be applied to the functional mechanics because this argument is based on the notion of individual trajectory. In the functional mechanics the state of the system is characterized by the distribution function and here the mean values might irreversibly tend to some limits without contradiction with the Poincaré theorem as it will be shown in the next section.

The Poincaré theorem is not applicable to the bunch of trajectories or even to two trajectories as it follows from the Lyapunov theory: if two points are situated in some small region of the phase space then they are not necessary come back to this region by moving along their trajectories.

7. Dynamics of a Particle in a Box

Dynamics of collisionless continuous medium in a box with reflecting walls is considered in [1,7,8]. This studied asymptotics of solutions of Liouville equation. In functional approach to mechanics, we interpret the solution of the Liouville equation as described the dynamics of a single particle.

Consider the motion of a free particle on the interval with the reflective ends. Using the method of reflections [16], the solution of the Liouville equation(8)

$$\frac{\partial \rho}{\partial t} = -\frac{p}{m}\frac{\partial \rho}{\partial q}$$

on the interval $0 \leq q \leq 1$ with the reflective ends we write as

$$\rho(q,p,t) = \sum_{n=-\infty}^{\infty} [\rho_0(q - \frac{p}{m}t + 2n, p) + \rho_0(-q + \frac{p}{m}t + 2n, -p)]\,, \qquad (34)$$

where it is assumed that the function ρ_0 has the Gaussian form (5).

One can show that for the distribution for coordinates

$$\rho_c(q,t) = \int \rho(q,p,t)dp \qquad (35)$$

one gets the uniform limiting distribution (pointwise limit):

$$\lim_{t\to\infty} \rho_c(q,t) = 1\,.$$

For the distribution of the absolute values of momenta $(p > 0)$

$$\rho_a(p,t) = \rho_m(p,t) + \rho_m(-p,t)\,,$$

where

$$\rho_m(p,t) = \int_0^1 \rho(q,p,t)dq\,,$$

as $t \to \infty$ we get the distribution of the Maxwell type (but not the Maxwell distribution):

$$\lim_{t\to\infty} \rho_a(p,t) = \frac{1}{\sqrt{\pi}b}[e^{-\frac{(p-p_0)^2}{b^2}} + e^{-\frac{(p+p_0)^2}{b^2}}]\,.$$

8. Conclusions

In this paper the functional formulation of classical mechanics is suggested which is based not on the notion of an individual trajectory of the particle but on the distribution function on the phase space.

It would be interesting to apply the functional approach to classical mechanics (i.e. the Liouville equation with Gaussian initial distribution for the particles) to various dynamical problems [18], in particular to the problems of the molecular dynamics [19].

Acknowledgements

The author expresses his sincere thanks to Masanori Ohya and Noboru Watanabe for the kind invitation to the very stimulating QBIC 2009 Conference and interesting discussions and to Ichiro Yamato for introduction to his approach to molecular dynamics. The work is also partially supported by grants NS-3224.2008.1, RFBR 08-01-00727-a.

References

1. Poincaré A. "Le mecanisme et l'experience", Revue de Metaphysique et de Morale 1, pp. 534-7 (1893); English translation, Stephen Brush, Kinetic Theory, vol.2, p.203.
2. Bogolyubov N.N., Problems of Dynamic Theory in Statistical Physics, [in Russian], OGIZ (1946).
3. Landau L.D., Lifshitz E.M. Statistical physics, part 1 (3 ed., Pergamon, 1980)
4. Zubarev D. N. Nonequilibrium Statistical Thermodynamics, New York, Consultants Bureau, (1974).
5. Ohya M. Some Aspects of Quantum Information Theory and Their Applications to Irreversible Processes. Rep. Math. Phys., vol. 27, pp. 19-47, 1989.
6. Prigogine I. Les Lois du Chaos. Flammarion. Paris. 1994.
7. Kozlov V.V. Temperature equilibrium on Gibbs and Poincaré, Moscow-Ijevsk (in Russian), 2002.
8. Kozlov V.V. Gibbs ensembles and nonequilibrium statistical mechanics, Moscow-Ijevsk (in Russian), 2008.
9. Accardi L., Lu Y. G. and Volovich I. Quantum Theory and Its Stochastic Limit. Berlin: Springer, 2002.
10. Feynman R. The character of physical law, A series of lectures recorded by the BBC at Cornell University USA, Cox and Wyman Ltd., London, 1965.
11. Volovich I. V. "Number theory as the ultimate physical theory", Preprint No. TH 4781/87, CERN, Geneva, 1987.
12. Volovich I.V. p-adic string. // Class. Quant. Grav. V. 4. P. L83–L87. 1987.
13. Vladimirov V.S., Volovich I.V., Zelenov E.I. p–Adic Analysis and Mathematical Physics. Singapore: World Scientific, 1994.
14. Khrennikov A. Yu. Non-Archimedean analysis: quantum paradoxes, dynamical systems and biological models, Kluwer Acad. Publishers, Dordreht/Boston/London, 1997.
15. Dragovich B., Khrennikov A. Yu., Kozyrev S. V. and Volovich I. V. On p -adic mathematical physics.// P-Adic Numbers, Ultrametric Analysis, and Applications. Vol. 1. N. 1. P. 1-17. 2009.
16. Vladimirov V.S. Equations of Mathematical Physics, Marcel Dekker, New York, 1971.
17. Pechen A.N., Volovich I.V. Quantum Multipole Noise and Generalized Quantum Stochastic Equations. Infinite Dimensional Analysis, Quantum Probability and Related Topics, Vol. 5, No 4 (2002) 441-464.

18. Inoue K., Ohya M. and Sato K. Application of chaos degree to some dynamical systems. Chaos, solitons and Fravtals, 11 (2000)1377-1385.
19. Ando T. and Yamato I. Basics of molecular simulation and its application to biomolecules. In: Quantum Bio-Informatics II, Eds. L. Accardi, W. Freudenberg and M. Ohya, World Sci. 2009, pp.228-241.

Quantum Bio-Informatics III
eds. L. Accardi, W. Freudenberg & M. Ohya

REGULATORY NETWORKS: INFERRING FUNCTIONAL RELATIONSHIPS THROUGH CO-EXPRESSION

DIERK WANKE*, ACHIM HAHN, JOACHIM KILIAN, KLAUS HARTER AND KENNETH W. BERENDZEN*

ZMBP Pflanzenphysiologie, Universität Tübingen
Auf der Morgenstelle 1, D-72076 Tübingen, Germany
**E-mail: dierk.wanke@zmbp.uni-tuebingen.de*
**E-mail: kenneth.berendzen@zmbp.uni-tuebingen.de*

Gene expression data not only provide us insights into discrete transcript abundance of specific genes, but contain cryptic information that can not readily be assessed without interpretation. We again used data of the plant *Arabidopsis thaliana* as our reference organism, yet the analysis presented herein can be performed with any organism with various data sources. Within the cell, information is transduced via different signaling cascades and results in differential gene expression responses. The incoming signals are perceived from upstream signaling components and handed to downstream messengers that further deliver the signals to effector proteins which can directly influence gene expression. In most cases, we can assume that proteins, which are connected to other signaling components within such a regulatory network, exhibit similar expression trajectories. Thus, we extracted a known functional network from literature and demonstrated that it is possible to superimpose microarray expression data onto the pathways. Thereby, we could follow the information flow through time reflected by gene expression changes. This allowed us to predict, whether the upstream signal was transmitted from known elements contained in the network or relayed from outside components. We next conducted the vice versa approach and used large scale microarray expression data to build a co-expression matrix for all genes present on the array. From this, we computed a regulatory network, which allowed us to deduce known and novel signaling pathways.

Keywords: Large scale gene expression data, co-expression matrix analysis, gene regulatory networks, signal transduction pathway, AtGenExpress

1. Introduction

During the last decade complete genome sequences became available for many species of all kinds of organism kingdoms, mainly those considered to be long known and well examined model organisms. However, progress in biotechnology allows us nowadays to surpass the restrictions of using non-model organisms. Affordable massive bulk genome sequencing approaches were the main driving forces and now it is possible for researchers start sequencing their organism of interest or go for transcript analysis by sequencing every mRNA template present in a cell.

While at present large scale studies on gene expression still require the information of a known genome sequence and the approximate number of genes present, this will be obsolete in the near future. However, entirely novel tools are urgently needed do curate on the enormous piles of information. Therefore, we need to exploit the information of present days model organisms to link the future data to. The deeper our understanding of our most studied organisms is, the better we will be able to exploit future's "data avalanche".

One of the best studied and understood eukaryotic model organisms to date is the flowering plant *Arabidopsis thaliana.* At the beginning of the new millennium, the DFG-funded Arabidopsis Functional Genomics Network (AFGN) AtGenExpress project has been initiated to uncover the transcriptome of Arabidopsis thaliana with the help of microarrays [1]. Microarray technology constitutes an effective means to get an insight into gene expression of all Arabidopsis genes simultaneously [2]. By now, the public has access to more than 1300 AtGenExpress gene expression profile datasets (http://www.uni-tuebingen.de/plantphys/AFGN/atgenex.htm).

Most informative for biologists and information scientists are those genes, which are co-expressed with other genes [3]. These are genes with highly similar expression trajectories over multiple experimental conditions[4,5]. More importantly, co-expressed genes are believed to constitute the best candidates that might likely fulfill a common function. Either they encode for proteins that act in the same signaling cascades, are enrolled in the same enzymatic pathways or represent interacting partners that form higher order complexes [5,6]. Thus, gene expression implies additional cryptic information that can not readily be assessed without additional processing and interpretation.

Within the cell, information is transduced via different signaling cascades and can result in differential gene expression responses. For example, environmental biotic or abiotic stimuli will at first act on the plant's surface, but its information will be conveyed to the inside and will finally be answered with a quantitative and qualitative response inside the cell [1]. These incoming signals are handed stepwise from upstream signaling components to downstream messengers and further on to effector proteins some of which, directly orchestrate distinct gene expression changes. In most cases, we have to assume that this information is not relayed in a linear fashion, but through complex regulatory networks. It has been shown that proteins, which are connected to other signaling components within such a regulatory network, can exhibit similar expression trajectories [6,7,8]. Recently, biologists and information scientists joint forces to exploit this linkage and to gain a new quality of meta-information.

Here, we give two examples of extracting biological information from the combined use of large scale gene expression data with regulatory network displays. First, we wanted to gain insight into where and when regulatory information operates which components of a known regulatory network. From this, we were able to deduce whether the signal is relayed by signaling elements contained within our network or acts from the outside on components within. Second, we retrieved regulatory network components based on co-expression analysis. Here, we assumed that functionally related genes will have similar expression trajectories and thus, enable us to deduce functional links between the elements contained in such a regulatory network.

2. Material and Methods

2.1. *Preprocessing of the Arabidopsis thaliana Gene Expression Data*

The Data from the AtGenExpress abiotic stress Data set [1] comprising the Affymetrix CEL file accession numbers at TAIR (http://www.arabidopsis.org): ME00325, ME00326, ME00327, ME00328, ME00329, ME00330, ME00338, ME00339 and ME00340 was used. From the heat stress data (ME00329) the cell culture results were excluded. The data was normalized with the GCRMA package [9] from Bioconductor (http://www.bioconductor.org). This contains background correction, normalization across arrays and calculation of probe level intensities as log2 values. The summary file was used for further analysis.

2.2. *Incorporate transcript regulation information into the Functional Regulatory Network*

Significant regulated genes/probsets were determined with the Limma package [10]. Default settings were used except that the p-value cutoff was set to 0.001. The resulting genes are regarded as regulated ignoring direction of change compared to the control. In this article we exemplarily focused on the UV-B stress results/analysis. Transcript regulatory information was incorporated into the functional regulatory network via Cytoscape (http://www.cytoscape.org).

2.3. *Identification and Graph construction of the Interaction Network from co-expression data*

Co-expressed genes for seed-genes were obtained from ATTED-II ver. c4.1 (http://atted.jp) [8]. A manual cut-off was set to a Pearson correlation of 0.45. A directed graph was made containing all genes (nodes) with co-expression as the edges using Cytoscape v.2.6.1 [11]. Co-expression tables from ATTED-II were converted to source (seed-gene) to target (co-expressed gene) nodes and imported into Cytoscape. The degree value was calculated using the Cytoscape Degree Sorted Circle layout and the Cytoscape Edge-weighted Spring Embedded layout for identifying cliques. The minimal hub value was chosen at a value to capture nodes with degree values in the top 5% (>19).

2.4. *Correlation Analysis and Statistics*

Pearson Correlations have been conducted either with Microsoft Excel or the GeneSpring GX software package from Agilent Technologies. Hypergeometric distribution from regulated gene lists has been computed with Microsoft Excel.

3. Results

3.1. *Imposing expression data on known networks*

One of the challenging tasks in the analysis of microarray expression data is the extraction of meaningful information. Most informative are co-expressed genes, which exhibit highly similar expression trajectories over different experimental conditions. It has previously been shown that gene products have a high probability to be involved in similar enzymatic pathways or are enrolled in the same signaling cascades when they are significantly co-expressed over many conditions [4,6,8].

A very simplified view of a signaling cascade is that it functions in a linear fashion (Fig. 1) and includes at least one receptor to perceive an incoming signal and a transcription factor to translate the information into gene expression changes.

The well studied example of such a simple linear signal transduction pathway is represented by the bacterial two-component system, where the changes in the activity state of the involved protein components is mediated by the delivery of a single phosphate residue [12]. In a mechanistic termed phosphorelay, the incoming signal is perceived by an upstream receptor Histidine-kinase and relayed to the downstream transcription factor Response Regulator. In this case, there is no amplification of the transferred information. In more complex eukaryotes, the principle of this kind of linear signaling pathway is retained as a modified form and is a multistep-two component system, which allows a more modulated response. Here, a second messenger protein serves as a link to shuttle the information between receptors and response regulators. However, the most frequent signaling pathways utilize several second messengers that can convey their signaling information upon activation by upstream components to more than one downstream target. First, amplification of the signaling information sometimes occurs, which allows the perception of very low amounts of upstream signals, i.e. hormones, to be sufficient for a full cellular response. Second, interconnectivity of multiple messengers ignites a complex response of recursive and iterative information flows that are necessary to translate a signal into a concerted expression response in a multicellular organism.

One of the best understood types of signaling cascades are those involving MAP kinases. A downstream MAP kinase (MPK) is activated by a MAP kinase kinase (MKK) that is positioned upstream. Accordingly, MKKs are activated by upstream MKK kinases (MAPKKK), and so on. At the very top of such a signaling pathway resides a receptor protein that ignited the activation cascade.

A well known example of such a MAP kinase signaling pathway in Arabidopsis thaliana is depicted in Figure 1, which is involved in the resistance to bacterial infections [13]. The receptor FLS2 perceives the presence of disease causing gram-negative bacteria and activates the downstream MAPKKK MEKK1, which in turn activates MKK4 and MKK5. Both, MKK4 and MKK5, can transfer the information to the downstream MPK3 and MPK6. Finally, this signal information is conveyed to WRKY transcription factors, which are common downstream transcription factors involved in MAP kinase signaling. In this example, the presence of bacterial effectors is perceived by FLS2 receptor and ends with transcriptional changes imposed by WRKY22 and WRKY29 transcription factors (Fig. 1).

The relationship between MAP kinases and WRKY transcription factors has been reported several times in literature [13,14,15]. Therefore, we

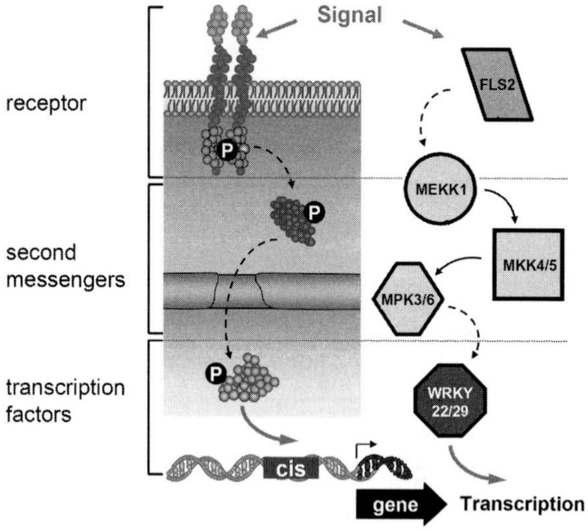

Figure 1. Scheme for a representative signal transduction cascade in eukaryotes. A signal is perceived by a receptor that translates the incoming information into chemical reactions, which will determine the activity state of downstream components. In many cases, this chemical information is transduced by mobile second messenger proteins to transcription factors, which subsequently will influence the gene expression of their respective target genes. On the right hand side, we depicted an example of a well characterized MAP kinase signaling cascade from *Arabidopsis thaliana* [13].

constructed a regulatory network from known physical, transitory and genetic interactions from literature. In total, 31 components have been identified and a regulatory network has been compiled (Fig. 2): it consists of two receptors, 18 second messengers (3 MKKKs, 7 MKKs, 8 MPKs) and 11 MAPK target proteins, including WRKY transcription factors. Depending on the information reported in the literature the components have been depicted at different 'levels'. Especially for the MAP kinase second messengers, a clear positioning within three levels could be assigned on the basis of their known downstream kinase activation [16].

The regulatory network depicted in Figure 2, is composed of any set of objects that are connected by some sort of relationship. These objects are called Nodes (vertices, or singular vertex) and the connections Edges. Undirected graphs are sets of unordered nodes and edges, whereas directed graphs have connections that are ordered (Fig. 3).

Once we consider that the MAP Kinases only phosphorylate their downstream targets, our regulatory network constructed from literature becomes

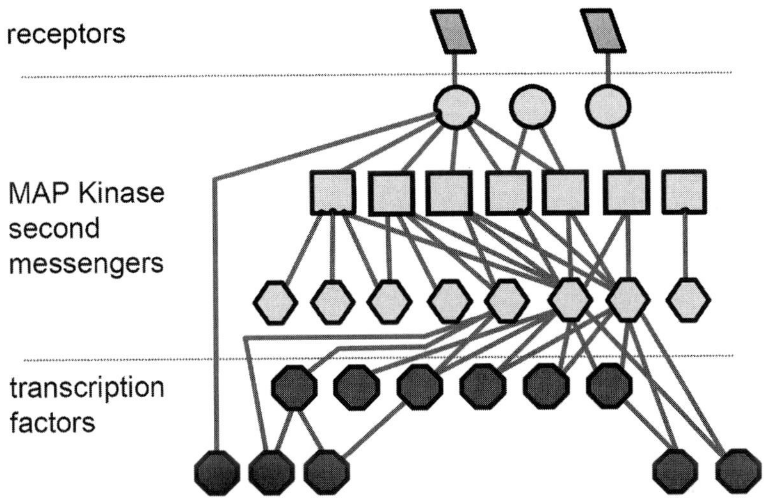

Figure 2. Schematic Display of a regulatory network built on functional Information. Known associations of MAP kinases, WRKY transcription factors and receptors were identified by literature searches. Relationships of the 31 components are shown by connecting lines. The components are shown as nodes of different level representing their functional placement in known regulating cascades. Dark Grey parallelograms: two known Receptors; Light Grey objects are parts of MAP kinase signaling: Circles: 3 MKKKs, Squares: 7 MKKs, Hexagons: 8 MPKs. Dark grey octagons: 11 MAPK target proteins.

a directed graph. It is obvious that the contained elements within the network posses different degrees of interaction. Some are directly linked with only one or a few other components, while several others form hubs where many lines of signaling merge from the upstream or split to the downstream partners.

We next wanted to gain insight into where and when regulatory expression information operates in our known regulatory network. In a first step, we exploited the AtGenExpress abiotic stress dataset and extracted those genes that exhibited a significant change in their expression state after stress application compared to untreated plants. Second, we chose the experiment with UV-B light irradiation, because we knew from our literature searches that the components of our regulatory network are not yet known to be involved in that kind of stress response. The UV-B irradiation has only been applied for 15 minutes before returning the plants to their previous state. On the one hand, this allows us to draw new conclusions of the involvement of these signaling components in that stress. On the

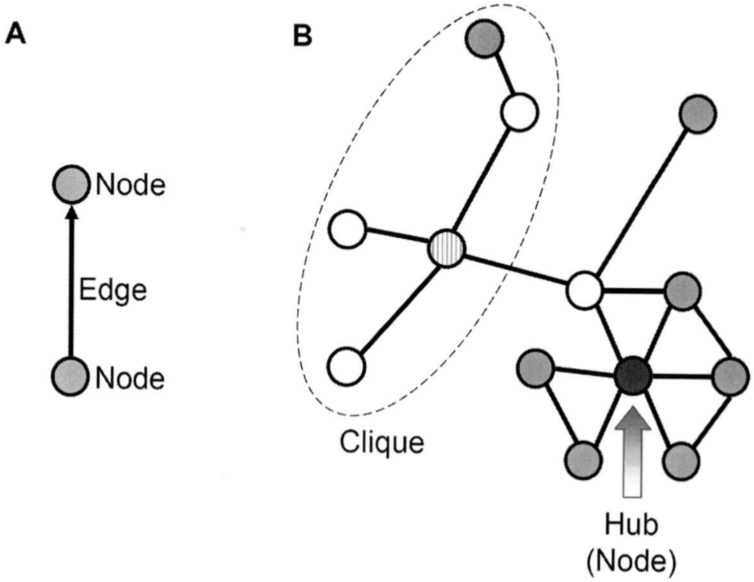

Figure 3. Graph Terminology. (A) Simple Directed Graph. (B) Schematic of an undirected graph. A Hub (the dark node) is a node that has most or many edges within a particular graph. A Clique is a subgraph which can be a complete graph in itself once removed from the full graph. A network neighborhood is the set of first nodes connected to any particular node. In this example, the network neighborhood of the white-stripped node in the center of the example clique is the white nodes.

other hand, and even more importantly, we wanted to be sure that our model regulatory network has been compiled without any precast structure and thus, is free of any prejudices that might bias our findings beforehand. Next, for the 31 genes present in the network, we determined if those genes were regulated (induced or repressed in expression) with respect to its control. If the transcripts of one of the genes were significantly regulated, the corresponding component object in our functional regulatory network has been highlighted by black fill-color (Fig. 3).

Figure 4 provides a schematic overview on the transcriptional involvement of all genes within the regulatory network and their responses upon UV-B irradiation. One can clearly see differences between the different measurements over time, but also between the genes and their level of involvement within the signaling cascade. Generally, a wavelike pattern of expression changes can be followed throughout the network.

In its initial phase of responses, the stress signal affects the genes within

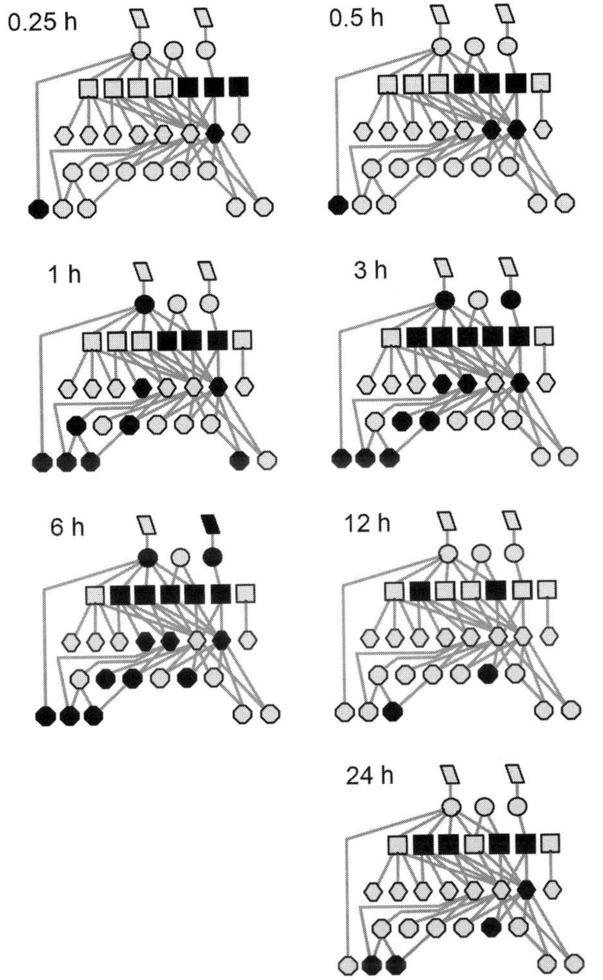

Figure 4. UV-B stress response signaling in the functional regulatory network. The known functional regulatory network (Fig. 2) has been superimposed with the regulatory information on UV-B regulated genes [1]. A schematic diagram has been depicted for each of the experimental points in time. Darkened nodes are genes that are regulated at that respective time point compared to the control.

the network at the second level of the MKKs first. All the genes that have already changed their expression state at as little as 15 min after the stress was applied are still differentially expressed at later time points.

At 1 h, after the stress was released, more downstream positioned components got involved in the signaling. One could conclude that the signal has finally reached the level of the WRKY transcription factors and that transcriptional reprogramming takes center stage. However, one upstream laying MAPKKK is differentially expressed, too. Besides the influence of the transcription factors, the involvement of this upstream MAPKKK might be one cause for the enrollment of more MAP kinase cascade components at the 3 h or 6 h time points.

The 3 h time almost reflects the maximum of responses seen in our functional network. All of those genes at any of the three levels of MAP kinase signaling that were not affected did not get involved at any of the other time points. Although they can be linked to our network by some functions, they are not affected by the UV-B light response and hence, not contained in its regulatory network. However, they might be of importance under other conditions.

Within our network, the maximum number of genes with expression responses was reached at the 6 h time point. Both upstream and downstream components of signaling cascades were involved. A transcriptional change of one of the two receptor genes is likely indicative for its function. Now, we could deduce the full response MAP kinase network of interchanging and merging signals that cascades and in some cases enrolls WRKY transcription factors as downstream effectors.

At the later time points, the majority of gene involvement contained within our functional network ceased. However, some genes still exhibited differential expression, especially the second level MKKs. As the second level messengers were also amongst the earliest genes that responded to the stress, one can speculate whether these genes are involved in priming and maintaining an alarmed state, which enables the cells to respond faster under recurring stresses. Taken together, we succeeded in the visualization of gene expression changes within our functional regulatory network and were able to follow them as a flow of signals through time. Moreover, we see an involvement of sets of genes that are enrolled at distinct points in time. A reason why we were able to make that observation might be a general enrollment of these gene sets under certain conditions together, which is reflected by their co-regulation.

To summarize our findings on the known regulatory network, we focused on only three time points to combine the inherent information of interaction between the components of the network with the paths taken by the flow of information based on transcriptional regulation. Connective lines linking the components (Fig. 2) were replaced with arrows that indicate the regulatory information, which acted on the respective genes (Fig. 5A). These arrows have two different qualities depending from where these acting signals might originate. White arrows indicate expression changes that have been imposed without a noticeable enrollment of upstream components while grey arrows point to downstream elements in cases where they might directly be affected by the regulation of the upstream components contained within our known functional regulatory network. By applying a strict chronological way analysis, grey arrows were only assigned if the downstream gene was functionally connected to the upstream component and both were regulated at the same time. To our surprise this simple assumption allowed a good differentiation between the signals relayed within the network's components and those influenced from the outside.

At the early 0.5 h time point, closest to the removal of UV-B irradiation, the rapid onset of responses acted mainly from the outside on our regulatory network through the second level MKKs and one downstream transcription factor. However, the MPKs at the lowest levels of MAP kinase signaling were already involved in the response, most likely due to the regulated upstream MKKs (Fig. 5A). Here, it is worth emphasizing the notion that the gene expression of only those components were affected, which have a functional relationship. On the one hand, this is indicative for the quality of our regulatory network; on the other hand, this again supports the idea of a certain functional relationship that is linked with a concerted responsiveness. We will come back to this idea later.

The 6 h time point is characterized by signals that were relayed from within the network and from the outside. Despite the enrollment of so many genes, one could follow one distinct train of signal that was initiated from the outside of the network, perceived by only one of the receptors and transduced via a discrete MAP kinase cascade, until integrating at only a few of the transcription factors (Fig. 5A). At 12 hours, the overall involvement of any of the components in the network is low. Only four genes are left with an altered gene expression response. They are not directly linked to each other and thus, no arrow can be assigned. All of them have been involved in responses at earlier time points. Therefore, gene expression of those genes might simply pursue a different kinetics and stay regulated over a prolonged period of time. This strategy would allow signaling information to be retained in a sustainable and self-preserving manner, a prerequisite for fine tuned and orchestrate signaling in a complex network.

To address the significance of our findings, we compared the number of regulated genes within our network (31) with all genes that are present on

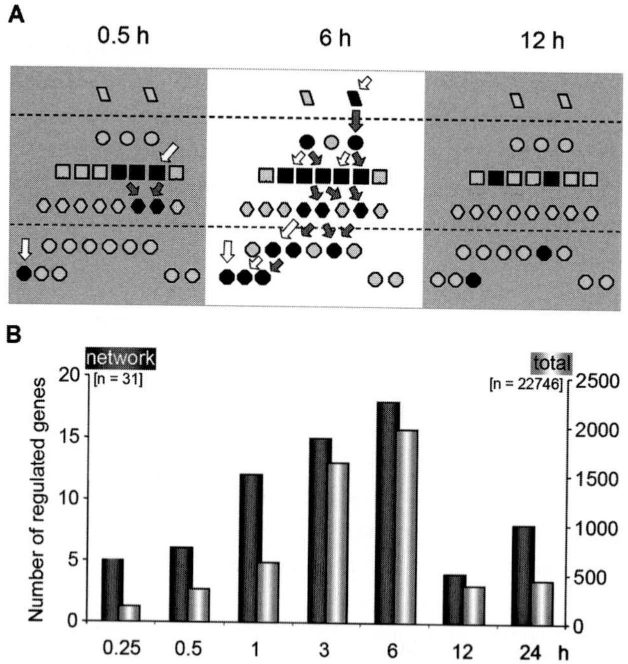

Figure 5. Signaling pathways throughout the network. (A) Signals relayed from outside the network and from within have been depicted. Three chronological points of the UV-B experiment displayed in Fig. 3 have been taken as examples: White arrows: expression changes that have been imposed without noticeable enrollment by components within the network. Grey arrows: downstream elements that might directly be affected by the regulation of the upstream components contained within the regulatory network. (B) Number of regulated genes after UV-B irradiation. The total number of responding genes in the shoot tissue is given for either our known functional regulatory network and for all regulated genes present on the chip [1]. Note that at time 6 h more than 50 % of the genes contained in the network are regulated, which is highly significant ($P(d) < 0.00001$)

the microarray (22746) [17] (Fig. 4B). Using Pearson correlation, we could support the visual impression that the number of genes follows the same trend line ($\varepsilon = 0.952$). However, both sets of data are completely different, when the sample size is included in the calculation. Our network consists of only 31 genes of which 17 (54 %) are regulated 6 h after the stress, while to total number of regulated genes in the entire experiment is 2244 (9 %) (Fig. 5B). Next, we tested our network for significant enrichment of regulated genes compared to all regulated genes using the hypergeometric distribution. All time points, except the 12 h time point, exhibited a highly significant difference from a random distribution of the number of regulated genes [P(d) $<$ 0.0001] . This is a strong evidence for the non-random composition and functional regulatory behavior in our network.

3.2. *Network expansion using co-expression*

We could demonstrate that the regulation of gene expression for genes enrolled in signaling cascades can be highly specific and comparable to the actual flow of information (Fig. 4). As a consequence, it should be possible to capture and expand the network by looking at gene expression data only, while we dismiss any true functional relationships.

Thus, we followed the conclusion by Bhardwaj *et al.* [18] that co-expression of genes is indicative for mutual relationship in signaling cascades or protein interaction. As a consequence of this assumption, we could expand the number of potential interaction partners if we could capture other genes that are co-expressed with the known network members. We next acquired all members of the MAP kinase families and WRKY transcription factors known in Arabidopsis thaliana and used them as 'seeds' in a large scale regulatory network approach. In order to reduce the complexity, only those genes that exhibited a certain co-expression in 1388 independent microarray experiments with our 'seed' gene list were retrieved from the ATTED-II database [8]. In order to reduce the complexity, we retrieved only co-regulated genes from the MAPK family members and their common signaling output transcription factors WRKYs [13,14,15,19,20,21,22] and other well described targets [19,23,24,25,26]. Because elements of the signaling and biosynthesis of the phytohormone ethylene have also been found to be frequent targets of MAPK signaling cascades [12], we decided to include them into our analysis. We refrained from implementing any algorithm for fusing different gene sets together as this would result in some genes being discarded; instead, the correlations are converted into a graph (Fig. 3). Thus, for each 'seed' gene, a set of co-expressed genes was obtained. This step was necessary to reduce computing time to a minimum, while on the other hand we reduced complexity to a meaningful context in terms of our 'seed' genes.

Figure 6A provides the entire co-expression graph for seed and target genes excluding those that were only connected to one node. Although it difficult to operate with the entire dataset and extract essential information from it, the number of total genes (5474) is indicative for a large number of processes our 'seed' genes can functionally be associated with. To extract those nodes that represent the major hubs (Fig. 6B) for regulatory signal transduction, only those with more than or equal to 20 edges were chosen for consecutive graph reduction analysis resulting in 216 hubs (Fig. 6C). This graph was then re-graphed to place nodes with shared connections close together and those with few connections far apart. In this way, one can easily identify which groups of genes, or cliques, have more connections between them (Fig. 6D). In our case, we were interested, whether we could identify novel, so far unknown components of MAP kinase / WRKY transcription factor regulatory networks. Therefore, we superimposed the most reduced regulatory network with our 'seed' genes and assigned distinct color and form styles for better comparability (Fig. 6D, 7).

Within the regulatory network groups of genes are formed that are tightly linked to each other and form cliques of nodes (Fig. 7A). On the basis of our computed regulatory network, we can immediately recognize several cliques, dominated by one huge and complex one in its center. This central clique might either act in a tightly controlled process with its members being highly co-expressed between all of them or need additional conditions to be meaningfully resolved. For this work, several interesting and well resolved regulatory cliques that possess remarkable gene compositions are explored here (Fig. 7B).

The first clique displays a linear network organization. It is characterized by two downstream transcription factors that receive information from three second upstream messengers, which presumably operate sequentially. In addition, we observed that two genes involved in ethylene signaling. The second clique constitutes a regulatory network that comprises the full set of upstream and downstream components of receptors, second messengers and transcription factors. It is comparable in size and composition to our known functional regulatory network. At the very top of the regulatory network, two novel receptor members of that respective pathway were identified. The third clique represents a small regulatory network that is composed of transcription factors only. As they all are co-expressed representatives of the most downstream component within our signaling cascade, it is possible that they are dependent on each others presence. This would be the case if they constitute interaction partners that affect the expression of their target genes as dimmers or multimers. Examples for that kind of function are e.g. bZIP transcription factors or Heat Shock Factors, which act as dimmers or trimers, respectively [27,28].

In the next step, we retrieved the gene lists for our representative cliques.

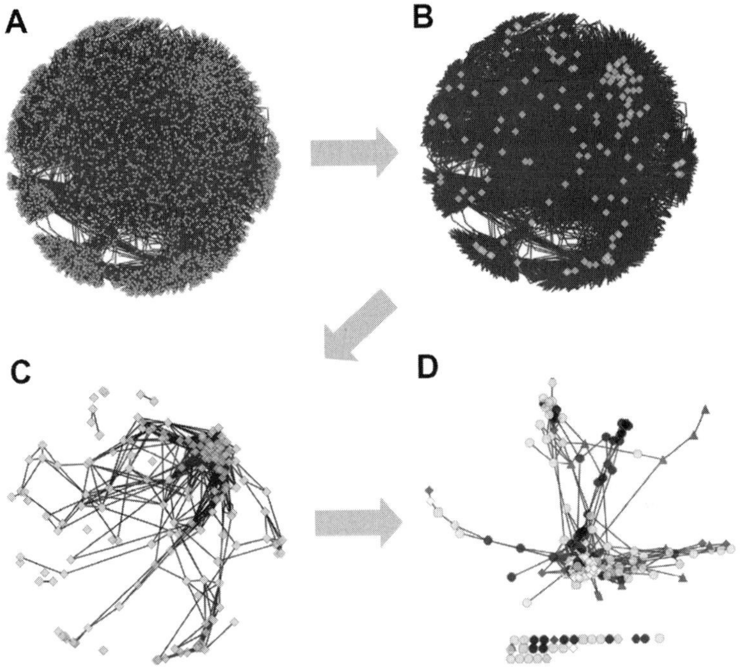

Figure 6. Graph reduction analysis on selected Hubs. (A) Entire co-expression network. Nodes with equal to or more that 20 edges (i.e. Degree > 19) were chosen as hubs for the graph reduction analysis. (B) The identified hubs are shown as large, grey nodes. (C) Displays the same co-expression network, with only the edges between hubs was retained. (D) Identical graph as depicted before, but sorted and with information added to the nodes. A spring-embedded alignment has been re-performed, marking all residual nodes that represent major hubs with the labeling scheme as has been introduced in Figure 1 and 2. Dark grey octagons: Transcription factors; Grey parallelograms: known Receptors; Grey triangles: proteins involved in ethylene signaling or biosynthesis; Diamonds: proteins, whose exact cascade position is not known; All other light grey objects represent second messengers.

To our big surprise, the first clique that was entirely built of transcription factors, consisting of WRKY factors only (Fig. 8, Clique 3). The 74 WRKY proteins from Arabidopsis can be classified into three major groups that split in seven subgroups total. Remarkably, with the exception of AtWRKY57, all the other WRKYs identified in that clique exclusively belong to group 2e (WRKY14, 35, 65 and 69) and 2b (WRKY09, 36, 61 and 72) [29]. This observation is supports the idea that these factors might interact within the cell to gain a higher level of modularity as had previously been reported for other factors [28]. The second clique contains well known

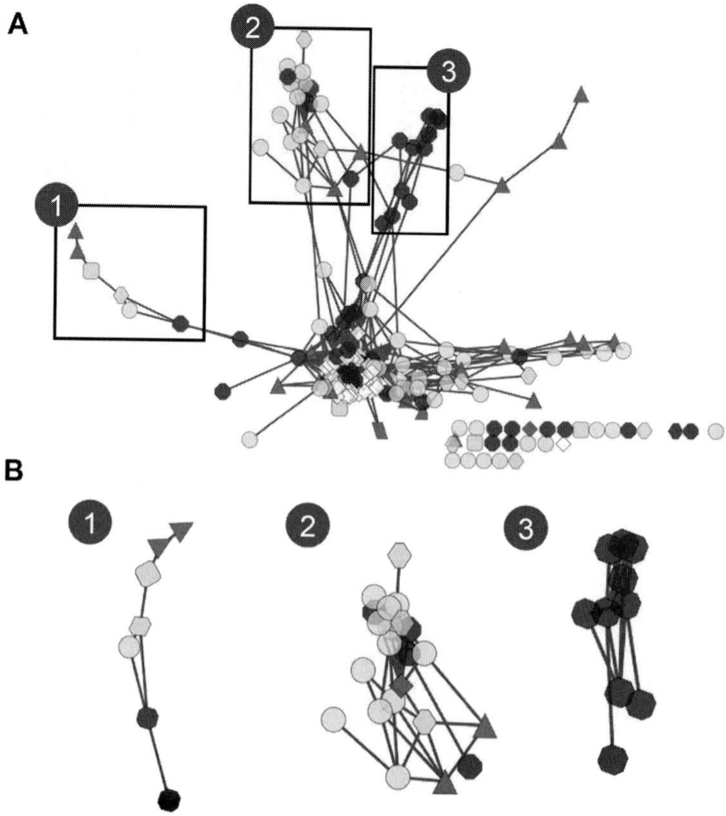

Figure 7. Overview of the extracted network and a Selection of Cliques. (A) Hubs and their edges between them extracted from the interaction network in Figure 6. The spring-embedded alignment places nodes sharing more edges or neighbors closer together and those that are not sharing any connections farther apart. Therefore, groups of nodes can be readily identified as cliques. (B) Three cliques were chosen as examples for further analysis are shown with their inter-connections. For comparability, the color- and shape- code has been retained from Figure 1 and 2. Dark grey octagons: Transcription factors; Grey parallelograms: known Receptors; Grey triangles: proteins involved in ethylene signaling or biosynthesis; Diamonds: proteins, whose exact cascade position is not known; All other light grey objects represent second messengers.

marker genes for the signaling of the phytohormone ethylene, ETO1 and EIN3 [30,31]. It has been shown before, that ethylene information is transduced involving MAP kinase cascades [26,32]. However, our data suggests a high level of possible interconnectivity between already known signaling components and those, we identified here. Two findings are especially novel

1

Degree	Gene Name	ID
104	ACS9	AT3G49700
189	EIL2	AT5G21120
193	MKK08	AT3G06230
196	MPK10	AT3G59790
161	Raf45	AT3G50730
137	AtWRKY66	AT1G80590
54	AtWRKY67	AT1G66550

2

Degree	Gene Name	ID
233	EIN3	AT3G20770
192	ETO1	AT3G51770
256	RAN1	AT5G44790
260	MAPKKKK	AT1G70430
272	MAPKKKK	AT1G79640
199	MPK06	AT2G43790
198	MPK08	AT1G18150
23	MPK09	AT3G18040
281	Raf16	AT1G04700
224	Raf26	AT4G14780
135	Raf36	AT5G58950
251	Raf39	AT3G22750
277	Raf40	AT3G24720
287	Raf46	AT3G59830
40	ZIK01	AT3G51630
227	ZIK06	AT5G41990
75	ZIK07	AT1G49160
260	ZIK10	AT1G64630
212	AtWRKY02	AT5G56270
41	AtWRKY31	AT4G22070
245	AtWRKY34	AT4G26440
255	AtWRKY42	AT4G04450

3

Degree	Gene Name	ID
234	AtWRKY09	AT1G68150
186	AtWRKY14	AT1G30650
30	AtWRKY27	AT5G52830
133	AtWRKY35	AT2G34830
253	AtWRKY36	AT1G69810
57	AtWRKY57	AT1G69310
35	AtWRKY61	AT1G18860
244	AtWRKY65	AT1G29280
221	AtWRKY69	AT3G58710
126	AtWRKY72	AT5G15130

Figure 8. Gene list tables that are major nodes. Genes in the cliques from Figure 7. Degree is the number of edges, which connected these hubs with related genes in the entire co-expression interaction network shown in Figure 6A. The coloring code from the previous Figures has been adopted. Black: Transcription factors; Dark grey: ethylene signaling; Light grey; second messengers.

in Clique 2: firstly, the involvement of WRKY proteins in the downstream signaling have already been associated with senescence, an ethylene driven process in plants; secondly, a possible bypass of the classical model of the ethylene signaling pathway could be deduced, which possibly involves the two MAPKKKK-type 'receptors' [33,34]. These MAPKKKKs could integrate signals from other signaling pathways that have an influence on ethylene signaling, and / or they could modulate the Raf-like MAPKKK CTR1, an element of the classical ethylene signaling pathway [32].

Finally, the linearly organized Clique 1 is very similar to what has been depicted at the very beginning (Fig. 1). The signaling information is relayed from a receptor to a MAPKKK - here termed RAF45, transferred to a MAPKK - here termed MKK08 and then transduced to a MPK, which is MPK10 in this case. Now, MPK10 forwards that information to the downstream transcription factor pair WRKY66 and WRKY67, which will influence the expression of their respective target genes. For MPK10 it has already been shown that it can phosphorylate WRKY factors in vitro [15]. This means, within the linear network we again could identify two genes with a physiological function in the ethylene signaling pathway [30,35].

4. Conclusion

In the present work, we discuss the use of regulatory network analysis. We presented two ways that exploit on the cryptic information inherent in sets of co-expressed genes. We could show that information flow can be followed in functional networks over time, which is a prerequisite for the assignment of meaningful attributes to known components of networks that function as signaling cascades. We were able to demonstrate that large regulatory networks, which were computed on the basis of co-expression data, provide data that in some part represents the expected signal flow for this type of biological network structure. In cases where known signaling networks were identified as a subset in a clique, an unknown cascade can likely be extended by those genes identified in that clique. Nonetheless, the verification of the findings thus leads back to the functional characterization in the cell. Finally, the synergy between both computational and experimental methods for gene expression analyses will help to complete our views and will be essential to our understanding of other eukaryote organisms, including humans.

Acknowledgements

We thank Jochen Supper and Andreas Zell for their support in the development of microarray analysis algorithms. This work was supported by the Arabidopsis Functional Genomics Network (AFGN) of the DFG (HA2146/5-2) and by the DFG-funded AtGenExpress project.

References

1. J. Kilian, D. Whitehead, J. Horak, D. Wanke, S. Weinl, O. Batistic, C. D'Angelo, E. Bornberg-Bauer, J. Kudla and K. Harter, *Plant J* **50**, 347 (2007).
2. D. J. Lockhart, H. Dong, M. C. Byrne, M. T. Follettie, M. V. Gallo, M. S. Chee, M. Mittmann, C. Wang, M. Kobayashi, H. Horton and E. L. Brown, *Nat Biotechnol* **14**, 1675(Dec 1996).
3. M. B. Eisen, P. T. Spellman, P. O. Brown and D. Botstein, *Proc Natl Acad Sci U S A* **95**, 14863(Dec 1998).

4. J. Supper, M. Strauch, D.Wanke, K. Harter and A. Zell, *BMC Bioinformatics* **8**, p. 334 (2007).
5. D. Wanke, J. Kilian, J. Supper, K. W. Berendzen, A. Zell and K. Harter, *QPPQ: QUANTUM BIO-INFORMATICS - From Quantum Information to Bio-Informatics, Quantum Probability and White Noise Analysis* **21** (World Scientific, 2008), ch. The Analysis of Gene Expression and Cis-Regulatory Elements in Large Microarray Expession Datasets., pp. 294 – 314.
6. K. Aoki, Y. Ogata and D. Shibata, *Plant Cell Physiol* **48**, 381(Mar 2007).
7. M. Mutwil, J. Obro, W. G. T. Willats and S. Persson, *Nucleic Acids Res* **36**, W320(Jul 2008).
8. T. Obayashi, S. Hayashi, M. Saeki, H. Ohta and K. Kinoshita, *Nucleic Acids Res* **37**, D987(Jan 2009).
9. Z. Wu, R. Irizarry, R. Gentleman, F. Martinez-Murillo and F. Spencer, *Journal of the American Statistical Association* **99**, 909(DEC 2004).
10. G. K. Smyth, *Stat Appl Genet Mol Biol* **3**, p. Article3 (2004).
11. P. Shannon, A. Markiel, O. Ozier, N. S. Baliga, J. T. Wang, D. Ramage, N. Amin, B. Schwikowski and T. Ideker, *Genome Res* **13**, 2498(Nov 2003).
12. A. Hahn and K. Harter, *Plant Physiol* **149**, 1207(Mar 2009).
13. T. Asai, G. Tena, J. Plotnikova, M. R. Willmann, W.-L. Chiu, L. Gomez-Gomez, T. Boller, F. M. Ausubel and J. Sheen, *Nature* **415**, 977(Feb 2002).
14. Y. Miao, T. M. Laun, A. Smykowski and U. Zentgraf, *Plant Molecular Biology* **65**, 63(SEP 2007).
15. S. C. Popescu, G. V. Popescu, S. Bachan, Z. Zhang, M. Gerstein, M. Snyder and S. P. Dinesh-Kumar, *Genes Dev* **23**, 80(Jan 2009).
16. L.-P. Hamel, M.-C. Nicole, S. Sritubtim, M.-J. Morency, M. Ellis, J. Ehlting, N. Beaudoin, B. Barbazuk, D. Klessig, J. Lee, G. Martin, J. Mundy, Y. Ohashi, D. Scheel, J. Sheen, T. Xing, S. Zhang, A. Seguin and B. E. Ellis, *Trends Plant Sci* **11**, 192(Apr 2006).
17. J. Redman, B. Haas, G. Tanimoto and C. Town, *Plant Journal* **38**, 545(MAY 2004).
18. N. Bhardwaj and H. Lu, *J Bioinform Comput Biol* **7**, 1(Feb 2009).
19. E. Andreasson, T. Jenkins, P. Brodersen, S. Thorgrimsen, N. H. T. Petersen, S. Zhu, J.-L. Qiu, P. Micheelsen, A. Rocher, M. Petersen, M.-A. Newman, H. Bjorn Nielsen, H. Hirt, I. Somssich, O. Mattsson and J. Mundy, *EMBO J* **24**, 2579(Jul 2005).
20. C. Y. Kim and S. Zhang, *Plant J* **38**, 142(Apr 2004).
21. F. L. H. Menke, H.-G. Kang, Z. Chen, J. M. Park, D. Kumar and D. F. Klessig, *Mol Plant Microbe Interact* **18**, 1027(Oct 2005).
22. S. Pandey and I. Somssich, *Plant Physiol* pp.109(May 2009).
23. Y. Liu and S. Zhang, *Plant Cell* **16**, 3386(Dec 2004).
24. G. Merkouropoulos, E. Andreasson, D. Hess, T. Boller and S. C. Peck, *J Biol Chem* **283**, 10493(Apr 2008).
25. D. Ren, Y. Liu, K.-Y. Yang, L. Han, G. Mao, J. Glazebrook and S. Zhang, *Proc Natl Acad Sci U S A* **105**, 5638(Apr 2008).
26. S.-D. Yoo, Y.-H. Cho, G. Tena, Y. Xiong and J. Sheen, *Nature* **451**, 789(FEB 14 2008).

27. C. Lohmann, G. Eggers-Schumacher, M. Wunderlich and F. Schoffl, *Mol Genet Genomics* **271**, 11(Feb 2004).
28. F. Weltmeier, A. Ehlert, C. S. Mayer, K. Dietrich, X. Wang, K. Schutze, R. Alonso, K. Harter, J. Vicente-Carbajosa and W. Droge-Laser, *EMBO J* **25**, 3133 (2006).
29. E. Mangelsen, J. Kilian, K. W. Berendzen, U. H. Kolukisaoglu, K. Harter, C. Jansson and D. Wanke, *BMC Genomics* **9**, p. 194 (2008).
30. Q. Chao, M. Rothenberg, R. Solano, G. Roman, W. Terzaghi and J. R. Ecker, *Cell* **89**, 1133(Jun 1997).
31. K. E. Woeste, C. Ye and J. J. Kieber, *Plant Physiol* **119**, 521(Feb 1999).
32. J. J. Kieber, M. Rothenberg, G. Roman, K. A. Feldmann and J. R. Ecker, *Cell* **72**, 427(Feb 1993).
33. S. Robatzek and I. E. Somssich, *Plant J* **28**, 123(Oct 2001).
34. S. Robatzek and I. E. Somssich, *Genes Dev* **16**, 1139(May 2002).
35. H. S. Chae, F. Faure and J. J. Kieber, *Plant Cell* **15**, 545(Feb 2003).

Quantum Bio-Informatics III
eds. L. Accardi, W. Freudenberg & M. Ohya

ON ENTROPIES OF QUANTUM DYNAMICAL SYSTEMS

NOBORU WATANABE
Department of Information Sciences,
Tokyo University of Science,
Noda City, Chiba 278-8510, Japan
E-mail: watanabe@is.noda.tus.ac.jp

Classical dynamical entropy is an important tool to discuss coding theorems in classical information theory. Quantum dynamical entropy was first studied by Connes, Størmer and Emch. Thereafter, there have been many researches to construct or calculate the dynamical entropies for several models.
In this paper, we briefly review two formulations due to (i) Ohya and (ii) Kossakowski, Ohya and Watanabe. Some numerical computations of these entropies are carried for several states.

1. Introduction

In classical information theory, Kolmogorov [13] discussed the (classical) dynamical entropy with respect to a measure preserving transformation defined on a message space for finite partitions of the probability measure space. The classical coding theorems of Shannon are important tools to analyze communication processes which have been formulated by the mean dynamical entropy and the mean dynamical mutual entropy. The mean dynamical entropy represents the amount of information per one letter of a signal sequence sent from an input source, and the mean dynamical mutual entropy does the amount of information per one letter of the signal received in an output system. In quantum information theory, Emch [9] and Conne, Størmer [8] first studied a complexity of quantum dynamical processes. After that, Ohya [20,21] defined a $\mathcal{S}$- mixing entropy in general quantum systems, which was natural extension of the von Neumann entropy. Based on the $\mathcal{S}$ - mixing entropy, Ohya [22] introduced a mean entropy and a mean mutual entropy for the quantum dynamical systems. Conne, Narnhoffer and Thirring formulated a dynamical entropy (CNT entropy) [7] and several attempts to discuss the CNT entropy [6,11,29]. By using a finite operational partitions of unity, AF entropy [4] was introduced by Alicki and Fannes. In

C* and W* algebras, dynamical approximation entropy [33] was defined by Voiculescu in 1995. Through the quantum Markov process, AOW entropy [2,3] was introduced by Accardi, Ohya and Watanabe in 1997. Based on completely positive maps, KOW entropy [16] was defined by Kossakowski, Ohya and Watanabe in 1999.

In this paper, we will discuss two quantum dynamical entropies of the mean entropy and the mean mutual entropy by Ohya and the generalized AOW entropy by means of KOW entropy, and we calculate those dynamical entropies for some models.

2. Quantum channels

Let $(\mathcal{A}, \mathfrak{S}(\mathcal{A}), \alpha(G))$ (resp. $(\overline{\mathcal{A}}, \overline{\mathfrak{S}}(\overline{\mathcal{A}}), \overline{\alpha}(\overline{G}))$) be a C*-dynamical system of an input system (resp. an output system), where $\mathcal{A}$ (resp. $\overline{\mathcal{A}}$) is a C*−algebra or the set of all bounded operators $\mathbf{B}(\mathcal{H})$ (resp. $\mathbf{B}(\overline{\mathcal{H}})$) on a separable Hilbert space $\mathcal{H}$ (resp. $\overline{\mathcal{H}}$), $\mathfrak{S}(\mathcal{A})$ (resp. $\overline{\mathfrak{S}}(\overline{\mathcal{A}})$) is the set of all states on $\mathcal{A}$ (resp. $\overline{\mathcal{A}}$) or the set of all density operators on $\mathcal{H}$ (resp. $\overline{\mathcal{H}}$) and $\alpha(G)$ (resp. $\overline{\alpha}(\overline{G})$) is an automorphism of $\mathcal{A}$ (resp. $\overline{\mathcal{A}}$) representing the dynamics of the input (resp. output) system.

A map Λ^* from $\mathfrak{S}(\mathcal{A})$ to $\overline{\mathfrak{S}}(\overline{\mathcal{A}})$ is called a channel.

(1) Λ^* is called a linear channel if Λ^* holds

$$\Lambda^*\left(\sum_n \lambda_n \varphi_n\right) = \sum_n \lambda_n \Lambda^*(\varphi_n)$$

for any $\varphi_n \in \mathfrak{S}(\mathcal{A})$ and $\sum_n \lambda_n = 1$ with $\lambda_n \geq 0 \quad (\forall n)$.

(2) Λ^* is a completely positive (C.P.) channel if its dual map Λ from $\overline{\mathcal{A}}$ to $\mathcal{A}$ * satisfies

$$\sum_{i,j=1}^{n} A_i^* \Lambda(B_i^* B_j) A_j \geq 0$$

for any $n \in \mathbf{N}$, any $B_j \in \overline{\mathcal{A}}$ and any $A_j \in \mathcal{A}$, where the dual map Λ of Λ^* is given by $\Lambda^*(\varphi)(B) = \varphi(\Lambda(B))$ for any $\varphi \in \mathfrak{S}(\mathcal{A})$ and any $B \in \overline{\mathcal{A}}$. Almost all physical transformations can be described by the CP channel [12,17,18,23,27].

2.1. *Noisy optical channel*

Take $\mathcal{A} = B(\mathcal{H}_1)$ (resp. $\overline{\mathcal{A}} = B(\mathcal{H}_2)$), hence $\mathfrak{S}(\mathcal{A}) = \mathfrak{S}(\mathcal{H})$ (resp. $\overline{\mathfrak{S}}(\overline{\mathcal{A}}) = \mathfrak{S}(\mathcal{H}_2)$) is the set of all density operators on a Hilbert space

$\mathcal{H}_1$ (resp. $\mathcal{H}_2$), and let $\mathcal{B}$ (resp. $\overline{\mathcal{B}}$) be a C*-algebra on another Hilbert space $\mathcal{K}_1$ (resp. $\mathcal{K}_2$). Then the state space of the total system is described by $\mathfrak{S}(\mathcal{H}_n \otimes \mathcal{K}_n)$ $(n = 1, 2)$.

Noisy optical channel Λ^* with a noise state ζ was defined by Ohya and Watanabe [28] such as

$$\Lambda^*(\rho) \equiv tr_{\mathcal{K}_2} \Pi^*(\rho \otimes \zeta) = tr_{\mathcal{K}_2} V (\rho \otimes \zeta) V^*, \tag{1}$$

where $\zeta = |m_1\rangle \langle m_1|$ is the m_1 photon number state in $\mathfrak{S}(\mathcal{K}_1)$ and V is a mapping from $\mathcal{H}_1 \otimes \mathcal{K}_1$ to $\mathcal{H}_2 \otimes \mathcal{K}_2$ denoted by

$$V(|n_1\rangle \otimes |m_1\rangle) = \sum_{j=0}^{n_1+m_1} C_j^{n_1,m_1} |j\rangle \otimes |n_1 + m_1 - j\rangle, \tag{2}$$

where

$$\begin{aligned} C_j^{n_1,m_1} &= \sum_{r=L}^{K} (-1)^{n_1+j-r} \frac{\sqrt{n_1! m_1! j! (n_1 + m_1 - j)!}}{r!(n_1 - j)!(j - r)!(m_1 - j + r)!} \\ &\quad \times \alpha^{m_1 - j + 2r} \left(-\bar{\beta}\right)^{n_1 + j - 2r}, \end{aligned} \tag{3}$$

and $|n_1\rangle$ is the n_1 photon number state vector in $\mathcal{H}_1$, and α, β are complex numbers satisfying $|\alpha|^2 + |\beta|^2 = 1$. K and L are constants given by $K = \min\{n_1, j\}$, $L = \max\{m_1 - j, 0\}$. For the noisy optical channel Λ^*, we have the following theorem [35]

Theorem 2.1. *The noisy optical channel Λ^* with noise state $|m\rangle \langle m|$ is described by*

$$\Lambda^*(\rho) = \sum_{i=0}^{\infty} O_i V Q^{(m)} \rho Q^{(m)*} V^* O_i^*, \tag{4}$$

where $Q^{(m)} \equiv \sum_{l=0}^{\infty} (|y_l\rangle \otimes |m\rangle) \langle y_l|$, $O_i \equiv \sum_{k=0}^{\infty} |z_k\rangle (\langle z_k| \otimes \langle i|)$, $\{|y_l\rangle\}$ is a CONS in $\mathcal{H}_1$, $\{|z_k\rangle\}$ is a CONS in $\mathcal{H}_2$ and $\{|i\rangle\}$ is the set of number states in $\mathcal{K}_2$.

$\Pi^*(\cdot) = V(\cdot) V^*$ is called a generalized beam splitting. The noisy optical channel with a vacuum noise is called an attenuation channel introduced in [18]. Lifting $\mathcal{E}_0^*$ from $\mathfrak{S}(\mathcal{H})$ to $\mathfrak{S}(\mathcal{H} \otimes \mathcal{K})$ in the sense of Accardi and Ohya [1] is denoted by

$$\mathcal{E}_0^*(|\xi\rangle \langle \xi|) = |\alpha\xi\rangle \langle \alpha\xi| \otimes |\beta\xi\rangle \langle \beta\xi|. \tag{5}$$

$\mathcal{E}_0^*$ (or Π_0^*) is called a beam splitting. Based on liftings, the beam splitting was studied by Accardi - Ohya and Fichtner - Freudenberg - Libsher [10]. We introduced the quantum capacity in [24] and calculated the quantum capacity for the attenuation channel [25] and the noisy optical channel [26].

3. Quantum Entropy for General Quantum Systems

For a unital C*-dynamical system $(\mathcal{A}, \mathfrak{S}(\mathcal{A}), \alpha(G))$, we denote a weak* compact and convex subset of $\mathfrak{S}(\mathcal{A})$ by $\mathcal{S}$, for instance, $\mathcal{S}$ is given by $\mathfrak{S}(\mathcal{A})$, $I(\alpha)$ (the set of all invariant states for α) and $K(\alpha)$ (the set of all KMS states).

Every state $\varphi \in \mathcal{S}$ has a maximal measure μ pseudosupported on $ex\mathcal{S}$ (the set of all extreme points of $\mathcal{S}$) such that

$$\varphi = \int_{\mathcal{S}} \omega d\mu. \tag{6}$$

Since the measure μ giving the above decomposition is not unique in general, the set of all such measures giving the above decomposition is denoted by $M_\varphi(\mathcal{S})$. μ is unique if $\mathcal{S}$ is a Choquet simplex. The subset of $M_\varphi(\mathcal{S})$ is defined by

$$\begin{aligned} D_\varphi(\mathcal{S}) = \Big\{ \mu \in M_\varphi(\mathcal{S});\quad & \exists \mu_k \subset \mathbb{R}^+ \text{ and } \{\varphi_k\} \subset ex\mathcal{S} \\ s.t.\quad & \sum_k \mu_k = 1, \quad \mu = \sum_k \mu_k \delta(\varphi_k) \Big\}, \end{aligned} \tag{7}$$

where $\delta(\varphi)$ is the Dirac measure with respect to an input state φ, and a functional $H(\mu)$ is given by

$$H(\mu) = -\sum_k \mu_k \log \mu_k \tag{8}$$

for a measure $\mu \in D_\varphi(\mathcal{S})$. Then the entropy of a state $\varphi \in \mathcal{S}$ with respect to $\mathcal{S}$ (**$\mathcal{S}$-mixing entropy** [20,21]) is defined by

$$S^{\mathcal{S}}(\varphi) = \begin{cases} \inf\{H(\mu);\quad \mu \in D_\varphi(\mathcal{S})\} \\ +\infty \qquad\qquad\qquad \text{if } D_\varphi(\mathcal{S}) = \varnothing \end{cases} \tag{9}$$

and it depends on the set $\mathcal{S}$. It describes the amount of information of the state φ measured from the system $\mathcal{S}$. We denote $S^{\mathfrak{S}(\mathcal{A})}(\varphi)$ by $S(\varphi)$ if $\mathcal{S} = \mathfrak{S}(\mathcal{A})$. It is a natural extension of von Neumann entropy [15] to general C*-systems.

For an initial state $\varphi \in \mathcal{S}$ and a channel $\Lambda^* : \mathfrak{S}(\mathcal{A}) \to \mathfrak{S}(\mathcal{B})$, two compound states are

$$\Phi_\mu^{\mathcal{S}} = \int_{\mathcal{S}} \omega \otimes \Lambda^* \omega \, d\mu, \tag{10}$$

$$\Phi_0 = \varphi \otimes \Lambda^* \varphi. \tag{11}$$

The compound state $\Phi_\mu^{\mathcal{S}}$ represents the correlation between the input state φ and the output state $\Lambda^*\varphi$. The mutual entropy with respect to $\mathcal{S}$ and μ is given by

$$I_\mu^{\mathcal{S}} (\varphi \,;\, \Lambda^*) = S\left(\Phi_\mu^{\mathcal{S}},\, \Phi_0\right) \tag{12}$$

and the mutual entropy with respect to $\mathcal{S}$ is defined by Ohya [20,21] as

$$I^{\mathcal{S}} (\varphi \,; \Lambda^*) = \sup \left\{ I_\mu^{\mathcal{S}} (\varphi \,; \Lambda^*) \ ; \ \mu \in M_\varphi (\mathcal{S}) \right\}. \tag{13}$$

It is a natural extension of the quantum mutual entropy [18,19] to general C*-systems.

4. Quantum mean mutual entropy of K-S type

In this section, we briefly explain quantum mean entropy and quantum mean mutual entropy introduced by Ohya [22].

In quantum information theory, a stationary information source is described by a C^*triple $(\mathcal{A}, \mathfrak{S}(\mathcal{A}), \theta_{\mathcal{A}})$ with a stationary state φ with respect to $\theta_{\mathcal{A}}$; that is, $\mathcal{A}$ is a unital C^*-algebra, $\mathfrak{S}(\mathcal{A})$ is the set of all states on $\mathcal{A}$, $\theta_{\mathcal{A}}$ is an automorphism of $\mathcal{A}$, and $\varphi \in \mathfrak{S}(\mathcal{A})$ is a state over $\mathcal{A}$ with $\varphi \circ \theta_{\mathcal{A}} = \varphi$.

Let an output C^*-dynamical system be the triple $(\mathcal{B}, \mathfrak{S}(\mathcal{B}), \theta_{\mathcal{B}})$, and $\Lambda^* : \mathfrak{S}(\mathcal{A}) \to \mathfrak{S}(\mathcal{B})$ be a covariant c.p. channel: $\Lambda : \mathcal{B} \to \mathcal{A}$ such that $\Lambda \circ \theta_{\mathcal{B}} = \theta_{\mathcal{A}} \circ \Lambda$.

In this section we explain new functionals $S_\mu^{\mathcal{S}}(\varphi; \alpha^M)$, $S^{\mathcal{S}}(\varphi; \alpha^M)$, $I_\mu^{\mathcal{S}}(\varphi; \alpha^M, \beta^N)$ and $I^{\mathcal{S}}(\varphi; \alpha^M, \beta^N)$ introduced in [22,14] for a pair of finite sequences of $\alpha^M = (\alpha_1, \alpha_2, \cdots, \alpha_M)$, $\beta^N = (\beta_1, \beta_2, \cdots, \beta_N)$ of completely positive unital maps $\alpha_m : \mathcal{A}_m \to \mathcal{A}$, $\beta_n : \mathcal{B}_n \to \mathcal{B}$ where $\mathcal{A}_m$ and $\mathcal{B}_n$ $(m = 1, \cdots, M,\ n = 1, \cdots, N)$ are finite dimensional unital C^*-algebras.

For a given finite sequences of completely positive unital maps $\alpha_m : \mathcal{A}_m \to \mathcal{A}$ from finite dimensional unital C*-algebras $\mathcal{A}_m$ $(m = 1, \cdots, M)$ and a given measure μ of $\varphi \in M_\varphi(\mathcal{S})$, the compound state of

$\alpha_1^*\varphi, \alpha_2^*\varphi, \cdots, \alpha_M^*\varphi$ on the tensor product algebra $\bigotimes_{m=1}^{M} \mathcal{A}_m$ is given by [22,14]

$$\Phi_\mu^S(\alpha^M) = \int_{S(\mathcal{A})} \bigotimes_{m=1}^{M} \alpha_m^* \omega \, d\mu(\omega). \tag{14}$$

Furthermore $\Phi_\mu^S(\alpha^M \cup \beta^N)$ is a compound state of $\Phi_\mu^S(\alpha^M)$ and $\Phi_\mu^S(\beta^N)$ with $\alpha^M \cup \beta^N \equiv (\alpha_1, \alpha_2, \cdots, \alpha_M, \beta_1, \beta_2, \cdots, \beta_N)$ constructed as

$$\Phi_\mu^S\left(\alpha^M \cup \beta^N\right) = \int_{S(\mathcal{A})} \left(\bigotimes_{m=1}^{M} \alpha_m^* \omega\right) \left(\bigotimes_{n=1}^{N} \beta_n^* \omega\right) d\mu. \tag{15}$$

For any pair (α^M, β^N) of finite sequences $\alpha^M = (\alpha_1, \cdots, \alpha_M)$ and $\beta^N = (\beta_1, \cdots, \beta_N)$ of completely positive unital maps (c.p.u. maps for short) $\alpha_m : \mathcal{A}_m \to \mathcal{A}$, $\beta_n : \mathcal{B}_n \to \mathcal{A}$ from finite dimensional unital C^*-algebras and any extremal decomposition measure μ of φ, the entropy functional S_μ and the mutual entropy functional I_μ are defined by [22,14]

$$S_\mu^S(\varphi; \alpha^M) = \int_{S(\mathcal{A})} S\left(\otimes_{m=1}^{M} \alpha_m^* \omega, \Phi_\mu^S(\alpha^M)\right) d\mu(\omega), \tag{16}$$

$$I_\mu^S(\varphi; \alpha^M, \beta^N) = S\left(\Phi_\mu^S(\alpha^M \cup \beta^N), \Phi_\mu^S(\alpha^M) \otimes \Phi_\mu^S(\beta^N)\right), \tag{17}$$

where $S(\cdot, \cdot)$ is the relative entropy for a finite algebra. The relative entropy of two states was introduced in [31] for σ-finite and semifinite von Neumann algebras. Araki [5] and Uhlmann [32] extended this relative entropy for more general quantum systems [23].

For a given pair of finite sequences of completely positive unital maps $\alpha^M = (\alpha_1, \cdots, \alpha_M)$, $\beta^N = (\beta_1, \cdots, \beta_N)$, the functional $S^S(\varphi; \alpha^M)$ (resp. $I^S(\varphi; \alpha^M, \beta^N)$) is given by taking the supremum of $S_\mu^S(\varphi; \alpha^M)$ (resp. $I_\mu^S(\varphi; \alpha^M, \beta^N)$) for all possible extremal decompositions μ's of φ:

$$S^S(\varphi; \alpha^M) = \sup\{S_\mu^S(\varphi; \alpha^M); \ \mu \in M_\varphi(\mathcal{S})\}, \tag{18}$$

$$I^S(\varphi; \alpha^M, \beta^N) = \sup\{I_\mu^S(\varphi; \alpha^M, \beta^N); \ \mu \in M_\varphi(\mathcal{S})\}. \tag{19}$$

Let $\mathcal{A}$ (resp. $\mathcal{B}$) be a unital C^*-algebra with a fixed automorphism $\theta_\mathcal{A}$ (resp. $\theta_\mathcal{B}$), Λ be a covariant c.p.u. map from $\mathcal{B}$ to $\mathcal{A}$, and φ be an invariant state over $\mathcal{A}$, i.e., $\varphi \circ \theta_\mathcal{A} = \varphi$.

$$\alpha^N \equiv \left(\alpha, \theta_\mathcal{A} \circ \alpha, \cdots, \theta_\mathcal{A}^{N-1} \circ \alpha\right), \tag{20}$$

$$\beta_\Lambda^N \equiv \left(\Lambda \circ \beta, \Lambda \circ \theta_\mathcal{B} \circ \beta, \cdots, \Lambda \circ \theta_\mathcal{B}^{N-1} \circ \beta\right). \tag{21}$$

For each c.p.u. map $\alpha : \mathcal{A}_0 \to \mathcal{A}$ (resp. $\beta : \mathcal{B}_0 \to \mathcal{B}$) from a finite dimensional unital C^*-algebra $\mathcal{A}_0$ (resp. $\mathcal{B}_0$) to $\mathcal{A}$ (resp. $\mathcal{B}$), $\tilde{S}^{\mathcal{S}}(\varphi;\theta_{\mathcal{A}},\alpha)$, $\tilde{I}^{\mathcal{S}}(\varphi;\Lambda^*,\theta_{\mathcal{A}},\theta_{\mathcal{B}},\alpha,\beta)$ are given by

$$\tilde{S}^{\mathcal{S}}(\varphi;\theta_{\mathcal{A}},\alpha) = \limsup_{N\to\infty} \frac{1}{N} S^{\mathcal{S}}(\varphi;\alpha^N), \tag{22}$$

$$\tilde{I}^{\mathcal{S}}(\varphi;\Lambda^*,\theta_{\mathcal{A}},\theta_{\mathcal{B}},\alpha,\beta) = \limsup_{N\to\infty} \frac{1}{N} I^{\mathcal{S}}(\varphi;\alpha^N,\beta_\Lambda^N). \tag{23}$$

The functionals $\tilde{S}^{\mathcal{S}}(\varphi;\theta_{\mathcal{A}})$ and $\tilde{I}^{\mathcal{S}}(\varphi;\Lambda^*,\theta_{\mathcal{A}},\theta_{\mathcal{B}})$ are defined by taking the supremum for all possible $\mathcal{A}_0$'s, α's, $\mathcal{B}_0$'s, and β's:

$$\tilde{S}^{\mathcal{S}}(\varphi;\theta_{\mathcal{A}}) = \sup_{\alpha} \tilde{S}^{\mathcal{S}}(\varphi;\theta_{\mathcal{A}},\alpha), \tag{24}$$

$$\tilde{I}^{\mathcal{S}}(\varphi;\Lambda^*,\theta_{\mathcal{A}},\theta_{\mathcal{B}}) = \sup_{\alpha,\beta} \tilde{I}^{\mathcal{S}}(\varphi;\Lambda^*,\theta_{\mathcal{A}},\theta_{\mathcal{B}},\alpha,\beta). \tag{25}$$

The next theorem [22,14] shows that the fundamental inequality in information theory holds for $\tilde{S}^{\mathcal{S}}(\varphi;\theta_{\mathcal{A}})$ and $\tilde{I}^{\mathcal{S}}(\varphi;\Lambda^*,\theta_{\mathcal{A}},\theta_{\mathcal{B}})$.

Theorem 4.1. $0 \le \tilde{I}^{\mathcal{S}}(\varphi;\Lambda^*,\theta_{\mathcal{A}},\theta_{\mathcal{B}}) \le \min\{\tilde{S}^{\mathcal{S}}(\varphi;\theta_{\mathcal{A}}), \tilde{S}^{\mathcal{S}}(\Lambda^*\varphi;\theta_{\mathcal{B}})\}$.

4.1. *Computation of mean mutual entropy of K-S type for modulated states*

We here introduce the concept of ideal modulator and give some examples of the ideal modulator (IPAM, IPPM).

Let $\{a_1,\cdots,a_N\}$ be an alphabet set constructing input signals and $\{E_1,\cdots,E_N\}$ be the set of all one dimensional projections on certain a Hilbert space $\mathcal{H}_0$ satisfying (i) $E_n \perp E_m \quad (n \neq m)$, (ii) E_n corresponds to the alphabet a_n.

(1) Ideal modulated state of PAM (Pulse Amplitude Modulator) is given by

$$E_n^{(IPAM)} \equiv |n><n|,$$

where $|n><n|$ is the n photon number state on $\mathcal{H}$.

(2) Ideal modulated state of PPM (Pulse Position Modulator) is given by

$$E_n^{(IPPM)} \equiv E_0^{(IPAM)} \otimes \cdots \otimes E_0^{(IPAM)} \otimes E_d^{(IPAM)} \otimes E_0^{(IPAM)} \otimes \cdots \otimes E_0^{(IPAM)},$$

where $E_0^{(IPAM)}$ is a vacuum state and $E_d^{(IPAM)} = |d><d|$.

Some ideal modulators were used to compute the transmission efficiency by the mutual entropy ratio [34].

Now we calculate the mean mutual entropy of K-S type for some ideal modulated states (IPAM, IPPM) expressed by the photon number states as above.

Let $B(\mathcal{H}_0)$ (resp. $B(\overline{\mathcal{H}}_0)$) be the set of all bounded linear operators on a Hilbert space $\mathcal{H}_0$ (resp. $\overline{\mathcal{H}}_0$) and $\mathcal{A}_0$ (resp. $\mathcal{B}_0$) be a finite subset in $B(\mathcal{H}_0)$ (resp. $B(\overline{\mathcal{H}}_0)$).

Let $\mathcal{A}$ (resp. $\mathcal{B}$) be an infinite tensor product space of $B(\mathcal{H}_0)$ (resp. $B(\overline{\mathcal{H}}_0)$) represented by

$$\mathcal{A} \equiv \bigotimes_{i=-\infty}^{\infty} B(\mathcal{H}_0), \quad \mathcal{B} \equiv \bigotimes_{i=-\infty}^{\infty} B(\overline{\mathcal{H}}_0).$$

Moreover, let $\theta_{\mathcal{A}}$ (resp. $\theta_{\mathcal{B}}$) be a shift transformations on $\mathcal{A}$ (resp. $\mathcal{B}$) defined by

$$\theta_{\mathcal{A}}(\bigotimes_{i=-\infty}^{\infty} A_i) = \bigotimes_{i'=-\infty}^{\infty} A_{i'} \quad (i' = i-1) \quad \text{for any} \bigotimes_{i=-\infty}^{\infty} A_i \in \mathcal{A},$$
$$\theta_{\mathcal{B}}(\bigotimes_{j=-\infty}^{\infty} B_j) = \bigotimes_{j'=-\infty}^{\infty} B_{j'} \quad (j' = j-1) \quad \text{for any} \bigotimes_{j=-\infty}^{\infty} B_j \in \mathcal{B}.$$

Let α (resp. β) be the embedding map from $\mathcal{A}_0$ to $\mathcal{A}$, (resp. $\mathcal{B}_0$ to $\mathcal{B}$) given by

$$\alpha(A) = \cdots I \otimes I \otimes A \otimes I \otimes \cdots \in \mathcal{A}, \quad \text{for any } A \in \mathcal{A}_0,$$
$$\beta(B) = \cdots I \otimes I \otimes B \otimes I \otimes \cdots \in \mathcal{B}, \quad \text{for any } B \in \mathcal{B}_0.$$

We denote the set of all density operators on $\mathcal{H}_0$ (resp. $\overline{\mathcal{H}}_0$) by $\mathfrak{S}_0$ (resp. $\overline{\mathfrak{S}}_0$), and let $\mathfrak{S}$ (resp. $\overline{\mathfrak{S}}$) be the set of all states ρ on $\mathcal{A}$ (resp. $\overline{\rho}$ on $\mathcal{B}$).

The maps $\alpha^N_{(IM)}, \beta^N_{\tilde{\Lambda}(IM)}$ are given by

$$\alpha^N_{(IM)} \equiv (\alpha \circ \tilde{\gamma}_{(IM)}, \theta_{\mathcal{A}} \circ \alpha \circ \tilde{\gamma}_{(IM)}, \cdots, \theta_{\mathcal{A}}^{N-1} \circ \alpha \circ \tilde{\gamma}_{(IM)}),$$
$$\beta^N_{\tilde{\Lambda}(IM)} \equiv (\tilde{\gamma}_{(IM)} \circ \tilde{\Lambda} \circ \beta, \tilde{\gamma}_{(IM)} \circ \tilde{\Lambda} \circ \theta_{\mathcal{B}} \circ \beta, \cdots, \tilde{\gamma}_{(IM)} \circ \tilde{\Lambda} \circ \theta_{\mathcal{B}}^{N-1} \circ \beta),$$

where we took a special channel and modulator such that $\tilde{\Lambda} \equiv \bigotimes_{i=-\infty}^{\infty} \Lambda$ and $\tilde{\gamma}_{(IM)} \equiv \bigotimes_{i=-\infty}^{\infty} \gamma_{(IM)}$.

(I) IPAM Take a stationary initial state

$$\rho = \sum_m \mu_m \bigotimes_{i=-\infty}^{\infty} \rho_m^{(i)} \in \mathfrak{S}. \tag{26}$$

Let $\rho_m^{(i)} = \sum_{n_i=1}^{M} \lambda_{n_i}^{(m)} E_{n_i} \in \mathfrak{S}_0$ be a Schatten decomposition [30] of $\rho_m^{(i)}$. Then we have

$$\begin{aligned}
&\Phi_E(\alpha_{(IPAM)}^N) \\
&= \sum_{n_0=1}^{M} \cdots \sum_{n_{N-1}=1}^{M} \left(\sum_m \mu_m \prod_{k=0}^{N-1} \lambda_{n_k}^{(m)} \right) \left(\bigotimes_{i=0}^{N-1} E_{n_i}^{(IPAM)} \right), \qquad (27)
\end{aligned}$$

$$\begin{aligned}
&\Phi_E(\beta_{\tilde{\Lambda}(IPAM)}^N) \\
&= \sum_{n_0=1}^{M} \cdots \sum_{n_{N-1}=1}^{M} \left(\sum_m \mu_m \prod_{k=0}^{N-1} \lambda_{n_k}^{(m)} \right) \left(\bigotimes_{i=0}^{N-1} \Lambda^* E_{n_i}^{(IPAM)} \right). \qquad (28)
\end{aligned}$$

When Λ^* is the noisy optical channel with a fixed noise $\zeta = |r\rangle\langle r|$, we have

$$\Lambda^* E_{n_i}^{(IPAM)} = \sum_{j_i=0}^{n_i+r} |C_{j_i}^{n_i,r}|^2 F_{j_i}^{(IPAM)}, \qquad (29)$$

where $F_{j_i}^{(IPAM)} = |j_i >< j_i|$ is the j_i-photon number state in the output space $\overline{\mathfrak{S}}_0$ and the coefficient $C_{j_i}^{n_i,r}$ is given by the noisy optical channel Λ^* [28,27]. The compound states through a channel $\tilde{\Lambda}^*$ becomes

$$\begin{aligned}
&\Phi_E(\alpha_{(IPAM)}^N \cup \beta_{\tilde{\Lambda}(IPAM)}^N) \\
&= \sum_{n_0=1}^{M} \cdots \sum_{n_{N-1}=1}^{M} \left(\sum_m \mu_m \prod_{k=0}^{N-1} \lambda_{n_k}^{(m)} \right) \sum_{j_0=0}^{n_0+r} \cdots \sum_{j_{N-1}=0}^{n_{N-1}+r} \\
&\quad \times \left(\prod_{k'=0}^{N-1} |C_{j_{k'}}^{n_{k'},r}|^2 \right) \left(\bigotimes_{i=0}^{N-1} E_{n_i}^{(IPAM)} \right) \left(\bigotimes_{i'=0}^{N-1} F_{j_{i'}}^{(IPAM)} \right), \qquad (30)
\end{aligned}$$

$$\begin{aligned}
&\Phi_E(\alpha_{(IPAM)}^N) \otimes \Phi_E(\beta_{\tilde{\Lambda}(IPAM)}^N) \\
&= \sum_{n_0=1}^{M} \cdots \sum_{n_{N-1}=1}^{M} \left(\sum_m \mu_m \prod_{k=0}^{N-1} \lambda_{n_k}^{(m)} \right) \sum_{n'_0=1}^{M} \cdots \sum_{n'_{N-1}=1}^{M} \\
&\quad \times \left(\sum_{m'} \prod_{k'=0}^{N-1} \lambda_{n'_{k'}}^{(m')} \right) \sum_{j'_0=0}^{n'_0+r} \cdots \sum_{j'_{N-1}=0}^{n'_{N-1}+r} \left(\prod_{k''=0}^{N-1} |C_{j'_{k''}}^{n'_{k''},r}|^2 \right) \\
&\quad \times \left(\bigotimes_{i=0}^{N-1} E_{n_i}^{(IPAM)} \right) \left(\bigotimes_{i'=0}^{N-1} F_{j'_{i'}}^{(IPAM)} \right). \qquad (31)
\end{aligned}$$

Lemma 4.1. *For an initial state ρ in (26), we have*

$$
\begin{aligned}
& I_E(\rho; \alpha^N_{(IPAM)}, \beta^N_{\tilde{\Lambda}(IPAM)}) \\
&= \sum_{j_0=0}^{M+r} \cdots \sum_{j_{N-1}=0}^{M+r} \sum_{n_0=J_0}^{M} \cdots \sum_{n_{N-1}=J_{N-1}}^{M} \\
&\times \left(\sum_m \mu_m \prod_{k=0}^{N-1} \lambda^{(m)}_{n_k} \right) \left(\prod_{k'=0}^{N-1} |C^{n_{k'},r}_{j_{k'}}|^2 \right) \\
&\times \log \frac{\prod_{k=0}^{N-1} |C^{n_k,r}_{j_k}|^2}{\sum_{n'_0=J_0}^{M} \cdots \sum_{n'_{N-1}=J_{N-1}}^{M} \left(\sum_{m'} \mu_{m'} \prod_{k'=0}^{N-1} \lambda^{(m')}_{n'_{k'}} \right) \left(\prod_{k"=0}^{N-1} |C^{n'_{k"},r}_{j_{k"}}|^2 \right)},
\end{aligned}
$$

where $J_i \equiv \max\{j_i - r, 0\} \quad (i = 0, 2, \cdots, N-1)$.

Under the above lemma, we obtain the following theorem.

Theorem 4.2. *(1) For an initial state ρ in (26), we have the lower bound of $\tilde{S}(\rho;\, \theta_{\mathcal{A}},\, \alpha_{(IPAM)})$ such as*

$$\tilde{S}(\rho; \theta_{\mathcal{A}}, \alpha_{(IPAM)}) \geq \sum_m \mu_m S(\rho^{(0)}_m).$$

(2) Let $\mu_0 = 1, \mu_k = 0$ ($\forall k \geq 1$) and $\lambda^{(0)}_n = \lambda_n$ in (26). Then we have the following equalities:

$$
\begin{aligned}
& \tilde{S}(\rho; \theta_{\mathcal{A}}, \alpha_{(IPAM)}) \\
&= -\sum_{n=1}^{M} \lambda_n \log \lambda_n, \\
& \tilde{I}(\rho; \Lambda^*, \theta_{\mathcal{A}}, \theta_{\mathcal{B}}, \alpha_{(IPAM)}, \beta_{(IPAM)}) \\
&= \sum_{j=0}^{M+r} \sum_{n=J}^{M} \lambda_n |C^{n,r}_j|^2 \log \frac{|C^{n,r}_j|^2}{\sum_{n'=J}^{M} \lambda_{n'} |C^{n',r}_j|^2}.
\end{aligned}
$$

(II) IPPM For an initial state ρ in (26), we have the following compound states:

$$
\begin{aligned}
& \Phi_E(\alpha^N_{(IPPM)}) \\
&= \sum_{n_0=1}^{M} \cdots \sum_{n_{N-1}=1}^{M} \left(\sum_m \mu_m \prod_{k=0}^{N-1} \lambda^{(m)}_{n_k} \right) \left(\bigotimes_{i=0}^{N-1} \left(\bigotimes_{j=0}^{N-1} E^{(IPAM)}_{d\delta_{j,n_i}} \right) \right). \quad (32)
\end{aligned}
$$

$$
\begin{aligned}
&\Phi_E(\beta^N_{\tilde{\Lambda}(IPPM)}) \\
&= \sum_{n_0=1}^{M} \cdots \sum_{n_{N-1}=1}^{M} \left(\sum_m \mu_m \prod_{k=0}^{N-1} \lambda^{(m)}_{n_k} \right) \\
&\quad \times \sum_{\ell_0=0}^{d\delta_{1,n_0}+r} \cdots \sum_{\ell_{N-1}=0}^{d\delta_{M,n_{N-1}}+r} |C^{d\delta_{1,n_0},r}_{\ell_0}|^2 \cdots |C^{d\delta_{M,n_{N-1}},r}_{\ell_{N-1}}|^2 \\
&\quad \times \left(\bigotimes_{i=0}^{N-1} \left(\bigotimes_{j=1}^{M} F^{(IPAM)}_{\ell_{iM+j}} \right) \right) .
\end{aligned} \tag{33}
$$

$$
\begin{aligned}
&\Phi_E(\alpha^N_{(IPPM)} \cup \beta^N_{\tilde{\Lambda}(IPPM)}) \\
&= \sum_{n_0=1}^{M} \cdots \sum_{n_{N-1}=1}^{M} \left(\sum_m \mu_m \prod_{k=0}^{N-1} \lambda^{(m)}_{n_k} \right) \sum_{\ell_1=0}^{d\delta_{1,n_0}+r} \cdots \sum_{\ell_{N-1}=0}^{d\delta_{M,n_{N-1}}+r} \\
&\quad \times |C^{d\delta_{1,n_0},r}_{\ell_1}|^2 \cdots |C^{d\delta_{M,n_{N-1}},r}_{\ell_{N-1}}|^2 \\
&\quad \times \left(\bigotimes_{i=0}^{N-1} \left(\bigotimes_{j=1}^{M} E^{(IPAM)}_{d\delta_{j,n_i}} \right) \right) \left(\bigotimes_{i'=0}^{N-1} \left(\bigotimes_{j'=0}^{M} F^{(IPAM)}_{\ell_{i'M+j'}} \right) \right) .
\end{aligned} \tag{34}
$$

$$
\begin{aligned}
&\Phi_E(\alpha^N_{(IPPM)}) \otimes \Phi_E(\beta^N_{\tilde{\Lambda}(IPPM)}) \\
&= \sum_{n_0=1}^{M} \cdots \sum_{n_{N-1}=1}^{M} \sum_{n'_0=1}^{M} \cdots \sum_{n'_{N-1}=1}^{M} \left(\sum_m \mu_m \prod_{k=0}^{N-1} \lambda^{(m)}_{n_k} \right) \\
&\quad \times \sum_{\ell_1=0}^{d\delta_{1,n_0}+r} \cdots \sum_{\ell_{N-1}=0}^{d\delta_{M,n_{N-1}}+r} |C^{d\delta_{1,n_0},r}_{\ell_1}|^2 \cdots |C^{d\delta_{M,n_{N-1}},r}_{\ell_{N-1}}|^2 \\
&\quad \times \sum_{n'_0=1}^{M} \cdots \sum_{n'_{N-1}=1}^{M} \left(\sum_{m'} \mu_{m'} \prod_{k'=0}^{N-1} \lambda_{n'_{k'}} \right) \\
&\quad \times \left(\bigotimes_{i=0}^{N-1} \left(\bigotimes_{j=1}^{M} E^{(IPAM)}_{d\delta_{j,n_i}} \right) \right) \left(\bigotimes_{i'=0}^{N-1} \left(\bigotimes_{j'=1}^{M} F^{(IPAM)}_{\ell_{i'M+j'}} \right) \right) .
\end{aligned} \tag{35}
$$

Lemma 4.2. *For an initial state* ρ *in (26), we have*

$$
\begin{aligned}
& I_E(\rho; \alpha^N_{(IPPM)}, \beta^N_{\tilde{\Lambda}(IPPM)}) \\
&= - \sum_{n_0=1}^{M} \cdots \sum_{n_{N-1}=1}^{M} \left(\sum_m \mu_m \prod_{k=0}^{N-1} \lambda^{(m)}_{n_k} \right) \\
&\times \sum_{p=1}^{N} \sum_{\{q_1,\cdots,q_p\} \subset \{1,2,\cdots,N\}} \sum_{\ell_1=1}^{d+r} \cdots \sum_{\ell_p=1}^{d+r} |C^{d,r}_{\ell_1}|^2 \cdots |C^{d,r}_{\ell_p}|^2 \\
&\times \left(|C^{d,r}_0|^2 \right)^{N-p} \left(1 - |C^{d,r}_0|^2 \right)^p \log \left(\sum_{m'} \mu_{m'} \prod_{k'=0}^{p} \lambda^{(m')}_{n_{q_{k'}}} \right).
\end{aligned}
$$

Theorem 4.3. *(1) For an initial state* ρ *in (26), we have the lower bound of* $\tilde{S}(\rho;\, \theta_{\mathcal{A}},\, \alpha_{(IPPM)})$ *such as*

$$\tilde{S}(\rho; \theta_{\mathcal{A}}, \alpha_{(IPPM)}) \geq \sum_m \mu_m S(\rho^{(0)}_m).$$

(2) Let $\mu_0 = 1, \mu_k = 0\ (\forall k \geq 1)$ *and* $\lambda^{(0)}_n = \lambda_n$ *in (26). Then we have the following equalities:*

$$\tilde{S}(\rho; \theta_{\mathcal{A}}, \alpha_{(IPPM)}) = - \sum_{n=1}^{M} \lambda_n \log \lambda_n,$$

$$\tilde{I}(\rho; \Lambda^*, \theta_{\mathcal{A}}, \theta_{\mathcal{B}}, \alpha_{(IPPM)}, \beta_{(IPPM)}) = (1 - |C^{d,r}_0|^2) \tilde{S}(\rho; \theta_{\mathcal{A}}, \alpha_{(IPPM)}).$$

5. KOW dynamical entropy

In this section, we briefly review the definition of the KOW entropy according to [16].

For a normal state ω on $\mathbf{B}(\mathcal{K})$ and a normal, unital CP linear map Γ from $\mathbf{B}(\mathcal{K}) \otimes \mathbf{B}(\mathcal{H})$ to $\mathbf{B}(\mathcal{K}) \otimes \mathbf{B}(\mathcal{H})$ and a normal, unital CP map Λ from $\mathbf{B}(\mathcal{H})$ to $\mathbf{B}(\mathcal{H})$ and the identity map id on $\mathbf{B}(\mathcal{K})$, the transition expectation

$$E^{\Gamma,\omega}_{\Lambda}\left(\tilde{A}\right) = \omega\left((id \otimes \Lambda)\, \Gamma\left(\tilde{A}\right)\right), \quad \forall \tilde{A} \in \mathbf{B}(\mathcal{K}) \otimes \mathbf{B}(\mathcal{H}) \qquad (36)$$

and the lifting in the sense of Accardi and Ohya [1] can be given by

$$E_{\Lambda}^{*\Gamma,\omega}(\rho) = \Gamma^{*}(\tilde{\omega} \otimes \Lambda^{*}(\rho)), \quad \forall \rho \in \mathfrak{S}(\mathcal{H}), \tag{37}$$

where $id \otimes \Lambda$ is a normal, unital CP map from $\mathbf{B}(\mathcal{K}) \otimes \mathbf{B}(\mathcal{H})$ to $\mathbf{B}(\mathcal{K}) \otimes \mathbf{B}(\mathcal{H})$ and Λ^{*} is a quantum channel [12,17,18,23,27] from $\mathfrak{S}(\mathcal{H})$ to $\mathfrak{S}(\mathcal{H})$ with respect to an input signal state ρ and a noise state $\tilde{\omega}$. Based on the following relation

$$\begin{aligned} &tr_{(\otimes_1^n \mathcal{K}) \otimes \mathcal{H}} \Phi_{\Lambda,n}^{*\Gamma,\omega}(\rho)(A_1 \otimes \cdots \otimes A_n \otimes B) \\ &\equiv tr_{\mathcal{H}} \rho \left(E_{\Lambda}^{\Gamma,\omega} \left(A_1 \otimes E_{\Lambda}^{\Gamma,\omega} \left(A_2 \otimes \cdots A_{n-1} \otimes E_{\Lambda}^{\Gamma,\omega}(A_n \otimes B) \cdots \right) \right) \right) \end{aligned} \tag{38}$$

for all $A_1, A_2, \cdots, A_n \in \mathbf{B}(\mathcal{K}), B \in \mathbf{B}(\mathcal{H})$ and any $\rho \in \mathfrak{S}(\mathcal{H})$, a lifting $\Phi_{\Lambda,n}^{*\Gamma,\omega}$ from $\mathfrak{S}(\mathcal{H})$ to $\mathfrak{S}((\otimes_1^n \mathcal{K}) \otimes \mathcal{H})$ and marginal state

$$\rho_{\Lambda,n}^{\Gamma,\omega} \equiv tr_{\mathcal{H}} \Phi_{\Lambda,n}^{*\Gamma,\omega}(\rho) \in \mathfrak{S}(\otimes_1^n \mathcal{K}) \text{ and } \bar{\rho}_{\Lambda,n}^{\Gamma,\omega} \equiv tr_{\otimes_1^n \mathcal{K}} \Phi_{\Lambda,n}^{*\Gamma,\omega}(\rho) \in \mathfrak{S}(\mathcal{H}) \tag{39}$$

are given, where $\Phi_{\Lambda,n}^{*\Gamma,\omega}(\rho)$ is a compound state with respect to $\bar{\rho}_{\Lambda,n}^{\Gamma,\omega}$ and $\rho_{\Lambda,n}^{\Gamma,\omega}$ in the sense of [16,18] .

Definition 5.1. The quantum dynamical entropy with respect to Λ, ρ, Γ and ω is defined by

$$\tilde{S}(\Lambda; \rho, \Gamma, \omega) \equiv \limsup_{n \to \infty} \frac{1}{n} S\left(\rho_{\Lambda,n}^{\Gamma,\omega}\right), \tag{40}$$

where $S\left(\rho_{\Lambda,n}^{\Gamma,\omega}\right)$ is von Neumann entropy of $\rho_{\Lambda,n}^{\Gamma,\omega} \in \mathfrak{S}(\otimes_1^n \mathcal{K})$ defined by $S\left(\rho_{\Lambda,n}^{\Gamma,\omega}\right) = -tr \rho_{\Lambda,n}^{\Gamma,\omega} \log \rho_{\Lambda,n}^{\Gamma,\omega}$. The dynamical entropy with respect to Λ and ρ is defined as

$$\tilde{S}(\Lambda; \rho) \equiv \sup \{ \mathfrak{S}(\Lambda; \rho, \Gamma, \omega); \Gamma, \omega \}. \tag{41}$$

Formulation of Generalized AOW entropies by KOW entropy

In this section, we briefly explain the generalized AOW entropy based on the KOW entropy [16].

For a finite operational partition of unity $\gamma_1, \cdots, \gamma_d \in \mathbf{B}(\mathcal{H})$, i.e., $\sum_{i=1}^{d} \gamma_i^{*} \gamma_i = I$, and a normal unital CP map Λ from $\mathbf{B}(\mathcal{H})$ to $\mathbf{B}(\mathcal{H})$, transition expectation $E_{\Lambda}^{\gamma(0)}$ from $M_d^0 \otimes \mathbf{B}(\mathcal{H})$ to $\mathbf{B}(\mathcal{H})$ is defined by

$$E_{\Lambda}^{\gamma(0)}\left(\sum_{i,j=1}^{d} E_{ij}\otimes A_{ij}\right)\equiv\sum_{i=1}^{d}\Lambda\left(\gamma_i^{*}A_{ii}\gamma_i\right). \tag{42}$$

where $E_{ij}=|e_i\rangle\langle e_j|$ with normalized vectors $e_i\in\mathcal{H}$ $(i=1,2,\cdots,d)$, M_d $(d\le\dim\mathcal{H})$ in $\mathbf{B}(\mathcal{H})$ is the $d\times d$ matrix algebra and M_d^0 is a subalgebra of M_d consisting of diagonal elements of M_d. Then the quantum Markov state

$$\begin{aligned}
&\rho_{\Lambda,n}^{\gamma(0)}\\
&=\sum_{i_1,\cdots,i_n=1}^{d} tr_{\mathcal{H}}\rho\Lambda\left(W_{i_1i_1}\left(\Lambda\left(W_{i_2i_2}\left(\cdots\Lambda\left(W_{i_ni_n}\left(I_{\mathcal{H}}\right)\right)\right)\right)\right)\right)E_{i_1i_1}\otimes\cdots\otimes E_{i_ni_n}\\
&=\sum_{i_1,\cdots,i_n=1}^{d} p_{i_1,\cdots,i_n}E_{i_1i_1}\otimes\cdots\otimes E_{i_ni_n},
\end{aligned}\tag{43}$$

are obtained, where

$$\begin{aligned}
W_{ij}(A)&\equiv\gamma_i^{*}A\gamma_j,\quad A\in\mathbf{B}(\mathcal{H}),\\
W_{ij}^{*}(\rho)&\equiv\gamma_j\rho\gamma_i^{*},\quad \rho\in\mathfrak{S}(\mathcal{H}),\\
p_{i_1,\cdots,i_n}&\equiv tr_{\mathcal{H}}\rho\Lambda\left(W_{i_1i_1}\left(\Lambda\left(W_{i_2i_2}\left(\cdots\Lambda\left(W_{i_ni_n}\left(I_{\mathcal{H}}\right)\right)\right)\right)\right)\right)\\
&=tr_{\mathcal{H}}W_{i_ni_n}^{*}\left(\Lambda^{*}\cdots\Lambda^{*}\left(W_{i_2i_2}^{*}\left(\Lambda^{*}\left(W_{i_1i_1}^{*}\left(\Lambda^{*}(\rho)\right)\right)\right)\right)\right).
\end{aligned}$$

Therefore the generalized AOW entropy $\tilde{S}_{\mathcal{B}}^{(0)}(\Lambda;\rho)$ of Λ and ρ with respect to a finite dimensional subalgebra $\mathcal{B}\subset\mathbf{B}(\mathcal{H})$ is obtained by

$$\tilde{S}_{\mathcal{B}}^{(0)}(\Lambda;\rho)\equiv\sup_{\{\gamma_i\}\subset\mathcal{B}}\tilde{S}^{(0)}(\Lambda;\rho,\{\gamma_i\}), \tag{44}$$

where the dynamical entropy $\tilde{S}^{(0)}(\Lambda;\rho,\{\gamma_i\})$ is given by

$$\tilde{S}^{(0)}(\Lambda;\rho,\{\gamma_i\})\equiv\limsup_{n\to\infty}\frac{1}{n}S\left(\rho_{\Lambda,n}^{\gamma(0)}\right). \tag{45}$$

$\tilde{S}_{\mathcal{B}}^{(0)}(\Lambda;\rho)$ is equal to the AOW entropy [2] if $\{\gamma_i\}$ is PVM (projection valued measure) and Λ is given by an automorphism θ. Then one can obtain the following theorem [35]:

Theorem 5.1. *When ρ is given by $\rho = \lambda|0\rangle\langle 0| + (1-\lambda)|\xi\rangle\langle\xi|$ and Λ^* is the noisy optical channel with a coherent noise $|\kappa\rangle\langle\kappa|$ and parameters α, β satisfying $|\alpha|^2 + |\beta|^2 = 1$, the quantum dynamical entropy with respect to Λ, ρ and $\{\gamma_j\}$ is obtained by*

$$\tilde{S}^{(0)}(\Lambda; \rho, \{\gamma_j\}) = -\sum_{j,k} q_{k,j} q_j \log q_{k,j},$$

where $q_j = \lambda |\langle \beta\kappa, x_j\rangle|^2 + (1-\lambda)|\langle \alpha\xi + \beta\kappa, x_j\rangle|^2$ and

$$q_{k,j} = \nu_j^+ \left|\langle x_k, y_j^+\rangle\right|^2 + \left(1-\nu_j^+\right)\left|\langle x_k, y_j^-\rangle\right|^2,$$

$$|y_j^+\rangle = a_j^+ |\beta\kappa\rangle + b_j^+ |\alpha\xi + \beta\kappa\rangle, \; |y_j^-\rangle = a_j^- |\beta\kappa\rangle - b_j^- |\alpha\xi + \beta\kappa\rangle,$$

$$a_j^+ = \varepsilon_j^+ a_j, a_j^- = \varepsilon_j^- a_j, b_j^+ = \varepsilon_j^+ b_j, b_j^- = \varepsilon_j^- b_j,$$

$$\varepsilon_j^+ = \sqrt{\frac{\tau_j^2 + 2\exp\left(-\frac{1}{2}|\xi|^2\right)\tau_j + 1}{\tau_j^2 + 2\exp\left(-\frac{1}{2}|\alpha\xi|^2\right)\tau_j + 1}},$$

$$\varepsilon_j^- = \sqrt{\frac{\tau_j^2 + 2\exp\left(-\frac{1}{2}|\xi|^2\right)\tau_j + 1}{\tau_j^2 - 2\exp\left(-\frac{1}{2}|\alpha\xi|^2\right)\tau_j + 1}},$$

$$\nu_j^+ = \frac{1}{2}\left(1 + \exp\left(-\frac{1}{2}\left(1-|\alpha|^2\right)|\xi|^2\right)\right)\frac{1}{\left(\varepsilon_j^+\right)^2},$$

$$\tau_j = \frac{-(1-2\lambda)}{2(1-\lambda)\exp\left(-\frac{1}{2}|\xi|^2\right)} + (-1)^j \frac{\sqrt{1 - 4\lambda(1-\lambda)\left(1 - \exp\left(-|\xi|^2\right)\right)}}{2(1-\lambda)\exp\left(-\frac{1}{2}|\xi|^2\right)},$$

$$|b_j|^2 = \frac{1}{\tau_j^2 + 2\exp\left(-\frac{1}{2}|\xi|^2\right)\tau_j + 1}, |a_j|^2 = \tau_j^2 |b_j|^2, \bar{a}_j b_j = a_j \bar{b}_j = \tau_j |b_j|^2$$

References

1. L. Accardi and M. Ohya, Compound channels, transition expectation and liftings, Appl. Math, Optim., **39**, 33-59 (1999).
2. L. Accardi, M. Ohya and N. Watanabe, Dynamical entropy through quantum Markov chain, Open System and Information Dynamics, **4**, 71-87, (1997).
3. L. Accardi, M. Ohya and N. Watanabe, Note on quantum dynamical entropies, Rep. Math. Phys., **38**, 457-469, (1996).
4. R. Alicki and M. Fannes, Defining quantum dynamical entropy, Lett. Math. Physics, **32**, 75-82, (1994).
5. H. Araki, Relative entropy for states of von Neumann algebras, Publ. RIMS Kyoto Univ. **11**, 809–833, (1976).

6. M. Choda, Entropy for extensions of Bernoulli shifts, Ergodic Theory Dynam. Systems, 16, No.6, 1197-1206 (1996).
7. A. Connes, H. Narnhoffer and W. Thirring, Dynamical entropy of C*algebras and von Neumann algebras, Commun. Math. Phys., **112**, 691-719, (1987).
8. A. Connes and E. Størmer, Entropy for automorphisms of von Neumann algebras, Acta Math., **134**, 289-306, (1975).
9. G.G. Emch, Positivity of the K-entropy on non-abelian K-flows, Z. Wahrscheinlichkeitstheory verw. Gebiete, **29**, 241 (1974).
10. K.H. Fichtner, W. Freudenberg and V. Liebscher, Beam splittings and time evolutions of Boson systems, Fakultat fur Mathematik und Informatik, Math/ Inf/96/ **39**, Jena, 105 (1996).
11. T. Hudetz, Topological entropy for appropriately approximated C*-algebras, J. Math. Phys. 35, No. 8, 4303–4333 (1994).
12. R.S. Ingarden, A. Kossakowski and M. Ohya, *Information Dynamics and Open Systems*, Kluwer, 1997.
13. A.N. Kolmogorov, Theory of transmission of information , Amer. Math. Soc. Translation, Ser. **2**, **33**, 291 (1963).
14. N. Muraki and M. Ohya, Entropy functionals of Kolmogorov Sinai type and their limit theorems, Letter in Mathematical Physics., **36**, 327-335, 1996.
15. J. von Neumann, *Die Mathematischen Grundlagen der Quantenmechanik*, Springer-Berlin, 1932.
16. Kossakowski, A., Ohya, M. and Watanabe, N., Quantum dynamical entropy for completely positive map, Infinite Dimensional Analysis, Quantum Probability and Related Topics, **2**, No.2, 267-282, (1999)
17. M. Ohya, Quantum ergodic channels in operator algebras , J. Math. Anal. Appl., **84**, 318-328, (1981).
18. M. Ohya, On compound state and mutual information in quantum information theory, IEEE Trans. Information Theory, **29**, 770-774 (1983).
19. M. Ohya, Note on quantum probability, L. Nuovo Cimento, **38**, 402-404, (1983).
20. M. Ohya, Entropy transmission in C*-dynamical systems, Journal in Mathematical Analysis and Applications, **100**, 222-235 (1984).
21. M. Ohya, Some aspects of quantum information theory and their applications to irreversible processes, Rep. Math. Phys., **27**, 19-47 (1989).
22. M. Ohya, State change, complexity and fractal in quantum systems, Quantum Communications and Measurement, **2**, 309-320, (1995).
23. M. Ohya and D. Petz, *Quantum Entropy and its Use*, Springer, Berlin, 1993.
24. M. Ohya, D. Petz and N. Watanabe, On capacity of quantum channels, Probability and Mathematical Statistics, **17**, 179-196 (1997).
25. M. Ohya, D. Petz and N. Watanabe, Numerical computation of quantum capacity, International Journal of Theoretical Physics, **37**, No.1, 507-510 (1998).
26. M. Ohya and N. Watanabe, Quantum capacity of noisy quantum channel, Quantum Communication and Measurement, **3**, 213-220 (1997).
27. M. Ohya and N. Watanabe, *Foundation of Quantum Communication Theory (in Japanese)*, Makino Pub. Co., 1998.

28. M. Ohya and N. Watanabe, Construction and analysis of a mathematical model in quantum communication processes, Electronics and Communications in Japan, Part 1, **68**, No.2, 29-34 (1985).
29. Y.M. Park, Dynamical entropy of generalized quantum Markov chains, Lett. Math. Phys. **32**, 63-74, (1994)
30. R. Schatten, *Norm Ideals of Completely Continuous Operators*, Springer - Verlag, 1970.
31. H. Umegaki, Conditional expectations in an operator algebra IV (entropy and information), Kodai Math. Sem. Rep., **14**, 59-85 (1962).
32. A. Uhlmann, Relative entropy and the Wigner-Yanase-Dyson-Lieb concavity in interpolation theory, Commun. Math. Phys., **54**, 21–32, (1977).
33. D. Voiculescu, Dynamical approximation entropies and topological entropy in operator algebras, Comm. Math. Phys., **170**, 249, (1995).
34. N. Watanabe, Efficiency of optical modulations for photon number states, Quantum Probability and Related Topics, **6**, 489 (1991).
35. N. Watanabe, Some Aspects of Complexities for Quantum Processes, Open Systems and Information Dynamics, **16**, No.2&3, 293-304, 2009.

Quantum Bio-Informatics III
eds. L. Accardi, W. Freudenberg & M. Ohya

SIGNIFICANT IMPROVEMENT OF SEQUENCE ALIGNMENT CAN BE DONE BY CONSIDERING TRANSITION PROBABILITY BETWEEN TWO CONSECUTIVE PAIRS OF RESIDUES

TOSHIHIDE HARA*, KEIKO SATO AND MASANORI OHYA

Department of Information Sciences, Tokyo University of Science,
2641 Yamazaki, Noda City, Chiba, Japan
**E-mail: j6307701@ed.noda.tus.ac.jp*

It is important to look for accurate sequence alignment because biological sequence analysis depends on the quality of alignment. However it is still a difficult computational problem because of not only the computational complexity but also unsatisfactory objective functions for measuring alignment. Here, we show the significant improvement of sequence alignment can be done by introducing a new metric defined on compound systems of two sequences. In the benchmark tests by PREFAB4 and HOMSTRAD, our method gives higher accuracy than other methods such as Needleman-Wunsch, ClustalW2. Especially for the sequences with less than 30% sequence identity, our method improves the alignment accuracy significantly.

Keywords: Sequence analysis; Sequence alignment; Transition probability

1. Introduction

To perform biological sequence analysis, biologists need accurate tools for generating sequence alignment. For this reason, several methods have been developed: Needleman-Wunsch [1], OU Alignment [2], ClustalW [3] and so on. These methods give high accuracy results for closely-related homologous sequences with more than 45% identity, but do not give satisfactory results for distantly-related homologous sequences. Recently, we developed an accurate sequence alignment algorithm called MTRAP (pairwise sequence alignment by a new Measure based on TRAnsition Probability). Most of alignment methods are based on finding a path that gives the minimum value to the sum of differences (the maximum value to the sum of similarity-scores) for all residue pairs between two sequences. In addition, our method is modified to compute this sum of differences by introducing a quantity through the transition probability between consecutive pairs of residues. In this paper, we will explain our alignment algorithm and discuss its performance.

2. Transition-quantity between consecutive pairs

Our algorithm MTRAP takes the local structural information through the following "Transition-quantity". Using the data in BAliBASE3 structural alignment database [4], we formulate the Transition-quantity as below. Let Ω be the set of all amino acids, and Ω^* be the Ω with the indel (gap) "$*$": $\Omega^* \equiv \Omega \cup \{*\}$. Let $[\Omega^*]$ be the set of all sequences of elements in Ω^*. We call a node of Ω a residue and a node of Ω^* a symbol. In addition, let $\Gamma \equiv \Omega \times \Omega$ and $\Gamma^* \equiv \Omega^* \times \Omega^*$.

Let $A_i \in [\Omega^*]$ $(i = 1, \cdots, N)$ with its length L:

$$A_i = a_{i1} \cdots a_{iL},\ a_{ij} \in \Omega^*.$$

Given a set $S^{L,N} = \{A_1, \cdots, A_N\}$. Let $u_{ij}^k = (a_{ik}, a_{jk}) \in \Gamma^*$ be a symbol pair in the kth site (column). Let $C_{S^{L,N}} : \Gamma^* \times \Gamma^* \mapsto \mathbf{R}$ be a mapping which represents the observed frequency of the symbol pair v accompanied by the pair w on the set $S^{L,N}$ such that

$$C_{S^{L,N}}(v, w) \equiv \sum_{k=1}^{L-1} \sum_{i=1}^{N-1} \sum_{j=i+1}^{N} \delta_{v,u_{ij}^k} \delta_{w,u_{ij}^{k+1}},\ (v, w \in \Gamma^*),$$

$$\delta_{p,q} \equiv \begin{cases} 1, & p = q \\ 0, & p \neq q \end{cases}$$

Let $\mathcal{S}^N$ be the set of all sets $S^{N,L}$ above: $\mathcal{S}^N = \bigcup_L S^{N,L}$, and $\mathcal{S}$ be the set of all sets $\mathcal{S}^N$: $\mathcal{S} = \bigcup_N \mathcal{S}^N$. Let $e(v, w)$ be a transition probability from the symbol pair v to the pair w on $\mathcal{S}$ such that

$$e(v, w) \equiv \begin{cases} \dfrac{\sum\limits_{S \in \mathcal{S}} C_S(v, w)}{\sum\limits_{S \in \mathcal{S}} \sum\limits_{z \in \Gamma^*} C_S(v, z)}, & \sum\limits_{S \in \mathcal{S}} \sum\limits_{z \in \Gamma^*} C_S(v, z) > 0 \\ 0, & \text{otherwise} \end{cases}$$

for $v, w \in \Gamma^*$. Here, let $T : \Gamma^* \times \Gamma^* \mapsto \mathbf{R}$ be a mapping called "Transition-quantity" such that

$$T(v, w) \equiv \begin{cases} \dfrac{-\log e(v, w)}{\max_{z \in \Gamma^*} \{-\log e(v, z)\}}, & e(v, w) > 0 \\ 1, & \text{otherwise} \end{cases}$$

for $v, w \in \Gamma^*$.

Let us define a new measure (difference) between two amino acids. Given a mapping $\mathcal{W} : \Gamma^* \mapsto \mathbf{R}$ defined a substitution matrix $\mathcal{W} = (w(a,b))$ such as PAM [5], BLOSUM [6] and GONNET [7]. Let $\mathrm{M} : [\min_{\mathrm{u}\in\Gamma}\{\mathcal{W}(\mathrm{u})\}, \max_{\mathrm{u}\in\Gamma}\{\mathcal{W}(\mathrm{u})\}] \mapsto \mathbf{R}$ be a function with

$$M(t) \equiv \frac{\max_{u\in\Gamma}\{\mathcal{W}(u)\} - t}{\max_{u\in\Gamma}\{\mathcal{W}(u)\} - \min_{u\in\Gamma}\{\mathcal{W}(u)\}},$$

$$0 \le M(t) \le 1.$$

We often denote the function $M(\mathcal{W}(u))$ by $M(u)$.

3. New Algorithm

The MTRAP is an alignment algorithm by minimizing the value of a certain objective function based on the Transition-quantity and the measure between two amino acids. Here we show the definition of the function and a solution by dynamic programming as a style of OU alignment [2].

Let $\mathcal{A}$,$\mathcal{B}$ be two amino acid sequences such as

$$\mathcal{A} : a_1\, a_2 \cdots a_m,$$
$$\mathcal{B} : b_1\, b_2 \cdots b_n,$$

where $a_i,\ b_j \in \Omega$. Take the lattice point $P_k = (i_k, j_k),\ i = 1, \cdots, m\ j = 1, \cdots, n$. We call the sequence of the lattice points

$$R = \{P_1, P_2, \cdots, P_N\}$$

a "route" with an initial point $P_1 = (0,0)$ and a final point $P_N = (m,n)$ if the following conditions are met:

$$i_{k-1} \le i_k,\ j_{k-1} \le j_k,\ P_{k-1} \ne P_k \text{ for any } k\,(= 2, 3, \cdots, N).$$

Let α_R, β_R be maps from a route $R = \{P_1, P_2, \cdots, P_N\}$ to a set Ω^* such that

$$\alpha_R(P_k) = \begin{cases} a_{i_k} & i_k \ne i_{k-1}\ (k = 2, \cdots, N) \\ * & k = 1 \text{ or } i_k = i_{k-1}\ (k = 2, \cdots, N) \end{cases}$$

$$\beta_R(P_k) = \begin{cases} b_{j_k} & j_k \ne j_{k-1}\ (k = 2, \cdots, N) \\ * & k = 1 \text{ or } j_k = j_{k-1}\ (k = 2, \cdots, N) \end{cases}$$

and μ_R be a map from the route R to the set of all symbol pairs $\Gamma^* (\equiv \Omega^* \times \Omega^*)$ such that

$$\mu_R(P_k) = (\alpha_R(P_k), \beta_R(P_k))\,.$$

We call the following $\mathcal{A}^*$ and $\mathcal{B}^*$ the alignment of $\mathcal{A}$ and $\mathcal{B}$ by the route R:

$$\begin{aligned}\mathcal{A}^* &: \alpha_R(P_1)\,\alpha_R(P_2)\cdots\ \alpha_R(P_N)\,,\\ \mathcal{B}^* &: \beta_R(P_1)\,\beta_R(P_2)\cdots\ \beta_R(P_N)\,.\end{aligned}$$

Let $R(P)$ be the set of all routes with the final point P; that is,

$$R(P) = \{\{P_1, \cdots, P_k\};\, P_k = P\}\,.$$

Let us fix the following notations for the sequel discussion: (1) $\Gamma^{*-} \equiv \Omega^* \times \Omega$, (2) $\Gamma^{-*} \equiv \Omega \times \Omega^*$, (3) $\Gamma^{g-} \equiv \{*\} \times \Omega$, (4) $\Gamma^{-g} \equiv \Omega \times \{*\}$, (5) W_{open} is a gap "opening" cost; $0 \le W_{\text{open}} \le 1$, (6) W_{extend} is a gap "extending" cost; $0 \le W_{\text{extend}} \le W_{\text{open}}$ and (7) ε is a weight $(0 \le \varepsilon \le 1)$ of Transition-quantity for generally used costs (i.e., the mixture of usual difference M and our new Transition-quantity T). The difference between $\mathcal{A}$ and $\mathcal{B}$ by a route R is given by

$$d(R) = \sum_{k=2}^{N} d_s\left(P_{k-1}, P_k;\, R\right),$$

where d_s is a function from $\Gamma^* \times \Gamma^*$ to $\mathbf{R}$ such that

$$d_s\left(P_{k-1}, P_k;\, R\right) = \begin{cases} d_e\left(\mu_R\left(P_{k-1}\right), \mu_R\left(P_k\right)\right), & k \ge 3 \\ d_i\left(\mu_R\left(P_k\right)\right), & k = 2 \end{cases}$$

$$d_i(u) = \begin{cases} (1-\varepsilon)\, M(u)\,, & u \in \Gamma \\ (1-\varepsilon)\, W_{\text{open}}, & u \notin \Gamma \end{cases}$$

$$d_e(u_1, u_2) = \begin{cases} (1-\varepsilon)\, M(u_2) + \varepsilon T(u_1, u_2)\,, & u_1 \in \Gamma^*, u_2 \in \Gamma \\ (1-\varepsilon)\, W_{\text{open}} + \varepsilon T(u_1, u_2)\,, & \begin{cases} u_1 \in \Gamma^{-*}, u_2 \in \Gamma^{g-} \\ u_1 \in \Gamma^{*-}, u_2 \in \Gamma^{-g} \end{cases} \\ (1-\varepsilon)\, W_{\text{extend}} + \varepsilon T(u_1, u_2)\,, & \begin{cases} u_1 \in \Gamma^{g-}, u_2 \in \Gamma^{g-} \\ u_1 \in \Gamma^{-g}, u_2 \in \Gamma^{-g} \end{cases} \end{cases}$$

The degree of difference between $\mathcal{A}$ and $\mathcal{B}$ with respect to a final point P can be defined as

$$D(P) = \min_{R \in R(P)} \{d(R)\}\,.$$

Hence the degree of difference between $\mathcal{A}$ and $\mathcal{B}$ is

$$D_{\mathcal{AB}} = D\left(P = (m, n)\right).$$

We calculate $D_{\mathcal{AB}}$ by a dynamic programming technique as below. For a final point $P_k = (i,j)$ and a route $R = \{P_1, \cdots, P_k\} \in R(P_k)$, we have

$$P_k = (i,j) \text{ and } P_{k-1} = Q_1 \text{ or } Q_2 \text{ or } Q_3,$$

where $Q_1 = (i-1,j)$, $Q_2 = (i-1,j-1)$ and $Q_3 = (i,j-1)$. Therefore

$$R(P_k) = R_1(P_k) \cup R_2(P_k) \cup R_3(P_k)$$

with

$$R_l(P_k) = \{\{P_1, \cdots, Q_l, P_k\}\,;\, \{P_1, \cdots, Q_l\} \in R(Q_l)\}$$

for $l = 1,2,3$. Thus we obtain

$$\begin{aligned} D\,(P = (m,n)) &= \min_{R \in R(P)} \{d(R)\} \\ &= \min_{l=1,2,3}\ \min_{R \in R_l(P)} \{d(R)\} \\ &= \min_{l=1,2,3}\ \min_{\substack{R_2 \in R(Q_l) \\ R_1 \in R_l(P)}} \{d(R_2) + d_s\,(Q_l, P;\, R_1)\} \\ &= \min\{D_1(P), D_2(P), D_3(P)\}, \end{aligned}$$

where

$$D_l\,(P_k = (i,j)) \equiv \min_{\substack{R_2 \in R(Q_l) \\ R_1 \in R_l(P_k)}} \{d\,(R_2) + d_s\,(Q_l, P_k;\, R_1)\}$$

for $l = 1,2,3$.

Each point Q_l has three points Q_l^1, Q_l^2, Q_l^3 which possibly go to Q_l one step after. These points are precisely written as

$$\begin{bmatrix} Q_1^1 & Q_1^2 & Q_1^3 \\ Q_2^1 & Q_2^2 & Q_2^3 \\ Q_3^1 & Q_3^2 & Q_3^3 \end{bmatrix} = \begin{bmatrix} (i-2,j) & (i-2,j-1) & (i-1,j-1) \\ (i-2,j-1) & (i-2,j-2) & (i-1,j-2) \\ (i-1,j-1) & (i-1,j-2) & (i,j-1) \end{bmatrix}$$

when $Q_1 = (i-1,j)$, $Q_2 = (i-1,j-1)$, $Q_3 = (i,j-1)$.
The distances $D_l\,(P_k = (i,j))$ can be obtained from one step before by the following recursion relations:

$$
\begin{aligned}
D_l\left(P_k=(i,j)\right) &= \min_{R_1\in R_l(P_k)} \min_{p=1,2,3} \min_{\substack{R_3\in R(Q_l^p)\\ R_2\in R_p(Q_l)}} \\
&\quad \left\{d\left(R_3\right)+d_s\left(Q_l^p,Q_l;\,R_2\right)+d_s\left(Q_l,P_k;\,R_1\right)\right\} \\
&= \min_{R_1\in R_l(P_k)} \min_{p=1,2,3} \left\{D_p\left(Q_l\right)+d_s\left(Q_l,P_k;\,R_1\right)\right\} \\
&= \min_{\substack{p=1,2,3\\ R_1\in R_l(P_k)}} \left\{D_p\left(Q_l\right)+d_s\left(Q_l,P_k;\,R_1\right)\right\}
\end{aligned}
$$

for $l=1,2,3$. The values D_l of initial point and those of the edge points are assumed as

$$
\begin{aligned}
D_2\left((0,0)\right) &= 0, \\
D_l\left((0,0)\right) &= \infty \text{ for } l=1,3, \\
D_l\left((1,j)\right) &= \infty \text{ for } l=1,2,\ j=1,\cdots,n, \\
D_l\left((i,1)\right) &= \infty \text{ for } l=2,3,\ i=1,\cdots,m,
\end{aligned}
$$

Moreover for other special cases, the recursive relation of the edge points satisfies

$$D_1\left(P_k=(i,1)\right)=D_3\left(P_{k-1}\right)+d_s\left(P_{k-1},P_k;\,R\right)$$

for $R\in R_1\left(P_k\right)$,$i=1,\cdots,m$,

$$D_3\left(P_k=(1,j)\right)=D_3\left(P_{k-1}\right)+d_s\left(P_{k-1},P_k;\,R\right)$$

for $R\in R_3\left(P_k\right)$,$j=1,\cdots,n$.
This calculation is completed in mn steps.

4. Evaluation

To compare the performance of MTRAP with those of the most often used two programs: Needle and ClustalW2, we used benchmark pairwise alignment sequences obtained from HOMSTRAD [8] and from PREFAB4 [9], which contain structural alignments and have been used to check the accuracy of alignment. The details of these six programs are: (1) Needle, a global pairwise alignment program using Needleman-Wunsch algorithm [1] contained in EMBOSS package ver. 5.0.0 [10] and it uses BLOSUM62 substitution matrix as default and (2) ClustalW2 [3,11], a typical progressive multiple alignment method and it uses GONNET matrix series as default. These programs basically used default parameters and MTRAP uses the following parameter values: $W_{\text{open}} = M(-9)$, $W_{\text{extend}} = M(-0.1)$ and $\epsilon = 0.8$, which were determined by sensitivity analysis and optimization. We removed gap of a set of 630 pairwise alignments obtained from HOMSTRAD November 1, 2008 and a set of 1682 pairwise alignments obtained from PREFAB4, then we realigned the protein pairs using each program.

Alignment accuracy was calculated with the Match Ratio (MR) score. The MR score is defined as the ratio of the number of residue pairs in generated alignment (i.e., the alignment obtained by a specified algorithm such as MTRAP, Needle, etc.) to the total number of aligned residue pairs in the generated alignment. Here the number of residue pairs in generated alignment mean that the number of the pairs that there exist same pairs in the reference alignment. Let mathematically it defines as follows: Let A_1, A_2 indicate the two sequences with length L properly aligned, and let $a_{1k}, a_{2k} \in \Omega^*$ indicate the kth symbol in the sequences A_1, A_2 respectively. When $a_{ik} \neq *$, it is important to find the number of the site in the reference sequence corresponding to the symbol a_{ik}, whose numbers are denoted by n_{ik} for $i = 1, 2$. When $a_{ik} = *$, put $n_{ik} = 0$ $(i = 1, 2)$. Then the MR score is given as

$$\text{MR score} = \frac{\sum_{k=1}^{L} \Delta_{a_{1k},a_{2k}} \delta_{n_{1k},n_{2k}}}{\sum_{k=1}^{L} \Delta_{a_{1k},a_{2k}}},$$

$$\Delta_{\mathrm{x,y}} = \begin{cases} 1, & x \neq * \text{ and } y \neq * \\ 0, & x = * \ \text{ or } \ y = * \end{cases}.$$

When all pairs are correctly aligned, the score have a maximum value 1, and when no-pairs are aligned the score have a minimum value 0.

5. Results

Table 1 shows the accuracy comparisons of MTRAP with the Needle and the ClustalW2 by using 630 protein pairs in the HOMSTRAD. The similarity between the sequence alignment by each method and the reference alignment of HOMSTRAD was measured with the MR score. Suppose as usual that the reference alignment is the optimal alignment, the MR scores over all 630 pairs indicate that our method works well compared with other two methods. For instance, the MTRAP has more than 80% accuracy (e.g., 80.9% with PAM250 and 81.1% with BLOSUM62), and the Needle

Table 1. Average MR scores on HOMSTRAD

Matrix / Method	Sequence identity (%) 0-15%(25)	15-30%(207)	30-45%(173)	0-100%(630)
PAM250				
MTRAP	**0.381**	**0.640**	**0.871**	**0.809**
Needle	0.226	0.548	0.837	0.764
ClustalW2	0.234	0.528	0.817	0.748
BLOSUM62				
MTRAP	**0.389**	**0.641**	**0.874**	**0.811**
Needle	0.223	0.556	0.843	0.768
ClustalW2	0.276	0.585	0.861	0.784
GONNET250*				
MTRAP	**0.398**	**0.649**	**0.875**	**0.814**
ClustalW2	0.313	0.619	0.867	0.800

Note: Each value is computed by averaging the MR scores over all pairwise sequence alignments with sequence identity within the specified range. The number in parentheses denotes the number of alignments in each sequence identity range. For each sequence identity range, the best result in each substitution matrix is in bold.

*Needle does not support GONNET matrix.

has less than 80% accuracy (e.g., 76.4% with PAM250 and 76.8% with BLOSUM62). Moreover it is important to notice that for two sequences with less than 30% sequence identity, our method improves the alignment accuracy significantly. For instance, MTRAP with PAM250 matrix has 0.381 for 0-15% sequence identity and 0.640 for 15-30% sequence identity, and ClustalW2 with PAM250 matrix has 0.234 for 0-15% sequence identity and 0.528 for 15-30% sequence identity, respectively.

We further compared the alignment accuracy by the different programs with several substitution matrices for 1682 protein pairs in another reference database PREFAB4. The similarity between the sequence alignments and the reference alignment of PREFAB4 was measured with the MR score. Figure 1 shows the ratio of the average MR scores with MTRAP to that with each program. On each substitution matrix, all three of these programs were almost the same alignment accuracy when the sequence identity is more than 60%, whereas the ratio clearly shows that MTRAP has higher accuracy than other programs in decreasing the sequence identity within 0-60%. For instance, MTRAP and Needle with PAM120 have 0.339 and 0.152 for 0-20% sequence identity, and those with BLOSUM80 have 0.336 and 0.166, respectively. For alignments with sequence identity 0-20%, the average MR score of MTRAP is twice higher than that of Needle. Moreover, MTRAP outperforms ClustalW2 at the same range by 1.3, 1.2 and 1.05 times for PAM, BLOSUM and GONNET series, respectively.

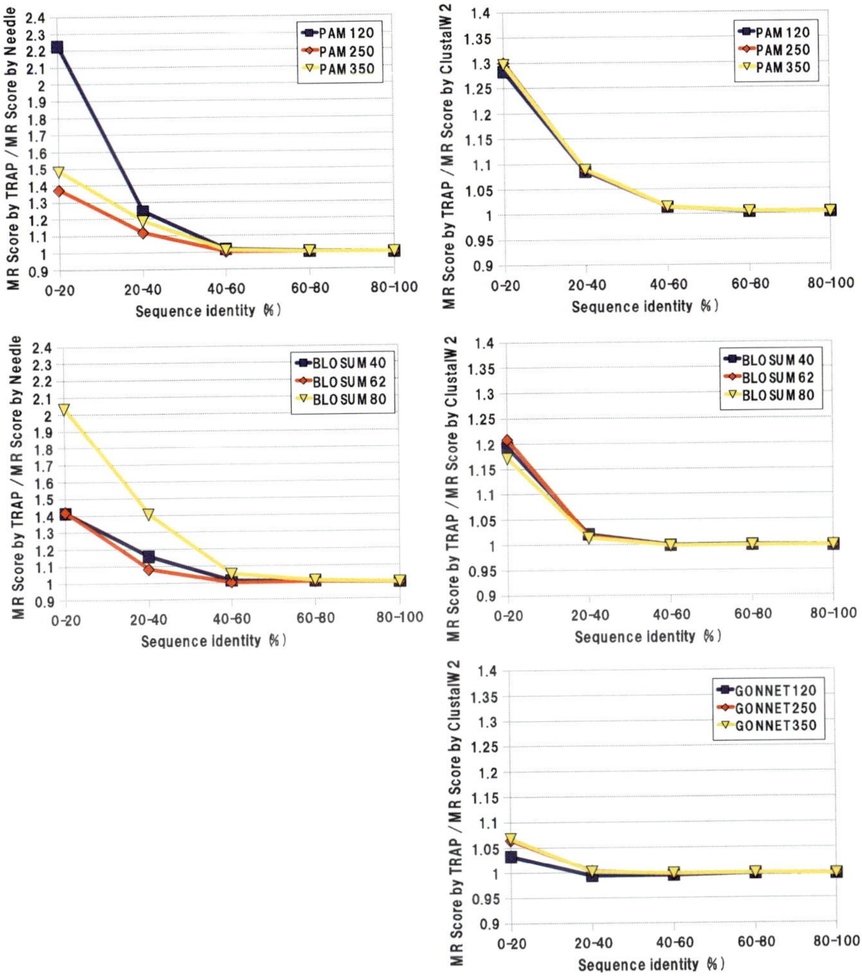

Figure 1. The ratio of the average MR scores on PREFAB4: The upper two figures show the ratio of the average MR score by MTRAP to that by Needle and the ratio of ours to that by ClustalW2, both for PAM substitution matrix. The middle two figures show the ratios for BLOSUM substitution matrix. The last figure shows the ratio for GONNET substitution matrix.

6. Conclusion

Here we indicated that our algorithm leads to a significant increase in alignment accuracy compared with other methods, and this is especially apparent in low identity range. We also indicated that MTRAP improves alignment accuracy across all identity range for any substitution matrix. That is, we can improve the alignment accuracy by using the measure which is determined by an adjusted substitution matrix and a transition probability-based matrix between two consecutive pairs of residues including gap-residue derived from structural alignments.

Pairwise sequence alignment is among the most important techniques to perform biological sequence analysis, and is fundamental to other applications in bioinformatics. Any multiple sequence alignment that is gradually built up by aligning pairwise sequences is essentially based on high-quality pairwise sequence alignments. By modifying existing multiple alignment algorithms based on our algorithm, the methods will significantly improve alignment accuracy.

References

1. S. B. Needleman and C. D. Wunsch, J. Mol. Biol. 48, 443(Mar 1970).
2. M. Ohya and Y. Uesaka, Information Sciences 63, 139 (1992).
3. J. D. Thompson, D. G. Higgins and T. J. Gibson, Nucleic Acids Res. 22, 4673(Nov 1994).
4. J. D. Thompson, P. Koehl, R. Ripp and O. Poch, Proteins 61, 127(Oct 2005).
5. M. O. Dayhoff, R. M. Schwartz and B. C. Orcutt, Atlas of Protein Sequence and Structure 5(3), 345 (1978).
6. S. Henikoff and J. G. Henikoff, Proc. Natl. Acad. Sci. U.S.A. 89, 10915(Nov 1992).
7. G. H. Gonnet, M. A. Cohen and S. A. Benner, Science 256, 1443(Jun 1992).
8. K. Mizuguchi, C. M. Deane, T. L. Blundell and J. P. Overington, Protein Sci. 7, 2469(Nov 1998).
9. R. C. Edgar, Nucleic Acids Res. 32, 1792 (2004).
10. P. Rice, I. Longden and A. Bleasby, Trends Genet. 16, 276(Jun 2000).
11. M. A. Larkin, G. Blackshields, N. P. Brown, R. Chenna, P. A. McGettigan, H. McWilliam, F. Valentin, I. M. Wallace, A. Wilm, R. Lopez, J. D. Thompson, T. J. Gibson and D. G. Higgins, Bioinformatics 23, 2947(Nov 2007).

Quantum Bio-Informatics III
eds. L. Accardi, W. Freudenberg & M. Ohya

A COMPUTATIONAL APPROACH TO EXPLORE PROTEIN TRANSLOCATION THROUGH TYPE III SECRETION APPARATUS*

THENMALARCHELVI RATHINAVELAN AND WONPIL IM†

Department of Molecular Biosciences and Center for Bioinformatics, The University of Kansas, 2030 Becker Drive, Lawrence, KS 66047, USA

Many Gram-negative bacteria initiate infections by injecting effector proteins into host cells through the type III secretion apparatus (TTSA) that is comprised of a basal body, a needle, and a tip. The needle channel is formed by the assembly of a single needle protein. To explore the export mechanisms of MxiH needle protein through the needle of *Shigella flexneri*, an essential step during needle assembly, we have performed steered molecular dynamics simulations in implicit solvent. Interestingly, the electronegative channel interior creates an energy barrier for MxiH to enter the channel, while the same may facilitate the ejection of the effectors into host cells. Structurally-known basal regions and ATPase underneath the basal region have also such electronegative interior, while effector proteins have considerable electronegative patches on their surfaces. Based on these observations, we propose a repulsive electrostatic mechanism for protein translocation through the TTSA. This mechanism is supported by the suggestion that an ATPase is required for protein translocation through these nanomachines, which may provide the energy to overcome the initial electrostatic energy barrier. A similar mechanism may be applicable to macromolecular channels in other secretion systems or viruses through which proteins or nucleic acids are transported.

1. Introduction

The type III secretion apparatus (TTSA) is a multi-protein molecular machine used by many Gram-negative bacteria to dock onto the eukaryotic host cell membrane and export effector proteins to initiate infections. One such Gram-negative bacterium, *Shigella flexneri*, causes shigellosis in humans and is responsible for over a million deaths each year [1]. The Centers for Disease Control and Prevention lists *Shigella* as a potential bioterrorism agent [2].

In general, the TTSA is composed of a basal body that spans the inner and outer bacterial membranes, an external needle, and a tip (Figure 1A) [3]. Although the atomic structure of the entire TTSA is not yet known, the three-

* This work was supported by institutional funding from the University of Kansas, Alfred P. Sloan Research Fellowship, NIH grants RR-017708, and National Science Foundation (NSF OCI-0503992) through TeraGrid resources provided by Purdue University to W.I.

† Email: wonpil@ku.edu

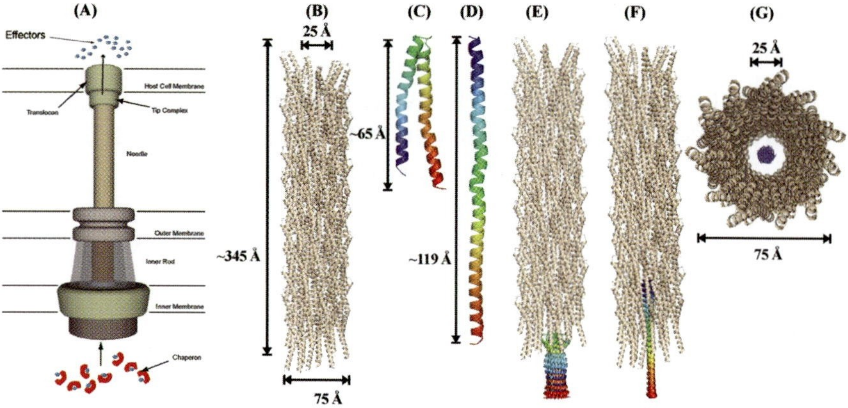

Figure 1 (A) Schematic representation of TTSA. (B) Structure of needle apparatus modeled with N-terminus facing inside the channel is viewed perpendicular to the channel axis. (C and D) Two MxiH conformations considered for the pulling simulations. (E and F) 18 Different starting configurations used in the pulling simulation of two-helix bundle and straight helix with each rotated 20° along the channel axis. The initial *Z* coordinates of the center of mass of the two-helix bundle and straight α-helix were -164 Å and -188 Å, respectively. (G) Perpendicular view of (E).

dimensional structures of several protein components involved in its formation are available for different Gram-negative bacteria. For instance, the structures of EscJ [4] and EscC [5] that are associated with the inner and outer membranes of the basal body in enteropathogenic *E. coli* are also proposed to form a ring-like structure based on X-ray and cryo-electron microscopy (EM) data. In addition, the needle of the *S. flexneri* TTSA, which is held by the basal body and protrudes from the bacterial surface, has been modeled recently based on a 16-Å cryo-EM density map [6]. The needle itself is a supramolecular complex that is formed by the assembly of approximately 120 copies of a needle protein around the needle channel axis. All the TTSA the needle protein structures solved to date have a helix-turn-helix motif which is attributed to a conserved PxxP motif at the turn region: MxiH from *S. flexneri* [6], PrgI from *S. typhimurium* [7], and BsaL from *B. pseudomallei* [8]. The structures of the tip proteins are also known, although the basis for their association with the needle is not known in atomic detail: IpaD from *S. flexneri* [9], BipD from *B. pseudomallei* [9] and LcrV from *Y. pestis* [10].

The ultimate aim in these research areas is to determine the structure of the entire type III secretion nanomachine and to understand the export mechanisms to facilitate vaccine and drug development against these bacteria. However, understanding how proteins are exported across the narrow needle channel poses a major challenge. In this context, we have investigated the underlying key microscopic forces responsible for protein export across the TTSA needle channel by performing steered molecular dynamics (SMD) simulations [11] to export the MxiH subunit through the needle of *S. flexneri*. It should be noted that MxiH itself needs to be exported through the growing needle during its formation. The export of MxiH is facilitated by pulling MxiH at a constant velocity through the modeled needle channel. Since the explicit treatment of the needle and solvent makes such simulations nearly intractable, we have utilized the implicit solvent models: the generalized solvent boundary potential (GSBP) [12] to describe the mean solvent-mediated potential from the rigid needle model, and the GBSW generalized Born model [13] to describe the solvation effects of the flexible pulled MxiH monomer. Based on the simulation results, we propose a repulsive electrostatic mechanism for MxiH transportation across the needle. We generalize this mechanism as a rationale for protein export through the TTSA based on the electrostatic surface of its components and the effectors.

2. Theory and Method

2.1 *Needle apparatus modeling*

A model of the *Shigella* needle was built with 64 subunits of MxiH (about a half of the biological needle) (Figure 1B) using the biomolecular simulation program CHARMM [14]. Following the helical parameters and the channel diameter of 25 Å determined by cryo-EM [6], the crystal structure of MxiH (PDB:2CA5) monomer was translated by 4.31 Å along the channel axis and rotated by 64.3° around the channel axis. The last five C-terminus residues as well as the first 19 residues of the N-terminus that were undetermined in the crystal structure were modeled as regular α-helices. Assuming that the needle channel is parallel to the Z-axis, the geometric center of the needle was located at $Z = 0$. The inter and intra molecular steric hindrance in the needle was removed by minimizing the entire needle with the GBSW generalized Born model [13] in CHARMM.

2.2 *Simulation details*

To export the MxiH monomer through the needle channel (Z-axis), we performed SMD simulations with a constant pulling speed using a harmonic restraint potential [11]. Since it is still unknown experimentally whether MxiH transports across the needle as a folded two helix bundle or an extended helix, we used two initial conformations of MxiH: a native-like folded two-helix bundle with the missing N- and C-terminal regions modeled as α-helices (Figure 1C), and an extended α-helix (Figure 1D). Starting conformations are positioned outside the needle as in Figures 1E and F.

The pulling speed was set to 15 Å/ns with a force constant of 150 pN/Å for the two-helix bundle and 100 pN/Å for the straight α-helix [100 pN corresponds to 1.44 kcal/(mol·Å)]. A higher force constant was chosen for the two-helix bundle systems to keep the pulling spring relatively stiff, so that MxiH trajectories would follow the reaction coordinate properly. The restraint force was applied to the center of mass of each MxiH monomer by modifying the AFM module [11] in CHARMM. The total simulation time to complete the

MxiH export was about 26 ns for the two-helix bundle and 25 ns for the extended α-helix.

Because the system size was too big for all-atom pulling simulations, we increased the computational efficiency by utilizing the implicit solvent models: the GSBP formalism [12] and the GBSW generalized Born model [13] (see "Theory" below). Based on the assumption of the rigid needle apparatus, the influence of the needle on exporting MxiH was approximated by the static electrostatic potential ($\phi^{\text{needle}}_{\varepsilon=80}$) from the needle as well as a repulsive core potential (U_{core}), Finally, the solvent effects on MxiH were treated by GBSW [13].

2.3 *Theory*

If 'A' and 'B' represent the needle apparatus and an MxiH monomer respectively, the total effective energy of the system can be expressed as [12],

$$E = U^{\text{AA}} + U^{\text{BB}} + U^{\text{AB}} + \Delta G_{\text{elec}} + \Delta G_{\text{np}} \tag{1}$$

U^{AA} and U^{BB} are the potential energy of 'A' and 'B'. U^{AB} is the interaction energy between 'A' and 'B', which is the sum of electrostatic ($U^{\text{AB}}_{\text{elec}}$) and van der Waals ($U^{\text{AB}}_{\text{vdW}}$) interactions. ΔG_{elec} and ΔG_{np} are the electrostatic and non-polar solvation free energy terms. The electrostatic free energy is given by [12],

$$\Delta G_{\text{elec}} = \frac{1}{2}\int d\mathbf{r}d\mathbf{r}'\rho_A(\mathbf{r})\,\mathrm{G}_{\text{rf}}(\mathbf{r},\mathbf{r}')\,\rho_A(\mathbf{r}') + \int d\mathbf{r}d\mathbf{r}'\rho_A(\mathbf{r})\,\mathrm{G}_{\text{rf}}(\mathbf{r},\mathbf{r}')\,\rho_B(\mathbf{r}') + \frac{1}{2}\int d\mathbf{r}d\mathbf{r}'\rho_B(\mathbf{r})\,\mathrm{G}_{\text{rf}}(\mathbf{r},\mathbf{r}')\,\rho_B(\mathbf{r}') \tag{2}$$

$\mathrm{G}_{\text{rf}}(\mathbf{r},\mathbf{r}')$ is the reaction field potential at $\mathbf{r}$ due to a unit charge at $\mathbf{r}'$. $\rho_{\text{A}}(\mathbf{r})$ and $\rho_{\text{B}}(\mathbf{r})$ are charge distributions in 'A' and 'B', respectively. The first term is the electrostatic solvation free energy of 'A', which is a constant and thus can be neglected because 'A' is kept fixed during the simulation. The second term is the coupling between the charges in 'A' and 'B'. This can be represented as the difference between the electrostatic interactions in the solvent dielectric environment and in vacuum, i.e., $\int d\mathbf{r}\,\rho_{\text{B}}(\mathbf{r})\left(\phi^{\text{A}}_{\varepsilon=80}(\mathbf{r}) - \phi^{\text{A}}_{\varepsilon=1}(\mathbf{r})\right)$, where $\phi^{\text{A}}_{\varepsilon=80}(\mathbf{r})$ is the electrostatic potential from 'A' in the solvent dielectric environment. Since the latter, $\int d\mathbf{r}\,\rho_{\text{B}}(\mathbf{r})\phi^{\text{A}}_{\varepsilon=1}(\mathbf{r})$, simply corresponds to the columbic interaction between 'A' and 'B' in vacuum ($U^{\text{AB}}_{\text{elec}}$), i.e.,

$$U^{\text{AB}}_{\text{elec}} + \Delta G_{\text{elec}} = \int d\mathbf{r}\,\rho_{\text{B}}(\mathbf{r})\phi^{\text{A}}_{\varepsilon=80}(\mathbf{r}) + \frac{1}{2}\int d\mathbf{r}d\mathbf{r}'\rho_{\text{B}}(\mathbf{r})\,\mathrm{G}_{\text{rf}}(\mathbf{r},\mathbf{r}')\,\rho_{\text{B}}(\mathbf{r}') \tag{3}$$

The static external potential $\phi^{\text{A}}_{\varepsilon=80}(\mathbf{r})$ can be calculated by solving the finite-difference PB equation [15]. Since it is independent of the instantaneous position of 'B', it is computed only once and used for the entire simulation [12].

However, the second term depends on the instantaneous position of 'B' and needs to be calculated every time-step during the simulation. But, solving the PB equations at every time-step will be computationally very expensive and hence, a generalized multipole approximation has been applied earlier [12]. However, in the present study, we simply used a generalized Born model ($\Delta G_{\mathrm{GB}}^{\mathrm{B}}$) to approximate the second term in Eq. (3), assuming that the reaction filed arising from the dielectric boundary of A is negligible.

The nonpolar solvation energy ΔG_{np} in Eq. (1) includes the formation of a cavity in the solvent as well as solvent-solute van der Waals interactions [16,17]. And, the van der Waals interaction between A and B ($U_{\mathrm{vdW}}^{\mathrm{AB}}$) includes the attractive ($U_{\mathrm{attractive}}^{\mathrm{AB}}$) and repulsive ($U_{\mathrm{repulsive}}^{\mathrm{AB}}$) components. Since 'A' is fixed, when 'B' approaches to 'A', ΔG_{np} will decrease and $U_{\mathrm{attractive}}^{\mathrm{AB}}$ will increase. For the sake of simplicity and computational efficiency, we assume that ΔG_{np} and $U_{\mathrm{attractive}}^{\mathrm{AB}}$ cancel each other and $U_{\mathrm{repulsive}}^{\mathrm{AB}}$ can be represented by a grid-based repulsive core potential U_{core}, i.e.,

$$\Delta G_{\mathrm{np}} + U_{\mathrm{vdW}}^{\mathrm{AB}} \approx U_{\mathrm{core}} \tag{4}$$

In terms of Eqs. (3) and (4), the total effective energy in Eq. (1) becomes

$$E = U^{\mathrm{BB}} + \Delta G_{\mathrm{GB}}^{\mathrm{B}} + \sum_{\alpha \in \mathrm{B}} q_\alpha \phi_{\varepsilon=80}^{\mathrm{A}}(\mathbf{r}_\alpha) + \sum_{\alpha \in \mathrm{B}} U_{\mathrm{core}}(\mathbf{r}_\alpha) \tag{5}$$

where q_α is the charge of atom α in 'B'. We used the GBSW generalized Born model [13] in CHARMM for $\Delta G_{\mathrm{GB}}^{\mathrm{B}}$ in Eq. (5).

Based on Eq. (5), the needle apparatus has been treated implicitly during the pulling simulations through the static electrostatic potential ($\phi_{\varepsilon=80}^{\mathrm{A}}$) arising from the needle protein charges as well as the grid-based core repulsive potential (U_{core}) arising from the shape of the needle channel. $\phi_{\varepsilon=80}^{\mathrm{A}}$ was calculated with a dielectric constant of 2 for the protein interior and 80 for the bulk solvent region with 150 mM salt concentration using the PBEQ module [16,18] in CHARMM. The optimized PB atomic radii have been used to setup the dielectric boundary [16]. The electrostatic potential was first calculated with a coarse grid (63 × 63 × 187 with a grid-spacing of 2.0 Å) centered on the entire needle apparatus. The result of the coarse calculation was then used to set the potential on the edge of a smaller box to perform second and third calculations using a finer grids (95 × 95 × 355 with a grid-spacing of 1.0 Å for second and 71 × 71 × 701 with a grid-spacing of 0.5 Å for the third) centered on needle apparatus. The PB radii augmented by the hydration radii layers were used to setup the molecular surface [18] by which a core repulsion potential map was built and stored on a grid (71 × 71 × 701 with a grid-spacing of 0.5 Å). It was set to be zero in all accessible region and 50 kcal/mol otherwise. The core repulsive energy and forces were calculated using the 3rd-order B-spline interpolation [19].

Although the present method is efficient to increase the computational speed, it has its own limitation too. For instance, the lack of explicit interactions between the needle and MxiH (U_{elec}^{AB} and U_{vdw}^{AB}) in the total energy may have influence in capturing the specific interaction between them. However, such limitation is unlikely to change the general conclusion of this study.

3. Results

3.1 *Electrostatics of the needle channel*

The electrostatic nature of the needle channel was first characterized by calculating the interaction energy between a chloride ion (Cl^-) and the electrostatic potential from the needle ($\phi_{\varepsilon=80}^{\text{needle}}$) at 150 mM salt concentration. As shown in Figure 2A, the resulting energy profile along the channel axis shows an energy barrier of about 3 kcal/mol for Cl^- to enter the channel due to the electronegative nature inside the needle channel (Figure 2C). As expected, the repulsion increases as the salt concentration decreases due to less salt screening. The energy barrier gradually increases from 3 kcal/mol to 10 kcal/mol when the salt concentration is decreased from 150 mM to 50 mM.

Although the cryo-EM data of the needle structure reveals that the inner diameter of the channel is about 25 Å [6], we examined the influence of different inner diameters on the electrostatic interaction by remodeling the needle with two different inner diameters (20 Å and 30 Å) [20] and recalculating $\phi_{\varepsilon=80}^{\text{needle}}$ at 150 mM. As shown in Figure 2B, the barrier reduces as the channel diameter increases (6 kcal/mol for 20 Å to 2 kcal/mol for 30 Å). Keeping in mind about such dependence of the electrostatic potential inside the needle channel, we used a salt concentration of 150 mM with a channel

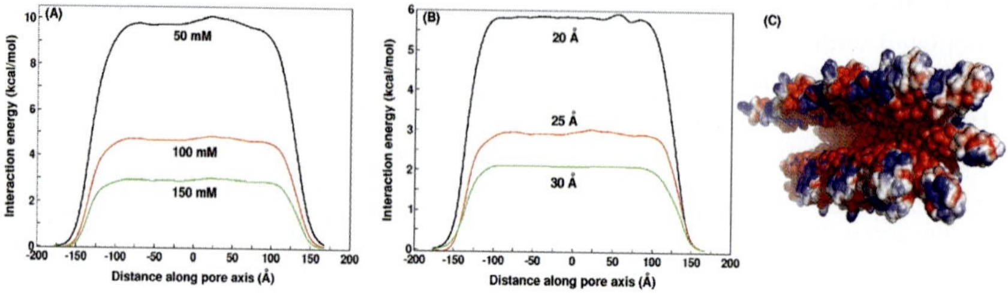

Figure 2 Effect of different salt concentration (A) and channel radius (B) on the energetics of Cl^- translation along the channel axis. (C) Cross sectional view (along the pore axis) of the needle apparatus of *S. flexneri* illustrating the surface electrostatics inside the pore. For the sake of clarity, some of the needle monomers in the front are not shown. The electrostatic surfaces are calculated by solving PBEQ module [15] in CHARMM and its pictorial representation is made using pymol [21]. Electrostatic scaling used for all the figures are -0.6 (red) to 0.6 kcal/(mol·e) (blue).

diameter of 25 Å based on the cryo-EM data in most pulling simulation studies and used other diameters for fewer simulations for a comparion purpose.

3.2 *Unfavorable energy barrier for MxiH export*

Although one can calculate the free energy profile using pulling simulation trajectories [22], our pulling speed (15 Å/ns) was too high and the number of trajectories were too small to reliably calculate the potential of mean force along the needle channel. Therefore, we calculated the mean interaction energy profile along the channel axis to explore the energetics of MxiH transport across the needle channel. As shown in Figure 3, while there is a flat potential for MxiH transport inside the channel, an energy barrier exists for MxiH to enter the needle, which, in fact, resembles the interaction energy profile between a chloride ion and the channel (Figure 2). It should be noted that the total net charge of MxiH is -4*e* based on the standard protonation state at pH 7. Although individual trajectories exhibit variations in the energy profiles (data not shown), the energy barrier is higher for the two-helix bundle (17 kcal/mol) than for the straight helix (10 kcal/mol) in the presence of 150 mM salt concentration (Figure 3). As expected, the barrier increases up to about 40 kcal/mol for both systems when the salt concentration is reduced to 50 mM.

We have examined the effects of MixH orientation, and channel diameter on the energy barrier. Several simulations that we have done with different starting conformations of the needle channel or MxiH monomer also reveal the presence of the electrostatic barrier, despite the marginal variation in the value of energy barrier (data not shown). For instance, pulling of the straight helix with C-terminal facing the needle channel yields essentially the identical energy barrier as in the case of N-terminal facing the needle channel. Second, the pulling simulations of straight helix (N-terminal facing the needle pore) and

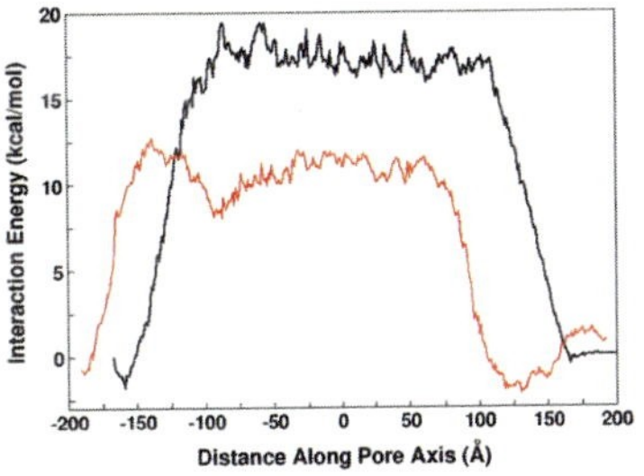

Figure 3 Mean interaction energy between the needle apparatus and the MxiH subunit along the channel axis: two-helix bundle (black) and straight helix (red).

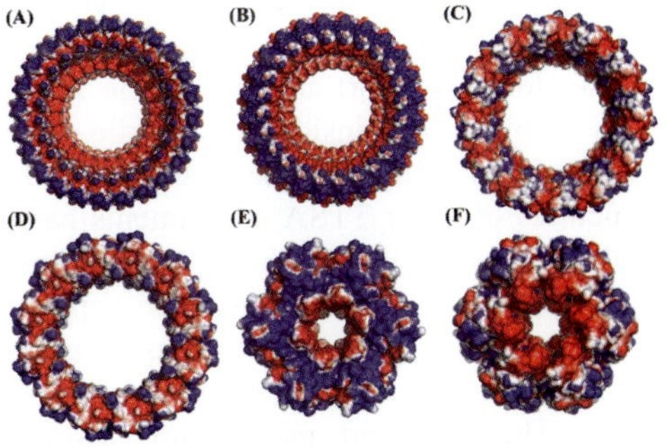

Figure 4 Surface electrostatics of the (A and B) EscJ, (C and D) EscC and (E and F) EscN. The electrostatic surfaces are calculated by solving PBEQ [15] in CHARMM and its pictorial representation is made using pymol [21]. Electrostatic scaling of -0.6 (red) to 0.6 kcal/(mol·e) (blue) is used.

two-helix bundle across the needle channel with 30 Å inner diameter [3] yields expected smaller energy barriers, 7 kcal/mol (straight helix) and 12 kcal/mol (two-helix bundle), than those with 25 Å inner diameter needle channel. Therefore, the presence of an energy barrier for MixH transportation across the needle is evident in all the cases. Assuming that the entropy contribution is similar or unfavorable even inside the TTSA compared to its exterior, we believe that the general shape of the free energy surface will be similar to the mean potential profile.

Interestingly, examination of the surface electrostatics of structurally-known TTSA components and effectors reveals similar overall electronegative features. The channels in the basal body (EscJ and EscC) [4, 5, 23] is highly electronegative (Figure 4 A-D), similar to the needle interior. EscN, the ATPase of *E. coli,* which is suggested to be anchored at the entrance of TTSA and acts as the inner-membrane recognition gate for TTSA chaperone-effector complexes [24], also possesses a highly electronegative channel interior (Figure 4 E, F). In addition, the monomeric tip proteins that sit on the top of the needle channel as well as almost all effectors that are known to pass through the TTSA contain electronegative patches on their surfaces (data not shown). Thus, our results suggest that the repulsive electrostatic interaction plays a key role in the export of the effectors through the TTSA.

3.3 *Repulsive electrostatic mechanism of protein export through TTSA*

Based on the energy barrier for MxiH export across the needle (Figure 3) and the electronegative nature of the TTSA components (Figure 4) as well as various effectors (data not shown), we propose a repulsive electrostatic mechanism for MxiH transportation through the TTSA channel. The highly electronegative nature of the channel creates an electronegative gradient outside the needle and thus develops a repulsive energy barrier for the electronegative effectors to enter into the channel. However, there is no (significant) energy barrier for the export as soon as the MxiH subunit enters inside the channel. This implies that inside the needle, the MxiH monomer exports through concentration-dependent passive diffusion. We also expect that the unknown structures of other TTSA components have the electronegative channel in line with the needle and the basal body. These results suggest that this secretion system has evolved to have such repulsive interactions because any strong attractive interactions could cause trapping of proteins inside the channel leading to obstruction of the channel.

The proposed mechanism raises an intriguing question about how effectors can enter the channel despite the presence of an electrorepulsive barrier at the entrance of the basal body. This may necessitate the need for an initial force to place the effectors into the apparatus. We envisage that this might be the reason that ATPase (which expected to sit at the bottom of TTSA and energize the export process) is required for placement of the proteins inside the TTSA channel in the Gram-negative bacteria [3], which may offer them the energy to overcome the barrier. Thus, one can believe that the directionality of the effectors through the TTSA is regulated by the ATPase. However, the presence of specific intracellular chaperones for almost all the effectors as well as the some of the available structure of effector-chaperone complexes clearly indicates that the directionality of the effectors towards the ATPase is regulated by those chaperones [3]. A similar electronegative gradient at the exit of the TTSA channel may facilitate the effective ejection of the effectors into the host cell because, as seen in the case of MxiH, the energetics for effectors to be outside the channel is much favored over those inside. Further, the presence of an electrorepulsive barrier could provide another layer of control to determine which proteins can be transported across the TTSA.

It is noteworthy that it was recently reported in the case of bacterial flagellar secretion apparatus that shares many commonalties with the TTSA [3], that FliI ATPase is needed for the initial placement of substrates into the flagellar gate [25]. It is believed that while the ATPase delivers the flagellar proteins to the nanomachine, the proton motive force may drive the progression of the proteins through the channel [26]. Thus, it is clear that the electrostatic gradient plays a role in the exportation process as evidenced by the results presented here.

4. Concluding Discussion

We have explored the mechanisms of effector translocation through the TTSA by performing SMD simulations to export the MxiH monomer, the needle protein of *S. flexneri* [6], through its needle (Figure 1). Because the system was too large for all-atom pulling simulations, we increased computational efficiency by using the GSBP formalism [12] to describe the mean solvent-mediated potential from the rigid needle model and the GBSW generalized Born model [13] to describe the solvation of the flexible pulled MxiH monomer. Such combination of the implicit solvent models may provide a computational approach to other supramolecular complex systems that are still intractable to use all-atom simulations.

The energetics deduced from the SMD trajectories indicates that, while the potential is quite flat for MxiH transport inside the channel, there is an energy barrier for MxiH to enter the needle (Figure 3), which is similar to the transportation of a chloride ion through a channel with a highly electronegative interior (Figure 2A, B). This energetic feature does not depend on the pulling speed, the orientation of pulled MixH, and the diameter of the channel interior. Detailed analyses of structurally-known TTSA components and effectors also reveal extensive electronegative patches on their surfaces (Figure 2C and 4A-F). Based on these observations, we propose a repulsive electrostatic mechanism for the protein translocation through the TTSA. The initial force required to surmount the energy barrier for proteins to enter into the TTSA may be offered by the ATPase as suggested experimentally [3]. A similar electronegative gradient at the exit of the channel may facilitate the effective ejection of the effectors into the host cell and thus, the electrorepulsive mechanism may control the proteins translocation across the TTSA. Based on the proposed mechanism, we speculate that the unknown structures of other TTSA components may have similar electronegative interior. Thus, this secretion system appears to have evolved this repulsive mechanism because any strong attractive interactions could cause trapping of proteins inside the channel, leading to obstruction of the protein channel. We expect that a similar mechanism may be applicable to macromolecular channels in other secretion systems or viruses through which proteins or nucleic acids are transported, e.g., protein channels that act as conduits during the DNA packing inside various virus capsids.

References

1. K. L. Kotloff, J. P. Winickoff, B. Ivanoff, J. D. Clemens, D. L. Swerdlow, P. J. Sansonetti, G. K. Adak and M. M. Levine, *Bull World Health Organ* **77**, 651 (1999).
2. L. D. Rotz, A. S. Khan, S. R. Lillibridge, S. M. Ostroff and J. M. Hughes, *Emerg Infect Dis* **8**, 225 (2002).
3. G. R. Cornelis, *Nat Rev Microbiol* **4**, 811 (2006).
4. C. K. Yip, T. G. Kimbrough, H. B. Felise, M. Vuckovic, N. A. Thomas, R. A. Pfuetzner, E. A. Frey, B. B. Finlay, S. I. Miller and N. C. Strynadka, *Nature* **435**, 702 (2005).
5. T. Spreter, C. K. Yip, S. Sanowar, I. Andre, T. G. Kimbrough, M. Vuckovic, R. A. Pfuetzner, W. Deng, A. C. Yu, B. B. Finlay, D. Baker, S. I. Miller and N. C. Strynadka, *Nat Struct Mol Biol* **16**, 468 (2009).
6. J. E. Deane, P. Roversi, F. S. Cordes, S. Johnson, R. Kenjale, S. Daniell, F. Booy, W. D. Picking, W. L. Picking, A. J. Blocker and S. M. Lea, *Proc Natl Acad Sci U S A* **103**, 12529 (2006).
7. Y. Wang, A. N. Ouellette, C. W. Egan, T. Rathinavelan, W. Im and R. N. De Guzman, *J Mol Biol* **371**, 1304 (2007).
8. L. Zhang, Y. Wang, W. L. Picking, W. D. Picking and R. N. De Guzman, *J Mol Biol* **359**, 322 (2006).
9. S. Johnson, P. Roversi, M. Espina, A. Olive, J. E. Deane, S. Birket, T. Field, W. D. Picking, A. J. Blocker, E. E. Galyov, W. L. Picking and S. M. Lea, *J Biol Chem* **282**, 4035 (2007).
10. P. Broz, C. A. Mueller, S. A. Muller, A. Philippsen, I. Sorg, A. Engel and G. R. Cornelis, *Mol Microbiol* **65**, 1311 (2007).
11. E. Paci and M. Karplus, *J Mol Biol* **288**, 441 (1999).
12. W. Im, S. Berneche and B. Roux, *J Chem Phys* **114**, 2924 (2001).
13. W. Im, M. S. Lee and C. L. Brooks, 3rd, *J Comput Chem* **24**, 1691 (2003).
14. B. R. Brooks, C. L. Brooks, 3rd, A. D. Mackerell, Jr., L. Nilsson, R. J. Petrella, B. Roux, Y. Won, G. Archontis, C. Bartels, S. Boresch, A.

Caflisch, L. Caves, Q. Cui, A. R. Dinner, M. Feig, S. Fischer, J. Gao, M. Hodoscek, W. Im, K. Kuczera, T. Lazaridis, J. Ma, V. Ovchinnikov, E. Paci, R. W. Pastor, C. B. Post, J. Z. Pu, M. Schaefer, B. Tidor, R. M. Venable, H. L. Woodcock, X. Wu, W. Yang, D. M. York and M. Karplus, *J Comput Chem* **30**, 1545 (2009).

15. W. Im, D. Beglov and B. Roux, *Comput Phys Commun* **111**, 59 (1998).
16. M. Nina, W. Im and B. Roux, *Biophys Chem* **78**, 89 (1999).
17. R. M. Levy, L. Y. Zhang, E. Gallicchio and A. K. Felts, *J Am Chem Soc* **125**, 9523 (2003).
18. N. K. Banavali, W. Im and B. Roux, *J Chem Phys* **117**, 7381 (2002).
19. W. Im and B. Roux, *J Mol Biol* **322**, 851 (2002).
20. F. S. Cordes, K. Komoriya, E. Larquet, S. Yang, E. H. Egelman, A. Blocker and S. M. Lea, *J Biol Chem* **278**, 17103 (2003).
21. W. L. DeLano, http://www.pymol.org (2002).
22. G. Hummer and A. Szabo, *Acc Chem Res* **38**, 504 (2005).
23. I. Andre, P. Bradley, C. Wang and D. Baker, *Proc Natl Acad Sci U S A* **104**, 17656 (2007).
24. R. Zarivach, M. Vuckovic, W. Deng, B. B. Finlay and N. C. Strynadka, *Nat Struct Mol Biol* **14**, 131 (2007).
25. T. Minamino and K. Namba, *Nature* **451**, 485 (2008).
26. J. E. Galan, *Nat Struct Mol Biol* **15**, 127 (2008).

Quantum Bio-Informatics III
eds. L. Accardi, W. Freudenberg & M. Ohya

CARBON NANOTUBES FOR BUILDING BLOCKS OF QUANTUM COMPUTING DEVICES

KOJI ISHIBASHI, SATOSHI MORIYAMA[1], TOMOKO FUSE[2], YUKIO KAWANO, AND TOMOHIRO YAMAGUCHI

Advanced Device Laboratory, RIKEN
2-1, Hirosawa, Wako, Saitama 351-0198, Japan

In this paper, we discuss possibilities to use single-wall carbon nanotubes (SWCNTs) as building blocks of quantum computing devices, particularly as quantum bits (qubits). To do this, electron transport properties of SWCNT quantum dots (QDs) are described, with experimental data that show artificial atom behaviors of the QD. It is shown that the SWCNT-QD behaves as a one-dimensional artificial atom with simple shell structures and energy scales associated with the artificial atom in the THz frequency range. The simple artificial atom feature would be useful for spin qubits. Another qubit type with normal materials is a charge qubit consisting of the coupled quantum dots. The experimental effort to fabricate it is also described.

1. Introduction

Quantum dots (QDs) are structures in which electrons are confined. The size of the quantum dot can be nm to μm, depending on the fabrication process used for the devices and temperature. For example, in the semiconductor quantum dots fabricated by conventional electron beam lithography technique with a typical size of submicron, the quantum dot behaviors have been studied below 1 Kelvin [1]. The main features of the quantum dots are confined quantum levels with a spacing of ΔE, and the single electron charging effect characterized by the charging energy for the single electron E_c ($=e^2/C_\Sigma$). When $\Delta E >> k_BT$, the dot is "quantum". When $E_c >> k_BT$, electrons can be charged on the dot one by one in the single electron transistor configuration [2].

The quantum dot is sometimes referred as an artificial atom in a sense that it resembles to natural atoms because, in both cases, electrons are confined in a small space. In the artificial atom, the confinement potential can be designed to some extent, a unique possibility for quantum computing devices [3]. In the next section, a carbon nanotube artificial atom is compared with the natural atoms as

1. Present address: National Institute of Materials Science (NIMS), 1-1, Namiki, Tsukuba, Ibaraki 305-0044, Japan
2. Present address: Kavli Institute of Nanoscience, Delft University of Technology, Lorenzweg, 2600GA Delft, The Netherlands

well as the semiconductor quantum dot. The artificial atom feature could be used for the quantum bit (qubit). For the normal state, a charge qubit consisting of the coupled quantum dot, and a spin qubit consisting of the single spin in the quantum dot are considered. The spin qubit may be particularly attractive in terms of the possible long coherence of the spin.

The quantum computing devices based on superconducting materials are extensively studied, and demonstration of the qubit manipulation and the CNOT gate, necessary requirements for realizing quantum circuit, has been done with those materials [4]. Although electrical access to the qubit composed of the superconducting loop with the Josephson junctions (Flux qubit) is easy, the coherence is limited by the charge fluctuation of the junction material. In contrast, the electrical access to the microscopic state is not easy, but, the manipulation of the single spin as an example of the microscopic state of normal materials is worth being pursued because of the possible longer coherence time [5].

2. Single electron transistor (SET) and artificial atom

2.1. *Single electron transistor*

The single electron transistor (SET) is widely used for investigation of the artificial atom. The structure and the equivalent circuit of the SET are shown in Fig.1(a). It is consisting of a quantum dot that is separated from source and drain electrodes through small tunnel junctions for possible current flow, and is connected to the gate electrode through the classical capacitance. Since the

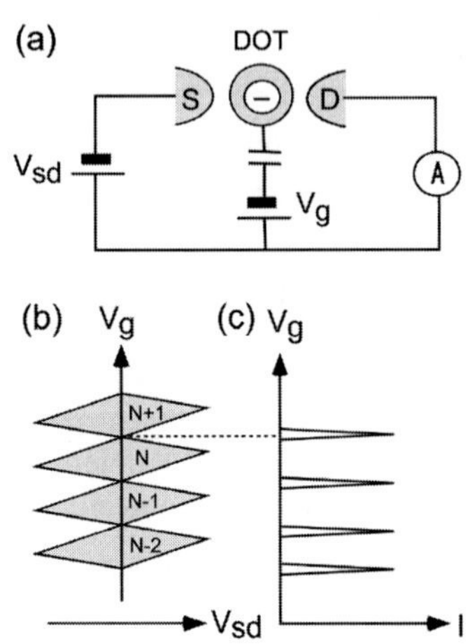

Fig.1 (a) Equivalent circuit of the SET (b) Coulomb diamonds (c) Coulomb oscillations

quantum dot is electrically floating, the chemical potential or the voltage of the dot is changed by the gate voltage (V_g). To flow the current, the source-drain voltage (V_{sd}) is applied.

The operation of the SET is determined by V_{sd} and V_g. Figure 1(b) shows a schematic gray scale plot of the conductance as function of V_{sd} and V_g, with a white color representing the zero conductance. In the diamond shaped regions, the conductance is zero, meaning that the current is blocked (Coulomb blockade). The important thing is that the number of electrons differs exactly by one between the adjacent diamonds. The charging energy, E_c, is necessary to increase the number of electrons by one, which energy is compensated by the gate voltage. This series of diamonds are called Coulomb diamonds. In the linear response regime with a small V_{sd}, the current flows at gate voltages where two diamonds meet, which occurs periodically as a function of the gate voltage (Fig.1(c)). The period is e/C_g, where C_g is the gate capacitance. The periodic current oscillations are called Coulomb oscillations [2].

2.2. *Artificial atom*

The above model of the SET, which does not take confinement levels in the quantum dot and the effect of the spin into account, is called the orthodox model [2]. When they are taken into account, the orthodox model needs to be modified [6]. The simplest modification would be that the charging energy or the addition energy is changed, depending on the number of electrons. When it is odd, the next electron can be charged in the level one electron has already occupied. But, the spin direction has to be opposite to the already occupying electron. In contrast, when the number of electrons is even, the next electron has to occupy one level above, because the level is already occupied by two electrons. In this case, the addition energy is $E_c+\Delta E$. This fact appears as an even-odd effect [7], and the size of the Coulomb diamonds changes alternately, depending on the

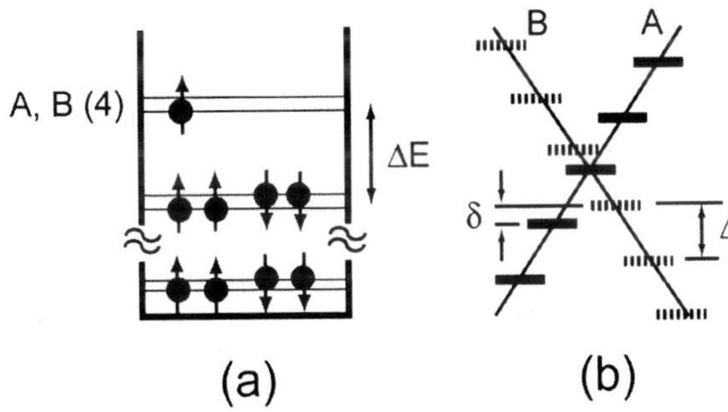

Fig.2(a) Four electron shell model in the one-dimensional hard wall potential (b) Two bands with confinement levels.δ, the misalignment of the levels in A and B bands, should be zero in the ideal case. ΔE is a level spacing determined by the confinement in a length direction.

number of electrons in the dot. In the practical quantum dot, the situation could be more complicated, with level degeneracy and electron-electron interactions. However, in the SWCNT artificial atom, the story is rather simple. In addition to the spin degeneracy considered above, the two-fold band degeneracy (Figure 2(b) has to be considered, so that each level can be occupied by four electrons, as shown in Figure 2. Because of the linear dispersion of E(k) in the carbon nanotube, the level spacing should be equal. In fact, the four electron shell structures are observed in the addition energy obtained from the distance between the adjacent Coulomb peaks in Figure 3(a) and (b) [8]. As seen in Figure 3(b), the addition energy has a peak for every four electrons, as expected. But, in another gate voltage range, a two-electron periodicity is observed. This may be explained by the broken two-fold band degeneracy which is produced by any kind of broken symmetry due to imperfections of the nanotube and impurities.

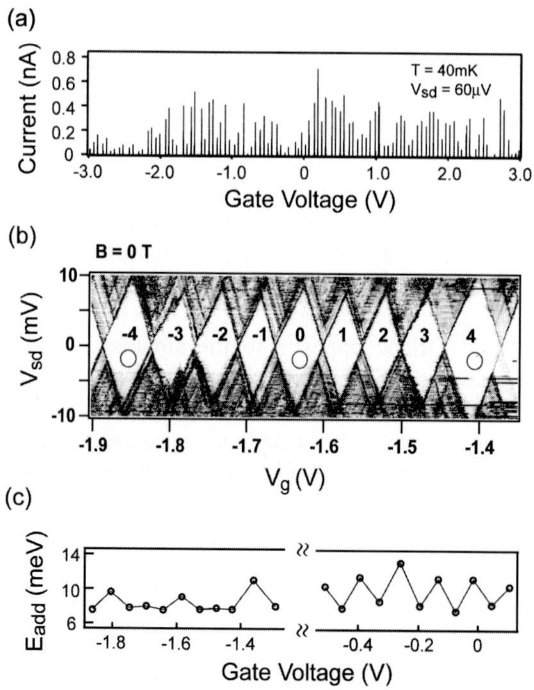

Fig. 3(a) Coulomb oscillations and (b) Coulomb diamonds of the SWCNT SET. Measurements have been carried out in the dilution refrigerator with a mixing chamber temperature of 40mK. (c) Calculated addition energy from the distances between adjacent Coulomb peaks.

Here, we like to compare the SWCNT artificial atom with a natural atom and the semiconductor artificial atom with vertical structures [9]. As mentioned before, the quantum dot is referred as an artificial atom in a sense that electrons are confined in a small space like a natural atom. But, for the quantum dot to be an artificial atom, it should have discrete quantum levels and clear shell structures that are specific to an atom and are determined by the confinement potential. In the Table 1, main features of the two artificial atoms are compared with a natural atom. In the natural atom, electrons are confined in the three-dimensional Coulomb potential, while, for the semiconductor artificial atom, they are confined in the two-dimensional harmonic potential, and for the SWCNT artificial atom, they are confined in the one-dimensional hard-wall potential. The size of the atom and the artificial atoms are very different that determines the energy scales associated with the atom and the artificial atoms. The addition energy defined in the artificial atom may correspond to the ionization energy for the natural atoms. For the natural atom, electrons escape to the vacuum, while for the artificial atom, electrons escapes to the electrodes that are separated with the dot by the tunnel barrier. The level spacing, ΔE, varies in a range from the X-ray to the infrared in the natural atom, while it is in the microwave range in the semiconductor artificial atom, and is in the THz range for the SWCNT artificial atom. The frequency range for the optical response of the atom and artificial atom is determined by these energies, so that we can expect the quantum THz response of the SWCNT artificial atom which we demonstrate later in this paper.

One important thing we should mention is that in the semiconductor artificial atom, the artificial atom feature is observable only when the number of

Table1: Comparison between a natural atom and artificial atoms

	Natural Atom	CNT Artificial Atom	Semiconductor Artificial Atom (Vertical Type)
	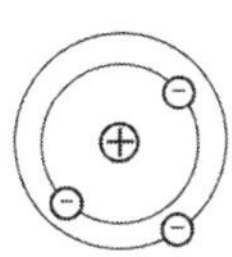	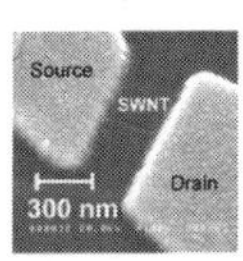 S. Moriyama, et al. Ref. 10	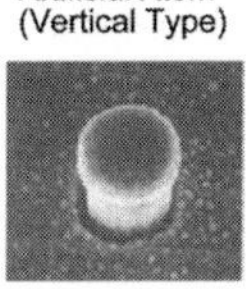 S. Tarucha et al. PRL 77, 3613 (1996)
Size	~ a few A	~ a few nm (diameter) ~ 100nm (length)	~ submicron (diameter) ~ 10nm (thickness)
# of electrons	1 ~ 100	0~10,000	0~10
Frequency of discrete levels	visible to X-ray	Submillimeter to THz	Microwave to millimeter
Charging energy	Ionization energy	> 10meV (1~10THz)	~1meV
Cofinement potential	3D Coulomb	1D hard wall	2D harmonic
Magnetic field effect	Zeeman effect	Zeeman effect	Orbital and Zeeman (when B is perpendicular to the disk)
Schematic picture of quantum states (Quantum states and degeneracy)	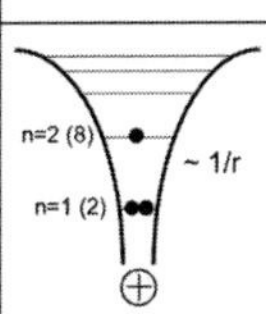	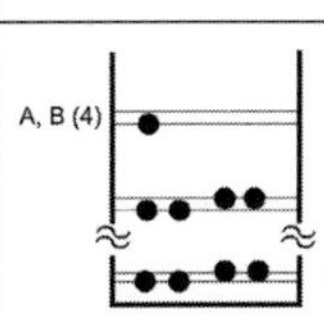	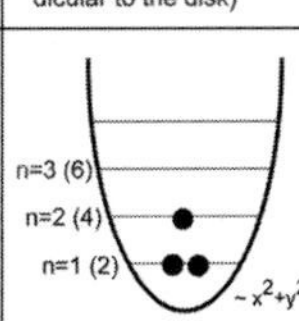

electrons are few, typically less than 10, which is in contrast to the SWCNT artificial atom where the artificial atom features are observable independent on the number of electrons. The reason for this is that for the semiconductor artificial atom, the level degeneracy increases as the quantum number increases, so that the effect is not easy to observe experimentally when the number of electrons is many (>~1,000). Besides, the parabolicity of the confined potential may be distorted in practical devices, especially as the energy becomes large, which lifts the degeneracy, and the electron-electron interaction may make the shell filling very complicated. In fact, the orthodox model that predicts the periodic Coulomb oscillation is very much applicable for the semiconductor quantum dots with many electrons. On the other hand, the degeneracy is always four, independent on the quantum number for the SWCNT artificial atom. Besides, the simple shell filling from the bottom to the top is possible because the level spacing, ΔE, is much larger than the electron interaction energies such as the on-site Coulomb interaction and the exchange interaction [8]. That is not the case for the semiconductor quantum dot.

2.3. *Energy level spectroscopy of the SWCNT artificial atom*

Energy level spectroscopy can be done in the Coulomb peak measurement with a larger source-drain voltage (V_{sd}) than ΔE. The basic idea of this is that in the specific Coulomb peak measurement, the current changes as soon as a new level comes into the bias window set by V_{sd}. To enhance the effect, the derivative of the current with respect to the gate voltage, dI/dV_g, is plotted in a color scale for the specific Coulomb peak as functions of the gate voltage and the magnetic field.

2.3.1. *One electron in a shell*

Figure 4(a) shows such a measurement for the Coulomb peak that appears for the number of electrons in a shell between n=0 and n=1. It is seen that there are two levels at the zero magnetic field (B=0). The two levels correspond to the two confinement levels belonging to the two bands, A and B. When the magnetic field is applied, each level splits into two, for both levels, with one going upward and the other going downward. The splitted levels change linearly, which suggests the Zeeman splitting of the single-particle state. The g-factor estimated from the slope is very close to 2, a value for the free electrons in vacuum.

2.3.2. *Two electrons in a shell –an interacting two-electron system-*

Figure 4(b) shows the case for two-electrons in a shell. It is found that each line corresponds to the following two-electron states. "e": two electrons occupy the Zeeman splitted two levels that belong to the A band. "f": |00> or |01>+|10>, "g": |11> or |01>+|10>, "h": |01>-|10> (0: spin-up, 1:spin-down) In this expression, |01>, for example, means that the spin-up electron in the A band and the spin-down electron in the B band. The lines, "f", and "g", originate from the triplet state of the two electron spins. However, in the present measurement scheme, we cannot distinguish |00> and |01>+|10> for the line "f" as well as |11> or |01>+|10> for the line "g". The lines, "f" and "g", that originate from the triplet state meet at B=0, a natural consequence of the three fold degeneracy of the triplet state at B=0. It is interesting to notice that the line, "h", runs above the line, "g", and can be attributed to the singlet state. The "h"-line do not meet the "f" and "g" lines at B=0, the fact that the singlet state has a larger energy by the amount of the exchange interaction energy (J) at B=0. From the separation at B=0, J~0.5meV which is much smaller than ΔE~6meV. The relation among the three numbers, ΔE>>J, ΔU (~0.4meV), where ΔU is an on-site Coulomb interaction energy, makes the simple shell filling possible in the SWCNT artificial atom.

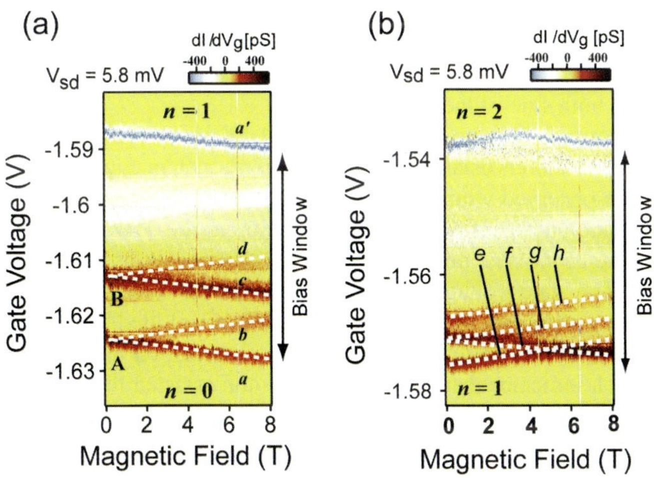

Fig.4 Energy spectrum in magnetic fields for (a) n=1 and (b) n=2

2.4. *Teraherz (THz) photon assisted tunneling (PAT)*

As mentioned in the subsection 2.3, the energy scales as an SWCNT artificial atom fall in the THz frequency range, for which reason we could expect the quantum response to the THz wave [10]. The THz frequency is an intermediate frequency between the light wave and the classical electromagnetic wave, and to study the response is interesting. The basic idea of the THz photon assisted tunneling is shown in Fig.5(c). Without the THz wave irradiation, the quantum dot is in Coulomb blockade condition. There is no available state in the bias window set by V_{sd}. However, if an electron in the dot absorbs the THz photon and tunnel out to the drain electrode, a new current path could be created. Therefore, a new current peak should appear at gate voltages where there is no

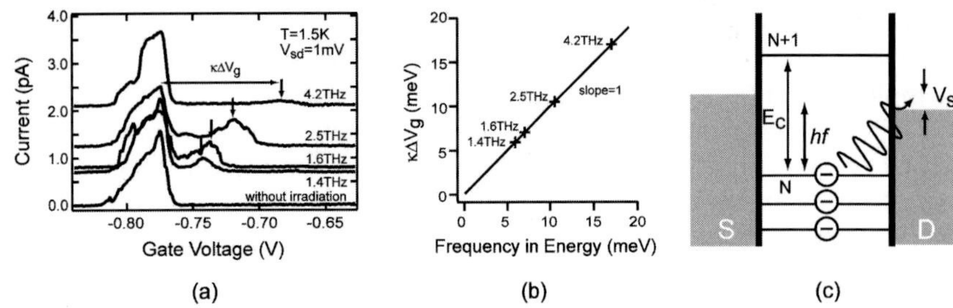

Fig.5 (a) Coulomb peaks under THz waves with different frequencies. (b) Energy distance between the main Coulom peak and the photon peak. (c) Schematic mechanism of the THz PAT that corresponds to the side peaks

current without the THz irradiation. This effect is analogous to the frequency modulation of the sinusoidal wave where frequency sidebands separated by *hf* appear on the both side of the main frequency.

The experiment has been carried out at 1.5K by irradiating the THz wave to the sample in the cryostat, and the result is shown in Figure 5(a) [11]. The side peaks are observed when the THz wave is irradiated on the right-hand side of the main Coulomb peak that exists even without the THz irradiation. Beside the appearance of the new peaks, the peak position moves as the THz frequency is changed. The gate voltage distances between the new peak and the main Coulomb peak, which is converted to the energy, are plotted as a function of the THz frequency, as shown in Figure 5(b). The plots fall on the line with a slope of unity, which indicates that the expected photon assisted tunneling (PAT) has indeed been observed in the sample. We should admit that all samples has not shown the THz-PAT, and some samples showed different response, which we do not discuss here [12, 13]. The reason may be that the samples did not have an appropriate antenna to receive the THz wave, so that the efficiency of the THz response could be very low.

Now, we like to discuss the condition of the THz-PAT in the quantum dot. For that to occur, there are following requirements. First, the charging energy, E_c, should be much larger than the THz frequency, *hf*, for the simple PAT process in Figure 5(c). In the present sample, E_c is about 20meV, while *hf* for 2.5THz is about 10meV, so that the requirement is met. Second, k_BT should be much smaller than *hf*, to avoid thermal smearing at the drain electrode. k_BT at 1.5K is about 0.2meV which is much smaller than *hf*. Third, the average escape time through the dot should be much longer than the period of the THz frequency, for an electron to feel the wave. The escape rate estimated from the current, Γ~I/e for the single electron transport, is about 10^8/s (0.1GHz) for ~1pA which is much smaller than the THz frequency. All the above requirements can be easily satisfied in the SWCNT quantum dot, and the THz-PAT may open a way to new kind of THz detectors with a spectroscopic capability.

3. On the possibility for quantum computing devices

3.1. *Spin qubit*

The electron spin qubit can be realized by setting one electron in the quantum dot, or preparing a single unpaired electron in the quantum dot with many electrons (Figure 6(a)). In the GaAs/AlGaAs quantum dots, the former strategy has been adopted [14]. By applying sufficient negative gate voltages to the appropriately designed quantum dot structure, one can squeeze the number of electrons to zero. In contrast, the latter is possible in the SWCNT when one electron exists in a shell. In the semiconductor SWCNT, the former is also possible [15], but, in practice, the chance to have a single quantum dot is lower probably because the semiconducting tube is not immune to the negative effects by imperfections and defects, as it is for metallic tubes. [16, 17, 18]

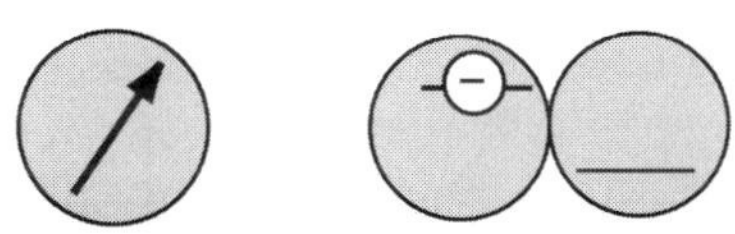

(a) Spin qubit (b) Charge qubit

Fig.6 Schematic picture of the spin qubit and the charge qubit

We can see this in Figure7. The lowest panel shows the Coulomb diamonds of the SWCNT-QD in the two-electron periodicity regime. In fact, the alternate sequence of the diamonds (large-small-large-small) is observed, and in the smaller diamonds, the number of electrons is odd, and in the larger diamonds, it is even. The important feature for the spin qubit is that in the smaller diamonds with the odd number of electrons, the single unpaired spin is realized. It is understood from the Coulomb peak shifts in magnetic fields. When the gate voltage is swept from the smaller to larger values, the number of electrons increases in the QD. The Coulomb peak that appears when the gate voltage pass from the even diamond to the odd diamond shifts linearly in the left direction, while the Coulomb peak that appears from the odd to even shifts linearly in the right direction (Fig.7(b)(c)). The mechanism of the behaviors is explained by using the top panels. For the former Coulomb peak, the new electron occupies the lower level of the Zeeman splitted two levels (left panel), while for the latter, Coulomb peak, the new electron occupies the upper level of the Zeeman splitted level. It should be noted that the single spin is realized in the left panel, as indicated by the dotted red circle. This is nothing more than the spin qubit. In terms of the requirement of the realization of the quantum computing [19], the present status of the carbon nanotube spin qubit is that the preparation the spin qubit and the initialization has been achieved. The next target is to coherently manipulate the spin qubit by using the electron spin

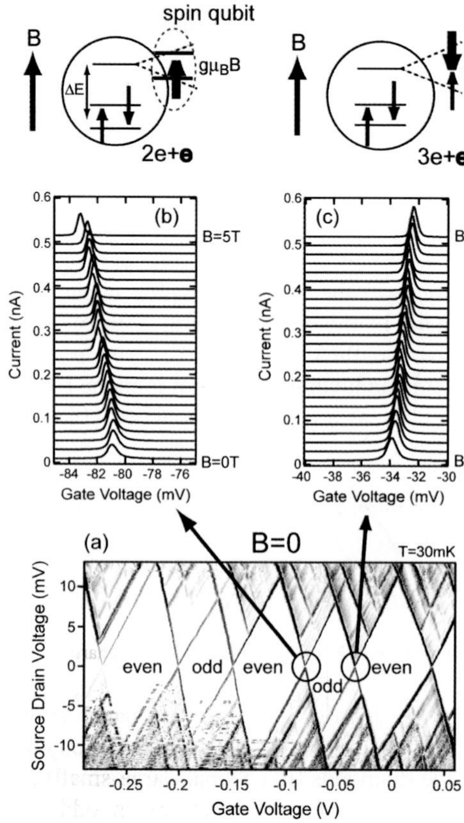

Fig.7 (a) Coulomb diamonds in the two-electron periodicity regime (b) and (c) the Coulomb peak shifts in magnetic fields for the two peaks. Top panel: The shell structures for the odd (left) and even (right) number of electrons in the dot. The single electron levels are Zeeman splitted.

resonance (ESR) technique.

Here, we discuss a little bit about the numbers for the ESR experiment. The Zeeman splitting at 5Tesla is 0.5meV with the obtained g-factor which is roughly 2. The energy corresponds to the frequency about 150GHz which corresponds to 0.6meV. Considering k_BT being roughly 0.3meV at 1K, we may expect the coherent manipulation even at liquid Helium temperatures. The electrical readout of the single spin has been demonstrated in the semiconductor quantum dot [20].

There are advantages of the SWCNT for the spin qubit. It should have a long coherence because the spin-orbit interaction is small, and the interaction with nuclear spin is small. Both of them are main causes of the decoherence in the compound semiconductor quantum dots [21]. The larger energy scales of the SWCNT artificial atom, compared with the submicron scale semiconductor quantum dots, make faster gate operation possible. The difficulties would be to get large interaction between the spin and the electromagnetic wave for the ESR -type manipulation. Furthermore, the primitive situation of the quantum dot fabrication process in the SWCNTs has to be improved.

3.2. *Charge qubit*

The charge qubit consists of the coupled quantum dots which have a well-defined zero-dimensional confined state. The coherent coupling, or the formation of the molecular state has been realized in the GaAs/AlGaAs semiconductor coupled quantum dots [22], and the coherent oscillations between the two levels in each quantum dot has been observed [23]. However, the oscillation decays rapidly, and the amplitude of the oscillations is small. For the charge qubit, the decoherence source would be acoustic phonon scatterings [24] and voltage fluctuations of the gates that is necessary to form the quantum dots with the split gate geometry.

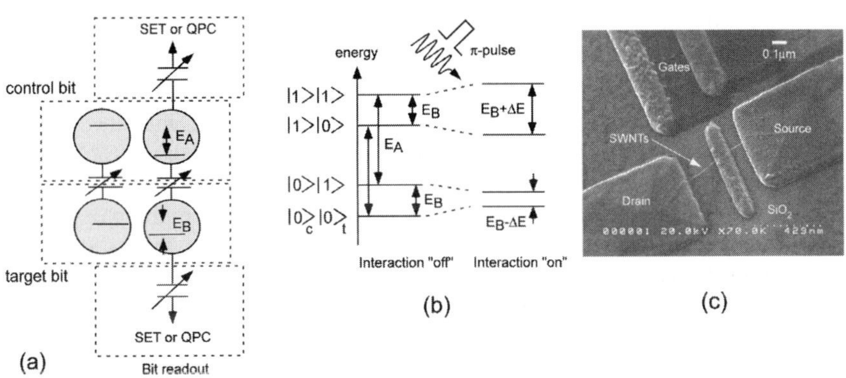

Fig.8 (a) Schematic configuration of the CNOT gate with charge qubits, (b) Energy diagram of the two-qubit system with and without the qubit interaction. (c) Scanning electron microscope image of the SWCNT coupled quantum dots

The control-NOT (CNOT) gate could be constructed, as shown schematically in Figure8(a) [25], where two qubits (coupled quantum dots) are faced with capacitive coupling which could be switched on and off. For the single qubit operation, the capacitive coupling between the qubits has to be switched off, and for the CNOT gate operation, it has to be switched on. The qubit state is read out either by the single electron transistor or by the quantum point contact (QPC) [26] which is a narrow constriction of the transport channel set in a tunneling regime. The faster readout could be possible with the RF-SET [27]. The CNOT operation mechanism is understood with a help of the total energy diagram shown in Figure8(b). Suppose that the qubit on the top is the control-qubit with the larger energy, while the qubit on the bottom is the target-qubit with the smaller energy. The qubit state with an electron on the left is defined |0>c, where "c" represent the control qubit, and vise versa. The energy diagram of the two-qubit system without the qubit coupling is shown in the left figure. When the qubit coupling is "on", the two qubit state with the two electrons in the diagonal arrangement would increase its energy much larger than the off-diagonal arrangement of two electrons, because of the Coulomb interaction between them. Then, by applying the π-pulse with an energy on resonance to the |10> and |11>, the CNOT operation is carried out. Due to the Coulomb interaction energy, the other transition between |00> and |01> does not occur, because it is out of resonance to the irradiated wave.

Here, we show our effort to fabricate coupled quantum dots with SWCNTs [28]. The advantage of the charge qubit with the SWCNT coupled quantum dots would be that the energy level spacing (ΔE) could be larger in the submillimeter to the THz range, which makes the gate operation frequency larger compared with the semiconductor based charge qubits that have a smaller ΔE. But, the fabrication of the coupled quantum dots with SWCNTs is not easy. Figure 8(c) is a fabricated coupled quantum dots with the individual SWCNT. The idea is that the single quantum dot in between the source and drain contacts is divided into two dots with a tunnel coupling, by depositing the narrow SiO_2 layer on top of the SWCNT. To see how it works, we compare the I-V curves with different gate voltages (Coulomb diamonds) for the SWCNT with and without the SiO_2 layer, as shown in Figure 9(a) and (b). For the SWCNT without the SiO_2 layer, the single quantum dot behavior with periodic diamonds along the gate voltage is observed. However, when the SiO_2 layer is deposited, the Coulomb diamonds have changed dramatically. There are two important features that suggest the coupled dot formation. First, the diamond shapes along the gate voltage become rather irregular, which is due to the stochastic Coulomb blockade that occurs for more- than-one dots with one gate voltage [29, 30]. Second, the negative differential conductance has been observed, which has never been observed in the single dot. This is most probably due to the resonant tunneling between confined states that are formed in each dot, which strongly suggests the formation of the coupled quantum dots. The problem of the present dots, the each dot cannot be controlled independently by the gate voltages, because the size of the gate is much larger than the coupled dots and they are not close enough to control the potential of each dot [31]. The device fabrication design should be improved to solve the problem [32].

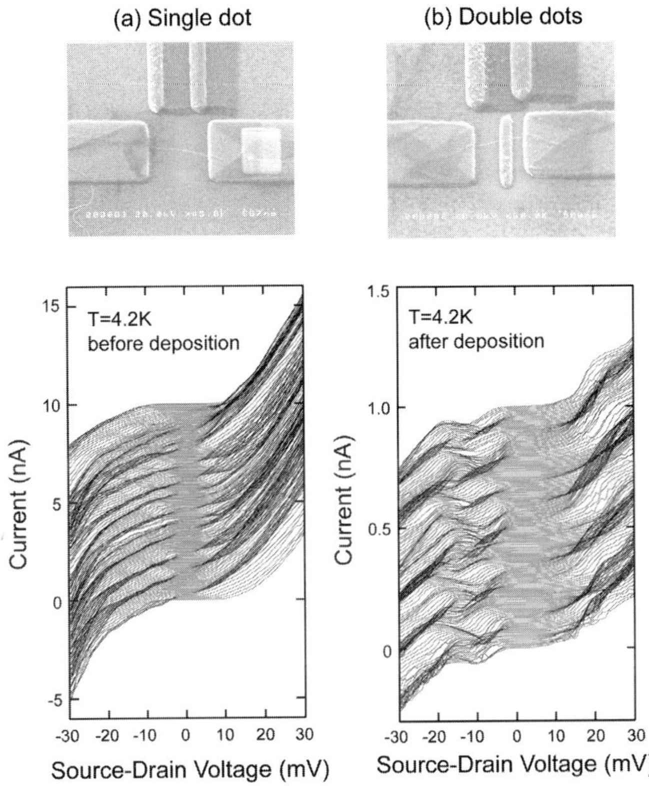

Fig.9 I-V curves with different gate voltages (a) without the SiO_2 layer and (b) with the SiO_2 layer. The measurements are essentially identical with the Coulomb diamond measurement

4. Summary

In this paper, we have shown the artificial atom behaviors of the SWCNT-QDs. They show the simple shell structures that accommodate four or two electrons in different gate voltage ranges. In magnetic fields, the single-particle levels are Zeeman splitted. When two electrons are in a shell, the quantum states of the interacting two electron system, the singlet and triplet states, have been observed in the energy spectrum as a function of the magnetic field. The simple shell structures occur, independent on the quantum number or the number of electrons. The energy scales associated with the artificial atoms are unique, which fall on the range in submillimeter to THz range. Focused on this, we have demonstrated the THz photon assisted tunneling (THz-PAT), a quantum response to the THz wave for the first time in the dot.

Based on the unique features of SWCNT artificial atoms, we have discussed on the possibility for the building block of the quantum computing devices or qubits. For qubits, we have considered the spin qubits and the charge qubits. Overall, the spin qubits in the SWCNT-QD appears to be more promising, because of the expected long coherence and the easy formation of the single spin even though there are many electrons.

But, there are many problems that have to be solved to make these in practice. The difficult problem would be the development of the reliable fabrication process specific to SWCNT. Besides, to realize the easy access to the single spin has to be realized.

References

[1] L. L. Sohn, L. P. Kouwenhoven and G. Schon, *Proc. of the NATO Advanced Study Institute on Mesoscopic Electron Transport, 1996* (Kluwer Academic Publishers, Dordrecht, 1997)

[2] D. V. Averin and K. K. Likharev: *Mesoscopic Phenomena in Solids*, eds. B. Altshuler, P. A.Lee and R. A. Webb (Elsevier Science Publishers, 1991) Chap.6

[3] Michael A. Nielsen and Isaac Chuang, "Quantum Computation and Quantum Information", Cambridge University Press 2000

[4] M. H. Devoret, A. Wallraff, and J. M. Martinis, cond.mat/0411174v1, 7, Nov. 2004

[5] K. C. Nowack, F. H. L. Koppens, Yu. V. Nazarov, and L. M. K. Vandersypen, Science, **318**, 1430 (2007)

[6] C. W. J. Beenakker, Phys. Rev. B **44**, 1646 (1991)

[7] D.C. Ralph, C.T. Black, M. Tinkham, Phys. Rev. Lett. **78**, 4087 (1997)

[8] S. Moriyama, T. Fuse, M. Suzuki, Y. Aoyagi, K. Ishibashi , Phys. Rev. Lett. **94**, 186806 (2005)

[9] S. Tarucha, D.G. Austing, T. Honda, R. J. vd. Hage and L. P. Kouwenhiven, Phys. Rev. Lett. **77**, 3613 (1996)

[10] T. Fuse, Y. Kawano, T. Yamaguchi, Y. Aoyagi, K. Ishibashi, Nanotechnology **18**, 044001 (2007)

[11] Y. Kawano, T. Fuse, S. Toyokawa, T. Uchida, K. Ishibashi, J. Appl. Phys. **103**, 034307 (2008)

[12] T. Fuse, Y. Kawano, M. Suzuki, Y. Aoyagi, K. Ishibashi, Appl. Phys. Lett. **90**, 013119 (2007)

[13] Yukio Kawano, Takao Uchida, and Koji Ishibashi, Appl. Phys. Lett. **95**, 083123 (2009)

[14] K. C. Nowack, F. H. L. Koppens, Yu. V. Nazarov, and L. M. K. Vandersypen, Science, **318**, 1430, 2007

[15] Pablo Jarillo-Herrero, Sami Sapmaz, Cees Dekker, Leo P. Kouwenhoven, Herre S. J. van der Zant, Nature **429**, 389, 2004

[16] T. Ando and T. Nakanishi, J. Phys. Soc. Jpn. 67, 1704 (1998)

[17] P. L. McEuen, M. Bockrath, D. H. Cobden, Y. -G. Yoon and S. G. Louie: Phys. Rev. Lett. **83** (1999) 5098

[18] M. Suzuki, K. Ishibashi, T. Ida, and Y. Aoyagi, *Jpn. J. Appl. Phys*. **40**, 1915 (2001)

[19] D. P. DiVincenzo, quant-ph/0002077

[20] J. M. Elzerman, R. Hanson, L. H. Willems van Beveren, B. Witkamp, L. M. K. Vandersypen,and L. P. Kouwenhoven, Nature, 430, 431 (2004)

[21] R. Hanson, B. Witkamp, L.M.K. Vandersypen, L. H. Willems van Beveren, J. M. Elserman, and L. P. Kouwenhoven

[22] T. H. Oosterkamp, T. Fujisawa, W. G. van der Wiel, K. Ishibashi, R. V. Hijman, S. Tarucha and L. P. Kouwenhoven, Nature, **395**, 873 (1998)

[23] T. Hayashi, T. Fujisawa, H. D. Cheong, Y. H. Jeong, and Y. Hirayama, Phys. Rev. Lett. **91**, 226804, 2003

[24] Toshimasa Fujisawa, Tjerk H. Oosterkamp, Wilfred G. van der Wiel, Benno W. Broer, Ramo'n Aguado, Seigo Tarucha, and Leo P. Kouwenhoven, Science, **282**, 932 (1998)

[25] K. Ishibashi, M. Suzuki, S. Moriyama, T. Ida, Y. Aoyagi, Superlattices and Microstructures, **31**, 141 (2002)

[26] J. M. Elzerman, R. Hanson, J. S. Greidanus, L. H. Willems van Beveren, S. De Franceschi, L. M. K. Vandersypen, S. Tarucha, and L. P. Kouwenhoven, _Phys. Rev. **B 67**, 161308 (2003)

[27] R. J. Schoelkopf, P. Wahlgren, A. A. Kozhevnikov, P. Delsing, and D. E. Prober, /Science/ **22**, 1238, 1998

[28] K. Ishibashi, M. Suzuki, T. Ida and Y. Aoyagi, Appl. Phys. Lett. **79**, 1864 (2001)

[29] I. M. Ruzin, V. Chandrasekhar, E. I. Levin and L. I. Glazman: Phys. Rev. B **45**, 13469 (1992)

[30] M. Kemerink and L. W. Molenkamp: Appl. Phys. Lett. **65**, 1012 (1994)

[31] K. Ishibashi, M. Suzuki, T. Ida and Y. Aoyagi, Physica E **13**, 782 (2002)

[32] N. Mason, M. J. Biercuk, C. M. Marcus, Science, **303**, 655, 2004

Quantum Bio-Informatics III
eds. L. Accardi, W. Freudenberg & M. Ohya

IN SILICO ANALYSIS FOR THE STUDY OF BOTULINUM TOXIN STRUCTURE

TOMONORI SUZUKI AND SATORU MIYAZAKI

Department of Medicinal and Life Science, Faculty of Pharmaceutical Sciences, Tokyo University of Science, 2641 Yamazaki, Noda 278-8510, Japan

Protein-protein interactions play many important roles in biological function. Knowledge of protein-protein complex structure is required for understanding the function. The determination of protein-protein complex structure by experimental studies remains difficult, therefore computational prediction of protein structures by structure modeling and docking studies is valuable method. In addition, MD simulation is also one of the most popular methods for protein structure modeling and characteristics. Here, we attempt to predict protein-protein complex structure and property using some of bioinformatic methods, and we focus botulinum toxin complex as target structure.

1. Introduction

In the post genome era, since it is generally assumed that function of a protein is closely linked to its three-dimensional structure, many research groups work on protein structure determination by X-ray crystallography or solution nuclear magnetic resonance (NMR). There are more than 58,000 structures registered in Protein Data Bank (PDB) as of 2009. However, since multidomain proteins and complex proteins are often difficult to crystallize and many are too large for NMR structure determination, the rate of structure determination is still low compared with sequence determination. On the other hand, a number of computational methods have been developed for the prediction of protein structures and interactions from genomic and structural information. We have been studying the structures of botulinum toxin, which is large multimeric protein complex consisting of several different components. Some of the component and complex structures have not yet been determined. Here, we describe the prediction of botulinum toxin and related protein structures.

Clostridium botulinum is anaerobic bacterium; which produces seven distinct serotypes (A-G) of neurotoxins (BoNT; 150 kDa) [1]. BoNT is synthesized as a single chain and associates with non-toxic proteins in culture supernatants. These include the non-toxic non-hemagglutinin (NTNHA; 130

kDa) and three types of hemagglutinin (HA) subcomponents, HA-70, HA-33 and HA-17 (70, 33 and 17 kDa, respectively), forming a large toxin complex (TC) held together by non-covalent binding.

Bacterial endogenous or exogenous proteases such as trypsin nick the single chain BoNT at about one third of its length from the N-terminus within a region inside a disulfide loop [2]. The BoNT is thereby split into a 50-kDa light chain (Lc) and a 100-kDa heavy chain (Hc) connected by disulfide binding. Trypsinized BoNTs from some strains of serotypes B, D and E are known to be fully activated forms exhibiting several-fold to a few hundred-fold intraperitoneal toxicity [3, 4]. In addition, NTNHA, which is also nicked at its N-terminal region, splits into 15 and 115 kDa fragments, while HA-70 is nicked at its N-terminal region, splitting into HA-22-23 (22-23 kDa) and HA-55 (55 kDa) [5, 6, 7]. With any subcomponents, nicking sites are oriented around the carboxyl side of Arg or Lys; thus the nicking proteases must be trypsin-like proteases (TLP). The protease responsible for nicking of BoNTs is closely related in toxicity and molecular modification of TC components.

The TLP were isolated from serotypes A, B, E and F [8, 9, 10], and in the case of only serotype A, the nucleotide sequence coding for the nicking protease was determined by genomic analysis [11]. On the other hand, the TLP of serotypes C and D have neither been purified nor characterized. Therefore, we and collaborators performed purification, enzymatic characterization and DNA sequencing of the TLP of serotypes C [12]. Using the sequence information of BoNT and TLP, we attempted the molecular modeling and docking simulation of interaction between the identified protease and the BoNT molecules [12]. In addition, we simulated the effect of nicking on the BoNT structure.

2. Modeling of BoNT and TLP structures

The modeled structures of TLP and BoNT produced by serotype C were constructed, using the program Discovery Studio v1.7 (Accelrys Software Inc., San Diego, CA, USA). For modeling of serotype C BoNT, serotypes A (PDB ID: 3BTA) and B (1EPW) BoNTs were used as template structures. On the other hand, as results of enzymatic assay and homology search of TLP, it was shown that the TLP was homolog of clostripain produced by *C. historyticum*. However, the structure of clostripain has not been determined. Therefore, for modeling of TLP, we used protein structures assigned to clan CD with clostripain as template structures; caspase-1 (1BMQ), caspase-2 (1PYO), caspase-3 (1PAU), caspase-7 (1K86), caspase-9 (1JXQ) and gingipain R (1CVR). The commonly conserved regions present in the TLP and template structure were determined by sequence homology and secondary structure alignment. Secondary structure prediction of the TLP was carried out using PSI-PRED [13]. Validation of the models was carried out using Ramachandran plot calculation computed with the Swiss-PdbViewer (available at http://swissmodel.expasy.org/spdbv/) after structure minimization.

As a result of modeling, the TLP catalytic domain models exhibited good geometry according to ramachandran plots with the most favored region containing 77.1% of the dihedral angle pairs and the additionally allowed region containing another 15.8% of the pairs. The catalytic residues of the protease (His172 and Cys 230) were located in the center of the deep pocket (Figure 1). On the other hand, the model for BoNT exhibited 84.9% in the favored region and 11.8% in the additionally allowed region. As shown in Figure 2, the loop region of the BoNT-specific nicking-site was exposed on the surface of the molecule with the Arg444 at the top, which is likely to be specifically attacked by the TLP [2].

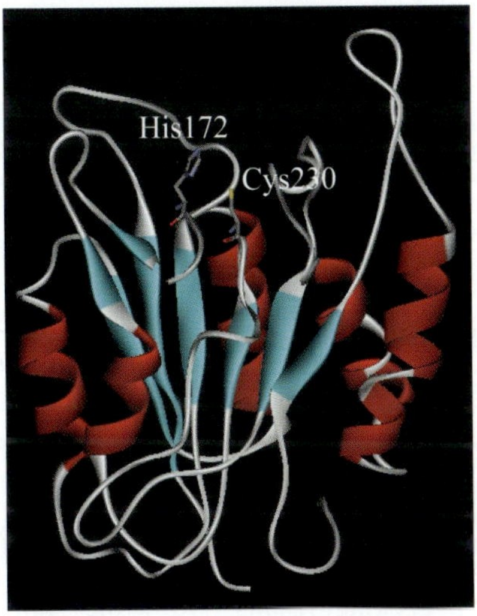

Figure 1. Predicted catalytic domain structure of serotype C TLP. His172 and Cys230 are catalytic dyad.

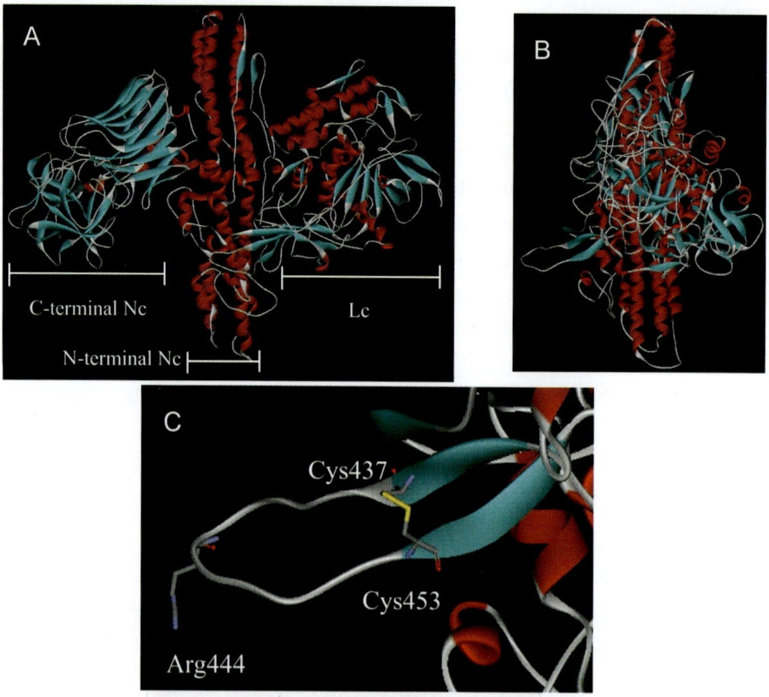

Figure 2. (A) Predicted structure of serotype C BoNT. (B) The structure with view from the Lc. (C) Close-up view of loop region.

3. Prediction of BoNT/TLP complex structure

We predicted the complex of modeled BoNT and TLP structures using the ZDOCK program [14]. The pdb files were processed using the ZDOCK blocking function before docking simulation. Catalytic residues and the surrounding region of the nicking protease were allowed to interact with the hydrolyzed residue (Arg444) and its surrounding region of BoNT. To facilitate the analysis of the docking results, we wrote short scripts to select only complexes in which Arg444 of BoNT made contact with catalytic residues of the protease. The resulting structures were investigated using the ZDOCK scoring function, contact distance of interacting residues and other protease/substrate complex data, such as trypsin and gingipain R.

We generated a model for the BoNT/TLP complex model to postulate specific nicking by the protease at the loop region of BoNT (Figure 3). The complex structure indicated that the guanidyl group of Arg444 of BoNT approached the vicinity of the carboxylate groups of Asp110 and Asp228 of the

TLP. Regarding the crystal structure of the complex of gingipain R and peptidic inhibitor, Eichinger *et al.* reported that the Arg side chain of inhibitor inserts into the S1 pocket of gingipain R, and its terminal guanidyl group makes a frontal 2N-2O salt bridge with the carboxylate group of Asp163 [15]. When the catalytic residues of gingipain R and Asp110 of our protease were superimposed, a very similar structure was obtained with an RMSD of 0.93 angstrom. Similar configuration in the neighborhood of the catalytic residues was also observed in the trypsin and peptide substrate complex [16]. From the analogy to this significant information, it is likely that Asp110 of the TLP characterized substrate specificity. Furthermore, Labrou *et al.* reported that Asp229 (clostripain numbering) could be responsible for substrate specificity, corresponding to Asp228 of our protease [17]. In the TLP, Asp110 and/or Asp228 were most likely responsible for substrate specificity, although further site directed mutagenic studies would be required to confirm this.

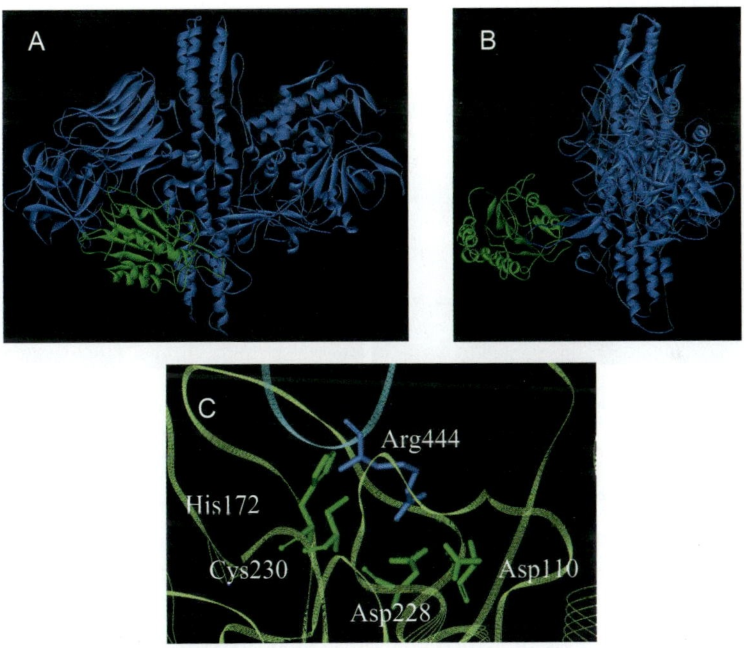

Figure 3. (A) Predicted structure of serotype C BoNT/TLP complex. (B) The structure with view from the Lc of BoNT . (C) Close-up view of binding region.

4. Simulation for effect of nicking on BoNT structure

As described above, BoNT/TLP complex structure was predicted. The purified TLP cleaves the loop region of the intact BoNT, and then the intact BoNT becomes nicked form [12]. Nicked BoNT is known to be fully activated forms with several-fold to a few hundred-fold intraperitoneal toxicity relative to the intact form [3]. The activated BoNT cleaves SNARE-proteins in nerve cell, resulting in animals paralyze. However, effect of cleavage on BoNT structure has been unknown. We predicted the effect of nicking for BoNT structure by MD simulation.

In the culture supernatant and nature, nicked C-St BoNT is deleted 5 residues at loop region [2]. We prepared two structures, intact BoNT and niced BoNT structures (Figure 4). To estimate an effect of nicking on BoNT, we performed MD simulation of these BoNT structures. B-factors of simulated results were analysed using PTRAJ of Amber program as flexibilities of intact and nicked BoNTs.

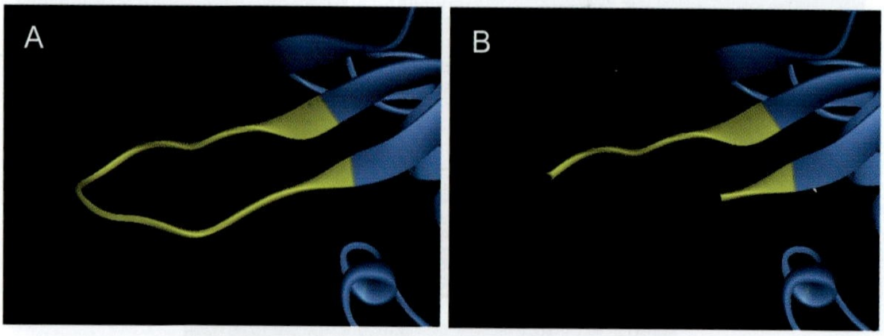

Figure 4. Close-up view of the loop region of intact (A) and nicked BoNT (B).

As a result of simulation, the nicked BoNT seemed to have nine flexible regions in comparison with intact form (Figure 5). This simulation result indicates the nick of loop region makes mainly Lc of BoNT flexibility. Lc of BoNT is catalytic domain, and cleaves SNARE protein at nerve cell. The complex structure of Lc of serotype A BoNT and SNAP-25 of SNARE protein was determined by crystal structure analysis [18]. The predicted flexible regions of nicked BoNT appear to be around the SNARE protein binding region. Thus far, we have known that nicking at loop region activates the BoNT, however effect of nicking on BoNT structure has been unclear. Our result leads to the prediction of effect of nicking on BoNT structure, although more analysis is required.

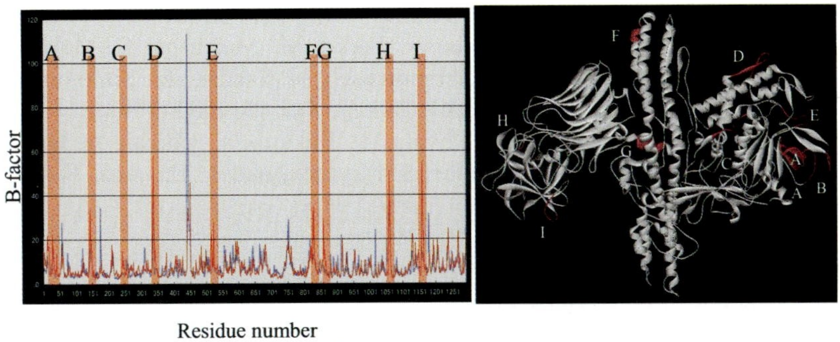

Figure 5. Predicted flexible regions of BoNT. (A) B-factor of intact (blue) and nicked (red) BoNT. (B) Highly flexible regions of nicked BoNT. The MD simulation was performed using MD program Amber 9.0 (force field; parm99, ensemble; NTP, SHAKE; off, time step; 1fs, solvate water box; TIP3P).

Acknowledgments

The authors thank Drs. T. Ohyama, T. Watanabe and K. Niwa, Tokyo University of Agriculture, for their in vitro experiments. The authors thank Miss Y. Sato for technical assistance.

References

1. H. Sugiyama, *Microbiol. Rev.* **44**, 419 (1980).
2. Y. Sagane, T. Watanabe, H. Kouguchi, H. Sunagawa, K. Inoue, Y. Fujinaga, K. Oguma and T. Ohyama, *J. Protein Chem.* **18**, 885 (1999).
3. I. Ohishi and G. Sakaguchi, *Infect. Immun.* **17**, 402 (1977).
4. T. Suzuki, H. Kouguchi, T. Watanabe, H. Hasegawa, T. Yoneyama, K. Niwa, A. Nishikawa, J. C. Lee, K. Oguma and T. Ohyama, *Protein J.* **26**, 173 (2007).
5. K. Oguma, K. Inoue, Y. Fujinaga, K. Yokota, T. Watanabe, T. Ohyama, K. Takeshi and K. Inoue, J. *Toxicol. Toxin Rev.* **18**, 17 (1999).
6. T. Watanabe, Y. Sagane, H. Kouguchi, H. Sunagawa, K. Inoue, Y. Fujinaga, K. Oguma and T. Ohyama, *J. Protein Chem.* **18**, 753 (1999).
7. Y. Sagane, T. Watanabe, H. Kouguchi, H. Sunagawa, S. Obata, K. Oguma and T. Ohyama, *Biochem. Biophys. Res. Commun.* **292**, 434 (2002).
8. M. L. Dekleva and B. R. DasGupta, *J. Bacteriol.* **172**, 2498 (1990).
9. B. R. DasGupta and H. Sugiyama, *Biochim. Biophys. Acta* **268**, 719 (1972).
10. A. Nakane, J. *Gen. Microbiol.* **107**, 85 (1978).
11. M. Sebaihia, M. W. Peck, N. P. Minton, N. R. Thomson, M. T. Holden, W. J. Mitchell, A. T. Carter, S. D. Bentley, D. R. Mason, L. Crossman, C. J. Paul, A. Ivens, M. H. Wells-Bennik, I. J. Davis, A. M. Cerdeño-Tárraga, C.

Churcher, M. A. Quail, T. Chillingworth, T. Feltwell, A. Fraser, I. Goodhead, Z. Hance, K. Jagels, N. Larke, M. Maddison, S. Moule, K. Mungall, H. Norbertczak, E. Rabbinowitsch, M. Sanders, M. Simmonds, B. White, S. Whithead and J. Parkhill, *Genome Res.* **17**, 1082 (2007).

12. T. Suzuki, T. Yoneyama, K. Miyata, A. Mikami, T. Chikai, K. Inui, H. Kouguchi, K. Niwa, T. Watanabe, S. Miyazaki and T. Ohyama, *Biochem. Biophys. Res. Commun.* **379**, 309 (2009).
13. D. T. Jones, *J. Mol. Biol.* **292**, 195 (1999).
14. R. Chen and Z. Weng, *Proteins* **47**, 281 (2002).
15. A. Eichinger, H. G. Beisel, U. Jacob, R. Huber, F. J. Medrano, A. Banbula, J. Potempa, J. Travis and W. Bode, *EMBO J.* **18**, 5453 (1999).
16. E. S. Radisky, J. M. Lee, C. J. Lu and D. E. Jr. Koshland, *Proc. Natl. Acad. Sci. USA.* **103**, 6835 (2006).
17. N. E. Labrou and D. J. Rigden, *Eur. J. Biochem.* **271**, 983 (2004).
18. M. A. Breidenbach and A. T. Brunger, *Nature* **432**, 925 (2004).

Quantum Bio-Informatics III
eds. L. Accardi, W. Freudenberg & M. Ohya

GENE DISCOVERY METHODS FROM LARGE-SCALE GENE EXPRESSION DATA

AKIFUMI SHIMIZU

School of Environmental Science, University of Shiga prefecture
Hikone, Shiga 522-8533, Japan

KENTARO YANO

School of Agriculture, Meiji University
Kawasaki, Kanagawa 214-8571, Japan

Microarrays provide genome-wide gene expression changes. In current analyses, the majority of genes on the array are frequently eliminated for further analysis just in order for computational effort to be affordable. This strategy risks failure to discover whole sets of genes related to a quantitative trait of interest, which is generally controlled by several loci that might be eliminated in current approaches. Here, we describe a high-throughput gene discovery method based on correspondence analysis with a new index for expression ratios [arctan (1/ratio)] and three artificial marker genes. This method allows us to quickly analyze the whole microarray dataset without elimination and discover up/down-regulated genes related to a trait of interest. We employed an example dataset to show the theoretical advantage of this method. We then used the method to identify 88 cancer-related genes from a published microarray data from patients with breast cancer. This method can be easily performed and the result is also visible in three-dimensional viewing software that we have developed. Our method is useful for revaluating the wealth of microarray data available from web-sites.

1. Introduction

Microarrays are widely used tools to monitor the expression levels of thousands to tens of thousands of genes in many organisms (1-3). In microarrays, genes showing two-fold relative expression levels at least or more than two standard deviations away from the mean are often considered to show significantly different expression from the control. And then these genes are selected for further analysis (4-6). This approach usually discards the majority of array genes just in order for computational effort to be affordable.

The expression levels of many genes show wide natural variation (7, 8). There is no theoretical basis whether there is a simple threshold to distinguish significant genes from the others (9). The considerable data elimination adopted in current approach causes a substantial loss of information and that is serious problem especially for analyses of quantitative traits.

Correspondence analysis (CA) is effective method for profiling gene-expression patterns without data elimination (10). CA calculations for the whole microarray dataset are not time-consuming, and find a low-dimensional projection of an originally high-dimensional data matrix (row [gene] and column [sample]). The scores (coordinates) of the first three components allow visual assessment of associations between genes and phenotypes in a three-dimensional subspace. Here, we describe a high-throughput gene discovery method based on CA with a new index for expression ratios and three artificial marker genes.

2. MATERIALS AND METHODS

2.1. *CA for microarray data*

CA was performed using the statistical software package R (http://cran.r-project.org/). CA with a new index was also performed by our developed software (10). A user-friendly java-based software has been developed and it will be released soon (11, 12). CA provide scores (coordinates) for genes and samples.

2.2. *Preparation of an example dataset*

An example dataset contains gene expression ratios for 516 genes and 100 samples. Of the 100 samples, 50 were phenotype A and 50 were phenotype B. There are five down-regulated genes (D1 to D5) and five up-regulated genes (U1 to U5) in phenotype A. Among the same phenotype, these 10 genes have the same expression ratios (Table 1). In addition, three housekeeping genes (HK1, HK2, and HK3) have the same expression ratios among all samples. The example dataset contains 500 genes unrelated to

Table 1 Expression ratios for an example dataset

Gene	Phenotype A	Phenotype B
D1	0.01	1.00
D2	0.10	1.00
D3	0.50	1.00
D4	1.00	2.00
D5	1.00	10.00
U1	1.00	0.01
U2	1.00	0.10
U3	1.00	0.50
U4	2.00	1.00
U5	10.00	1.00
HK1	100.00	100.00
HK2	10.00	10.00
HK3	1.00	1.00
ExtraGene1	0.00	100.00
ExtraGene2	100.00	0.00
ExtraGene3	1.00	1.00

phenotype (Unrelated1 to Unrelated500). The expression ratios for the unrelated genes were randomly selected from the published microarray data (13) (data not shown). This example dataset includes three artificial marker genes.

2.3. *Preparation of breast cancer expression data*

The published microarray data (13) includes 24,481 genes and 117 samples. Two samples and 457 genes that had more than two missing values were eliminated from the dataset. For the remainder, the missing values, at most two, were replaced by the average expression ratio for the same phenotype. Out of 115 samples, 62 samples were from patients that developed metastases within five years after their initial diagnosis (poor prognoses), and 53 samples were from patients that remained free of disease for at least five years after diagnosis (good prognoses).

2.4. *A new index for gene expression ratios*

We used the new index for gene expression ratios (arctan[1/ratio]) (10). It is the arctangent (inverse tangent) of the reciprocal intensity ratio and it ranges from 0 to 90 degrees. When the ratio is equal to one, the index is 45 degrees. As the gene is repressed or induced, this index increases or decreases from 45 degrees, respectively.

2.5. *Significant distances in a low-dimensional projection*

We used a confidence area for a location of a point (plot) of genes in a low-dimensional projection obtained by CA to identify up/down-regulated and housekeeping genes. For an i-th row (gene) in a contingency table, a 95% confidence area is defined as a confidence circle centered at the location of the gene in the two-dimensional subspace (14), where the radius is $\sqrt{\chi^2/K_{i\bullet}}$, the value of the statistic χ^2 with two degrees of freedom (d.f.) is 5.99 at a 0.05 significance level, and $K_{i\bullet}$ is the total of the elements in the i-th row. The d.f. of χ^2 are equal to the number of dimensions in the subspace.

3. RESULTS

3.1. *Three artificial marker genes to identify genes associated with a trait*

We added three artificial genes (ExtraGenes) to the dataset to classify all genes on the array. Assuming that there are two phenotypes (A and B) of a quantitative trait, genes on the array can be classified into the following four categories: (a) genes specifically expressed in either phenotype, (b) genes up- or down-regulated between the two phenotypes, (c) genes up- or down-regulated that are unrelated to the phenotypes, and (d) housekeeping genes that show constant levels of expression in all samples. The genes related to the phenotypes are included in categories (a) and (b). To classify all genes, we used three ExtraGenes (ExtraGene1, ExtraGene2 and ExtraGene3) to the dataset (Table 1). ExtraGene1 and ExtraGene2 assist in the discovery of genes related to the phenotypes and housekeeping genes (categories a, b, and d) as described below. ExtraGene3 shows the same ratio (1.0) in all samples and also aids in the identification of housekeeping genes (category d). Consequently, genes related to the phenotypes can be obtained.

3.2. *A line segment to identify up/down regulated genes*

For the example dataset, we performed CA using the new index (arctan [1/ratio]). As expected, projections of genes into the first two-dimensional subspaces with the cumulative contribution ratios 60.0% show up- and down-regulated genes have negative and positive scores in Factor1, respectively (Figure 1). The up- and down-regulated genes lie on a line segment between ExtraGene1 and ExtraGene2. We call this line segment the UDL (up/down line). Moreover, all housekeeping genes and ExtraGene3 lie in the center of the UDL, which is the origin of the subspace. Locations of genes unrelated to phenotypes are random and independent of the UDL.

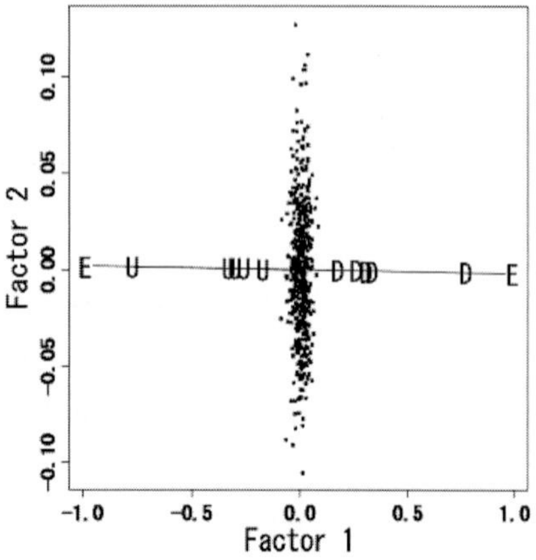

Figure1. CA result for an example dataset

3.3. *Analysis of breast cancer data*

We next applied CA with the new index to published human breast cancer microarray data (13). The three ExtraGenes were also added to the dataset. We calculated seven-dimensional scores to 24,027 genes and the 115 samples. This process takes only approximately ten seconds. Thus, CA requires considerably less time for calculation than hierarchical clustering.

3.4. *Up/down regulated genes in significant regions*

As shown in Figuer 1, a UDL was determined as a line segment between ExtraGene1 and ExtraGene2 (Figure 2a). Genes up- or down-regulated between good and poor prognosis samples and housekeeping genes are expected to lie on the UDL. However, the locations of these genes can statistically deviate from the UDL as well as biometric data generally deviate from the expected value.

For the breast cancer data, we applied the confidence areas to a seven-dimensional space. The value of the statistic χ^2 with seven d.f. at a significance level of 0.05 is 14.0671. The significant distance from an ExtraGene becomes $\sqrt{\chi^2 / \sum_{i=1}^{n} f_i}$,

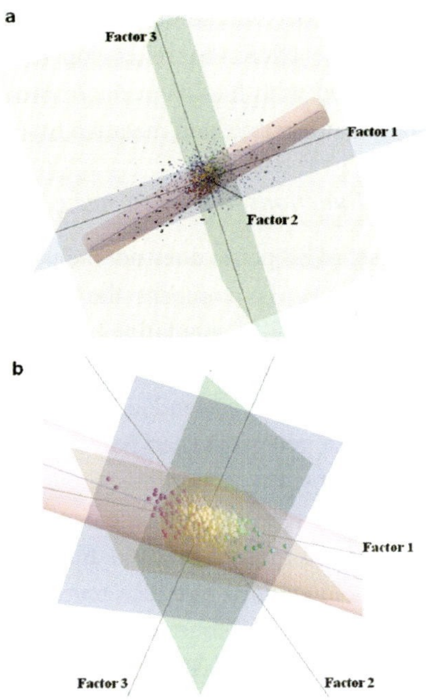

Figure2. CA result for breast cancer data.

where n is the number of samples and f_i is the new index (arctan[1/ratio]) for the ExtraGene of the i-th sample. Consequently, the significant distance from ExtraGene1 is 0.0502 because $\sum_{i=1}^{n} f_i = 90 \times 62$. Similarly, the significant distances from ExtraGene2 and ExtraGene3 are 0.0543 and 0.0521, respectively.

The significant distance from ExtraGene1 was used as the significant distance from the UDL because they are nearly equal. Up/down-regulated and housekeeping genes were located inside the confidence area of UDL with a 95% probability. We call this confidence area the up/down region (UDR). In the first three-dimensional subspace, the UDR is visualized as a cylindrical shape (Figure 2a). Using this UDR, we estimated that 544 genes are up/down-regulated or housekeeping genes.

3.5. *Detection of 88 genes related to breast cancer*

It is expected that housekeeping genes cluster around the position of ExtraGene3. Housekeeping genes were defined as those that have less than the significant distance from ExtraGene3. A statistical test with a 95% significant distance from ExtraGene3 is also available. This significant region forms as a spherical space in the three-dimensional subspace (Figure 2b). Here, we call this region the HKR (housekeeping region). Consequently, we detected 88 genes associated with the diagnosis of breast cancer. Out of the 88 genes, 45 and 43 genes had positive and negative first-axis coordinates, respectively.

3.6. *Functions of the detected genes*

The set of 88 genes does not include any marker genes identified in the previous report (13). To compare the biological functions of the two gene sets, we investigated GO annotations. The result shows that GO terms related to cell cycle (e.g., cell cycle and division) and apoptosis (e.g., I-kappaB kinase/NF-kappaB cascade and induction of programmed cell death) are highly significant for the 88 genes detected here. These biological processes are well-known to be affected by the activities of oncogenes. The majority of GO terms for the previously reported 70 genes are related to cellular development (e.g., cellular growth and morphogenesis) and DNA metabolism (e.g., DNA metabolism and strand elongation). We also investigated MeSH terms assigned for the 88 genes shown here. The fourteen MeSH terms are significantly related to neoplasms. Together with the GO terms, these MeSH terms suggested that the 88 genes detected here are related to cancer.

The biological functions of 35 of the 88 genes identified here have been investigated in previous studies. There is no detailed information on the function of the other 53 genes in public databases or published reports. Based on the published reports on the 35 previously studied genes, 18 of them are oncogenes, candidate target genes for tumor therapy, or genes with known carcinogenic functions.

4. DISCUSSION

We describe here a method for gene discovery from microarray data. Our method, CA with a new index for expression ratios coupled with the inclusion of ExtraGenes, allows us to define a UDL, UDR and HKR, which assist in the detection of genes related to the phenotype of interest. Although the confidence regions used here for UDR and HKR are defined for a contingency table, the application shows good classifications of genes. Our method also dramatically reduces the calculation time, and it is effective at predicting the phenotype based on the gene expression profile.

Using this method, we detected 88 prognostic marker genes from a published human breast cancer dataset (13). Veer et al. (13) selected 4968 genes from the 24,481 genes on this array, from which 70 marker genes were identified using a three-step supervised classification method. Both the 70 candidate genes identified in this previous report and the 88 genes detected here show up/down-regulation between poor and good prognosis samples. The 88 genes identified here do not include any of 70 previously identified genes, but it does include known cancer-related genes. Especially, gene associated with breast cancer, such as tumor suppressers, NF-κB activators and genes associated with estrogen receptor-α (ERα) were identified.

We suspect that the discrepancy between this work and the previous report (13) arises from the current tendency to over-reduce the number of genes for further analyses. Genes that do not have more than two-fold differences in expression ratios and P values less than 0.01 are commonly excluded. The threshold in the previous work would have eliminated 84 of the 88 genes detected here. Furthermore, the candidate genes identified in the previous report are located outside the UDR, although some are close. This result implies that the candidate genes that were identified from the 4968 threshold-selected genes include those that are the closest to the UDR (Figure 2a). Generally, the expression ratios show greater variation at lower expression levels. Yang et al. (15) suggested that even less than two-fold expression levels can be significant. We also believe that a method that assesses the expression ratios and/or P values of detected genes from the whole microarray dataset would be better than the current method, which detects candidate genes from those selected using the threshold. Differences in the selection of genes using the UDR, which can vary according to the significant distance (significant level), may also explain the difference between the two sets of candidate genes.

As the information for samples increases, the number of phenotypes may increase to more than two. Our method can be extended to more than two phenotypes. In the current study, we prepared two ExtraGenes for the two phenotypes, wherein the ExtraGene is specifically expressed in either phenotype. When there are more than two phenotypes, ExtraGenes specific to each phenotype would be created to detect the genes related to the trait.

Acknowledgments

This work was supported in part by Grants-in-Aid for Special Research on Priority Areas (19043015, 21024010), Scientific Research (B) (20380022) from the Ministry of Education, Culture, Sports, Science and Technology of Japan (MEXT). This research was also supported in part by Grants-in-Aid for Exploratory Research (19651084) from the Japan Society for Promotion of Science (JSPS).

References

1. T. S. Tanaka, T. Kunath, W. L. Kimber, S. A. Jaradat, C. A. Stagg, M. Usuda, T. Yokota, H. Niwa, J. Rossant and M. S. Ko, *Genome Res.* **12**, (2002).
2. T. R, Hughes, M. Mao, A. R. Jones, J. Burchard, M. J. Marton, K. W. Shannon, S. M. Lefkowitz, M. Ziman, J. M. Schelter, M. R. Meyer, et al, *Nature Biotechnol.* **19**, 342 (2001)
3. K. Suwabe, G. Suzuki, H. Takahashi, K. Shiono, M. Endo, K. Yano, M. Fujita, H. Masuko, H. Saito, T. Fujioka, F. Kaneko, T. Kazama, Y. Mizuta, M. Kawagishi-Kobayashi, N. Tsutsumi, N. Kurata, M. Nakazono and M. Watanabe, *Plant Cell Physiol.* **49**, 1407 (2008).
4. A. Mueller, J. O'Rourke, J. Grimm, K. Guillemin, M. F. Dixon, A. Lee and S. Falkow, *Proc. Natl. Acad. Sci. USA*, **100**, 1292 (2003).
5. J. M. Sperger, X. Chen, J. S. Draper, J. E. Antosiewicz, C. H. Chon, S. B. Jones, J. D. Brooks, P. W. Andrews, P. O. Brown and J. A. Thomson, *Proc. Natl. Acad. Sci. USA* **100**, 13350 (2003).
6. Y. Zhang, C. Ma, T. Delohery, B. Nasipak, B. C. Foat, A. Bounoutas, H. J. Bussemaker, S. K. Kim and M. Chalfie, *Nature*, **418**, 331 (2002).
7. V. G. Cheung, L. K. Conlin, T. M. Weber, M. Arcaro, K. Y. Jen, M. Morley and R. S. Spielman, *Nature Genet.* **33**, 422 (2003).
8. W. P. Kuo, T.K. Jenssen, A. J. Butte, L. Ohno-Machado and I. S. Kohane, *Bioinformatics* 18, 405 (2002).
9. J. Quackenbush, *Nature Rev. Genet.* **2**, 418 (2001).
10. K. Yano, K. Imai, A. Shimizu, T. Hanashita, *Nucleic Acids Res.* **34**, 1532 (2006).
11. K. Yano et al., Japanese Patent No.4302466 granted on May 1, 2009
12. K. Yano and A. Shimizu, Japanese Patent pending (No. 2009-063273), 2009
13. L. J. van 't Veer, H. Dai, M. J. van de Vijver, Y. D. He, A. A. Hart, M. Mao, H. L. Peterse, K. van der Kooy, M. J. Marton and A. T. Witteveen, *Nature* **415**, 530 (2002).
14. L. Lebart, A. Morineau and K. M. Warwick, Multivariate descriptive statistical analysis: correspondence analysis and related techniques for large matrices. translated by Berry, E. M. Wiley: New York. (1984)
15. I. V. Yang, E. Chen, J. P. Hasseman, W. Liang, B. C. Frank, S. Wang, V. Sharov, A. I. Saeed, J. White, J. Li, et al, *Genome Biol.* **24**, research0062 (2002)

International Conference

March 11-14 2009

Quantum Bio-Informatics Center

QBIC 2009

BI Towards New Paradigm for QI

Main Session

March 11, Wednesday 8:50-18:00
March 12, 13, and 14, 9:00-18:30
at Conference Room in Canal Hall
Noda Campus, Tokyo University of Science
Noda City, Chiba Japan

Satellite Session

March 6 and 7, 9:00-17:00
at Suwa Campus, Tokyo University of Science
Chino City, Nagano Japan

Poster Session

March 11, Wednesday 12:30-13:30
March 13, Friday 12:30-13:30
at Canal Hall

Banquet

March 12, Thursday 18:30-
at Cafeteria(2F) in Canal Hall

Speakers (Tentative)

L. Accardi (Roma II University, Italy)
R. Belavkin (Middlesex University, UK)
D. Chruscinski (Nicolaus Copernicus University, Poland)
K-H. Fichtner (Friedrich Schiller University Jena, Germany)
L. Fichtner (Friedrich Schiller University Jena, Germany)
W. Freudenberg (Brandenburg Techn. University Cottbus, Germany)
T. Hida (Emeritus Professor, Nagoya University, Japan)
W. Im (The University of Kansas, USA)
K. Ishibashi (RIKEN Advanced Science Institute, Japan)
T. Itoh (National Inst. of Agrobiological Sciences, Japan)
A. Jamiolkowski (Nicolaus Copernicus University, Poland)
J. Jurkowski (Nicolaus Copernicus University, Poland)
A. Khrennikov (University of Växjö, Sweden)
A. Majewski (University of Gdańsk, Poland)
M. Michalski (Nicolaus Copernicus University, Poland)
H. Nakano (NTT Basic Research Laboratories, Japan)
I. Ojima (Kyoto University, Japan)
D. Petz (Budapest Univ. of Tech. & Economics, Hungary)
M. Regoli (Roma II University, Italy)
Si Si (Aichi Prefectural University, Japan)
J. Skolnick (Georgia Institute of Technology, USA)
O. G. Smolyanov (Moscow State University, Russia)
L. Streit (BiBoS, University Bielefeld, Germany)
Y. Sugita (RIKEN Theoretical Biochemistry Lab, Japan)
I. Volovich (Steklov Mathematical Institute, Russia)
D. Wanke (University of Tübingen, Germany)
K. Yano (Meiji University, Japan)
Some QBIC members in Tokyo University of Science... and more

Venue

Noda Campus
Tokyo University of Science
Noda-city, Chiba, Japan

http://www.noda.tus.ac.jp/qbic/qbic.html

Purpose

The main aim of QBIC and the conference is to create a new paradigm synthesizing Quantum Information and Bio-Informatics based on efforts by active researchers traversing various fields of Mathematics, Physics, Information and Life Science.

Organizer

M.Ohya, Chief (Tokyo University of Science, Japan)
H. Sakata (Tokyo University of Science, Japan)
Y. Togawa (Tokyo University of Science, Japan)
I. Yamato (Tokyo University of Science, Japan)

Advisory Committee

L. Accardi (Roma II University, Italy)
W. Freudenberg (Brandenburg Techn. Univ. Cottbus, Germany)
T. Hida (Emeritus Professor, Nagoya University, Japan)
H. Kamimura (Tokyo University of Science, Japan)
I. Ojima (Kyoto University, Japan)

Local Committee

K. Kuchitsu (Tokyo University of Science, Japan)
S. Miyazaki (Tokyo University of Science, Japan)
H. Takayanagi (Tokyo University of Science, Japan)
M. Takeda (Tokyo University of Science, Japan)
S. Tomizawa (Tokyo University of Science, Japan)
N. Watanabe (Tokyo University of Science, Japan)

Contact Information

M. Kuroda, Secretary
Quantum Bio-Informatics Center
Tokyo University of Science
Noda City, Chiba 278-8510 Japan
Tel:+81-4-7124-1501 ext.5249,3319
Fax:+81-4-7124-7130
E-mail:qbic09@rs.noda.tus.ac.jp

The QBIC (Quantum Bio-Informatics Center) project from 2006 to 2011 is sponsored by Japanese Ministry of Education and Tokyo University of Science